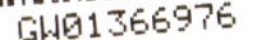

COMPUTING WITH BIOLOGICAL METAPHORS

COMPUTING WITH BIOLOGICAL METAPHORS

EDITED BY

Ray Paton

Department of Computer Science
University of Liverpool
Liverpool
UK

CHAPMAN & HALL
London · Glasgow · Weinheim · New York · Tokyo · Melbourne · Madras

Published by Chapman & Hall, 2–6 Boundary Row, London SE1 8HN, UK

Chapman & Hall, 2–6 Boundary Row, London SE1 8HN, UK

Blackie Academic & Professional, Wester Cleddens Road, Bishopbriggs, Glasgow G64 2NZ, UK

Chapman & Hall GmbH, Pappelallee 3, 69469 Weinheim, Germany

Chapman & Hall USA, One Penn Plaza, 41st Floor, New York NY 10119, USA

Chapman & Hall Japan, ITP-Japan, Kyowa Building, 3F, 2-2-1 Hirakawacho, Chiyoda-ku, Tokyo 102, Japan

Chapman & Hall Australia, Thomas Nelson Australia, 102 Dodds Street, South Melbourne, Victoria 3205, Australia

Chapman & Hall India, R. Seshadri, 32 Second Main Road, CIT East, Madras 600 035, India

---

First edition 1994

Typeset in 10/12 Palatino by Best-set Typesetter Ltd., Hong Kong
Printed in Great Britain by the University Press, Cambridge

ISBN 0 412 54470 9

A catalogue record for this book is available from the British Library

Library of Congress Catalog Card Number: 94-70980

∞ Printed on permanent acid-free text paper, manufactured in accordance with ANSI/NISO Z39.48-1992 and ANSI/NISO Z39.48-1984 (Permanence of Paper).

CONTENTS

CONTRIBUTORS

MICHAEL ARBIB
Center for Neural Engineering,
University of Southern California,
Los Angeles, USA.

THOMAS BÄCK
Department of Computer Science,
University of Dortmund,
Germany.

AVIV BERGMAN
Department of Biological Sciences,
Stanford University,
California, USA.

HUGUES BERSINI
Université Libre de Bruxelles,
Belgium.

GAIL A. CARPENTER
Center for Adaptive Systems,
Boston University, Boston,
Massachusetts, USA.

ROBERT J. COLLINS
Alife Laboratory,
Department of Computer Science,
UCLA, Los Angeles, USA.

YUVAL DAVIDOR
Department of Applied Mathematics and
Computer Science,
The Weizmann Institute,
Israel.

STEPHEN GROSSBERG
Center for Adaptive Systems,
Boston University, Boston,
Massachusetts, USA.

PATRICK HAMILTON
Neurosurgical Clinic,
University of Bonn,
Germany.

WILLIAM D. HAMILTON
Department of Zoology,
University of Oxford,
UK.

MIKE HOLCOMBE
Department of Computer Science,
University of Sheffield,
UK.

IAN HORSWILL
AI Lab,
MIT,
Cambridge, Massachussetts, USA.

BERNARDO A. HUBERMAN
Xerox Palo Alto Research Center,
Palo Alto,
California, USA.

GEORGE KAMPIS
Department of Theoretical Chemistry,
University of Tubingen,
Germany.

JOHN R. KOZA
Department of Computer Science,
Stanford University,
California, USA.

BERNARD MANDERICK
AI Lab., Vrije University,
Brussels, Belgium.

PEDRO C. MARIJUÁN
Seminario Interdisciplinario,
University of Saragossa,
Spain.

JAMES M. NYCE
Center for Information and Communication Sciences,
Ball State University,
Muncie, IN and
Department of Computer and Information Sciences,
Linköping University,
Sweden

RAY PATON
Department of Computer Science,
University of Liverpool,
UK.

GEOF STANIFORD
Department of Computer Science,
University of Liverpool,
UK.

W. RICHARD STARK
Department of Mathematics,
University of South Florida,
Tampa, USA.

BRIAN H. SUMIDA
Department of Zoology,
University of Oxford,
UK.

DENIS THIEFFRY
Laboratoire de Génétique,
Université Libre de Bruxelles,
Belgium.

RENÉ THOMAS
Laboratoire de Génétique,
Université Libre de Bruxelles,
Belgium.

FRANCISCO VARELA
CREA, Ecole Polytechnique,
Paris, France.

FRANK T. VERTOSICK, JR.
West Penn Center for Neuro-oncology,
Pittsburgh, Pennsylvania, USA.

G. RICK WELCH
Department of Biological Sciences,
University of New Orleans,
USA.

CHRIS WINTER
British Telecom Research,
Martlesham Laboratories,
Ipswich, UK.

GERSHOM ZAJICEK
Medical Computer Unit,
Hebrew University of Jerusalem,
Israel.

PREFACE

> *We are always looking for metaphors in which to express our ideas of life, for our language is inadequate for all its complexities. Life is a labyrinth. But a labyrinth is a static thing, and life is not static. Life is a machine. But machines do not repair themselves, nor do they reproduce themselves. Life is a laboratory, a workshop. But it is a workshop in which a thousand processes go on within a single microscopic cell, all crossing and intercrossing each other, and it is a workshop which is constantly multiplying and producing its like.*
>
> *It is but a metaphor. When we speak of ultimate things we can, maybe, speak only in metaphors. Life is a dance, a very elaborate and complex dance. The physiologist cannot consider the dance as a whole. That is beyond his experimental power. Rather he isolates a particular corner or a particular figure. His conception of the dance, as thus derived, is imperfect in itself and, moreover, in obtaining it he has disturbed the very pattern of the dance.*
>
> Singer, C. (1959). *A Short History of Scientific Ideas to 1900.* Clarendon Press, Oxford, p. 498.

This book, like the biological structures it describes, has been the product of evolutionary processes. It started out as a newsletter for the Association for Computing Machinery Special Interest Group on Biomedical Computing. The topic for the special edition of the newsletter was Biologically Motivated Computing (BMC). My aim in producing that newsletter edition was to interest the SIGBIO readership in BMC by looking at a broad range of topics and to show how ideas are being transferred between the disciplines of biology and computing. The newsletter contained eight short articles and was 50 pages or so in length.

After being advised to produce a book, I wanted to provide a rich source of material on a broad range of topics. However, the perceptive reader will find that although the book ranges from cells to ecosystems and from anthropology to zoology, there are important themes which pervade the whole. My motivation in approaching potential contributors and in editing the whole was always to enthuse readers about this rapidly developing area rather than provide a manual on one topic. Hence, the topics covered are broad and the contributors come from many academic backgrounds and many different parts of the world. No single approach to the study of the complex biological and computational systems is advanced.

ACKNOWLEDGEMENTS

I would like to thank all the contributors for their encouraging comments, for reviewing each others' material and for providing feedback on their own chapters. Thanks also to Eliane Wigzell and the sub-editorial staff of Chapman & Hall for their support throughout the project, and to Pat McCarren of ACM for his advice when the idea of a book was first getting off the ground. Finally, I would like to thank Geoff Kendall for help with proof-reading and Chris, my wife, for help with proof-reading and index compilation.

INTRODUCTION TO COMPUTING WITH BIOLOGICAL METAPHORS 1

1.1 BACKGROUND AND CONTEXT

Biological form and function have long inspired and motivated people to design and construct new and often very exciting artefacts. We need go no further than mention the inspiration for the original Daedalus project to identify the antiquity of this enterprise and, moving to the present century, list such achievements as novel designs for aircraft wings (slots, winglets, high lift devices, variable flight geometry) or the mechanical designs for new materials such as Velcro (based on a plant burr, e.g. burdock) and bullet-proof vests (based on spider silk) or gene and enzyme biotechnologies, and so on. It is also important to appreciate that as the engineering sciences have developed so the biosciences have greatly benefited. This is the case in understanding the mechanical design of organisms or the way subcellular systems process information (ranging from the electronic model of the nerve cell action potential to the idea of cellular metabolism as a parallel distributed processing system). Ideas that biological systems are like machines or laboratories or chemical plants have been important metaphors over a number of centuries (Paton 1992).

It has also been the case that scientists have come up with solutions to certain problems only to find that the biological world got there first. This is the case with bat ears and radar antennae, infrared imaging systems, sonar echolocation and the plethora of naturally occurring chemical substances that have proved to be vital in the pharmaceutical industry. So, given this general picture of common ground between biology and engineering, it should come as no great shock that the biological and computational sciences can mutually benefit from each other. We see this mutualism in the emergence of cybernetics and systems theory especially from the 1940s onwards.

If biologists were asked to assess the value of computing to their field of study they would realize that they use computational ideas in their description of biological systems and processes which are expressed in the language of information, signals, codes, cybernetics, input–output, machine thinking, biological components as processors, and so forth. This computational language has enriched biology. Indeed, for some it has become the dominating metaphor as, for example:

> If you want to understand life, don't think about vibrant, throbing gels and oozes, think about information technology.
>
> Dawkins, R. (1986). *The Blind Watchmaker*. Penguin, Harmondsworth, p. 112.

For others it is not as simple as this, as biological complexity requires a number of metaphors, for example:

> No one knows or can express what an organism is. The best we can do is to try to say what an organism is like.
>
> Rosen, R. (1986). Causal Structures in Brains and Machines, *Int. J. General Systems*, **12**, 107.

Biology has greatly enriched computing. This is the major motivation for producing this book. We see it in the development of evolutionary computation, artificial neural networks, immune algorithms, artificial life, cellular automata and other ideas, many of which are discussed in this book.

Over the last decade there has been a considerable expansion of interest in the interchange of ideas between biology and computing. Numerous conferences have been organized, books published and research journals started. The purpose of this book is to help with this expansion, by bringing together in one volume contributions from many disciplines. In so doing it is hoped that at least three goals will to some degree be achieved:

1. to extend biological and computational theory;
2. to develop practical tools and systems at the interface of biology and computing;
3. to promote meaningful dialogue between disciplines.

In saying this it is important to distinguish two basic approaches, namely, biologically motivated computing and computationally motivated biology. In the former, biology provides sources of models and inspiration for computing, and in the latter computing provides sources of models and inspiration for biology. As a result, biologically motivated computing can provide:

1. the inspiration for engineering designs;
2. information about similar kinds of system to the intended design;
3. computational devices (i.e., engineering with biological materials).

Computationally motivated biology can provide source models for:

1. clarifying fuzzy ideas;
2. suggesting hypotheses about how biosystems may or may not process information;
3. providing a powerful simulation environment for testing models.

This said, the two approaches overlap so much that it is not always easy or useful to separate them.

1.2 THE STRUCTURE OF THIS BOOK AND PREVIEW OF THE CONTENTS

The book has been organized into five parts: cells, tissues, genetics, ecology and theoretical and conceptual issues. It could have been organized in a number of other ways, for example around integrative themes. However, a perceptive reader would soon find so many interrelations across themes that the sections would seem arbitrary. Table 1.1 shows some of the main themes in the book.

1.2.1 CELLS

Mike Holcombe examines some ways of transferring ideas between algebraic machines, VLSI and cellular metabolism. He introduces the idea of an abstract machine called an *X*-machine and initially applies it to the specification of a word processor interface and then to Krebs' cycle. From here he goes on to look at other levels of machine organization in his computational cell and discusses concurrent processing in an enzyme control net machine. Using the example of a VLSI chip design for controlling a robot arm, he discusses an algebraic approach to networking cellular processing sites in an idealized cell. The chapter concludes with a brief discussion of emerging ideas concerned with spatial organization and hierarchical relations in computational systems.

René Thomas and Denis Thieffry provide an overview of the former's work on the application of logical networks to understanding biological systems. In particular René and Denis focus on applications to logical networks of genes although their approach can be generalized to other systems. They show how this method can be used to reveal relations between feedback loops and steady states in biological systems. In particular they seek to provide a means of modelling interactions between feedback loops of a system rather than feedback between elements.

Rick Welch provides a biophysicist's perspective on some of the computational capacities of the cell. He describes two metaphors to do with field theory and computation and applies them to a description of the living state of the cell. A major focus of his chapter is the computational capacities of enzymes and he looks at the nature of the driving force behind the computational processing associated with enzyme catalysis. The view that an enzyme is a molecular machine is presented and he describes how the physicochemical nature of the cellular microenvironment is more complicated than any bulk-phase modelling. A computational account of the machine-like capacities of macromolecular configurations in these microenvironments is then made.

Pedro Marijuán continues with the theme of enzyme systems but this time as networks of molecular automata. He discusses how cellular networks (which could also be extended to include operons and molecular channels) can be modelled and identifies three collective, emergent properties: self-organization, self-reshaping and self-modification. He goes on to consider the computational society in the cell and postulates a multidisciplinary area of study focused on the artificial cell. In order to achieve this long-term goal Pedro promotes the need for the integrative study of bioinformation.

Chris Winter discusses aspects of possible developments in molecular computation which have some basis in biological systems. After reviewing different kinds of biocomputer architectures that could be fabricated (given substantial developments in technology) he goes on to consider a number of specific details such as data transport and molecular logic elements. From here he looks at fabrication techniques for manipulating small numbers of individual molecules in the production of thin films. Then follows a description of a possible biocomputer system which could be developed. The chapter concludes with a discussion of some prospects for the emergence of nanobiotechnology.

Patrick Hamilton's chapter on computing dendritic growth describes an experimental system for simulating dendritic growth. The general approach of mutating L-systems is well documented and Patrick has developed an interactive CAD-like platform for building dendritic trees. Users of this system are able to define strings which code for trees in a graph-grammar fashion and subject them to change using a number of operators. A number of examples of the kinds of trees that can be generated are provided.

1.2.2 TISSUES

Michael Arbib examines some of the ways the brain can be used as a metaphor for sixth generation computing such as cooperative computation, perceptual robotics and learning. In order to present his case he reviews schema theory as an approach to functional analysis. From here the functionality of a biological neuron is described but with an emphasis on some currently unexplored applications to neural computing. He then provides a model of the brain as an action-oriented somatotopically organized computer and applies this approach to *Rana computatrix* – the frog that computes. This leads on to issues associated with competition, cooperation and self-organization. The final part of the chapter brings together many of the ideas previously discussed to show how sixth generation computers could be designed.

The chapter by Gail Carpenter and Stephen Grossberg on ART 3 is a re-publication of an article which first appeared in *Neurocomputing*. It has been included in this book because it is a very good example of how a biological source has provided many direct computational features. The authors have developed a model which implements parallel search of compressed or distributed pattern recognition codes in a neural network hierarchy based on adaptive resonance theory (ART). The search process emerges when computational properties of the chemical synapse such as transmitter accumulation, release, inactivation and modulation are embedded within an ART architecture called ART 3. Formal analogues of ions such as Na^+ and Ca^{2+} control nonlinear feedback interactions that enable presynaptic transmitter dynamics to model the postsynaptic short-term memory of a pattern recognition code.

Some of the connections between the nervous and immune systems are discussed by Frank Vertosick who applies parallel distributed processing (PDP) ideas to immune network modelling. Given that PDP networks can be characterized by their capacity to recognize, analyse and extrapolate patterns, Frank compares some aspects of immune and nervous system complexity. A number of questions are raised about learning in immune networks, the role of cytokines and spatiotemporal factors. The immune systems has some fixed components (e.g., lymph nodes) and a fluid phase (e.g., in the plasma). This kinds of hybrid network of fixed and fluid phases is discussed and a generalized framework for describing a large number of systems from cell to ecosystem is suggested.

Hugues Bersini and Francisco Varela provide a detailed analysis of how an understanding of the dynamics and metadynamics of immune networks can provide a rich source of ideas for computing and in particular an additional viewpoint on evolutionary-based systems. The immune network source has been developed into a methodology for adaptive and distributed problem solving. Of particular importance to this work is the application of immune recruitment as a mechanism for generating individuals within a population which are very similar to the best individuals which are already present. Following a review of a biological model of immune networks, the general principles behind the design methodology, which include Q_learning, immune recruitment and population-based memory, are discussed. Two applications of the methodology are reported – recruitment mechanism for the cart-pole problem and population-based memory for the robot in a maze problem.

This section has so far looked at some commonly occurring biological sources, namely, nervous and immune systems. The final two chapters of this section introduce new examples of tissue-level computation albeit in different ways and for different purposes. W. Richard Stark introduces and develops the concept of a non-neural artificial tissue based on distributed information processing via gap junctions. After reviewing some aspects of the basic biology of these structures Richard presents a framework for describing the kinds of information which gap junctions could process and the ways in which anonymous communication is achieved in a network of processing units. The abstract tissue model is described in terms of a communication graph and implications of the model for distributed information processing are developed. A number of biological issues are addressed concerned with pattern formation and coordination in pulsing tissues.

Gershom Zajicek reports on the application of simulations based on cellular automata to biomedical domains in which tissues stream. He begins with a explanation of the nature of streaming tissues based on his own empirical research and his application of the concept of chaos to medical understanding. A streaming system, the crypt-villus unit, is then described and from this Gershom develops an abstract model called a tissue unit. The biological details of this model are then discussed and a simpler simulation model, the tissue automat, is introduced. The results of a number of simulations of this system are then reported dealing with cellular differentiation, acute and chronic pathology, carcinogenesis and embryogenesis.

1.2.3 GENETICS AND EVOLUTION

Thomas Bäck provides a comparison of the different approaches to the design of three major kinds of evolutionary algorithm, namely, evolutionary programming, evolution strategies and genetic algorithms. For each type Thomas reviews the evaluation and representation of fitness, initialization of the population, mutation, recombination, selection and a general description of the conceptual algorithm. A summary of the major similarities and differences between these algorithms is then given together with a comparison between the representation of information in biological and artificial systems. The chapter concludes with a basic guide to the literature on this subject.

The next two chapters provide a valuable insight into the ways biologists and computer scientists can apply themselves to some common problems. Rob Collins introduces us to the idea of artificial (simulated) evolution which provides a bottom-up approach for tackling a problem in evolutionary biology: the evolution of sexual reproduction. Following a comparison between natural and artificial evolutionary systems, Rob describes a simulation system called Parasite which has been developed to simulate macroevolutionary processes in an artificial environment. This system was used to test the hypothesis that parasites present their host species with a changing and challenging environment over long time periods for which the ability to mix genes through segregation and recombination (due to sexual reproduction) will be an important selective feature. His results suggest that parasites bring about selection for higher recombination rates in a simple host–parasite simulation.

Brian Sumida and Bill Hamilton transfer a number of biological ideas to the enhancement of evolutionary algorithms and provide an excellent example of how biological details can be applied to artificial evolution. They describe a genetic algorithm based on an evolutionary model which adds parasites and structured populations to its configuration and was applied to the travelling salesman problem (TSP). It is shown that the addition of parasites or the structuring of the population or both of these together are significantly better at solving TSP than when they are absent. A number of issues are raised including the effects of random genetic drift on small demes, interdemic migration and parasite pressure on the evolutionary search process.

John Koza discusses the recently developed genetic programming approach to evolutionary computation in which computer programs are bred to solve a wide variety of problems. This approach applies a number of neo-Darwinian ideas such as reproduction of the fittest programs. John describes the way in which the genetic programming approach has been applied to the 'painted desert' problem and shows how the evolved program exhibits emergent and cooperative behaviour.

Aviv Bergman reports on the development of neural-network-based adaptive signal recognition systems which are designed using genetic algorithms. Aviv describes the nature of these encoders and explains why the use of evolutionary techniques for their design was adopted. From this the results from a number of experiments which tested the genetic system are presented including effects of changing environment, noise, large population size and parasites. A distinctive feature of this work is that it does not seek the 'most fit' encoder but rather seeks subpopulations of encoders which have become specialized to what is the equivalent of ecological niches.

1.2.4 ECOLOGY

Yuval Davidor adopts an ecologically motivated approach to evolutionary computation. His purpose is to make systems which make the use of evolutionary algorithms more robust. The ECO GA (ECOlogical Genetic Algorithm) framework combines GAs with some features of cellular automata where the dynamics of the system are governed by local interaction rules. However, unlike cellular automata, the rules in the ECO environment are not based on topology, but rather on simplified evolutionary information processing rules. Yuval describes his approach and then gives some results of its application to a high-dimensional quadratic assignment problem and two job shop scheduling problems.

Bernardo Huberman shows how distributed computational systems can be modelled as communities of concurrent processes and presents an ecological analysis of such systems which takes account of distributed control, asynchrony, resource contention and a great amount of communication between agents. The computational ecology of a distributed system can also be characterized as having many interactions but incomplete knowledge. He shows that stability can be achieved by making chaos a transient phenomenon. The possibility of achieving global stability through local controls is demonstrated by using fitness mechanisms related to evolutionary game theory.

The final two chapters of this section look at some biological influences on the specification of computational agents. Ian Horswill discusses an approach to the definition of computational agents and their environments by focusing on the important notions of habitat description and specialization as optimization. He then goes on to describe the development of a vision-based system which detects when people nod their head. A robot called Polly is then described which could make use of this nod detector. Ian concludes his chapter with a discussion about the design of computational agents and their context (habitat) sensitivity.

Geof Staniford describes the way biological and sociological metaphors have acted as sources for the architectural specifications of independent communicating agents in a distributed AI system for supporting cooperative authorship. Geof discusses how ethological ideas have helped to clarify the hierarchical and network structures in his system in which different categories of agent can be characterized by different amounts of informaton processing capability. The simplest agent class exhibits tropistic behaviour whereas the most sophisticated have considerable problem-solving capacities. Differences between communicating agents and their environment are very important in understanding the overall operation of this system.

1.2.5 THEORETICAL AND CONCEPTUAL ISSUES

This section begins with a study of selectionist categorization by Bernard Manderick. Beginning with a discussion of the bootstrap problem for autonomous agent design Bernard shows how a selectionist approach can overcome some of the difficulties. Three selectionist systems are compared and contrasted: neo-Darwinian models of evolution in a population, clonal selection theory applied to the immune system and the theory of neuronal group selection as applied to the brain. The cognitive capacities of these systems are considered and Bernard then focuses his attention on the design of an autonomous cognitive creature, the selectionist automaton. It is important to bear in mind that this chapter provides a number of underpinning principles for other chapters in several other sections.

George Kampis presents an analysis of living systems which reveals both the limits of the computational metaphor in biology as usually defined, and a way of overcoming the limitation based on the self-modifying behaviour of biosystems. An important component of his argument concerns the problem of representing physical systems as computers in that the former cannot be reduced to the latter. He argues that although certain computational systems exhibit a multidimensional modality, they are basically one-dimensional or serial. George uses the idea of a shifting reading frame to show how a biological system can interact with its changing environment and then goes on to describe an abstract machine which can read a knotted tape (shifting reading frame). This leads on to the development of self-modifying component systems which generalize known properties of macromolecular systems. Some computer models of component systems are produced and the results of simulations using these models are presented.

Jim Nyce provides a historical reflection on the cultural context of the Macey Conferences on cybernetics and how a dominant metaphor emerged related to the digital computer. He argues that the analogy set did not simply provide a common language between participants (i.e., filled in a vocabulary), it also provided the means for validating the analogy set (i.e., had an ontological capacity). From here Jim looks at some of the work of Vannevar Bush in relation to his Memex machines. Although never built, these mechanical analog machines provide another insight into the nature of computation, one which in many ways is all but forgotten. This chapter provides a valuable analysis of the way certain metaphors can become firmly established in scientific thought.

Ray Paton seeks to provide a summary of some of the key themes developed in the book and listed in Table 1.1. A number of conceptual issues are addressed beginning with a framework for making transfers of ideas between biology and computing. The results of this analysis are applied to ways of thinking about information processing in neural systems. A fuller application of the analytical scheme is then described involving an examination of the concept of information as variously used when modelling biological and computational systems. A general description of some of the features of a computational non-neural tissue is then given. The chapter concludes with a plea for researchers in this area to strike a balance between the needs for both a common language and a plurality of views.

1.3 AN INTEGRATIVE VIEW OF THE BOOK

There are many paths through this book and many connections between chapters. Table

Table 1.1 Some of the integrative themes in this book

Theme	*Some relevant chapters*
Automata	Thomas and Thieffry, Holcombe, Marijuan, Stark, Davidor, Kampis
Biological sources	Arbib, Carpenter and Grossberg, Bersini and Varela, Collins, Sumida and Hamilton, Kampis
Computational sources	Thomas, Welch, Vertosick, Zajicek
Selectionism	Manderick, Bersini and Varela, Bäck, Koza
Distributed computation	Holcombe, Marijuan, Arbib, Vertosick, Stark, Huberman, Staniford
Engineering applications	Winter, Bersini and Varela, Koza, Bergman, Davidor, Horswill
Medical applications	Hamilton, Zajicek
Nature of bio-information	Marijuan, Stark, Kampis, Paton
Impact of metaphors	Welch, Nyce, Paton

1.1 gives some key themes and some of the chapters which consider them in some way.

This book is both multidisciplinary and multinational. Many different views by people from many different backgrounds are expressed. At this point in the discussion about models and biocomputational issues it is worth reflecting on Hesse's analysis of analogies and models in science (Hesse 1963). In her study she distinguished two modelling approaches. One produces formal, mathematical or logical descriptions and is highly abstract; the other is imaginative, less precise and often analogical in nature. Both types of modelling are found in this book because both types are accepted as valid within their sphere of application. All of the contributions have a common thread whether the focus is biologically motivated computing or computationally motivated biology. There is a great deal of common language; the challenge is whether the lingua franca of biocomputation will remain at the level of common words or will progress to a stage where there can be common meanings.

Finally, before the reader embarks on his or her exploration of this broad topic, it may be pertinent to consider the value of getting hold of a dictionary. Consider the following list:

Biology	crypt-villus unit, tropism, Ito cell, Krebs cycle
Computing	finite state machine, cellular automata, data type, Petri net
Artificial intelligence	connection weight, schema, autonomous agent, LISP

If you are unable to define all the terms on the right it is suggested you will need the kind of dictionary listed on the left.

REFERENCES

Hesse, M. (1963). *Models and Analogies in Science.* Sheed and Ward, London.

Paton, R. C. (1992). Towards a Metaphorical Biology, *Biology and Philosophy* **7**, 279–94.

PART ONE
CELLS

FROM VLSI THROUGH MACHINE MODELS TO CELLULAR METABOLISM 2

Mike Holcombe

2.1 INTRODUCTION

The principal philosophy for this work is the representation of some aspects of cellular metabolism as the computation of suitably defined data. The main benefit is the opportunity to make use of some of the theoretical models of general, concurrent computing systems in the hope that: the computational models may enlighten the biochemical theory; and the analysis of successful parallel biochemical systems may enlighten the theory of parallel computing.

It is not a new insight to try and relate chemical activity with data processing, and the idea of comparing a cell, or even an organism, with a computer has been around for many years. The work of McCulloch and Pitts, Arbib and Rosen have all followed this direction. The relationship between biology and computer science has been quite a fruitful one and has helped in the development of many concepts. The ideas associated with simple models of neural nets have been influential in the theory of computing, for example, and in this chapter we look at other relationships between the two subjects. The recent advances in the theoretical foundations of concurrent computing are likely to be of interest to theoretical biologists attempting to understand complex concurrent systems. We have been interested in developing these connections further, in particular to see what modern theoretical models of computation might say about the theoretical basis for some aspects of biology. The underlying philosophy for much of this work can be expressed as follows. We attempt to relate:

molecules with information

and

metabolic activity with computation.

An early attempt to model metabolic processing was due to Krohn, Langer and Rhodes (1967) who used finite state automata (machines) to model basic metabolic reactions. Their principal aim was to utilize a mathematical theorem which described how any such machine could be decomposed into a collection of very elementary machines; reset machines which consist of machines with two states only and machines that are constructed from basic algebraic constructs called simple groups. The use of a computer algorithm to help with this decomposition was feasible and there was a hope that the information would be useful in trying to understand some of the biochemistry involved.

Machine theoretic models have also been used by other authors (Welch and Kell 1986) to describe and model certain aspects of cellular (and multicellular) activity. In many cases the system is considered to be an input–output system with a set of internal states which change in the light of environmental inputs and produce external outputs. A simple illustration can be given where the

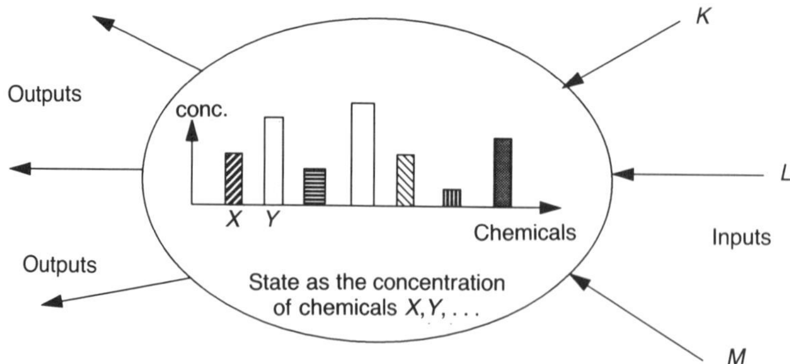

Figure 2.1 A cell as a machine.

internal state of the machine is defined by the concentrations of various chemicals, X, Y, and so on, as shown in Figure 2.1.

Formalizing this work is achieved by using the algebraic theory of state machines which has been an important aspect of the theory of computing (Holcombe 1982).

Definition. A (finite) sequential machine consists of a quintuple:

$$M = (S, A, B, F, G)$$

where

1. S is a finite set of states,
2. A is a finite set of input symbols,
3. B is a finite set of output symbols,
4. $F: S \times A \rightarrow S$ is the next state funtion,
5. $G: S \times A \rightarrow B$ is the output function.

The interpretation is this. Given a system in state $s \in S$ with an incoming input $a \in A$, then, at the next time instant, the state changes to $F(s, a) \in S$ and an output $G(s, a) \in B$ is generated. A pictorial interpretation of a sequential machine is given in Figure 2.2.

In many cases the theory of these machines makes it is possible temporarily to ignore the role of the outputs and concentrate on the state transitions in the model and to then 'stick back' the outputs at a later stage of the analysis. An example of this is the modelling of metabolic activity such as the Krebs cycle. This is an example of an approach that can be set into the context of the 'biocybernetics' or machine theory of biology discussed by Ji (1991). Here metabolic machines are defined and organized into a hierarchy of increasing sophistication based on functionality. We, too, consider a hierarchical approach based on the nature of the interaction between models of different types of metabolic activity.

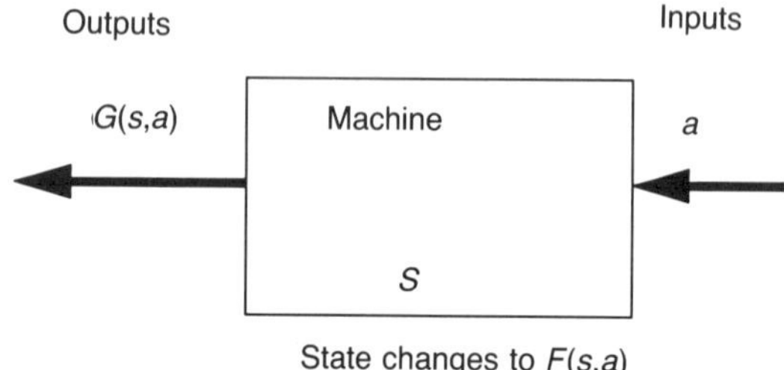

Figure 2.2 A pictorial interpretation of a sequential machine.

Our basic idea is to consider a hierarchy of algebraic models or machines that describe certain organizational and computational features to be found in the biochemical behaviour of cellular systems. This, inevitably, leads us into a more sophisticated class of algebraic structures which provide us with more powerful modelling capabilities. One important motivation for the work is the realization that it is important to integrate architectural features with the functional descriptions of cellular processing. It is not sufficient to assume that the cell is a 'bag containing a soup of interreacting chemicals', we must take account of the fact that much of the processing is located at specific sites within a cell. In order to develop the ideas associated with a hierarchy of metabolic machines we must first consider some mathematical definitions.

2.2 THE MATHEMATICAL BASIS

There are a number of different types of machine that can be useful in the mathematical theory of metabolic organization and computation. The most general setting that we have considered is the *X*-machine of Eilenberg (1974). This machine, in its simplest form consists of a fundamental data type, *X*, which is the basic environment in which computations are carried out. The functions or relations which are the mechanisms which 'do' the computations are defined as operating on *X*, we identify a set, Φ, of such relations on *X*. A set of possible internal states of the system, *Q*, are postulated and a structure within the state space is defined by a relation $F: Q \times \Phi \rightarrow Q$. If $q, q_1 \in Q$ and $q_1 = F(q, \phi)$ then there will be an arc from q to q_1 labelled by ϕ ($\phi \in \Phi$):

$$q \xrightarrow{\phi_x} q_1$$

Thus the function or relation *F* will define a labelled, directed graph with nodes from *Q* and labels from Φ.

Essentially an *X*-machine looks like a finite state machine with one important exception; the labels of the arcs are functions that operate on a fundamental data type, *X*. (Actually the labels could be relations in theory; in practice we usually use functions.)

The way in which a computation is carried out is defined as follows:

1. Select a start state (usually there is only one) and an initial value for the data type *X*.
2. Construct a path from the initial state to a terminal state (in many examples every state is considered a terminal state and so all we are doing there is defining a path through the state space).
3. The path labels form a sequence of functions that operate on *X*.
4. We apply them in turn to the initial value of *X* and end with a final value of *X*.

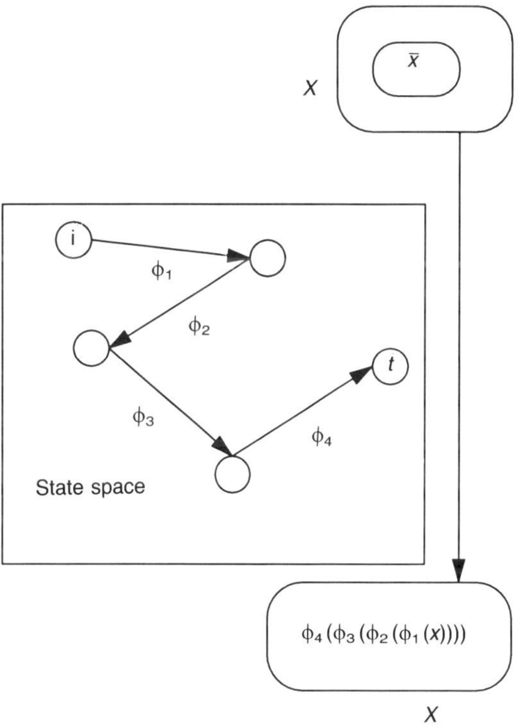

Figure 2.3 A typical *X*-machine computation.

The transformation from the initial value $\bar{x} \in X$ to the final value $\underset{\sim}{x} \in X$ represents the computation carried out by the machine. A specimen computation in a general *X*-machine is illustrated in Figure 2.3.

One benefit of the model is the clear distinction that it makes between the control state of the system and the state of the data being processed at any particular moment, which is given by a value from the construct in Figure 2.3. This is reflected in the way in which computations are defined by sequences of functions that are applicable in certain

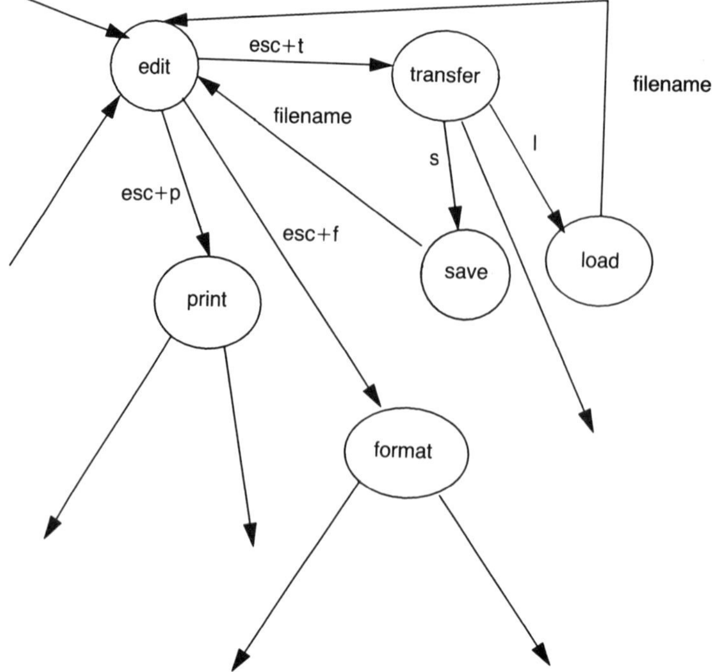

Figure 2.4 Part of the interface of a popular word processor.

states only and to data satisfying specific conditions. Another benefit is that this model provides a fully general computational model as powerful as the Turing model (Eilenberg 1974).

A simple example of the kind of data type X that could be involved is obtained if we consider part of the interface of a word processing system. There is a type Document which describes the content and format of a document in the system. The treatment of the two types of state, the system state which defines which function is available, for example '(file) transfer', 'edit', 'print', and so on (indicated in Figure 2.4) or the data state which is the exact form of the document at that moment is both clear and convenient. In some cases the state of the control is determined from aspects of the screen display; for example, a highlighted menu item, an open window in the foreground; and the state of the data is also indicated, in part, by the display, such as the contents of a window or area of screen which might be text in the case of a document or an annotated diagram in the case of a diagram editor. The role of the Display is thus crucial as is its link with the two state types.

The machine in Figure 2.4 has the set of internal states $S = \{$edit, transfer, print, save, format, load, . . .$\}$ and various directed arcs that link some of these states.

The machine computes over the following data type X. First let Char denote the set of all characters including space, punctuation, end_of-line, and so on. Let Font denote the set of possible font types $\{$bold+italic+times, standard+straight+courier, and so on$\}$.

A document will consist of a number of paragraphs with the cursor located somewhere in the middle of the document. Let Para = Seq[Char $\times$ Font] denote the set of

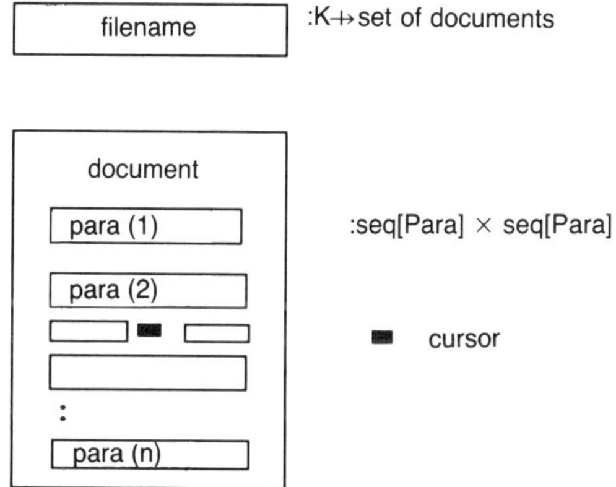

Figure 2.5 The fundamental data type of the system given in Figure 2.4.

all sequences of character-font pairs. The Document type is then a pair of sequences of Para with the interpretation that the cursor is positioned at the end of the first sequence of paragraphs. The filing system consists of a partial function from K to the set of documents, where K is the set of filenames. The Data type X is then the set consisting of pairs of items, the filing function and the actual document. It is described mathematically as:

$$[\text{K} \nrightarrow \text{set of documents}] \times [\text{seq[Para]} \times \text{seq[Para]}]$$

which corresponds to the diagram in Figure 2.5.

The specific functions labelling the arcs of the machine are of the form 'esc+t', 'filename', and so on. These define various functions that affect the state of X. Some require user input in the form of strokes, mouse movements and clicks.

A computation is defined with reference to a prescribed initial state and a set of terminal states. Given the particular data state $x \in X$ a path is selected, possibly nondeterministically, from the initial state to a terminal state. This path determines a sequence of path labels derived from the relation F and the elements of this sequence are taken from the set Φ. By composing the relations used as labels of the path we can form a relation defined on X which can then be applied to the initial data 'state' x to obtain the result of the computation. This is a somewhat simplified description of these machines and the full model involves encoding and decoding relations to allow for the computation of more general types of computable relation.

These machines generalize most of those previously studied and do provide a general framework in which to discuss computational issues. There is a range of types of machine worthy of study including finite state automaton, push-down machines and Turing machines and some of these provide suitable frameworks in which to discuss many metabolic processes. The main weakness of these machines is in the modelling of massively parallel computational systems and here we use, instead, a related concept called a hybrid data net. The complete theory of X-machines seeks to provide a framework within which we try to describe the computational theory of any structure. The results from this theory will be useful in defining a computational model of the function and organization of many levels of metabolic activity.

2.3 THE COMPUTATIONAL ORGANIZATION OF THE CELL

2.3.1 THE BASIC LEVEL MACHINE – SIMPLE METABOLIC COMPUTATION

An example might explain the idea. Consider the classic Krebs cycle pathway. We can identify a number of molecular components: substratal components such as citrate, *cis*-aconitate; imported cofactor components like acetyl Co-A, NAD; and the resulting output components such as Co-A, NADH.

There are clearly other important components including enzymes. In this model we assume that sufficient enzymes are available for the reactions to take place. A substrate such as oxaloacetate reacts with an 'input' such as acetyl Co-A to form an 'output' Co-A and a new state – citrate.

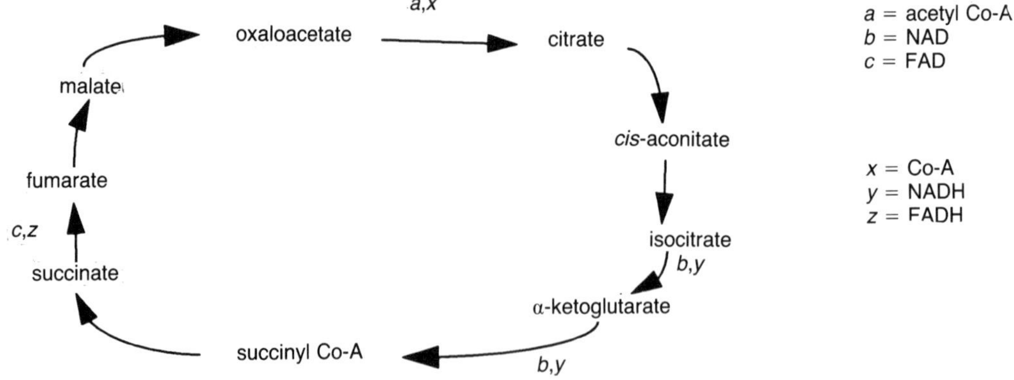

Figure 2.6 The Krebs cycle as a state machine.

The organization of this computation, whereby the cofactor molecules imported into this machine from other parts of the system drive the processing of the cycle at a molecular level, can be thought of as a simple finite state machine. The basic concepts are those of internal state represented by the substratal components; system inputs represented by the imported cofactors; and system outputs represented by the output components.

The basic organization is depicted in Figure 2.6 which interprets the important steps of the cycle as state transitions labelled by the input–output relationships. Such machines can be regarded as very simple examples of *X*-machines.

In Figure 2.6

$$\text{oxaloacetate} \xrightarrow{a,\ x} \text{citrate}$$

means that the input when the system is in state oxaloacetate gives rise to the output x and the next state citrate and so on. It may not seem immediately obvious that this type of machine is a special case of an *X*-machine but this is in fact the case and is described in detail in Eilenberg (1974). Essentially the set X is defined to be the set of all pairs of strings of input and output symbols,

$$X = \Sigma^* \times \Gamma^*$$

where $\Sigma = \{a, b, c\}$ and $\Gamma = \{x, y, z\}$. The functions on the arcs systematically remove symbols individually from the beginning of the input string and place them on the end of the output string, thus the function between the states oxaloacetate → citrate is of the form

$$(al, m) \to (l, mx) \quad \text{where } l \text{ is a string from } \Sigma^* \text{ and } m \text{ is a string from } \Gamma^*.$$

Further development of this simple basic state machine idea is possible and is considered in Ji (1991), Holcombe (1990, 1991) and, originally, in Krohn, Langer and Rhodes (1967).

We can then look at the mathematical theory of these machines to see what it might tell us about the biochemical system. For example, the decomposition theory for finite state machines depends on the fact that any finite state machine can be turned into a transformation semigroup in a natural way.

We consider the set of states, S. Each input string $a_1, a_2, \ldots, a_n$ defines a function $f(a_1, a_2, \ldots, a_n)$: $S \to S$. (We note that $\#S$ is finite so there are only a finite number of such functions possible; however there are an infinite number of possible input strings.) We identify those functions that can occur as functions defined by strings. This set of

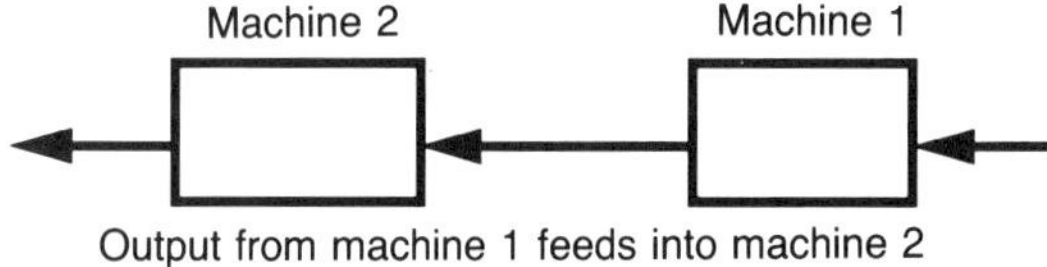

Figure 2.7 The series connection of two machines.

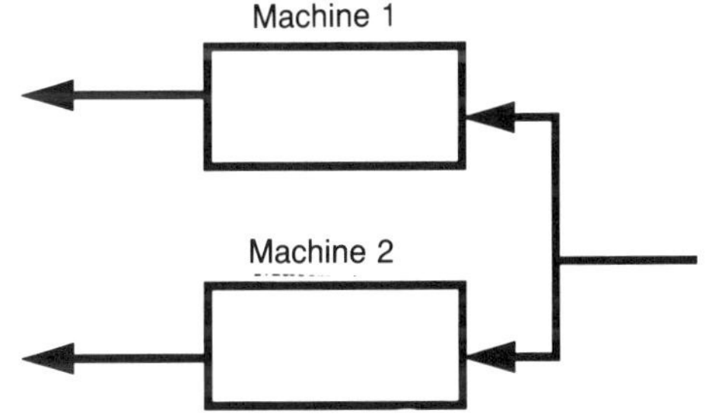

Figure 2.8 The parallel connection of two machines.

functions is a semigroup – that is, it is a finite set with a binary operation (function composition).

Let T be the set of all functions on S that arise from strings. Then for any two such elements t, t_1 from T we can construct $t_0\ t_1$ which is also a member of T. (Recall that t, t_1: $S \rightarrow S$ and so $t_0 t_1(s) = t_1(t(s))$.) The pair (S, T) forms what is called a transformation semigroup. Various theorems can now be examined.

2.3.2 DECOMPOSITION THEOREMS

Any transformation semigroup can be replaced by a collection of simpler transformation semigroups connected together in suitable ways. Effectively the two principal ways of connecting together two transformation semigroups originate from series connections (wreath products) as shown in Figure 2.7 and parallel connections (direct products) as shown in Figure 2.8.

The simple machines can be constructed from finite simple groups (all of these are now known) and simple aperiodic semigroups (essentially machines with a very trivial structure; the state transition functions map the entire set of states onto a single state (Holcombe 1982)).

We might now try to see what the Krebs cycle example tells us. It can be decomposed into wreath products of these simple groups and aperiodic semigroups, but the resulting algebraic constructs are difficult to interpret biologically.

The component machines are constructed from cyclic groups of order 2 and 3; and two aperiodic semigroups of orders 2 and one of order 3. The hope is that the algebra can inform the biology as summarized in Figure 2.9.

Further machine models can be built to try and describe the metabolic pathways but the situation is not entirely straightforward; we cannot just put together state machine models of different metabolic pathways using the constructs mentioned above (wreath and direct products). There is also the problem of how the overall system is controlled; we need to address the inherent parallelism of

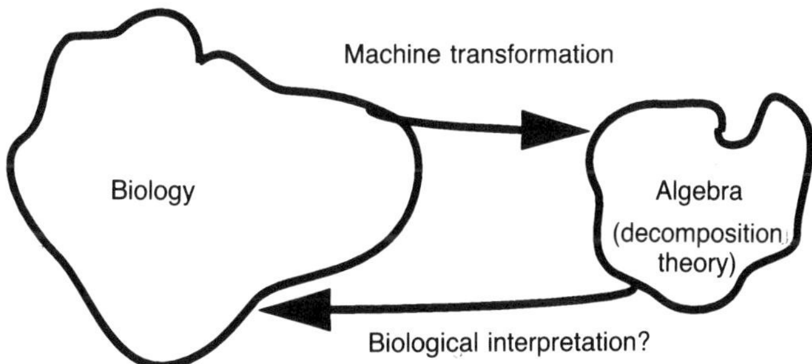

Figure 2.9 An approach to the algebraic analysis of biology.

the system and this can be achieved with another series of models at a higher level of abstraction.

2.3.3 HIGHER LEVEL METABOLIC MACHINES AND CONCURRENCY

The machines that have been discussed so far are quite effective for modelling the 'bottom level' metabolic processing. However they suffer from a number of limitations:

1. They cannot model concurrent processing very well.
2. They rapidly become very complex and difficult to analyse.
3. Their computational capability is not powerful enough to handle many sequential processing situations.

The possible solution to these difficulties involves the development of hierarchical models that involve fully general computations. This is an area of study that is discussed further in Ji (1991) where a hierarchy of machines is described.

Here, we will approach the problems of complexity and concurrency by considering different levels of metabolic processing and their interrelationships. At the bottom level we have the 'simple' machines that model specific metabolic reactions such as our example of the Krebs cycle. Central to these machines are assumptions about, for example, the availability of enzymes which enable the different stages of the reactions to take place. To model this vital control aspect we can use a type of model that is positioned at a higher level than the foregoing. We use a Petri net (Reisig 1985) for the modelling of some of the concurrent enzyme control processing that might be found in a general cell. The individual state machines representing the metabolic subsystems such as Krebs, are represented as 'transitions' which are related to other transitions by virtue of sharing some product or result and the whole system is controlled by resources, such as the availability of enzyme catalysts which enable transitions (basic reactions) to operate. Thus certain reactions – which might produce enzymes – are produced by certain transitions and these enzymes are represented by tokens (dots within a place) which are donated to various other transitions. Thus we see from Figure 2.3 that reactions t_1 and t_3 have received their enzymes or other inputs from elsewhere in the system and can then operate. When they have completed a computation they then send their products, which might be further enzymes, to the places 3 and 4 which control the transitions t_2, t_4 and t_5. Only when t_5 receives both its resources is it enabled to happen. This then feeds on – or back – to other parts of the complex network of interacting reactions. This type of system – we call it an enzyme control net – provides a higher level model which, from a control point of view, sits on top the basic level processing machines. Such a net is shown in Figure 2.10.

At the next level we may try to model the way in which the enzyme control system is itself controlled by further complex interreactions involving DNA transcription and RNA synthesis and translation. I think that it is likely that the hierarchy of machines will involve many more different types of machine. Some, probably the lower level ones, may be state machines of the type we have looked at, others may be described using net-like systems, and so on. With this plethora of machine types it is useful to try to relate them in some mathematical way. The fundamental property of the *X*-machine is that it is possible to regard all of these as special cases of the *X*-machine and thus we have potential for a substantially unified theory for the subject (Holcombe 1990, 1991). However, the benefits of *X*-machines do not end there. As we mentioned before, we wish to relate processing with processors, that is, function with structure. The next section demonstrates how *X*-machines can achieve this.

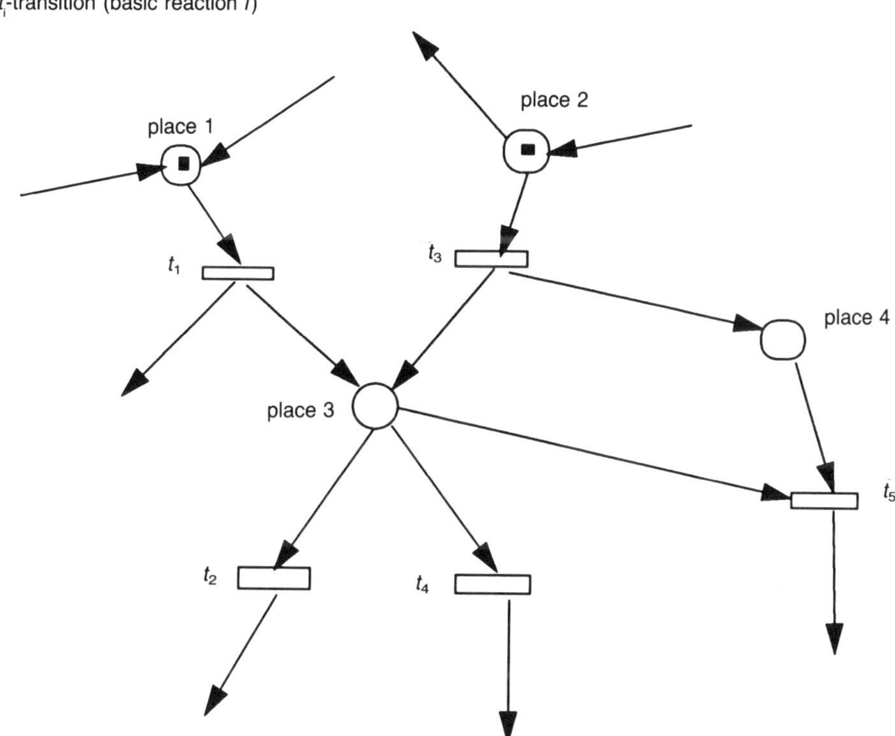

Figure 2.10 An enzyme control net.

2.3.4 AN ARCHITECTURAL MODEL

The basic problem with these models, as has already been mentioned, is their treatment of the cell as a 'bag of chemicals' with no account taken of their internal structure. Our next model tries to address this failing.

The origins of this model are in the field of VLSI – the silicon level implementations that lie at the heart of most modern computing systems. A simple example may demonstrate the idea. The system described in Figure 2.11 is of a chip that controls the behaviour of a robot arm (Holcombe and Pang, in press). We may describe the operation of this computational system as follows.

The information describing the last position of the arm is fed in at the input and stored temporarily in the register IN. The registers I and D contain particular constants that describe the nature of the path required for the robot head to traverse. A register is a particular device that can store information – in these cases binary numbers containing 8 or 16 digits as appropriate. The controller organizes the way in which the computation in the chip is carried out; it switches on and off various components – registers, adders, and so on – and thus allows data to be processed and passed through the system. Finally the appropriate information is sent out of the OUT register and this controls the next position of the robot head.

It was a simple matter to construct an *X*-machine model of this device which told us explicitly what was happening and when. We will look at a few stages in this process, bearing in mind that our proposal is to use a simple version of this type of model to try and model the architectural basis of the

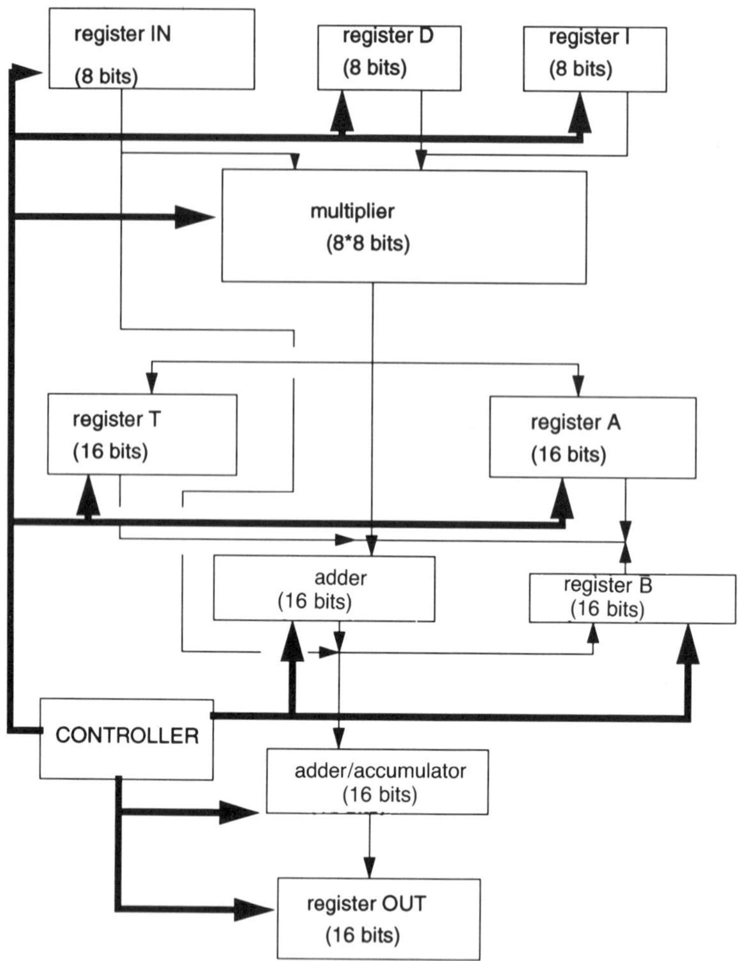

Figure 2.11 A chip architecture.

processing in a cell. The appearance in many biological examples of organelles and substructures in cells that play a similar role to registers – temporary stores of substances (information) and adders, simple functional processing units (perhaps metabolic machines (Ji 1991)) – is possibly an insight worth sharing. The detailed modelling of a specific biological system using this approach is underway.

If we look at Figure 2.12 we see a very simple model of the dynamics of the chip.

It is convenient to represent an *X*-machine using a combination of a diagram to indicate

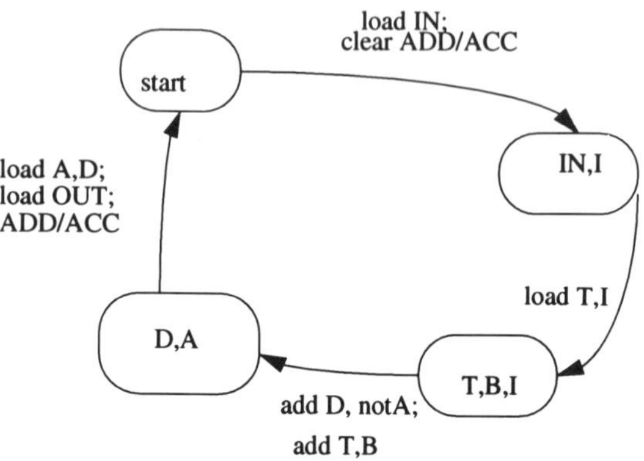

Figure 2.12 An *X*-machine model of Figure 2.4.

the possible states and the state transitions together with labels on these transitions which represent functions on the data type X. The data type X must also be defined formally. We have to identify an appropriate candidate for the data type X and in this case a sensible definition, in the spirit of Eilenberg's canonical representation (Eilenberg 1974), is the set

$$X = [\mathrm{OUT}] \times [\mathrm{A}] \times [\mathrm{B}] \times [\mathrm{D}] \times [\mathrm{T}] \times [\mathrm{I}] \times [\mathrm{ACC}] \times [\mathrm{IN}].$$

Here the notation [A] means the set of all possible data values held in register A, and which in our case are all possible 16 bit values, and similarly for [B], [D] and so on. The X-machine will be carrying out computations on this set of data values which will represent the flow of data from the input to the output and each cycle of the machine will correspond to a cycle of the algorithm.

The next issue is the identification of the states of the machine. A sensible choice is to decide which registers are enabled, at any moment of time. In Figure 2.12 the states are indicated by ellipses and the enabled registers in that state are listed inside the ellipse.

We need to define the functions that label the transitions in the diagram. The functions are described by their effect on the data type X, so each function maps

$$\mathrm{OUT} \times \mathrm{A} \times \mathrm{B} \times \mathrm{D} \times \mathrm{T} \times \mathrm{I} \times \mathrm{IN} \times \mathrm{ACC} \times \mathrm{IN}$$

into

$$\mathrm{OUT} \times \mathrm{A} \times \mathrm{B} \times \mathrm{D} \times \mathrm{T} \times \mathrm{I} \times \mathrm{IN} \times \mathrm{ACC} \times \mathrm{IN}$$

Thus we have the function

$(\_,\_,\_,\mathrm{D},\_,\mathrm{I},\_,\_,x) \rightarrow (\_,\_,\_,\mathrm{D},\_,\mathrm{I},x,\_,x)$
[representing the function LOAD IN]
$(\_,\_,\_,\mathrm{D},\_,\mathrm{I},\_,x,\_) \rightarrow (\_,\_,\_,\mathrm{D},\_,\mathrm{I},\_,0,\_)$
[CLEAR ADD/ACC]
$(\_,\_,\_,\mathrm{D},\_,\mathrm{I},\_,\_,\_) \rightarrow (\_,\_,\_,\mathrm{D},\mathrm{I},\mathrm{I},\_,\_,\_)$
[LOAD T,I]
$(\_,\_,y,\mathrm{D},x,\mathrm{I},\_,z,\_) \rightarrow (\_,\_,y+x,\mathrm{D},x,\mathrm{I},\_,z,\_)$
[LOADB, (T+B)]

and so on.

The details are in Holcombe and Pang (in press), where this machine was refined further. One point about the refinement is that we have to remove some of the concurrent actions carried out by functions and replace them with simpler functions at the expense of increasing the number of states and introducing a set of possible paths through the system. From our experience with this VLSI example, it seems likely that such a model can be analysed in a reasonably simple way. For example, we were able to functionally verify that the control algorithm of the chip satisfied the requirements. Thus we might expect to be able to reconstruct the overall mathematical activity of the system from a knowledge of its individual component behaviours.

2.3.5 APPLICATION TO CELLULAR PROCESSING SYSTEMS

The question here is what this has to do with biology. Let us consider a simple, idealized cellular architecture as depicted in Figure 2.13. The general information flows are indicated using the thick arrows and are meant to be indicative of the sort of functional relationships that might apply in different processing sites. The analogy with the VLSI model is strengthened when we consider some of the sort of processing that might be involved. For example we have simple metabolic processes that might correspond to the behaviour of a register, that is, information (molecules) might be stored temporarily at a location prior to being further transformed by another processing site; we might have sites performing fairly simple processing akin to that of an adder which operates when its input places contain appropriate values; and then there are structures like multipliers that operate continuously. The overall model of the system could then be determined with reference to the data types involving the various active sites and the functions operating on the system described at this level.

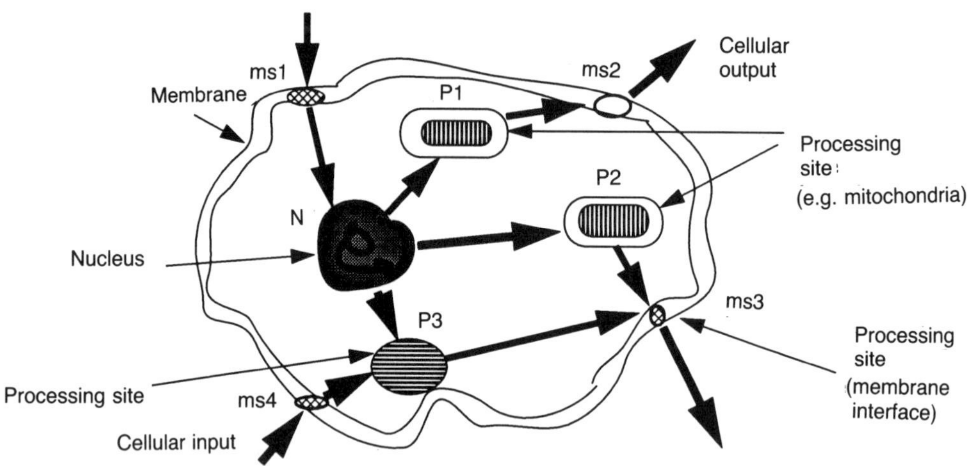

Figure 2.13 A hypothetical cell indicating active sites.

The fundamental data type will be of the form:

$$X = \text{INms1} \times \text{INms4} \times [\text{MS1}] \times [\text{MS2}] \times [\text{MS3}] \times [\text{MS4}] \times [\text{N}] \times [\text{P1}] \times [\text{P2}] \times [\text{P3}] \times \text{OUTms2} \times \text{OUTms3}$$

where each element defines the type of values that the site can deal with and would be defined in a formal manner following our previous work (Holcombe 1990; Eilenberg 1974), INms1 and INms4 are the input data types at ms1 and ms4 respectively and OUTms2 and OUTms3 are the output data types at ms2 and ms3 respectively.

The next stage is to identify when the various sites are active and from this we construct a top level model of the system using the specific processing functions that determine what happens to the data.

In Figure 2.13 we note the existence of functions indicated by the arrows. Thus we can list the functions and the appropriate data types. As a convention let us indicate input data types and output data types to sites as follows. For site K we define IN(K) and OUT(K) to be the appropriate types and COMP(K): IN(K) $\nrightarrow$ OUT(K) to be the specific processing function (which might be a partial function allowing for processing to be undefined under certain data conditions). The processor then stops until the situation is resolved.

It should be remarked that, although discrete models were the inspiration for this approach, we are making no assumptions about the type of the functions and types being considered here. What we do need is some concept of the start and the end of processing activity by each component site. This may raise issues, later, of time and its role in these models. It is possible to augment the X-machine model with time information but we will not deal with this added complication here. An example dealing with this is to be found in Mezhoud and Holcombe (1992).

Returning to our hypothetical cell, we note that in some cases there are relationships between types. For example, the environmental inputs and outputs are: INms1, INms4, OUTms2, OUTms3, where

IN(ms3) = [P3] × [P2],
IN(P3) = [ms4] × [N],
IN(P1) = [N],
and so on.

The component functions are described thus:

COMP(ms1): INms1 ⇸ [ms1],
COMP(ms2): [P1] ⇸ OUTms2,
COMP(ms3): [P2] × [P3] ⇸ OUTms3,
COMP(ms4): INms4 ⇸ [ms4],

COMP(N): [ms1] ⇸ [N],
COMP(P1): [N] ⇸ [P1],
COMP(P2): [N] ⇸ [P2],
COMP(P3): [ms4] × [N] ⇸ [P3],

The system functions are all defined on *X* and operate on the elements of *X* as follows:

FUNms1: (il,_,a,_,_,_,_,_,_,_,_) → (il,_,il,_,_,_,_,_,_,_,_);
FUNms4: (_,i2,_,_,_,a,_,_,_,_,_,_,_) → (_,i2,_,_,_,i2,_,_,_,_,_);
FUNN: (_,_,a,_,_,_,_,_,_,_,_) → (_,_,_,_,_,_,b,_,_,_,_)
where COMP(N)(a) = b ∈ [N];
FUNP1: (_,_,_,_,_,_,a,_,_,_,_) → (_,_,_,b,_,_,_,_,_,_,_)
where COMP(P1)(a) = b ∈ [P1];
FUNP2: (_,_,_,_,_,_,a,_,_,_,_) → (_,_,_,_,_,_,_,_,b,_,_,_)
where COMP(P2)(a) = b ∈ [P2];
FUNP3: (_,_,_,a,_,_,b,_,_,_,_) → (_,_,_,_,_,_,_,_,_,c,_,_)
where COMP(P3)(a, b) = c ∈ [P3];
FUNms2: (_,_,_,_,_,_,_,a,_,_,_,_) → (_,_,_,_,_,_,_,_,_,_,b,_)
where COMP(ms2)(a) = b ∈ OUTms2;
FUNms3: (_,_,_,_,_,_,_,_,a,b,_,_) → (_,_,_,_,_,_,_,_,_,_,_,c)
where COMP(ms3)(a, b) = c ∈ OUTms3;

Note that we have, at ms3, an example of 'fan-in', whereby the data paths diverge at N and come back together at ms3. This is a rather undesirable feature in VLSI design; it is not clear if it occurs naturally in cellular metabolism and, if so, what consequences it has.

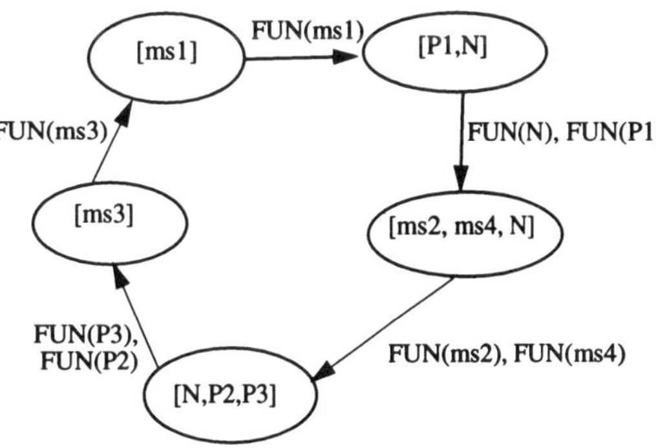

Figure 2.14 A possible state space describing the functional dynamics of the cell.

The processing is determined by what sites are active and what functions are processing and we can postulate an abstract scenario for such a system in Figure 2.14. However, we are not claiming this to be a realistic example; it is just an indication of the possibilities of the method.

It would be possible, once the specific details of the various functions were known, to simulate the situation of such a system operating and to analyse various aspects of its behaviour. Let [P1, N, ms3] mean that each of the sites P1, N, ms3 are active and processing. From this we can detect a certain amount of simple parallelism. For example, there is simultaneous processing by N and P2 and this can lead to a refinement of the state space by introducing new states and interleaving the functions as described in the VLSI example. A refinement of the model in Figure 2.14 is shown in Figure 2.15. This new model expands the concurrency and demonstrates how we can replace the parallel processing by sequential processing at the expense of requiring a larger collection of paths through the state. This might assist in modelling some aspects of the system. Other aspects, such as time factors, will need other types of model and will be an interesting area for further

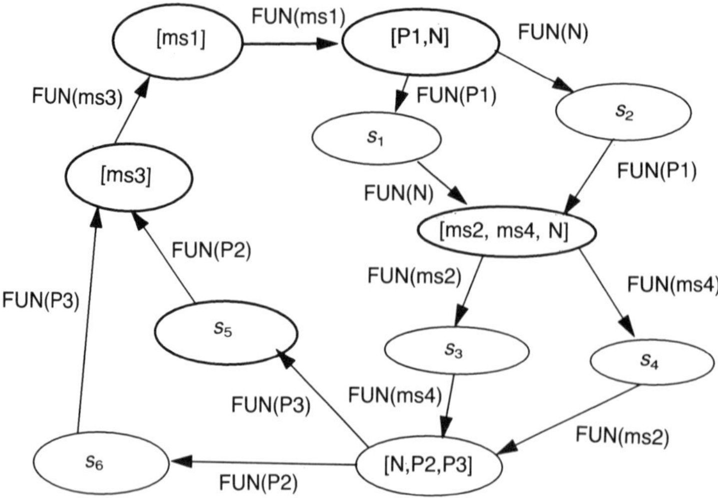

Figure 2.15 A possible refined state space (s_1, s_2, . . . , s_6 are new states introduced during the refinement).

study. Naturally, the state space for a real cell will be an order of magnitude more complex.

2.4 CONCLUSIONS AND FURTHER WORK

The issue of what happens in more detail at each active site can be explored by using the same technique, albeit at a lower level. Thus an individual mitochondrion can be modelled more explicitly, identifying specific sites of active processing and building a machine model whose computational function equals the higher level function (such as COMP(P1)) identified in the model described in the last section.

We have discussed some preliminary ideas for overcoming the problems associated with previous algebraic models of cellular metabolism which tend to ignore the explicit connection between processing and physical location. The proposed model will not provide a complete picture, however, and we still maintain that there is a need for a hierarchy of machine-type models to describe various aspects of these complex biochemical systems. It is clear from software engineering, where systems of a similar order of complexity have to be described and analysed, that a number of different models are required, each capturing a particular 'view' of the system at a suitable level of abstraction.

These ideas need to be properly evaluated and the next stage is to try and apply this theory to a real example. One promising area is photosystem II in noncyclic photophosphorylation. The principal architecture of the model is based around three subsystems:

1. water oxidization;
2. electron-generation;
3. a photon–electron transducer system.

We hope to report on this investigation later.

It is perhaps too early to say whether these models may enlighten our knowledge and understanding of the biochemistry of a cell, but it might well be possible to use this approach to construct symbolic simulators of these complex systems. Conversely, the organization of successful, fault-tolerant, concurrent systems, such as we find in biology, will provide some inspiration for the design of more sophisticated and effective computational devices. It may even be possible to harness the natural computing properties of some biological systems to construct hybrid machines. The story is still unfolding and conclusions must await much more study.

REFERENCES

Eilenberg, S. (1974). *Automata, Languages and Machines*, Vol. A. Academic Press, London.

Holcombe, M. (1982). *Algebraic Automata Theory*. Cambridge University Press.

Holcombe, M. (1990). Towards a formal description of intracelluar biochemical organization. *Comp. Math. Applic.* **20**, 107–15.

Holcombe, M. (1991). Mathematical models of cell biochemistry. In *Molecular Theories of Cell Life and Death* (ed. S. Ji). Rutgers University Press, NJ.

Holcombe, M. and Pang, R. (in press). Formal specification and verification of a VLSI device – a reverse engineering case study.

Ji, S. (1991). Biocybernetics: a machine theory of biology. In *Molecular Theories of Cell Life and Death* (ed. S. Ji). Rutgers University Press, NJ.

Krohn, K., Langer, R. and Rhodes, J. L. (1967). Algebraic principles for the analysis of a biochemical system. *J. Theor. Biol.* **116**, 399–426.

Mezhoud, B. and Holcombe, M. (1992). Formal specifications and verification of a hard-time user interface. Department research report, University of Sheffield, UK.

Reisig, W. (1985). *Petri Nets*. Springer, Berlin.

Welch, G. R. and Kell, D. B. (1986). Not just catalysts – molecular machines in bioenergetics. In *The Fluctuating Enzyme* (ed. G. R. Welch). Wiley, New York, pp. 451–92.

DEVELOPING A LOGICAL TOOL TO ANALYSE BIOLOGICAL REGULATORY NETWORKS 3

René Thomas and Denis Thieffry

3.1 INTRODUCTION

Several years ago, one of us (René Thomas) was invited by the great topologist René Thom to deliver a seminar at the Institut des Etudes Supérieures in Bures-sur-Yvette, near Paris. After the discussion, René Thom told him that the logical method he had just described could only have been developed by a biologist. Knowing the slightly mischievous spirit of René Thom, several interpretations are conceivable; the most favourable one can be taken as a justification for including the present chapter in this book on the impact of biology on informatics.

Our group has been involved for twenty years or so with the following challenge: given a network of elements which can interact with each other positively or negatively, derive the possible dynamics of the system from the graph of its interactions. The origin of the problem was biological: after many years of experimental work on the genetics of temperate bacteriophages, it was realized that even in such simple living beings, one has to consider sets of elements (gene products, etc.) which form complex networks of interactions. These networks, which can be symbolized by oriented, signed graphs, are usually too intricate to be directly understandable; more specifically, one can describe them verbally by saying that this gene exerts a positive control on that gene and so on, but the possible dynamics of the system are far from obvious and, in fact, turn out often to be perfectly anti-intuitive.

Facing this situation, the senior author decided to try developing a method appropriate for this type of analysis. He immediately (a) realized that the probability of success was low and (b) was convinced that, if he succeeded, the method would be usable not only in genetics but in other fields in and outside biology as well. Although the first prognostication was pessimistic, we succeeded. The second prognostication turned out to be correct insofar as the method has since been applied in fields like genetics (Thomas and D'Ari 1990; Thieffry and Thomas, in press; Thieffry, Verdin and Colet 1993), immunology (Kaufman 1988), climatology (Nicolis 1982) and psychiatry (Ciompi, Ambühl and Dünki 1992). Nevertheless, in agreement with the statement of René Thom, we now realize that most specific features of the method we developed have been deeply impregnated by the biological reality.

This paper is not a technical one. All the information concerning the practical use of our logical method can be found in Thomas and D'Ari (1990), Thomas (1991), Thieffry, Colet and Thomas (1993), Thomas, Thieffry and Kaufman (1994), and the formal basis of our present approach is developed in

Snoussi and Thomas (1993). In this chapter, we shall describe characteristic aspects of our logical method with a major aim of this book in mind: how did biological considerations influence the development of a method which can also be used outside biology?

In order to reach this goal, we shall first give the minimal information required to understand our problems and the general way adopted to solve them. Then we shall discuss the interest of using a logical description, that is, a description in which variables have a finite (and in fact low) number of values: 0 and 1 wherever tolerable, otherwise 0, 1, 2, . . .

Several aspects of the logical description will be discussed:

1. asynchronous vs. synchronous;
2. the use of variables with more than the two classical (0 and 1) values where required;
3. the introduction of logical parameters (Snoussi 1989) which give a proper weight to each term of a logical expression;
4. the introduction of the thresholds as logical values;
5. the logical identification of all steady states;
6. the concept of loop-characteristic states;
7. the consideration of a set of interacting loops rather than a set of individual interactions.

As we will see, the generalized logical method recently developed sheds a new light on the relation between feedback loops and steady states. In doing so, it helps in the understanding of the logical basis of regulation, and particularly the specific roles of the two types of feedback loops in biology. On the other hand, from a much more general viewpoint, it should provide a significant contribution to the qualitative (more specifically, topologic) view of complex systems described by sets of ordinary nonlinear differential equations. The major novelty consists of treating systems as sets of interacting feedback loops rather than as sets of individual interactions.

3.2 REGULATORY NETWORKS

In order to function properly, a complex system must somehow be informed of the level of at least some of its elements, and have ways to take this information into account. Regulations may be defined as the constraints which adjust the rate of production of the elements of a system to the state of the system itself and of relevant environmental variables. Biological systems usually comprise complex networks of interacting elements whose wheels are feedback loops (see Section 3.2.2).

3.2.1 THE SHAPE OF REGULATORY INTERACTIONS

In biology, and probably also elsewhere, most regulatory interactions are nonlinear. Consider a positive regulator involved in the expression of a gene. For increasing concentrations of the regulator, we have an increasing rate of expression of the gene; however, this relation eventually levels out, if only because other factors limit the rate of expression of the gene. On the other hand, one generally observes that the regulator is almost inefficient unless its concentration exceeds a minimal value. The relation between the concentration of a regulator and its effect is thus doubly nonlinear: there is a threshold of concentration below which the regulator is inefficient and a boundary value of the rate of expression of the regulated gene. This results in sigmoid curves (Figure 3.1(a)).

The most obvious way to treat complex regulatory networks uses sets of differential equations. The fact that many interactions are nonlinear enormously complicates the situation, because in most cases there is no analytical solution. One might be tempted to use a 'linear caricature' (Figure 3.1(b)) but this turns out to be acceptable only in the close vicinity of the steady states; elsewhere, the

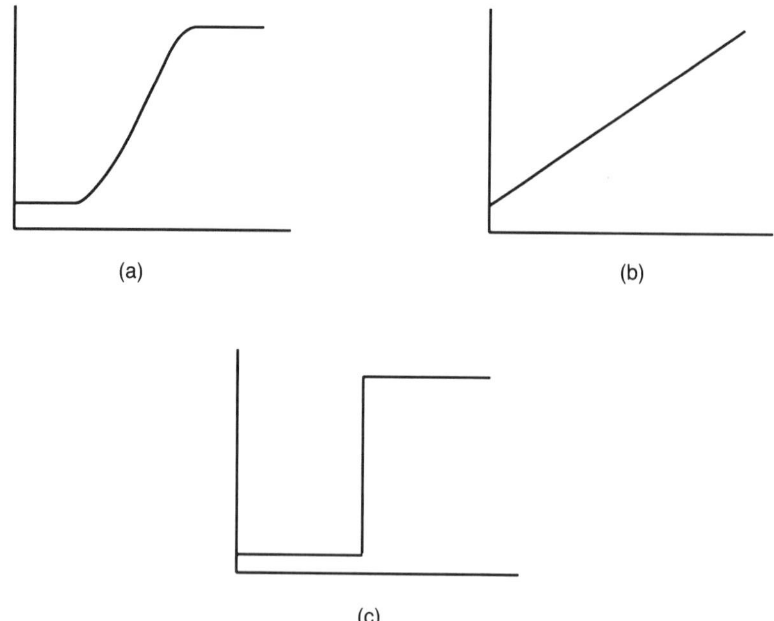

Figure 3.1 A sigmoid curve and its caricatures: (a) sigmoid curve; (b) linear caricature; (c) infinitely nonlinear caricature (step function).

complex dynamics vanishes, showing in fact that the nonlinear character of the interactions is not merely a boring complication, but also an essential feature of regulatory systems. One can also take the diametrically opposite attitude and use instead an infinitely nonlinear caricature in which the actual sigmoids are represented by step functions (Figure 3.1(c)). This is the basis of the so-called 'logical' description, in which one uses variables with a small number of values (in the simple cases, only two: 0 and 1). It has become clear (Glass and Kauffman 1973; Thomas and D'Ari 1990) that the essential dynamical behaviour of sigmoid systems displays a low sensitivity to the slope of the sigmoids. In other words, there is a wide range of steepness of the sigmoids (including the limit case of a step function) within which the global behaviour of our systems remains essentially the same. Since it is much easier to analyse systems whose interactions are step functions, the 'logical' caricature is at the same time convenient and justified.

3.2.2 FEEDBACK LOOPS

One speaks of feedback loops* when elements of a system interact in a cyclic way, as for example if gene x exerts a control on the expression of gene y which exerts a control on the expression of gene z, which in turn influences the expression of gene x. A feedback loop is the way a system can measure the levels of its elements and take them into account; for example, one type of loop tends to stabilize the levels of its elements near supposedly optimal values. For this reason, one can say that feedback loops are the wheels of regulation.

The structure and functions of feedback

*The word 'loop' is unfortunately used not only with different but sometimes opposite meanings. In graph theory, the general word used to describe a cyclic set of interactions is 'circuit'; 'loop' refers to a direct action of an element on itself, in other words, to a one-element circuit. At the other extreme, Tyson (1975) uses 'loop' for circuits with at least three elements. Biologists use the word 'feedback loop' for circuits of any length, and here we adopt this terminology.

loops are discussed at length, for instance in Thomas and D'Ari (1990). Let us simply mention here some essential points:

1. In a feedback loop, each element exerts a direct action on one element only (its follower) and is acted on directly by one element only (its predecessor). However, if one considers not only the direct but also the indirect effects, one can tell that each element acts on all the elements of the loop including itself.
2. In any loop, either each element exerts on itself a positive action (activation), and accordingly one speaks of a **positive loop**; or each element exerts on itself a negative action (inhibition), and accordingly one speaks of a **negative loop**.
3. Whether a loop is positive or negative depends on the **parity** of the number of **negative** interactions in the loop; if even, one deals with a positive loop, if odd, with a negative loop.
4. Negative loops generate **homeostasis**, with or without oscillations; the best examples are the thermostat and the Watt regulator.
5. Positive loops generate **multistationarity**; insofar as multistationarity is the biological modality of differentiation, positive loops have a crucial role in differentiation.
6. As the essential function of negative loops is to generate homeostasis and the essential function of positive loops is to generate multistationarity, we say that a loop is **functional** if the parameter values are such that it can actually exert its function (homeostasis if negative, multistationarity if positive).

3.3 LOGICAL DESCRIPTION OF REGULATORY NETWORKS

3.3.1 LOGICAL DESCRIPTION: GENERAL

In simple cases, we use a Boolean description; the logical variable x has the value 1 if the (real) value $\boldsymbol{x}$ exceeds the threshold value $\boldsymbol{s}$ (x 'present'), and $x = 0$ if $\boldsymbol{x} < \boldsymbol{s}$ (x 'absent'). For example, when we know that a process depends on the fulfilment of one or more of the conditions – presence of x, absence of y – we write $x + \bar{y}$ (x or not y). If both conditions are required we write $x \cdot \bar{y}$ (x and not y). The symbol '+' is the logical sum (inclusive or), the symbol '·', the logical product (and).

3.3.2 LOGICAL DESCRIPTION: ASYNCHRONOUS VS. SYNCHRONOUS

In order to introduce time into the logical description, the classical approach consists of giving the situation at time $t + 1$ as a function of the situation at time t:

$$x_{t+1} = f(x, y, z, \ldots)_t, \quad \text{etc.}$$

This can be described by a state table in which the left column gives the whole repertoire of the possible states at time t; for each state at time t, the right column gives the state at time $t + 1$, that is, the 'next' state. Clearly, in this so-called 'synchronous' description, each state has a single possible follower; one can converge from various states to a common follower, but one cannot diverge from one state to two or more potential followers. Let us describe in this way the very simple four-variable system:

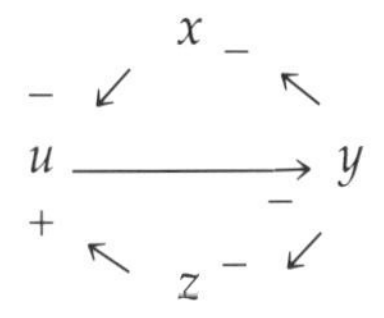

x → u (−), u → y (−), y → x (−), y → z (−), z → u (+) or

$$\begin{aligned} x_{t+1} &= \bar{y}_t \\ y_{t+1} &= \bar{u}_t \\ z_{t+1} &= \bar{y}_t \\ u_{t+1} &= \bar{x}_t + z_t \end{aligned} \qquad (3.1)$$

(see Table 3.1).

We see in Figure 3.2 that some states may converge to a common follower but each state has a single follower. In addition, some states differ from their follower at the level of more than one variable. For example, the follower of 0000 is 1111; thus, the description assumes that in this particular case all four variables will commute from 0 to 1 in perfect synchronism. This is why this description is called synchronous.

These aspects of the synchronous descrip-

Table 3.1 State table of system (3.1) in the synchronous description. State 1011 is circled because the next state is also 1011; it is thus stable

$(x, y, z, u)_t$	$(x, y, z, u)_{t+1}$
0000	1111
0001	1011
0010	1111
0011	1011
0100	0101
0101	0001
0110	0101
0111	0001
1000	1110
1001	1010
1010	1111
(1011)	1011
1100	0100
1101	0000
1110	0101
1111	0001

tion are not readily acceptable to a biologist. On the one hand, certain states must have more than one potential follower in order to permit the description of differentiative events. On the other hand, when a system has received orders to synthesize more than one gene product, there is no reason whatsoever why all of them would reach an efficient concentration simultaneously.

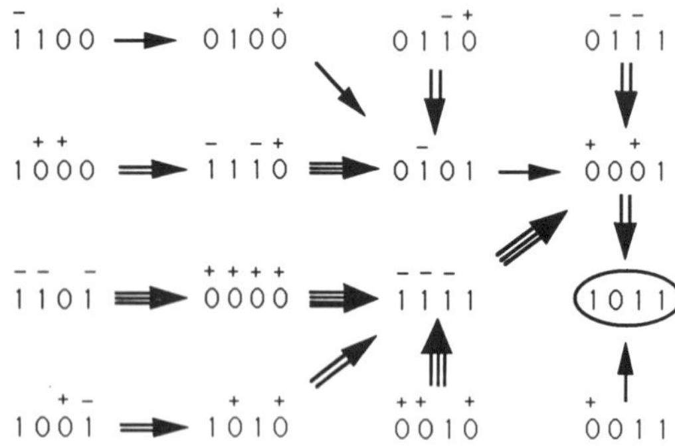

Figure 3.2 Graph of sequences of states of system (3.1) in the synchronous description. Each state has only one possible follower and, in the present case, the unique final state is 1011.

These serious difficulties disappear if instead of writing $x_{t+1} = f(x, y, z, \ldots)_t$ we write that at any time $X = f(x, y, z, \ldots)$, in which X (the 'image' of x) is a prospective value of x; if X differs from x, x has an order to adopt the value X, and it will do it after a suitable delay t_x unless there has been a counter-order in the meantime. Instead of (3.1), we now have:

$$\begin{aligned} X &= \bar{y} \\ Y &= \bar{u} \\ Z &= \bar{y} \\ U &= \bar{x} + z \end{aligned} \tag{3.2}$$

The state table is the same as above, except for the replacement of the 'next state' $(x, y,$

Table 3.2 Stable table of system (3.1) in the asynchronous description. Each difference between a state (vector $x\ y\ z\ u$) and its image (vector $X\ Y\ Z\ U$) means that there is an order to change the value of a variable. For example, state 1010 has an image 1001. There is thus an order for variable z to switch from 1 to 0 and for variable u to switch from 0 to 1. Instead of describing the situation by 1010/1001, we write in a more compact way $10\overset{-}{1}\overset{+}{0}$. Note that in the table the +s and −s are redundant but convenient

x, y, z, u	X, Y, Z, U
$\overset{+}{0}\ \overset{+}{0}\ \overset{+}{0}\ \overset{+}{0}$	1 1 1 1
$\overset{+}{0}\ 0\ \overset{+}{0}\ 1$	1 0 1 1
$\overset{+}{0}\ \overset{+}{0}\ 1\ \overset{+}{0}$	1 1 1 1
$\overset{+}{0}\ 0\ 1\ 1$	1 0 1 1
$0\ 1\ 0\ \overset{+}{0}$	0 1 0 1
$0\ \overset{-}{1}\ 0\ 1$	0 0 0 1
$0\ 1\ \overset{-}{1}\ \overset{+}{0}$	0 1 0 1
$0\ \overset{-}{1}\ \overset{-}{1}\ 1$	0 0 0 1
$1\ \overset{+}{0}\ \overset{+}{0}\ 0$	1 1 1 0
$1\ 0\ \overset{+}{0}\ \overset{-}{1}$	1 0 1 0
$1\ \overset{+}{0}\ 1\ \overset{+}{0}$	1 1 1 1
(1 0 1 1)	1 0 1 1
$\overset{-}{1}\ 1\ 0\ 0$	0 1 0 0
$\overset{-}{1}\ \overset{-}{1}\ 0\ \overset{-}{1}$	0 0 0 0
$\overset{-}{1}\ 1\ \overset{-}{1}\ 0$	0 1 0 1
$\overset{-}{1}\ \overset{-}{1}\ \overset{-}{1}\ 1$	0 0 0 1

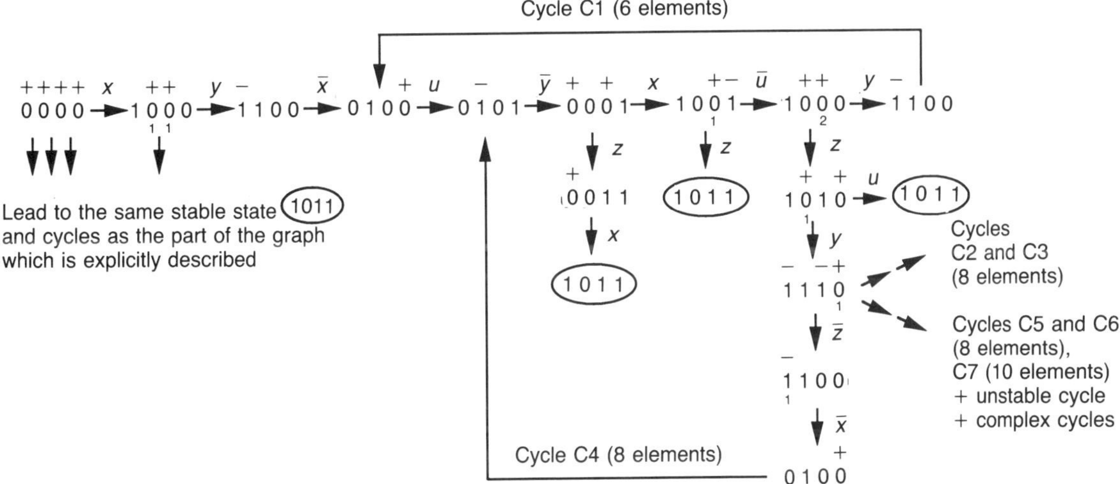

Figure 3.3 Partial graph of sequences of states of system (3.1) in the asynchronous description. The essential result is that the system can end up in the stable state 1011 or in a periodic trajectory whose exact profile (cycle C1, . . . , C7, etc.) depends on the time delays. When the same system is described by nonlinear differential equations one reaches the same conclusion: there are two attractors, one of which is a stable node and the other can be a limit cycle whose precise shape depends on the parameter values. In the graph, we use subscripts which draw attention to the fact that orders may have already been given one, two or more steps ahead. For example, states $\underset{11}{\overset{++}{1000}}$ and $\underset{2}{\overset{++}{1000}}$ differ because in the first case variables y and z have received the order to change their value at the former step, whereas in the second case variable y has just received the order and variable z has received it already two steps ahead. This is why the string between $\underset{11}{\overset{++}{1000}}$ and $\underset{2}{\overset{++}{1000}}$ is not considered a cycle.

$z, \ldots)_{t+1}$ by the 'image' $X\,Y\,Z$. The essential difference is that if the image of the state vector 0000 is 1111, the next state is usually 1000, 0100, 0010 or 0001 depending on which of the variables first changes its logical value.* In this so-called asynchronous description, the graph of the sequences of states of the system described above becomes much more complex. Part of this graph is given in Figure 3.3. Suffice it to mention here that in practice it gives the system a choice between two attractors, one of which is a stable state and the other a logical cycle (whose exact sequence displays a wide variety). Note that this behaviour fits nicely with the differential description of the same system. See also the analysis of a much more complex system in Thomas and D'Ari (1990, appendix 4).

At the beginning, some people considered that an asynchronous description (a) would be too heavy to be tractable and (b) would generate 'anything'. It rapidly became obvious that this description (a) is perfectly practicable and (b) instead of giving 'anything', gives results which are indeed richer than those of the synchronous description but, in contrast with the former, fit well with the differential description.

3.3.3 MULTILEVEL LOGICAL VARIABLES: WHY AND WHEN?

As ably remarked by Van Ham (1979), the purpose of using multilevel logical variables is not to imitate better and better the continu-

*Double, triple or even quadruple commutations are by no means excluded but considered unlikely compared with the single commutations.

ous description by introducing more and more intermediate steps: rather, one should introduce additional logical levels only when it is required for qualitative reasons. But when? What are the criteria?

The answer turns out to be extremely simple in most cases; if an element acts at more than one place in the system, there is no reason why these various actions would be effective in the same range of concentration, and consequently we use more than one threshold. More concretely, if an element has n distinct actions, we use n thresholds and describe it with an $(n + 1)$ level variable.

In practice, following Van Ham, we use: (a) a multilevel logical variable x which takes the values 0, 1, 2, . . . depending on whether the real value is below the first threshold ${}^{1}s$, between the first and the second ${}^{2}s$, etc.; (b) in parallel, a set of Boolean variables ${}^{1}x$, ${}^{2}x$ whose value (0 or 1) tells whether or not we are above the threshold considered.

In many cases (as, for instance, when one writes a state table, or when one describes the system in the space of the variables) it is more convenient to use the multivalued logical variable. However, it often happens that what we are interested in is simply whether or not we are above a given threshold; in such cases, it is convenient to use the Boolean variable corresponding to that threshold. In addition, the use of the set of Boolean variables allows one to keep the Boolean formalism and yet treat variables as multivalued. For example, the logical equations:

$$\begin{aligned} X &= {}^{1}\bar{y} \\ Y &= {}^{1}x + {}^{2}y \end{aligned} \tag{3.3}$$

although purely Boolean, tell that gene X is on iff y is below its first threshold, and that gene Y is on iff x exceeds its (only) threshold or y exceeds its second threshold. These logical equations can also be symbolized by the matrix:

$$\begin{pmatrix} 0 & -1 \\ +1 & +2 \end{pmatrix}$$

This system results in the state table and graph of the sequences of states:

$x\ y$	$X\ Y$
$\overset{+}{0}\ 0$	1 0
$0\ \bar{1}$	0 0
$0\ \bar{2}$	0 1
$1\ \overset{+}{0}$	1 1
$\bar{1}\ 1$	0 1
$\bar{1}\ \bar{2}$	0 1

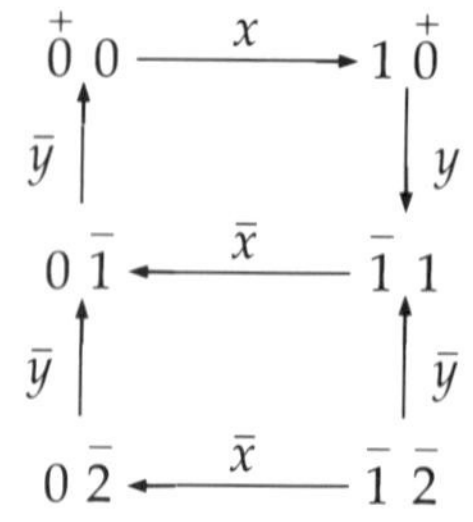

$\overset{+}{0}\,0 \xrightarrow{x} 1\,\overset{+}{0}$; $1\,\overset{+}{0} \xrightarrow{y} \bar{1}\,1$; $\bar{1}\,1 \xrightarrow{\bar{x}} 0\,\bar{1}$; $0\,\bar{1} \xrightarrow{\bar{y}} \overset{+}{0}\,0$; $\bar{1}\,\bar{2} \xrightarrow{\bar{y}} \bar{1}\,1$; $\bar{1}\,\bar{2} \xrightarrow{\bar{x}} 0\,\bar{2}$; $0\,\bar{2} \xrightarrow{\bar{y}} 0\,\bar{1}$

As suggested above, it is convenient here to use the Boolean variables ${}^{1}x$, ${}^{1}y$ and ${}^{2}y$ in the logical equations and the multivalued logical variables x, y in the state table and in the graph of the sequences of states.

3.3.4 LOGICAL PARAMETERS

The crucial concept of logical parameters was introduced by Snoussi (Snoussi 1989; Snoussi *et al.* in Thomas and D'Ari 1990). Basically, the idea consists of ascribing a proper weight to each term of our logical expressions.

The utility of logical parameters becomes obvious if one considers a system like the three-element negative loop:

$x\ y\ z$	$X\ Y\ Z$
$\overset{+}{0}\,0\,0$	1 0 0
$0\,0\,\bar{1}$	0 0 0
$\overset{+}{0}\,\bar{1}\,\overset{+}{0}$	1 0 1
$0\,\bar{1}\,1$	0 0 1
$1\,\overset{+}{0}\,0$	1 1 0
$\bar{1}\,\overset{+}{0}\,\bar{1}$	0 1 0
$1\,1\,\overset{+}{0}$	1 1 1
$\bar{1}\,1\,1$	0 1 1

$$\begin{aligned} X &= \bar{z} \\ Y &= x \\ Z &= y \end{aligned} \tag{3.4}$$

The 'naïve' logical description predicts the cyclic trajectory:

$$\begin{array}{ccccc} \overset{+}{0}\,0\,0 & \xrightarrow{x} & 1\,\overset{+}{0}\,0 & \xrightarrow{y} & 1\,1\,\overset{+}{0} \\ \bar{z}\uparrow & & & & \downarrow z \\ 0\,0\,\bar{1} & \xleftarrow{\bar{y}} & 0\,\bar{1}\,1 & \xleftarrow{\bar{x}} & \bar{1}\,1\,1 \end{array}$$

However, it is well known from differential analysis that, depending on the parameter values, the (unique) steady state of such a system can be a focus (in which case there is indeed a periodic trajectory, stable if the focus is unstable, damped if the focus is stable) or a stable node (in which case the trajectory is not periodic). Thus, in this case, the naïve logical description privileges one type of behaviour of the system and lacks generality.

Let us now write instead of (3.4):

$x\,y\,z$	$X\ Y\ Z$
0 0 0	K_1 0 0
0 0 1	0 0 0
0 1 0	K_1 0 K_3
0 1 1	0 0 K_3
1 0 0	K_1 K_2 0
1 0 1	0 K_2 K_3
1 1 0	K_1 K_2 K_3
1 1 1	0 K_2 K_3

$$\begin{aligned} X &= K_1\bar{z} \\ Y &= K_2x \\ Z &= K_3y \end{aligned} \tag{3.5}$$

in which the Ks are logical parameters which in the present case can each take the value 1 or 0 according to the significance of the term considered. We see that instead of a unique state table we have now a set of state tables, each characterized by a defined combination of values of the Ks. Where all three Ks = 1, we fall into the situation treated by the naïve logical description (and we have a periodic behaviour), but if any of the Ks = 0, we have instead a stable logical state and no cycle. Depending on which K is (are) nil, we have a different stable state and its character fits nicely with the differential description.

This is a simple illustration of the fact that logical parameters introduce at the same time a flexibility which was absent from the 'naïve' description and a much greater generality.

For a more general description of logical parameters, let us come back to system (3.3). The Boolean expression ${}^1x + {}^2y$ has the same value 1 if either or both of the conditions ${}^1x = 1$, ${}^2y = 1$ (in other words, $\boldsymbol{x} > {}^1\boldsymbol{s}$ $\boldsymbol{y} > {}^2\boldsymbol{s}$) is fulfilled. Let us now write the semi-Boolean expression:

$$k_3\,{}^1x + k_4\,{}^2y$$

in which ${}^1x$ and ${}^2y$ are Boolean variables as above, k_3 and k_4 are reals and + is the algebraic (not the logical) sum. Clearly, this expression can take either of four (real) values:

0	(if ${}^1x$ and ${}^2y = 0$)
k_3	(if ${}^1x = 1$ and ${}^2y = 0$)
k_4	(if ${}^1x = 0$ and ${}^2y = 1$)
$k_3 + k_4$	(if ${}^1x = 1$ and ${}^2y = 1$)

In fact, in the logical description, what we are interested in is not the real value but its location in the logical scale of the variable, in other words the logical value. This is why we apply the discretization operator d_x or d_y which transforms a real into a logical value; the expression

$$d_y(k_3\,{}^1x + k_4\,{}^2y) \tag{3.6}$$

can take only the values 0, $d_y(k_3)$, $d_y(k_4)$ or $d_y(k_3 + k_4)$, and if we write $d_y(k_3) = K_3, \ldots,$ $d_y(k_3 + k_4) = K_{34}$, expression (3.6) can take only the values 0, K_3, K_4 or K_{34}. Moreover, since the Ks are logical constants, their value can be only 0, 1, 2, . . . , n (n is the upper value of the variable considered).

Let us come back to system (3.3) and apply the generalization just mentioned. (3.3) becomes:

$$\begin{aligned} X &= d_x(k_2\,{}^1\bar{y}) \\ Y &= d_y(k_3\,{}^1x + k_4{}^2y) \end{aligned} \tag{3.7}$$

in which we see that x acts at only one level. For this element, we thus consider a single threshold and we treat it as a binary variable; thus x, X and K_2 can only have the values 0 or 1. In contrast, y acts at two levels. For this element, we thus consider two thresholds and we treat it as a three-valued variable; y, Y, K_3, K_4 and K_{34} can thus have the values 0, 1 or 2. This gives the state table:

$x\,y$	$X\;Y$
0 0	K_2 0
0 1	0 0
0 2	0 K_4
1 0	K_2 K_3
1 1	0 K_3
1 2	0 K_{34}

While in the 'naïve' description we had unique values of X and Y for each state of the system, here we have several possibilites according to the values of parameter K_2 (0 or 1) and of parameters K_3, K_4 and K_{34} (0, 1 or 2). In fact, each combination of values of the logical parameters corresponds to a typical qualitative situation. Figure 3.4 gives four examples, from which it can be seen that the structure considered can give a choice between two stable logical states, a choice between a stable logical state and a logical cycle, or a logical cycle only (also a single stable state, not shown). This variety is not surprising, because the system comprises a positive loop, which can generate multistationarity, and a negative loop which can generate a cycle; depending on the parameter values either, both or none of the loops

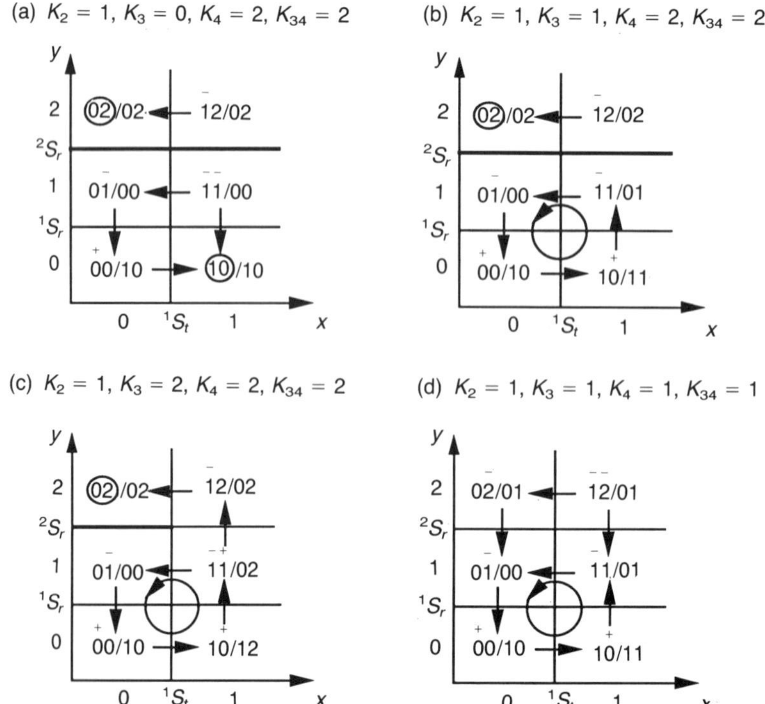

Figure 3.4 State tables of system (3.6) for various sets of values of the logical parameters.

will be functional, thus generating or not multistationarity or oscillations.

A final point concerning the use of logical parameters. It is especially convenient to use, whenever possible, logical expressions which are sums of terms each comprising only one variable; in particular, in this description the network can be described by a matrix of interactions, each term a_{ij} of which deals with the effect of variable j on variable i. Of course, it is by no means excluded to use other types of logical expression, for example, $X = d_x(k^1x \cdot {}^2y)$. We simply would like to point out that such a product can be written perfectly well in the canonic form $X = d_x(k_1\,{}^1x + k_2\,{}^2y)$. It suffices to use proper parameter values; in the present case, one can use $K_1 = 0$, $K_2 = 0$ and $K_{12} = 2$, which means that one condition (${}^1x$, or ${}^2y$) is not sufficient by itself but ${}^1x$ and ${}^2y$ is.

3.3.5 LOGICAL IDENTIFICATION OF ALL STEADY STATES

Without entering into details (but see Thomas and D'Ari 1990; Thomas 1991; Snoussi and Thomas 1993) the fit between this generalized logical description and the differential one is perfect when the differential description uses step interactions and remains surprisingly good when one uses sigmoid interactions (provided the sigmoids are not too flat).

However, our original logical description (both 'naïve' and 'generalized') visualized only part of the steady states which can be identified by the differential description. This is in fact easily understandable for the following reason. In our original logical description, we defined logical stable states as those for which the state vector ($x\,y\,z\ldots$) and its image ($X\,Y\,Z\ldots$) are identical. In such cases, there is no command to change the value of any variable (no + or − superscript) and the situation will not change in the absence of a major perturbation; this is why we call these states 'stable'. It was soon realized that these logical stable states correspond to stable nodes in the differential description; other steady states seen in the differential description (and even some of the stable nodes) were not seen in the logical description. The reason is simple: a steady state can be located at the level of one or more thresholds. Remember, however, that in the logical description we had so far considered the real situations $x < s$ described by the logical x = 0, and $x > s$ described by x = 1, but not the borderline situation $x = s$. Thus, we could not identify the steady states located on one or more thresholds as logical states. In fact, to solve the difficulty, it suffices to include the thresholds in the logical description. We now write:

(logical description)		(real description)
x = 0	if	$x < {}^1s$
x = ^{1}s	if	$x = {}^1s$
x = 1	if	${}^1s < x < {}^2s$
x = ^{2}s	if	$x = {}^2s$ etc.

We thus now have 'regular' states, corresponding to the classical logical states and 'singular' states, in which the value of one or more variables is a threshold value. Note that a state located on one threshold is at the junction between two adjacent regular states, and more generally a state located on n thresholds is at the junction between 2^n adjacent regular states.

This generalization also obliged us to generalize the definition of logical steady states. In our description, the logical steady state equations are simply the relations $x = X$, $y = Y$, $z = Z, \ldots$, and we define a logical steady state as any logical state (regular or not) whose image is consistent with the logical steady state equations.

We already know how to calculate the image of a regular state. The image of a singular state can be computed by comparing the images of appropriate pairs of adjacent regular states.

Consider the simple case of a system whose variable x is located on, say, threshold ${}^2s$. In the subspace of variable x, the singular state $x = {}^2s$ is sandwiched between the regular states $x = 1$ and $x = 2$. Let K_1 be the image of state $x = 1$ and K_2 the image of $x = 2$. For reasons of continuity, the image of $x = {}^2s$ is comprised between K_1 and K_2. Thus, for state $x = {}^2s$, the situation is consistent with the steady state equation $x = X$ if ${}^2s$ is comprised in the interval $]K_1, K_2[$; if K_1 and K_2 are such that it is indeed the case, we say that variable $x = {}^2s$ is steady (and write $X = {}^2s$).

For a more complete description, see Thomas (1991); suffice it to mention here that once this generalization is introduced, all the steady states found in the differential description can be identified on logical grounds. As a matter of fact, the logical identification of steady states has become so easy that for some time we usually begin with the logical identification and only afterwards turn to the differential description, when it is felt useful.

3.3.6 THE CONCEPT OF THE LOOP-CHARACTERISTIC STATE

As mentioned above, we consider a positive loop functional if it actually confers multistationarity to its elements, and a negative loop as functional if it actually generates homeostasis. It is invariably found that when a feedback loop is functional there is a steady state at (or near, if instead of step functions we have (not too flat) sigmoids) the intersects of its thresholds. Consider, for example, a functional loop in which variable x acts (on y) above its second threshold ${}^2s$, variable y acts (on z) above its first threshold ${}^1s$ and variable z (on x) above its third threshold ${}^3s$; there is a steady state at the point ${}^2s{}^1s{}^3s$ (i.e., $x = {}^2s$, $y = {}^1s$, $z = {}^3s$).

This leads to the concept of the loop-characteristic state: with each feedback loop (or reunion of usually disjoint loops) one can associate a loop-characteristic state, which is located at the threshold values involved in the loop. When the loop is functional, its characteristic state is steady, and vice versa. It was soon observed (Thomas 1991) and subsequently formally demonstrated (Snoussi and Thomas 1993) that: (a) among the singular states of a system, only loop-characteristic states can be steady; (b) when a (singular) state is loop-characteristic, there exist parameter values for which it is steady.

These properties of loop-characteristic states introduce a surprisingly simple relation between feedback loops and steady states. Moreover, the logical analysis of complex systems is greatly simplified. A four-variable system with three-level (0, 1 and 2) variables has 625 logical states (81 regular, 544 singular). So far we have had to scan each state of the system for steadiness. Now, we identify the feedback loops, memorize their characteristic state and, among all the singular states, check only those which are loop-characteristic for steadiness.

Needless to say, when a loop involves only part of the variables of the system, its characteristic state must be considered in the subspace of the variables which actually take part in the loop.

In system (3.3), the negative loop involves both variables; its characteristic state is ${}^1s{}^1s$. When the loop is functional, ${}^1s{}^1s$ is a steady state of the system (and vice versa). In contrast, the positive loop of this system concerns only variable y. Its characteristic state can be symbolized $-{}^2s$, in which the dash indicates that the value of x remains open. Let us consider states $0\,{}^2s$, sandwiched between the regular states 01 and 02, and $1\,{}^2s$, between 11 and 12. From the state table 3.3, we extract:

$x\,y$	$X\,Y$
0 1	0 0
0 2	0 K_4

and

$x\,y$	$X\,Y$
1 1	0 K_3
1 2	0 K_{34}

Thus, in the domain $x = 0$, the condition for the positive loop to be functional is $^2s \in [0, K_4]$, that is, $K_4 = 2$. Since variable x is steady ($x = 0$, $X = 0$), the condition ensures not only that the positive loop be functional (and variable y multistationary) in the region $x = 0$, but also that $0\,^2s$ be a steady state of the system. In the domain $x = 1$, the conditions for the positive loop to be functional are $K_3 > 1$ and $K_{34} = 2$. If these conditions are fulfilled, variable y will be multistationary in region $x = 1$; however, state $1\,^2s$ of the system cannot be steady, because variable x is not steady in this region ($x = 1$, $X = 0$). These predictions can be checked by looking at the state tables of Figure 3.4. Note that, in this table, multistationarity is visualized by a heavy line (a 'separatrix') which is not crossed by any arrow. In subtable 3.4(c), the conditions of stationarity of state $-^2s$ are fulfilled for $x = 0$ (since $K_4 = 2$) but not for $x = 1$ (since $K_3 = 2$). Accordingly there is a separatrix showing multistationarity of variable y in region $x = 0$ but not in $x = 1$. In subtables 3.4(a) and 3.4(b), the conditions of steadiness of $-^2s$ are fulfilled in both regions $x = 0$ and $x = 1$; accordingly there is a separatrix showing that y is multistationary in the whole space of the variables. Finally, in subtable 3.4(d), $-^2s$ can be steady neither in region $x = 0$ (because $K_4 \neq 2$) nor in region $x = 1$ (because $K_{34} \neq 2$); accordingly, there is no trace of multistationarity.

3.3.7 CONSIDERING A SET OF INTERACTING LOOPS RATHER THAN A SET OF INDIVIDUAL INTERACTIONS

The word 'metaphor' belongs to the title of this book and to several of its chapters. We thus feel allowed to use here metaphors to depict situations found in biology. In many cases, we are accumulating data which permit us to understand the mechanism of individual processes in more and more molecular detail, but the more detail there is, the more difficult it is to have a general view of the situation. We are to some extent in the situation of someone who is studying all the teeth of the mechanism of a clock individually. Even though it may be necessary to check the operation of each tooth of the mechanism, it is at least as necessary to examine (a) the wheels of which the teeth are the components and (b) the interactions between the wheels. Instead of looking at the interactions between the individual teeth, we look at the interactions between the wheels. More strictly speaking, instead of considering the interactions between the elements of a system, we consider the interactions between the feedback loops constituted by these elements.

As an example, let us consider the two graphs of interactions of Figure 3.5 and the corresponding matrices. The first matrix comprises eight individual interactions and six feedback loops. At first sight, it is extremely complex in the absence of a formal analysis. However, it is soon realized that in this particular case all the feedback loops are negative. In the absence of any positive feedback loop we do not expect any trace of multistationarity; whatever the parameter values chosen, there can be only one steady state. Insofar as each functional feedback loop generates a steady state near the junction of its thresholds, this means that in this system not more than one feedback loop (or reunion of disjoint loops) can be functional for a given set of parameter values. Thus, we know already that, depending on the para-

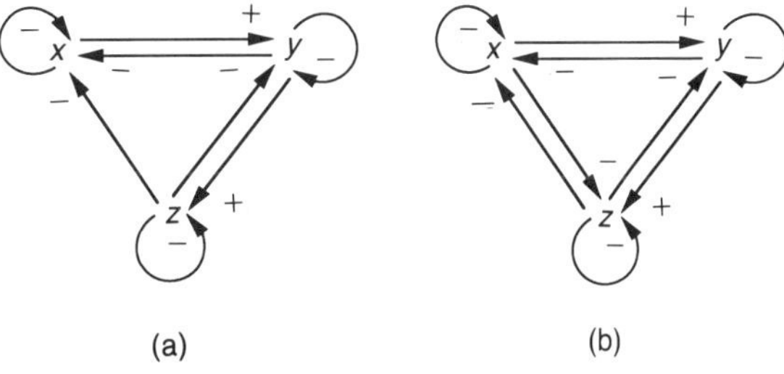

Figure 3.5 Two three-element regulatory networks differing only by one interaction.

meter values, the unique steady state of the system will be either a regular state (if none of the loops is functional) or the characteristic state of the only loop which is functional.

The system of Figure 3.5(b) differs from the preceding one only by an additional negative interaction. In this case, there is one positive loop. If the parameters are such that this loop is not functional, we are led back to the preceding situation. If it is functional, we know already that the system has three steady states, one of which is characteristic of the positive loop.

More generally, given a regulatory network, we first identify the feedback loops (or unions of disjoint loops) and their characteristic states. For each feedback loop, we compute the range of parameter values which make the loop functional in part or all of the space of the variables. Now, the problem of the complex interactions between the loops reduces to the analysis of the compatibilities between these constraints; we know in which conditions several loops will function simultaneously from a simple confrontation of the constraints operating on the individual loops. In this way, we can determine whether and in what range of parameter values a system can adopt a complex dynamic combining multistationarity and homeostatic behaviour.

We have developed computer programs which fully automate the dynamical analysis of complex regulatory models. Depending on the concrete knowledge about the system one wishes to analyse, one can choose between the following options:

1. Entering the logical structure of a system ('matrix of the interactions') and the parameter values ('matrix of the parameters'), one can identify the steady states, their location and their nature (stable or unstable nodes, focus, saddle point, etc.).
2. Given only the logical structure of the system (no predefined values for the logical parameters), one can compute which constraints on the parameters have to be respected for a feedback loop to be functional, and what the possibilities of coexistence of two or more functional loops are. To do that, on the basis of the matrix of interactions, our program finds the elementary loops which constitute the network, finds their characteristic states, calculates the steadiness constraints and checks their compatibilities.
3. Finally, if one has pre-existing (experimental or logical) indications concerning the parameter values, one can use them in order to simplify the analysis of the feedback loops and their combinations.

In addition to use as a tool for analysing concrete biological systems (see Section 3.1), our generalized logical method has considerably helped us in identifying general principles of biological regulation (and regulation in general), and it may be efficient for the qualitative analysis of a wide range of system of differential equations.

REFERENCES

Ciompi, L., Ambühl, B. and Dünki, R. (1992). Shizophrenie und chaostheorie. Methoden zur untersuchung der nicht-linearen dynamik komplexer psycho-sozio-biologischer systeme. *System Familie* **5**, 133.

Glass, L. and Kauffman, S. A. (1973). The logical analysis of continuous non-linear biochemical control networks. *J. Theor. Biol.* **39**, 103.

Kaufman, M. (1988). Role of multistability in an immune response model: a combined discrete and continuous approach. In *Theoretical Immunology* (Part One) (ed. A. Perelson). SFI Studies in the Sciences of Complexity. Addison-Wesley, Redwood City, CA, p. 199.

Nicolis, C. (1982). A Boolean approach to climate dynamics. *Quart. J. R. Met. Soc.* **108**, 707.

Snoussi, E. H. (1989). Qualitative dynamics of piecewise-linear differential equations: a discrete mapping approach. *Dyn. Stability Syst.* **4**, 189.

Snoussi, E. H. and Thomas, R. (1993). Logical identification of all steady states: the concept of feedback loop characteristic states. *Bul. Math. Biol.* **55**, 973.

Thieffry, D. and Thomas, R. (in press) Dynamical behaviour of biological regulatory networks. II. Immunity control in temperate bacteriophages.

Thieffry, D., Colet, M. and Thomas, R. (1993). Formalization of regulatory networks: a logical method and its automatization. *Mathematical Modelling and Scientific Computing* **2**, 144.

Thieffry, D., Verdin, M. and Colet, M. (1993). Regulation of HIV expression: a logical analysis. In *Mathematics Applied to Biology and Medicine* (ed. J. Demongeot and V. Capasso). Wuerz, Winnipeg, Canada, p. 291.

Thomas, R. (1991). Regulatory networks seen as asynchronous automata: a logical description. *J. Theor. Biol.* **153**, 1.

Thomas, R. and D'Ari, R. (1990). *Biological Feedback*. CRC Press, Boca Raton, FL.

Thomas, R., Thieffry, D. and Kaufman, M. (1994). Dynamical behaviour of biological regulatory networks. I. Biological role and logical analysis of feedback loops.

Tyson, J. (1975). Classification of instabilities in chemical reaction systems. *J. Chem. Phys.* **62**, 1010.

Van Ham, P. (1979). How to deal with more than two levels. *Lect. Notes Biomath.* **29**, 326.

THE COMPUTATIONAL MACHINERY OF THE LIVING CELL 4

G. Rick Welch

The essence of science is to uncover patterns and regularities in nature by finding algorithmic compressions of observations.

Paul Davies (1992)

4.1 INTRODUCTION

The philosophical train of 'linear' thinking in the western (Greek) tradition, most recently via the track of logical positivism, has generated an ever deepening mathematical abstraction of natural phenomena into refined algorithmic compressions of sensory data. Underlying this historical march of science, many analogues and metaphors have been invoked to rationalize the order of nature. Since the mid nineteenth century, two powerful constructs have come to play encompassing roles: **field** and **computer**. The mathematical lawfulness of the world around us, as interpreted by theoretical physics, is now firmly entrenched in the imagery of these two entities. The ability of humankind to predict (and, from a biologically utilitarian standpoint, control) the behavior of natural processes has been greatly enhanced by the application thereof.

The epistemological and ontological thrust of contemporary physics is advancing the unification of scientific knowledge on a grand cosmic scale. The realm of biology is being pulled, with increasing urgency, into this unity crescendo. The import of relativity and quantum physics suggests that the logical quest of science is 'circular', rather than 'linear'. Humankind – the sentient observer of nature – has a biological essence, which imbues an active biotic (anthropic) character into the physical view of the cosmos (Barrow and Tipler 1986). Accordingly, rather than adopting a philosophical bent that operates to reduce isolated biological phenomena to mere physical mechanisms, I take a physicalist view that seeks to characterize the functioning of biological systems by way of actual analogies and metaphors from pure physics. Pursuant thereto, it is natural to ponder an intrinsic role of the 'field' and the 'computer' in the living state.

Although overshadowed by today's reductionistic molecular-genetic focus, the annals of biology hold a rich and colorful history of embracement of the 'field' metaphor, dating to the mid 1800s when it was concretized in theoretical physics (reviewed in Welch (1992)). Could it be that the affinity is self-imposed? The 'circularity' in the advancement of science would appear to bring western thought into direct confrontation with eastern philosophy, that is, with the position that the living organism perceives (and orders) its external environment as a 'field' because life, itself, is a field; we look into nature and see our own reflection (Siler 1990).

The point of origin of the metaphorical circularity lies perhaps in ancient Greek thought itself. From Adler's *The Great Ideas: A Lexicon of Western Thought* (Adler 1992), under the entry of 'world' one finds the following three metaphors: organism, society, and machine. The first of these is the most primordial, as crystallized from the Greek (e.g., Platonic) view of the world as an animated whole; life and nature were seen as one and the same. The idea of the world as a society under divine law seems to have been grafted later from Judeo-Christian influence. The machine picture has a long historical thread, dating to early Greek (e.g., Democritean) concern over the principle of causation. The birthright of modern physics has been to attempt the deanimation of Nature according to the machine metaphor; while modern biology continues, intrinsically, to employ all three metaphors – albeit taking a dialectical-materialistic view of life's hierarchy.

Over the years, many workers have addressed the biological applicability of metaphors/analogues from information theory and from computer theory. Recently, Paton (1992) brought into comparative focus the polytypical, metaphorical nature of biological thought, especially as regards computational models. Paton (1993) has drawn attention, in particular, to the comparative views of the living cell as a 'society', as a 'field', and as a 'text'. The relational character of the text metaphor offers some novel insights, for example, as to the emergent properties attendant on the hierarchy of life-forms.

It was the nineteenth-century English inventor, Charles Babbage, who provided modern vision for the metaphorical view of the physical world as a gigantic computer, with the notion that the laws of physics and computable mathematics form a closed circle of existence. There is, indeed, an increasing awareness in science of the link between physical proccesses and computation; whereby the dynamical evolution of a physical system converts INPUT data into OUTPUT data via some kind of intermediary COMPUTATION. Here, I will discuss the possible isomorphism of this construct in the biological realm, specifically at the level of cellular metabolism.

4.2 THE ENIGMA OF BIOLOGICAL ORGANIZATION: THE METAPHOR REVEALED

One of the hallmarks of life is 'organization'. This feature is evident at all levels of the biological hierarchy, from the level of the cell to that of socio-ecosystems (Miller 1978). Teleonomic questions of the 'How?' and the 'Why?' of the manifested organization have permeated biological thought since the writings of Aristotle. To many biologists, simple realization of the idea of structure-function duality seems to quench the inquisitive thirst here. However, such cognizance is myopic and arbitrary; it is based on case-specific physiological reasoning and, in the final analysis, sidesteps the philosophical issue. Moreover, attempts by biologists in modern times to objectify (and, alas, quantify by the Cartesian alembic of theoretical physics) the principle of organization have led to considerable factionalization. (For relevant discussion, see the older works of the eminent mathematical biologists, L. von Bertalanffy, A. Lotka and N. Rashevsky.)

In my opinion (shared by others), a true physicalist insight into the essence of biological organization can only be gained by a holistic approach to science, by taking into account the circumstance that living systems are part and parcel (and, in fact, an epiphenomenon) of the physical evolution of the cosmos (Davies 1988). Moreover, I would argue that the issue of 'organization' must be approached from a 'process' perspective, as opposed to the 'substance' picture which dominates the science of biology today. Taking such a view, the paradigms of 'dissipative structure' 'synergetics', 'chaos', 'complexity', among other things, have yielded penetrating analysis of the question of how organization arises and perpetuates itself (Nicolis and Prigogine 1989; Kauffman 1993).

It is my contention that access to the question of the 'Why?' of biological organization demands a marriage of theoretical physics and physiology (or, perhaps, a reunion thereof – see Welch, 1987). Among the fruits of this union, there ensue the following lines of reasoning. First, one must appreciate the hierarchical symmetry spanning the living world, from the cellular to the socio-ecosystem level (Rashevsky 1938; Miller 1978). That is to say, despite the difference in size and the varying jargon used to describe the system components at the different levels, the vital processes and working relations among components at each level of the biological hierarchy manifest an invariant order and a covariant mathematical representation akin to that in theoretical physics (Welch 1987, 1992). Second, taking lessons from the history of physics, a central epistemological metaphor is required for a physicalist characterization of the system dynamics, to serve as a guide in the mathematical depiction of the spatiotemporal behavior; herein lies the importance of such ideas as the 'field' and the 'computer'. Third, a kinship to theoretical physics demands a superlative principle, which provides some kind of directionality (purpose?) to the dynamical evolution of the system. For a physical field, the formalism of Lagrangian dynamics, with its attendant principle of least action, comes to bear; while, for computer technology, the notions of dissipationless computation and maximal information inclusion enter the scene. It should be noted that superlative principles are fraught with metaphysical accoutrements, whether it be in physics (Yourgrau and Mandelstam 1968) or in biology (Welch 1992).

Elsewhere (Welch 1992, 1994), I have reviewed the arguments that, by virtue of the macroscopic ensemble character of living systems (at all hierarchical levels), the theoretical principles of the thermodynamics of irreversible processes (using, for example, the Rayleigh–Onsager 'dissipation function') provide an epistemological physicalist bridge between pure physics and the 'field' quality of life. Armed therewith, one may then approach the 'Why?' of biological organization in a manner analogous to that in pure physics. On this plane of reckoning, the classical biophilosophical ratiocination, which focuses on substantific organization, pales in significance. In physics, one does not ask why organization exists; one just accepts organization as a verity of nature, attaches a Cartesian-analytic metaphor (namely, field) thereto, and proceeds to ponder the mathematical essence of the process behavior therein. In biology, as well as in physics, it is the functional process, in the end, that rationalizes the structure.

Now, let us halt from any further slide into metaphysical abysm and attempt to reify the metaphorical issue of 'biocomputation' at the level of the living cell.

4.3 BIOCOMPUTATION IN THE LIVING CELL: THE ENZYME

The computer metaphor befits the function of the living cell at various levels, from that of the cell in its entirety through to the hierarchy of subprocesses operative therein (cf. Paton 1993). The fundamental event generator in cellular metabolism is the enzyme – a wondrous proteinaceous microcolloid, many of whose properties still elude the trenchant analysis of physical chemists. The textbook rationale for the existence of enzymes is to speed up a chemical reaction, ostensibly by lowering the Arrhenius activation-energy barrier. Beneath the simplistic superfice of this most basic metabolic event lies a layer of complexity.

First of all, most enzymes enhance the rate of their respective reactions, not merely by lowering the activation barrier, but by providing an altered (albeit more energetically favorable) mechanistic path from substrate to product molecular states. For the catalytic process *per se*, conformational substates within the macromolecular enzyme-substrate complex serve as identifiable computational logic elements. An individual enzyme reaction, though, cannot perform macroscopic work on the surroundings. Constrained by the thermodynamic character of the given chemical reaction, an enzyme speeds up the process in the forward and reverse directions to the same degree. Thus, if left alone (i.e., not linked to other processes), the enzyme will, in time, just equilibrate the substrate-product concentration pools, whereby higher order computational functions (e.g., metabolic pathways) require the cell to couple whole enzyme reactions *per se* as the individual logic elements.

Consider the simple monomolecular reaction, $S \leftrightarrow P$, interconverting substrate (S) and product (P), catalyzed by an enzyme (E) in the following fashion: $E + S \leftrightarrow ES \leftrightarrow EP \leftrightarrow E + P$. Once the substrate has become bound to the enzyme (the 'input' stage), the 'computation' involves the generation of localized Gibbs free-energy events necessary to drive the enzyme substrate (ES) complex through the $ES \rightarrow EP$ transition state, followed by release of product (the 'output'). Such computation is far from Boolean, alas, owing to thermal influence. The macroscopically observed conformation of a protein dissolved in solution corresponds to a mixture of a large number of instantaneous (microscopic) conformational states in thermal equilibrium. Any measured property of the system (e.g., the catalytic turnover constant for the $ES \rightarrow EP$ transition) is a weighted average over all such states. Free-energy linkage ('computation') within this proteinaceous system is based ultimately on the equilibrium fluctuational character of the conformational microstates (Welch 1986). The transduction of internal energy and heat exchange with the surroundings are interwoven, as part and parcel of the fluctuational interaction of the protein and the solvent.

Most enzymes are much larger in size (by a factor of 10 or so) than their respective substrates. As to the question of why enzymes are so big, the traditional rationale is that the remainder of the protein macromolecule serves as a mere scaffolding for the active-center geometry. It has now become apparent that the large-scale structure of the protein is designed to provide a local, specific solvent medium for a given chemical reaction, wherein the combined chemical and protein subsystems engage in a fluid and variable exchange of free energy, facilitating the entrance of the bound chemical system into its transition state (McCammon and Harvey 1987; Welch 1986). Accordingly, the protein matrix serves as an intermediary, a deterministic mediator, between a localized chemical reaction coordinate and the surrounding phase. One begins to appreciate why such a large organized macromolecular design is required for the selective and rapid action of enzymes. The dynamical $ES \rightarrow EP$ transition demands more than just an energetic fluctuation; it also requires a specific configuration in the statistical-mechanical phase space of the protein-conformational variables. The protein structure serves as a filter, in selecting and focusing anisotropically a narrow band-width of the thermal noise relevant to the particular chemical reaction.

Protein structure contains a number of elements which suggest modes of energy transduction. Obvious possibilities are regions of local secondary structure, namely, α-helix and β-structure. These have been implicated by many workers (reviewed in Welch (1986), in generating local electric fields, protonation/deprotonation events, proton semiconduction, vibrational excitation, among other things. Also, hydrogen-bond networks – within single proteins and among conjoined proteins – are of great interest as possible conduits for protochemical processes in enzyme action. Transient gaps (faults) in the internal bonding arrangement can elevate locally the free energy of the system – electrostatically and mechanically, as well as protonically. Catalytic functions would, then, depend on precise, anisotropic fault configurations (Lumry and Gregory 1986). Such faults can arise thermally in a protein dissolved in aqueous solution, in conjunction with binding/relaxation of bound water, fluctuating proton-transfer processes, charge-density fluctuations at the surface, etc. Inside the protein, these faults can migrate, for example, by proton hopping (Nagle and Tristram-Nagle 1983).

What, then, is the driving force on the 'computational' process in enzyme catalysis? Because of its size, the bulk reservoir controls the situation through its energy-level density (its entropy) and forces upon the system the canonical distribution of states – within the anisotropic, internal constraints of the folded globular protein. Because of the great degeneracy of protein-conformational substates, the computational aspect of enzyme catalysis is essentially a directed diffusion under a bias (Kamp *et al.* 1988). The enzyme-protein functions as a thermal-equilibrium chemodynamical machine (Welch and Kell 1986).

In a whole metabolic pathway, the component enzymes themselves become the individual logic-elements – each behaving in the statistical-mechanical mode just described. If the pathway operates in a bulk solution environment supplied with a (steady-state) nonequilibrium substrate source (and product sink), the reaction-diffusion coupling of the sequentially acting enzyme components engenders a kinetic irreversibility for the pathway flux.

It is an empirical fact that many enzymes of intermediary metabolism *in situ* are located in organized, structured states – often associated with such nonequilibrium energy sources as local electric fields and local proton gradients (Welch and Berry 1983). As detailed elsewhere (Welch and Kell 1986), within these cellular microenvironments, enzyme action may have rather bizarre qualities, compared to the familiar bulk solution situation, and may be more akin to 'molecular machines' driven in a cyclical manner by an external flow of energy.

4.4 ENZYME ORGANIZATION AND HIGHER ORDER BIOCOMPUTATION

The simplistic view of the cell as a homogeneous, isotropic 'bag' of metabolites and enzymes is now obsolete. Living cells, particularly the larger eukaryotic cells, are replete with infrastructure (Porter and Tucker 1981). This structure encompasses an extensive membranous reticulation, as well as a variform microstructure permeating the hyaloplasmic space (the so-called 'ground substance') of the cell. The latter region is laced with a dense array of proteinaceous cytoskeletal elements and an interstitial 'microtrabecular lattice' (Clegg 1984). Calculations of protein concentrations associated with cytomatrix structures indicate high, crystal-like local density of protein molecules (Sitte 1980). It appears that cytomembranous and cytoskeletal elements have evolved to function as effective 'protein collectors' in the operation of the cellular machinery (Sitte 1980; Porter and Tucker 1981).

Accumulating evidence shows that the majority of enzymes of intermediary metabolism function *in vivo* in organization with the particulate structures, and numerous thermodynamic and kinetic advantages have been attributed thereto (reviewed in Srere 1987; Welch 1985*a*; Welch and Clegg 1986). Some metabolic processes (e.g., electron-transport phosphorylation) are linked permanently to structure, while others exhibit defined variability and a biphasic *modus operandi*. For example, with glycolysis in skeletal muscle, there is a bifurcation of enzyme locale (and of the kinetic properties of the respective enzymes as well) between the cytosol and the cytomatrix (namely, myofilaments), with the partitioning between bound and soluble forms being regulated *in vivo* according to the physiological state of the muscle (Masters 1981). Such evidence attests to the circumstance that cytomatrix surfaces represent the business site of much (perhaps most) cellular metabolism. Accordingly, the coupled reaction-diffusion metabolic flow has a local topographic character more akin to a two- (or, in some cases, one-) dimensional form. The enzyme catalysts, interacting with the cytomatrix, become an integral part of the very organization.

A grasp of the physicochemical nature of the microenvironments in these organized, surface states is of paramount importance to the understanding of the thermodynamic properties of cellular metabolism. Empirical evidence thereon is, at present, quite meager; it is clear, though, that these metabolic microenvironments differ drastically from the kind of bulk-phase solution defined *in vitro* (Siegbahn *et al.* 1985; Westerhoff and Welch 1992). In some of the structured systems, metabolite molecules are effectively channelled in a vectorical fashion from enzyme to enzyme in a reaction sequence, thereby preventing the intermediates from equilibrating with the bulk phase and potentially maintaining some degree of control over the energy states (chemical potentials) thereof. Theoretical models have provided some insight into the essence of long-range energy-transduction modalities potentially extant in the organized regimes *in vivo* (reviewed in Welch 1992; Welch and Kell 1986). In these organized states, the macromolecular matrix of the enzyme couples the bound substrate electromechanochemically – and nonthermally – to ambient nonequilibrium bioenergetic sources. There may exist a coherent 'pumping' of electromechanochemical states analogous to that in the operation of laser systems (Astumian *et al.* 1987; Del Giudice *et al.* 1986; Fröhlich 1986; Kamp *et al.* 1988).

Long-range, mobile protonic states are finding increasing relevance to many organized cellular processes, stemming from Peter Mitchell's pioneering suggestion as to their role in electron-transfer phosphorylation (Mitchell 1979). This kind of energy continuum is emerging as a unifying theme in cell metabolism. Consideration of the roles of mobile protons in enzyme structure, function and evolution has led to suggestions that externally derived high-energy protons may function in the modulation of enzymatic events in organized states intracellularly (reviewed in Welch and Kell 1986).

Hydrogen-bond structures in proteins might also couple to other energy continua in organized states. A likely source would be the electric fields at the surfaces of cytological particulates. A mode of operation here was expounded by Fröhlich (1986), who conjectured that coherent dipole excitations of proteins (entailing phonon/soliton-like modalities) should play an important role in their biological activity. Over the years, theoretical calculations have supported this claim (Del Giudice *et al.* 1986), although the direct experimental evidence is thus far limited.

Superimposed on these designs may be long-range electronic semiconduction states, which might be important in some enzymes – especially in organized states *in vivo*. Various modes of electronic coupling between enzyme and substrate have been offered as theoretical possibilities. Interaction of electronic and nuclear degrees of freedom is well known in solid-state physics. The potential catalytic roles of such modes in enzyme action have been widely discussed (reviewed in Welch and Kell 1986). Very recent experimental evidence (Sucheta *et al.* 1992), for example, suggests that a mitochondrial electron-transport enzyme behaves in a manner resembling the current-voltage character of a tunnel diode.

Long-range electric-field effects on enzyme action have been discussed by a number of workers (reviewed in Fröhlich and Kremer 1983; Welch and Berry 1983; Pethig and Kell 1987). Westerhoff *et al.* (1986) and Astumian *et al.* (1987) have developed realistic models for membrane-transport proteins driven by an electric field, involving direct coupling of the field to dipoles (e.g., α-helices) in the protein structure. Noninvasive dielectric analysis of cellular membranous processes by Kell *et al.* (1988) and by Woodward and Kell (1991) has yielded exciting findings on the coupling of enzyme action to electric fields *in situ*.

There is still much to learn about the physicochemical properties of enzyme catalysts, especially regarding their operation in the living cell. Which of the aforementioned theoretical energy-transduction modalities will prove relevant remains to be seen. It has become apparent, though, that the enzyme serves as a local field transducer in the events of chemical catalysis, and that many of the structured microenvironments *in vivo* are energized by nonequilibrium sources. Such a design is, of course, maintained at the expense of oxidative catabolic processes in the cell. Interestingly, recent observations (Gregory and Berry 1992) with hepatocytes (an oft-used cell type in metabolic studies) show that a significant portion of the O_2-consumption has nothing to do with production/utilization of ATP – the central, diffusible energy currency of the cell. What is the role of this residual oxidative energy generation? The answer, perhaps, is to be found in the surface metabolic field.

4.5 'BIOMOLECULAR ENERGY MACHINES' AND COMPUTER TECHNOLOGY

During the early 1970s, the late C. W. F. McClare (1971, 1974) drew attention to the fact that far-from-equilibrium bioenergetic systems, owing to their molecular size, pose unique thermodynamic problems. For instance, how does the molecular transfer of energy in subcellular processes add up to produce a macroscopic effect? Pursuing the issue, he coined the term 'molecular energy machine', to describe a single enzyme molecule that, in a cyclical fashion, acts to couple the energy released by one form of reaction (e.g., the chemical energy of ATP hydrolysis) to an otherwise unfavourable reaction without, during the lifetime of the cycle, being exposed to the macroscopic (and thermalizing) environment. The concept of a 'molecular energy machine' entails the storage (or 'pumping') of energy over a sufficient period of time such that useful work can be performed with it. As McClare realized (and as is obvious from Section 4.3), such a *modus operandi* requires organization of the enzymic components, with an intrinsic modality for efficient coupling of energy sources with specific molecular work functions in order to minimize the heat exchange during the energy transduction processes. As reviewed elsewhere (Blumenfeld 1981; Welch and Kell 1986; Schneider 1991), the biological relevance of 'molecular energy machines' has attracted considerable attention in the 20 years or so since McClare's writings; and there are many biological examples where this construct is potentially applicable. More evidence thereof will emerge, as biologists shed the empirical prejudices ingrained by bulk solution methods of analyzing enzyme action *in vitro*, in favor of holistic *in situ* approaches.

Distinct parallels exist between the utilitarian advances of humankind in computer technology and the evolutionary development of the enzymatic machinery of cell metabolism (Paton 1993). A design concern in both worlds is the control of information flow, efficiency and heat. Heat is produced where/when a particle in the system must choose between two or more available states. A molecular computation coupled to an output, in the presence of thermal noise, must dissipate some energy as heat. The theoretical foundation of this issue dates to the (in)-famous 'Maxwell's demon', as rationalized by L. Szilard and L. Brillouin. To be exact, an energy, $E \geqslant k_B T \cdot \ln(2)$ (where k_B is the Boltzmann constant and T is the absolute temperature), must be dissipated per 'bit' of information gained in a molecular computation (namely, one which is coupled to an 'output') (Leff and Rex 1990). The way, then, to improve the efficiency of operation (i.e., decrease the heat production) is to reduce the accessibility of the system to particle states arising from the surrounding thermal bath. The lower bound on heat generation (in the total absence of noise) would be a pure Boolean logic. As detailed elsewhere (Bennett 1988; Fredkin and Toffoli 1982; Schneider 1991), the 'computation' portion of a device operation can, in theory, be executed with virtually 100% efficiency; it is the linkage to the outside world (namely, the 'output') which, Szilard and Brillouin proved, demands at least some energy dissipation.

According to the conventional wisdom in biochemistry, one measure of the evolutionary 'perfection' of individual enzymes is the attainment of a mechanistic equienergy profile of all enzyme-bound intermediate states on the ES ↔ EP course, entailing an efficient dynamical matching of the fluid protein matrix and the localized chemical subsystem. There are numerous documented examples of this degree of perfection, including the ATP synthase that couples ATP synthesis to the electronmotive proton gradient during oxidative phosphorylation (Kamp *et al*. 1988).

Such evolutionary 'perfection' applies to the level of supraenzyme organization as well. Within the structured multienzyme systems that channel intermediary metabolites, numerous potential modalities exist for modulating the Gibbs energy profile of the intermeshed chemical and protein systems (Welch and Keleti 1981). Moreover, as the changes for the enzyme-bound states are linked to conformational-mechanical modes in the protein molecule, the energy profile of these states can be adjusted by the aforementioned nonequilibrium energy sources in the organized states *in vivo* (Welch and Kell 1986).

Thus there is the distinct probability that many of the intermediary metabolic processes in cellular microenvironments are much more efficient (and, therefore, generate less heat) than reckoned by the conventional, bulk-phase methods of Gibbs free-energy measurement of isolated metabolic processes (Welch 1985*b*, 1993). The macromolecular configurations in these microenvironments must have machine-like computational properties, in order to execute such a coherent energy transduction. A pure Boolean logic for such biocomputation is a distinct theoretical possibility (Schneider 1991).

In computer technology, of course, there is great concern with dissipationless computation (Bennett 1988; Leff and Rex 1990). In particular, we note a device, the 'Fredkin gate', which is a conservative logic gate that executes a Boolean function in a reversible, 100% efficient manner (Fredkin and Toffoli 1982). It entails a billiard-ball scheme with baffles, which is strikingly analogous to the mechanochemical features of molecular channelling in organized enzyme schemes. Metabolic channelling, in effect, reduces the number of spurious 'outputs' in the system, thereby expanding the configurational domain of the useful 'computation' part of the metabolic machinery (Westerhoff and Welch 1992).

The area of computer technology is rich in metaphor and analogy for biology. An amalgam of ideas is fiting, when we realize that the modern theoretical views of field, machine, computer, organism and physiology, as well as the conception of work, heat and efficiency, arose from a common crucible of natural philosophy in the mid nineteenth century (Welch 1987, 1992). Thus, in one sense, today's quest for the unity of science is just a rediscovery of the past.

REFERENCES

Adler, M. J. (1992). *The Great Ideas: A Lexicon of Western Thought*. Macmillan, New York.

Astumian, R. D., Chock, P. B. and Tsong, T. Y. (1987). Absorption and conversion of energy from dynamic electric fields by membrane proteins: electro-conformational coupling. *Studia Biophysica* **119**, 123–30.

Barrow, J. D. and Tipler, F. J. (1986). *The Anthropic Cosmological Principle*. Oxford University Press.

Bennett, C. H. (1988). Dissipation, information, computational complexity, and the definition of organization. In *Emerging Syntheses in Science*, Vol. 1 (ed. D. Pines). Addison-Wesley, New York, pp. 215–33.

Blumenfeld, L. A. (1981). *Problems of Biological Physics*. Springer, Heidelberg.

Clegg, J. S. (1984). Properties and metabolism of the aqueous cytoplasm and its boundaries. *American Journal of Physiology* **246**, R133–R151.

Davies, P. (1988). *The Cosmic Blueprint*. Simon & Schuster, New York.

Davies, P. (1992). *The Mind of God*. Simon & Schuster, New York.

Del Giudice, E., Doglia, S., Milani, M. and Vitiello, G. (1986). Collective properties of biological systems. In *Modern Bioelectrochemistry* (ed. F. Gutmann and H. Keyzer). Plenum, New York, pp. 263–87.

Fredkin, E. and Toffoli, T. (1982). Conservative logic. *International Journal of Theoretical Physics* **21**, 219–53.

Fröhlich, H. (1986). Coherence and the action of enzymes. In *The Fluctuating Enzyme* (ed. G. R. Welch). Wiley, New York, pp. 421–49.

Fröhlich, H. and Kremer, F. (eds.) (1983). *Coherent Excitations in Biological Systems*. Springer, Heidelberg.

Gregory, R. B. and Berry, M. N. (1992). Stimulation by thyroid hormone of coupled respiration and of respiration apparently not coupled to the synthesis of ATP in rat hepatocytes. *Journal of Biological Chemistry* **267**, 8903–8.

Kamp, F., Welch, G. R. and Westerhoff, H. V. (1988). Energy coupling and Hill cycles in enzymatic processes. *Cell Biophysics* **12**, 201–36.

Kauffman, S. A. (1993). *The Origins of Order*. Oxford University Press.

Kell, D. B., Astumian, R. D. and Westerhoff, H. V. (1988). Mechanisms for the interaction between nonstationary electric fields and biological systems. *Ferroelectrics* **86**, 59–78.

Leff, H. S. and Rex, A. F. (eds.) (1990). *Maxwell's Demon: Entrophy, Information, Computing*. Adam Hilger, Bristol.

Lumry, R. and Gregory, R. B. (1986). Free-energy management in protein reactions. In *The Fluctuating Enzyme* (ed. G. R. Welch). Wiley, New York, pp. 3–190.

Masters, C. J. (1981). Interactions between soluble enzymes and subcellular structure. *CRC Critical Reviews of Biochemistry* **11**, 105–43.

McCammon, J. A. and Harvey, S. C. (1987). *Dynamics of Proteins and Nucleic Acids*. Cambridge University Press.

McClare, C. W. F. (1971). Chemical machines, Maxwell's demon, and living organisms. *Journal of Theoretical Biology* **30**, 1–34.

McClare, C. W. F. (1974). Resonance in bioenergetics. *Annals of the New York Academy of Science* **227**, 74–97.

Miller, J. G. (1978). *Living Systems*. McGraw-Hill, New York.

Mitchell, P. (1979). Compartmentation and communication in living systems. *European Journal of Biochemistry* **95**, 1–20.

Nagle, J. and Tristram-Nagle, S. (1983). Hydrogen-bonded chain mechanisms for proton conduction and proton pumping. *Journal of Membrane Biology* **74**, 1–14.

Nicolis, G. and Prigogine, I. (1989). *Exploring Complexity*. Freeman, New York.

Paton, R. C. (1992). Towards a metaphorical biology. *Biology and Philosophy* **7**, 279–94.

Paton, R. C. (1993). Some computational models at the cellular level. *BioSystems* **29**, 63–75.

Pethig, R. and Kell, D. B. (1987). Dielectric properties of biological systems. *Physiology in Medicine and Biology* **32**, 933–70.

Porter, K. R. and Tucker, J. B. (1981). The ground substance of the living cell. *Scientific American* **244**, 56–67.

Rashevsky, N. (1938). *Mathematical Biophysics*. Dover, New York.

Schneider, T. (1991). Theory of molecular machines. *Journal of Theoretical Biology* **148**, 83–123, 125–37.

Siegbahn, N., Mosbach, K. and Welch, G. R. (1985). Models of organized multienzyme systems: use in microenvironmental characterization and in practical application. In *Organized Multienzyme Systems* (ed. G. R. Welch). Academic Press, New York, pp. 271–301.

Siler, T. (1990). *Breaking the Mind Barrier: The Art-science of Neurocosmology*. Simon & Schuster, New York.

Sitte, P. (1980). General principles of cellular compartmentation. In *Cell Compartmentation and Metabolic Channeling* (eds. L. Nover, F. Lynen and K. Mothes). Elsevier/North-Holland, Amsterdam, pp. 17–32.

Srere, P. A. (1987). Complexes of sequential metabolic enzymes. *Annual Reviews of Biochemistry* **56**,

89–124.

Sucheta, A., Ackrell, B. A. C., Cochran, B. and Armstrong, F. A. (1992). Diode-like behaviour of a mitochondrial electron-transport enzyme. *Nature* **356**, 361–2.

Welch, G. R. (ed.) (1985*a*). *Organized Multienzyme Systems*, Academic Press, New York.

Welch, G. R. (1985*b*) Some problems in the usage of Gibbs free energy in biochemistry. *Journal of Theoretical Biology* **114**, 433–46.

Welch, G. R. (ed.) (1986). *The Fluctuating Enzyme*. Wiley, New York.

Welch, G. R. (1987). The living cell as an ecosystem: hierarchical analogy and symmetry. *Trends in Ecology and Evolution* **2**, 305–9.

Welch, G. R. (1992). An analogical 'field' construct in cellular biophysics: history and present status. *Progress in Biophysics and Molecular Biology* **57**, 71–128.

Welch, G. R. (1993) Bioenergetics and the cellular microenvironment. *Pure and Applied Chemistry* **65**, 1907–14.

Welch, G. R. (1994). Frontiers in the mathematical description of 'biothermokinetic' phenomena. In *Frontiers in Modern Biothermokinetics* (eds. J. P. Mazat, R. Ouhabi, M. Rigoulet and S. Schuster). Plenum, New York, pp. 3–6.

Welch, G. R. and Berry, M. N. (1983). Long-range energy continua in the living cell: protochemical considerations. In *Coherent Excitations in Biological Systems* (ed. H. Fröhlich and F. Kremer). Springer, Heidelberg, pp. 115–40.

Welch, G. R. and Clegg, J. S. (eds.) (1986). *The Organization of Cell Metabolism*. Plenum, New York.

Welch, G. R. and Keleti, T. (1981). On the 'cytosociology' of enzyme action *in vivo*: a novel thermodynamic correlate of biological evolution. *Journal of Theoretical Biology* **93**, 701–35.

Welch, G. R. and Kell, D. B. (1986). Not just catalysts – molecular machines in bioenergetics. In *The Fluctuating Enzyme* (ed. G. R. Welch). Wiley, New York, pp. 451–92.

Westerhoff, H. V. and Welch, G. R. (1992). Enzyme organization and the direction of metabolic flow: physicochemical considerations. In *Current Topics in Cellular Regulation*, Vol. 33 (eds. E. R. Stadtman and P. B. Chock). Academic Press, New York, pp. 361–90.

Westerhoff, H. V., Tsong, T. Y., Chock, P. B., Chen, Y. D. and Astumian, R. D. (1986). How enzymes can capture and transmit free energy from an oscillating electric field. *Proceedings of the National Academy of Sciences (USA)* **83**, 4734–8.

Woodward, A. M. and Kell, D. B. (1991). Dual-frequency excitation: a novel method for probing the nonlinear dielectric properties of biological systems and its application to suspensions of *Saccharomyces cerevisiae*. *Bioelectrochemistry and Bioenergetics* **25**, 395–413.

Yourgrau, W. and Mandelstam, S. (1968). *Variational Principles in Dynamics and Quantum Theory*, 3rd edn. Saunders, Philadelphia.

ENZYMES, AUTOMATA AND ARTIFICIAL CELLS 5

Pedro C. Marijuán

5.1 INTRODUCTION: THE INFORMATIONAL ANALYSIS OF THE CELL

The simplest way to postpone most of the controversies related to the thorny scientific subjects of 'information' and 'life' is to declare first of all an engineer's point of view – with the explicit promise to abandon it when it becomes inadequate. This is not intended to be a pejorative statement, as Scarrot (1986) has noted when dealing with these same subjects: science and engineering are quite distinct but they are interdependent and maintain a symbiotic relationship.

Like other contributions to this book, this chapter explores the conceptual relationship between two worlds widely separated: the world of cells (and by extension of living beings) and the world of computers (information processing sciences and technologies). In a pragmatic scientific-engineering style, it may happen, if the relationship is fertile in the direct sense, that the biological, biochemical, biomathematical, molecular biological and biophysical knowledge of the cell, gained following well-honoured scientific approaches, might share company with a new approach following the sciences and technologies developed around the computer. In the opposite sense, new computer developments could be gained through the examination of the 'biological style' of computation. Paton (1992) cogently develops this subject and adds more complex conceptual dimensions.

In this chapter we are mainly interested in the first possibility. We want to examine the cell, looking for a new scientific perspective for the study of living systems. In the light of this goal we have decided to take as our object system the simplest cellular entity capable of autonomous life, the prokaryotic cell (in a sort of heuristic, or Occam's razor elimination of superfluous complexity). In the engineering style, we should consider the cellular system just as another 'processing machine', an unconventional computational artefact whose components and functionality are relatively unknown to us, so that we are committed to their investigation.

When we open a prokaryotic cell, what we basically find is: first, a tight membrane, with a number of components and substructures, completely surrounding the system; second, inside the membrane, an intricate network of interconnected chemical reactions controlled by a multitude of enzymes; third, a set of many diverse nucleic acids, especially a long DNA molecule, more or less coiled.

This seems to be the basic 'hardware' inside the unknown microscopic computer we are committed to analyse; at the same time we can realize that, scientifically, it is upon these three kinds of components that many other information-like theories and synthetical strategies have been developed. For example, membranes have been taken as models for boundaries and 'closure' of the system (Brown 1969), or as 'frontiers' and interfaces (Margalef 1980). The accumulation of information in the DNA was carefully studied by Gatlin (1972) and the enzymic-computing view, as we will see later, has been advanced mainly by Sugita, Rosen and Conrad. Historically, illustrious pioneers like D'Arcy Thompson, Rashevsky and Schrödinger opened new ways of thinking about the operation of the whole ensemble.

The concrete analysis in the next sections will proceed as follows. First we detect the active elements of the system (enzymes and proteins), the controlling part of the cell; afterwards we formalize the automata function performed by these active elements, and study its general application to channels and other cellular elements; additional factors of complexity are taken into account concerning automata formalisms, particularly the extent to which they may 'capture' the function of the enzyme. Second, when analysing the collective properties of enzyme networks, we proceed to apply three conceptual categories: self-organization, self-reshaping (reconfiguration) and self-modification. We shall talk about the computational core of the cell; in this respect DNA appears not only as a hereditary material, but as a computing mechanism too. Third, we introduce the idea of 'artificial cells', and some general goals and possibilities are suggested for these fields; the informational analysis of the cell shows itself as a powerful multidisciplinary paradigm. Finally, the development of these informational studies transcends the engineering perspective and leads to reflection on the integrative panorama of the sciences. A brief philosophical corollary is sketched in this respect on the relationship between information, life sciences and other disciplines; the possibility of a general 'information science' (and a particular bioinformation discipline) is considered.

5.2 THE AUTOMATA FUNCTION OF THE ENZYME

5.2.1 APPROACHES TO ENZYMES

The reason for focusing on enzymes and proteins is fairly similar to the distinction made in electronics between transistors (the active components of a circuit) and resistors, capacitors, inductances and so on (the passive ones). In both cellular and electronic systems we have to centre the logical study on the active components that control the dynamics of the ensemble – the ones capable of changing their chemical or electrical potential depending on other variables of the system. Curiously, as was eagerly remarked by Monod, Jacob and other early workers in the field, both enzymes and transistors share similar families of curves. In fact, the influence of electronics and engineering ideas was pervasive in the formative period of modern enzymology, and heavily marked the analytical approaches (for references, see Marijuán 1991). As Changeux (1965, p. 295) writes:

> the cell is a mechanical microcosm: a chemical machine in which the various substructures are interdependent and controlled by feedback systems quite similar to the systems devised by engineers who specialize in control theory . . . The elementary machines of the cellular factory are the biological catalysts known as enzymes.

In the analysis of enzymes we have to include many other complexities. Nowadays the body of scientific literature describing the function performed by enzymes extends over four main disciplinary groups: molecular biology, biochemistry, theoretical biology and biophysics. Other related fields have rapidly advanced towards maturity and deserve consideration: molecular dynamics (as a separated branch of biophysics), computational biology (coming from molecular biology), molecular computing and some areas of artificial life. There is a fast-moving mosaic of interrelated fields, not always clearly connected.

In our approach to enzymes, we can start by surveying a basic computational view, capable of communicating the mainstream of biochemical control theory with other realizations in molecular computing and theoretical biology. In this sense Marijuán and Westley (1992) have retaken one of the simplest classical models: the molecular automata approach started by Sugita (1961, 1963), and developed by Rosen (1967, 1979), Conrad (1972, 1985), Okamoto *et al.* (1980, 1989) and others. It stands in a middle common ground between the two worlds we are relating, the computer and the cell; it will orient our approach to enzymes.

5.2.2 THE AUTOMATON MODEL OF A REGULATED ISOMERASE

In a simplified biochemical perspective, the application of automata theory to enzymic action follows naturally. As Westley (1969, ch. 12) writes:

> the business of enzymes, whether viewed teleologically as synthetic or degradative in function, is clearly the making and breaking of bonds, i.e., the electronic alteration of substrates. The mechanism is not always predominantly electronic at all steps, but the result has an electronic character.

The discrete character of these electronic enzymic operations, their cyclic disposition, their disjunctive nature (0 or 1), their constitutive stochasticity, are general characteristics highly suitable for an automaton model (Kraut 1988).

In the chosen case here, a regulated isomerase, the enzyme action represents one of the simplest cases of allosteric regulation. (If an isomerase-catalysed reaction with a single substrate and a single product, where the form a changes into the isomer a^*, can be considered the simplest formal enzyme mechanism, so also might the scheme given

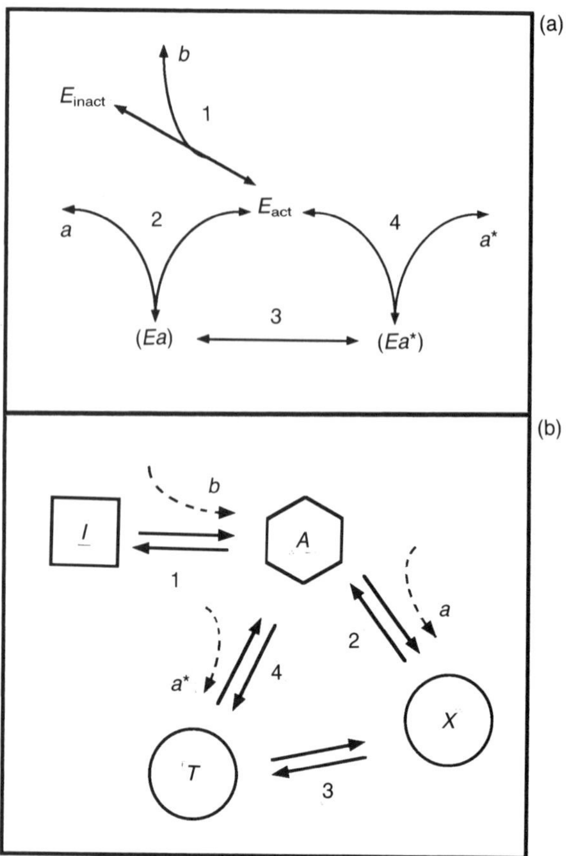

Figure 5.1 (a) Formal mechanism of an isomerase regulated by the activator b. The substrate and product are a and a^*; (Ea) and (Ea^*) represent the corresponding enzymatic complexes with these substances. (b) Qualitative representation of the enzyme's action. The states I, A, X and T correspond with E_{inact}, E_{act}, (Ea) and (Ea^*).

in Figure 5.1 be considered the simplest regulated isomerase mechanism.) The inactive enzyme, represented as (I), binds the effector b and changes to the active form (A); in this state it can form either an enzyme-substrate complex with a, represented as (X), or an enzyme-product complex with a^*, (T); these two complexes can either interconvert or discharge a and a^*.

The following rate equations constitute a mathematical description of the system:

Table 5.1 Logical table of the automaton

States		*Inputs*								*Outputs*	
Previous	*Next*	a	b	a^*	k_{-1}	k_{-2}	k_3	k_{-3}	k_4	a	a^*
I	I	/	0	/	/	/	/	/	/	0	0
I	A	/	1	/	/	/	/	/	/	0	0
A	I	0	/	0	1	/	/	/	/	0	0
A	A	0	/	0	0	/	/	/	/	0	0
A	X	1	/	0	0	/	/	/	/	0	0
A	T	0	/	1	0	/	/	/	/	0	0
X	A	/	/	/	/	1	0	/	/	1	0
X	T	/	/	/	/	0	1	/	/	0	0
X	X	/	/	/	/	0	0	/	/	0	0
T	A	/	/	/	/	/	/	0	1	0	1
T	X	/	/	/	/	/	/	1	0	0	0
T	T	/	/	/	/	/	/	0	0	0	0

$$\frac{dI}{dt} = -k_1 b \cdot A + k_{-1} A$$

$$\frac{dA}{dt} = k_1 b \cdot I - k_{-1} A + k_{-2} X - k_2 a \cdot A + k_4 T - k_{-4} a^* \cdot A$$

$$\frac{dX}{dt} = k_2 A + k_{-3} T - k_{-2} X - k_3 X$$

$$\frac{dT}{dt} = k_3 X + k_{-4} a^* \cdot A - k_{-3} T - k_4 T$$

As R. Rosen has proved (1979), one can create an automaton regulated by activation and inhibition rules which can parallel the behaviour of any system of rate equations. Even more, there is a sense in which automata formulations of activation-inhibition patterns have more generality than rate equations (and very often, in biological systems the quantitative observation of the former is the only way to formulate the latter).

In this case, the four states I, A, X, T, with the state variables b, a, a^*, k_{-1}, k_{-2}, k_3, k_{-3}, k_4 (all of them binary, and b, a, and a^* including the value of the constants k_1, k_2, k_{-4}), together with the transition table shown in Table 5.1, can faithfully reproduce, by means of a Monte Carlo simulation, the dynamics of the system (Marijuán and Westley 1992). For the introduction of the appropriate string of binary inputs into the system, a whole set of transition probabilities has to be calculated (and recalculated, as the automaton completes successive iterations in one or another sense). These probabilities correspond to the formula:

$$P_i = 1 - \exp(-k_i \cdot \Delta t / \Delta a^*)$$

Δt and Δa^* being scale parameters corresponding to time and concentration.

Then, the simple counting of the iterations completed by the automaton gives the net production of a^* (or a) in the steady state; but the counting needs more sophistication in the transient state, when the states of the enzyme have not yet been populated according to the equilibrium constants. Let us point out too that the model requires a certain initial 'tinkering' with its scale parameters and the initial variables in order to obtain a meaningful evolution by the system. (In many cases, just slight changes either in the kinetic constants or in the initial concentration of modulator and substrate cause the apparent breakdown of the system, and force the scale of the approach to change.)

In summary, after introducing a set of initial

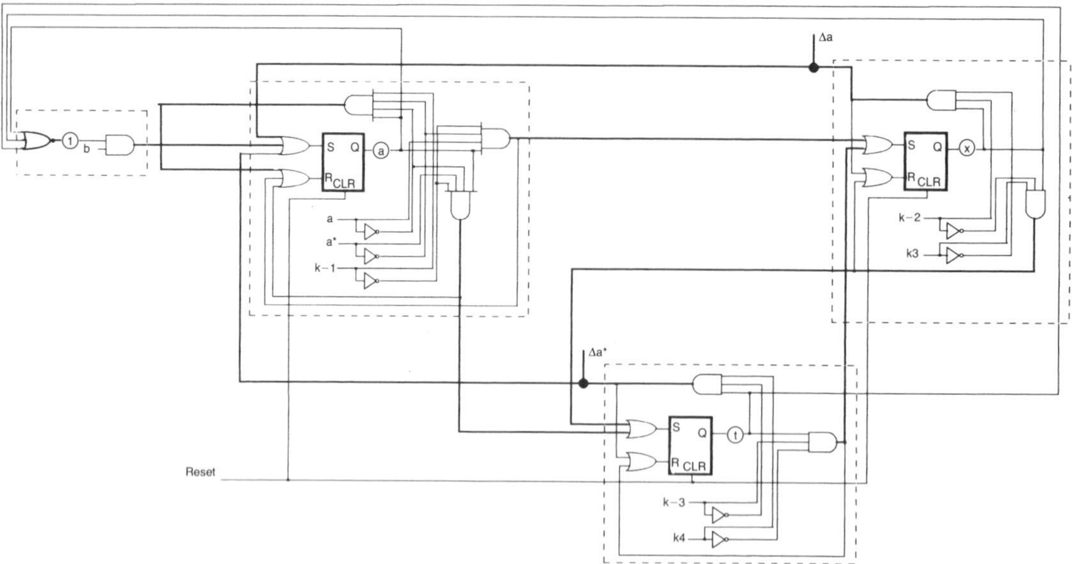

Figure 5.2 Equivalent circuit with flip-flops and logic gates. The initial state *I* is represented by a logic gate NOR; the other three states (indicated by a small circle) correspond to the three flip-flops; the thick interconnections stand for state transitions commands. It has to be emphasized that this logic circuit, like the automaton model, contains reversible activation and the capability to operate in reverse direction. (The circuit has been made by Tomás Pollán, CPS-VZ.)

variables into the system (as if a 'biochemical problem' were posed to the automaton) there will be the development of all the transitions, outputs and whole cycles performed by the automaton under the influence of both its internal (k_i) and external (a, b, a^*) variables and, depending on the related transition probabilities, a set of final conditions will be achieved. It represents the answer that the system gives to the initial conditions, the solution produced by the automaton to the biochemical problem introduced by the initial variables. A more detailed description and some other consequences (that do not always follow from the differential equations) are given in Marijuán and Westley (1992). Additionally, let us mention that this enzyme dynamics corresponding to the transition table shown in Table 5.1 can be represented by an equivalent digital circuit with flip-flops and logic gates, as shown in Figure 5.2. This representation has been proposed (Capstick *et al.* 1992) as an advantageous tool in order to facilitate the exploration of enzyme networks and bionetworks from the electronic engineering perspective.

5.2.3 OTHER AUTOMATA MODELS FOR EMZYMES, OPERONS AND CHANNELS

The automaton idea may go well beyond this initial application to a regulated isomerase. Most of the control mechanisms that appear in the reviews of enzymic mechanisms seem capable of the automaton approach: competitive inhibition, noncompetitive inhibition, acompetitive inhibition, substrate inhibition, allosterism, many cases of homotropic and heterotropic kinetics, molecular cofactors, and even covalent modification (Herrera 1991) could be modelled by following the procedure sketched in the previous section. At this level of generality, the language of enzymic graphs produces isomorphic results with automata theory; we will return to the subject later in this section, and also when dealing with enzyme networks. Additionally, this theory can be applied to operon and gene networks regulation (which obviously constitute the final regulatory step in the control of enzymes). This has been done in recent models by Tchuraev (1991) and Thomas (1991); the work of Kauffman on the emergence of statistical properties from Boolean networks, as a model for the organization of gene networks, can be regarded as a classic (1974, 1985). A great variety of other biological automata models, intracellular or multicellular ones, can be found in different reviews on the subject (Stahl 1966; Sampson, 1976; Weisbuch 1986). As we have already mentioned, Capstick *et al.* (1992) have developed the digital representation of small enzyme pathways (nonregulated irreversible ones) by means of automata theory and equivalent digital circuits, with flip-flops and logic gates.

With respect to membrane channels, the availability of detailed patch-clamp and voltage-clamp recordings, as far as molecular analysis has gone, have opened a promising new field for automata concepts. It has been discussed whether deterministically chaotic instead of stochastic models could constitute a more realistic approach for understanding channel behaviour (Liebovitch and Toth 1991); however, kinetic and energetic models (like the three-state model for inactivation of voltage dependent in channels, as discussed by Jones (1991) seem to fit better with stochastic behaviours in the line of the previous model. Subsequently, automata-like views have already served to anticipate interesting experimental results in channels. Modelling the dynamics of scarce populations of postsynaptic channels at a synapse, by means of transition probabilities and Monte Carlo simulations (following a philosophy very similar to the model previously explained, but implementing it into a complex populational system), Faber *et al.* (1992) report the occurrence of intrinsic variability due to stochastic channel behaviours. Synaptic responses are usually modelled by a series of coupled differential equations that do not take intrinsic channel variations into account; however, the model of Faber *et al.*, extracting the individual history of a channel, shows a series of unpredictable state transitions, including multiple openings similar to those found in the experimental single-channel records. This flicker, which cannot be predicted with analytical solutions, prevents saturation occurring at central synapses where quanta are small. These results have been experimentally tested in some neurons of the CNS, and are contributing to refining present views on synapse saturation in nervous systems (Faber *et al.* 1992).

Another aspect relates channels to automata. Why have channels been developed in families involving a bewildering variety of slight changes? K^+, Na^+, Ca^{2+}, Cl^- and other channels appear in families of dozens or even more constituent members (e.g., 85 K^+ channels), all of them following a common pattern of design but showing slight kinetic differences. Studies in a new field derived from cellular automata, distributed robotic systems (DRS), could help with the answer. These DRSs consist of very simple computational agents capable of realizing sensing operations and 'swarm intelligence' (Hackwood and Beni 1992). The application of the Jaynes–Shannon maximum entropy principle (namely, that the optimum choice of probability distribution is the one that makes no other assumptions about the prior information – the one that maximizes the entropy) to the elements of the DRSs allows the design of optimal sensing configurations by sets of value-specific sensing units. Given a total number of units, we can calculate how many units of each type should be present so that the swarm maximizes the entropy of the probability distribution and therefore is most efficient at measuring, recognizing and so on, the values of a concrete variable (Beni and Hackwood 1991). The problem that the cell has to face is that every channel, like our previous automata model, effectively breaks down its dynamics outside a narrow margin, and then the cell has to arbitrate an appropiate entropic distribution of channels in order to cover the range of variability of the corresponding variables. But how can this distribution be done by the cell?

In a fine review (Miller 1991), where 1990 is consecrated as the *annus mirabilis* of potassium channels, the author explains the solution discovered in the shaker gene of *Drosophila*. This gene codes for a voltage-dependent K^+ channel, or rather for a multiplicity of these channels, through promiscuous alternative splicing to form a family of shaker isochannels. Deletions of increasing length between residues 23 and 40 progressively speed up the inactivation rate, providing a natural rationale for the seemingly continuous gradation between fully inactivating, weakly inactivating, and noninactivating K^+ channels found in nature. This is an elegant evolutionary solution to create a tuned population of automata sensors.

In summary, voltage channels and other gated channels seem to provide a good opportunity to put to test the validity of DRSs, cellular automata and other automata models in the cellular realm in order to capture the variety of actual measurements (voltage, concentrations, mechanical forces, etc.) performed within the cell by distributed populations of these molecules.

5.2.4 TO WHAT EXTENT DOES AUTOMATA THEORY CAPTURE THE FUNCTION OF THE ENZYME?

We have already said that automata theory can parallel the behaviour of any system of rate equations in enzyme kinetics (Rosen 1979), and perhaps it fares a bit better (with the help of Monte Carlo simulations) in cases at the border of the law of mass action. But some more realistic intracellular factors have to be included in this discussion: first of all, the formation of aggregates and clusters, and particularly the channelling phenomenon; second, the complex role that water plays; third, the aging, oxidation and degradation problems in enzymes and proteins (and their turnover); fourth, the wondrous complexity that emerges in the dynamics of the collective networks; fifth, the special character that the logical function of the enzyme shows at the subatomic dimension.

Current automata approaches are rather inadequate to overcome the obstacles numbered one to four; arriving at number five it seems that they definitely fail. Let us very briefly consider these subjects.

The first obstacle needs no comment here (elegantly reviewed by Welch and Clegg (1987), Welch (1992) and by Srere (1988). The oxidation, aging and degradation problems have been recently reviewed by Stadtman (1992); special gene networks have been found inside the cell to organize these destructive phenomena into a coherent dynamics, a sort of creative destruction (like in economic systems). However, many aspects are poorly understood and defy quantitative modelling. With respect to water, Szent Györgyi (1957, 1968) had already warned about the strong influence of its crystalline, dielectric, electromagnetic and solvent peculiarities in the intracellular world ('water, the mother and matrix of life'); but now it seems that water has to be considered even as a ligand in allosterism. Colombo *et al.* (1992) show that as four oxygens bind to haemoglobin, so do 60 additional water molecules; and far from being futile the water cycle these molecules perform appears necessary for the sequential conformational changes in the haemoglobin; in the case of cytochromes, the water cycle is part of the coupled process of moving protons through an associated membrane (as reviewed by Rand (1992)). The importance of water for the whole cell has put evolutionary pressures on channels too – and it seems that some families of mechanosensitive channels have evolved towards sensing the solvent (Kung *et al.* 1990); these authors advance very interesting views on different families of channels supporting the evolution of 'solute senses' versus 'solvent senses' in the neurons of multicellular animals.

All these molecular details seriously compromise the integrity of our model and warn about the superficiality of the approach (let us leave the network complexity for the next section). However, the discussion at the subatomic level takes us to a series of theoretical questions (mainly the Turing–Church thesis) elaborated by the early practitioners of artificial intelligence, and later on widely discussed in theoretical biology, molecular computing, artificial life and so on.

Following M. Conrad (1992), there appears a substantial difference between the switching function provided by a logic gate in a digital computer and the switching provided by enzymes and proteins. The first one corresponds to a (statistically) picturable quantum phenomenon. But the pattern recognition process that underlies enzyme switching transitions corresponds to a hypercomplex energetic exploration performed collectively by numerous electrons, nuclei, atoms and so on, working in parallel to explore a succession of partial energetic wells by means of tunnelling and other complex quantum effects; that is, behind many of the '1s' and '0s' of our automaton there is a statistically nonpicturable quantum process. It includes in general the subprocesses of substrate recognition – even at a distance – attraction to the active site, docking, transformation and ejection. So, when an enzyme computes it does something far more powerful and advanced than a group of switches realizing Boolean functions into a digital computer. The enzyme is a pattern recognition machine. We can effectively tap these pattern recognition capabilities, applying them to information processing in artificial systems (as cells do in natural systems), but cannot meaningfully formalize and specify their internal dynamics into a 'user manual' as we can in conventional computers and Turing machines: they are structurally nonprogrammable devices (Conrad 1992). The separability between hardware and software is not possible here, and consequently the idea of both program and logical form become invalid. Therefore it seems that the Turing–Church thesis about the equivalence between computation and dynamics cannot be substantiated in cellular systems. In theoretical biology, R. Rosen (1985*a*) has widely discussed the acritical assumption of this thesis in current analyses of biological and neuronal systems, showing the epistemological confusions introduced about natural law, causality, modelling and the material realization of formalisms. This is a point where theoretical biology and molecular computing contradict current research assumptions in artificial life, based, like artificial intelligence, on the validity of the Turing–Church thesis.

The problems related to the networking of enzymes and the collective functions that emerge (at the macroscopic level) will receive our attention in the next section. We will have to superimpose some extra computational categories upon the biochemical dynamics these enzyme networks generate, particularly upon their integration with the DNA world.

5.3 ENZYME NETWORKS: THE 'SOCIETY OF ENZYMES'

5.3.1 THE ARCHITECTURE OF THE SYSTEM

Any prokaryote contains about 2000 different enzymes and proteins; the elemental architecture that interconnects their information processing is in principle the network of chemical reactions of the cell. Therefore our pragmatic approach to the cellular 'computer' should produce a long list of enzymes together with their substrates and products. In 1956, in the rush to discover new enzymes, the International Union of Biochemistry or-

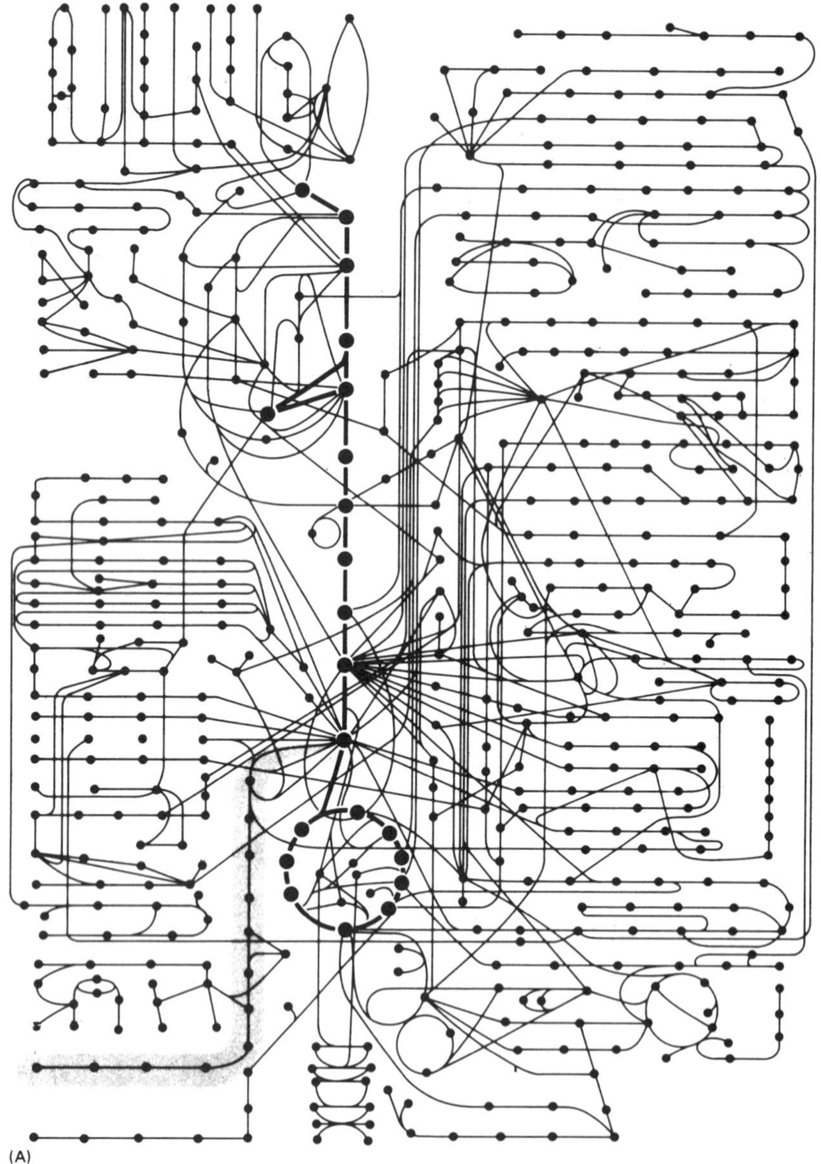

Figure 5.3 The network of chemical reactions of the cell, taken from Alberts *et al.* (1983).

ganized a committee to establish a standard classification, with precise chemical definition, nomenclature and number for every discovered enzyme; we should do something similar with our model cell. We could graphically represent the results in a complete chemical map, or better still by using some sort of simplified network. (This has been done by Alberts *et al.* (1983). See Figure 5.3 for the intermediary metabolism of carbohydrate, amino acid, lipid and nitrogen base metabolism of the cell.)

In Figure 5.3 we can perceive about 500 related metabolic reactions (enzymes would label the arcs, and the black dots correspond to the intermediate products). Srere (1988) has made cogent remarks on the connectivity of this network; for example, about 80% of the metabolic intermediates shown have just one use in the cell, so the formation of sequential multienzyme complexes to channel reactions has many advantages. But let us note that this graph does not represent important populations of enzymes and proteins, such as receptors, channels and membrane enzymes, DNA and RNA related enzymes and proteins, the enzymes responsible for the covalent modification of other enzymes and proteins (converter enzymes). Neither does the graph represent two other important factors: the control system of activations and inhibitions superimposed on the network of reactions, and the multiple connections of these phenomena with gene expression.

Trying to introduce a minimum of informational sense into this multiplicity of intervening factors, we will group them into three conceptual categories: self-organization, self-reshaping and self-modification. These terms have been taken in part from the field of distributed robotic systems (DRS), with some conceptual changes (Marijuán 1991; Beni and Hackwood 1991). We have already said that this field works with rather special cellular automata that solve 'sensing problems', not much different from the problems that the cellular system has to confront.

5.3.2 THE SELF-ORGANIZING PROPERTY

The network of enzymes and chemical reactions depicted in Figure 5.3 is not viable without the superimposition of a tight control architecture. Quite a few substrates and products intervene in a variety of direct feedbacks upon their own enzymes, and also in the modulation (as inhibitors and activators) of regulated enzymes in other branches; together these connections stabilize the evolution of the system, allowing for its permanence. In this respect, the studies known as control theory, either in the simplified form of flux coefficients (Kacser 1983), or in a more elaborate analysis (Crabtree and Newsholme 1987), represent the conventional biochemical cybernetic approach to the stability problem in the branches and bifurcation points of the network. This approach has been criticized for its limitations concerning both nonlinear phenomena and the analysis of extensive pathways (Discussion Forum 1987); however the general properties of the control architecture have not received very much attention either.

It can be argued that the development of a wide system of feedbacks – of an appropriate number of interconnections between the logical tables of our automata – is necessary in order to bring a sort of superresonance capacity to the entire system. Macroscopically, this allows the permanence of steady states of fluxes, or the emergence of constant oscillatory behaviours. In the cellular 'computer', paraphrasing a suggestion made by Stonier (1992, p. 179) addressed to create autonomous electronic computers:

> probably the best approach is to create feedback loops which result in harmonious resonances when the system is performing its tasks well . . . one might create a series of resonating [enzymic] circuits, [to which] the laws of physics [and biochemistry] will favor returning to a stable state if the system has been disturbed. By having numerous, interdependent resonating subsystems and feedback loops, a [cellular] computer would search for a state which brings harmony – that is, a state of super-resonance – to the entire system.

In an enzymic system with this organization, we can tap both the introduction of disturbances and the appearance of state changes into some sort of macroscopic information processing system (either natural or artificial). This emergent self-organization property, in the Prigogine sense of a switching capacity between different attractors (Nicolis and Prigogine 1989), seems a minimum basis to start to talk about information processing.

Together with John Westley, I have pointed out that the sheer dimension of the enzymic system of the cell suggests the emergence of rather complex attractors (Marijuán and Westley 1992), similar to the ones that emerge for Boolean networks randomly connected (as the classical studies carried by S. A. Kauffman (1974, 1985) have shown for simplified genetic systems). Then the presence of supernumerary control systems based mainly on inhibition (forcing functions), and the presence of more regulated enzymes in most areas of metabolism than can possibly be needed for the purpose, would not only be a mere fossil record of control evolution (Ottaway 1988) but a positive condition in order to differentiate a richer collection of stable metabolic attractors for the system.

Indeed a rudimentary form of biological information and symbolism emerges when we count with this basic system, a membrane bag with a collection of numerous enzymes and channels endowed with appropiate control mechanisms. Because of the rather arbitrary conformation of control circuits and resulting attractors, some strategic substances can be used as rudimentary signals, their presence or their release being able to change the basin of attraction (the state) in the same system or in a neighbouring one. This interpretation is in accordance with the metabolic code hypothesis framed by Tomkins (1975): the possibility of coupling metabolic states between different enzymic systems might be considered as the basis of biological semiotics, rather than the existence of a coding relationship between enzymes and DNA (see Emmeche (1991) for a recent review on different biosemiotic theories).

5.3.3 THE SELF-RESHAPING PROPERTY (SELF-RECONFIGURATION)

A necessary condition for the advance of further complexity in any information-driven system is to free its agents from noninformational constraints in the development of their specialized functions. This is what ATP and the associated phosphates mean for the enzymic systems: independence from the particular energy barriers of their specific reactions, or what the invention of money means for the producers of primitive societies evolving towards specialization: independence from barter and immediate exchange. This is the underlying reason for the frequent comparison between ATP and 'currencies' (Welch 1987, 1992). New dimensions of organizational complexity can be incorporated into the system.

The information processing power – problem-solving capacity – of the bag of specialized enzymes can be dramatically increased by the development of the self-reshaping (or self-reconfigurating) property, which consists in the permanent, though reversible, alteration of enzyme structures and functions. By a variety of (enzymic) means, the logical tables of the individual enzymes can be altered, and with them the whole connectivity and the 'logical shape' of the ensemble becomes significantly different. Among the procedures are phosphorylation, dephosphorylation, glycosylation, adenylation, and methylation of side chains, modification of precursors and so on.

Most of the information processing of the cell, and indeed the appearance of genuine signals, are dependent on the converter enzymes which perform these transforming operations upon other enzymes and proteins (notably, kinases for phosphorylation, and phosphatases for dephosphorylation; some kinases act upon other kinases too). These processes are crucial in the tuning of receptors to compensate for the changing level of incoming signals, in the scaling of intracellular signals between micro- and macroscopic dimensions (amplification cascades or, vice versa, amortiguating ones), in the integration of many partial subprocesses into unitary ones, in the elaboration of complex responses, and so on. More concretely, the continuum of experimental and conceptual developments related to the self-reshaping property can be traced from the methylation of receptors in bacteria (Koshland 1986) to the specialized properties of the families of signal-transduction proteins (Koshland *et al.* 1982; Stock 1987; Stock *et al.* 1990) and to the concrete computational role of scale changing between micro-, meso- and macroscopic levels by the cell (Conrad 1984, 1990).

At this stage appear the second messengers, small molecules acting as a variable 'flag' or link, which transmit the integrated metabolic status of the cell to responsive metabolic enzymes, though the information they transmit may be very different between one phylum and another (e.g., cyclic AMP means, in bacteria, glucose starvation; but in animals it functions as a mere transmitter of signals between receptors and amplifier kinases via G proteins – see Ottaway 1988). The enormous number of signalling pathways discovered in recent years, with the corresponding cohort of converter and converted enzymes and proteins (in the hundreds), stands in contrast with the scarce number of second messengers (only half a dozen or so). A generalized cross-talk between the existing signalling pathways seems to facilitate to an extraordinary extent the evolution of new ones (Bray 1990); conversely, the low figure of second messengers is in accordance with their integrative role within the cellular problem-solving mechanisms (Conrad 1984, 1990).

The metabolic code hypothesis sketched by Tomkins (1975), with its distinction between symbols and domains (the term 'symbol' refers to a specific intracellular effector molecule which accumulates when a cell is exposed to a particular environment; the domain of a symbol refers to all the metabolic and genetic processes controlled by the symbol) becomes fully fledged through the new complexity of these self-reshaping enzymic systems. The development of this hypothesis would allow us to connect these views with other recent integrative concepts in physiology, endocrinology, and neurobiology on the origin and evolution of intercellular communication (Barrington 1980; Le Roith *et al.* 1986; Fujita 1988; Campbell 1991).

5.3.4 SELF-MODIFICATION AND CELLULAR 'INTELLIGENCE'

Though a considerable body of information processing capabilities is provided by enzymic self-organization and self-reshaping (a basic 'soft-hardware' plus its reconfigurations), the cell contains a supplementary device in order to multiply its computational performance. Through the DNA (and RNA) worlds the cell can modify its own population of enzymes and proteins on a vast scale, subsequently incorporating new sets of basic soft-hardwares.

Computationally, the DNA does not only constitute a reservoir mechanism for heredity and for the self-construction of the system, but a problem-solving device continuously used: it constitutes the main source of adaptation to the demands posed by the internal and external environment (Marijuán 1991). Recent heterodox views in artificial intelligence have emphasized the manipulation of adaptative information-processing mechanisms before environmental changes as the benchmark of intelligence (Brooks 1989; Beer 1990). In this sense, the networks of enzymes, endowed with a separated DNA reservoir for modifying their own population, constitute the 'computational core' that supports the intelligence of the cellular system.

The cell has developed a peculiar computational style. The 'creative destruction' (like in economic systems) that characterizes the society of enzymes implies a fundamental constraint in relation to the rate of protein synthesis (around 10 000 peptide bonds per second in prokaryotes). Not only do the DNA and RNA worlds contain extensive control architectures in order to select the adequate genes and carefully tune their expression and translation rates; the limited rate of protein synthesis also influences the control design of the metabolic pathways and many other cellular factors (Brown 1991). In general, these constraints have forced the cellular system to adopt a parsimonious computational strategy. The plethora of natural phenomena that impinge on the cell has to be largely ignored by the system. The cell cannot take care of its infinite universe of details, except in the relatively few crucial points that matter. (And it only matters what the cell's computational core has managed to control during the evolutionary process in order to keep the integrity of the system: a few specifics about the performance or impact of such natural phenomena.) There we find specific chemical networks, specialized enzymes, modulators and signalling pathways, genome architectures and so on. For example, the viscosity of the membrane is carefully controlled by changing the composition of lipids in most eukaryotic cells and some prokaryotes. But other prokaryotes do not care; they do not control (compute) the viscosity variable, and as a result they are unable to thrive in many environments – they cannot solve the 'viscosity problem' that has been ignored in the evolutionary process. Somehow, the nature of the network is what counts about what is computed or not. The global result is that this strange social processing entity is extremely well suited to evolve, to specialize and to play social games (e.g., symbiosis, multicellularity, ecosystem integration).

The formalization of this crucial informational property of self-modification, how activations of gene expression relate to the collective functions that appear in the cytoplasm, goes well beyond the scope of this article and beyond any particular automata models for operons and gene control. The theoretical problem was anticipated by Rashevsky (1954) when he postulated a relational biology; Rosen (1985*b*) has developed these notions elaborating the concepts of metabolism/repair systems and anticipatory systems (see other theoretical developments in Welch 1987, 1992). Perhaps this whole subject constitutes one of the most intriguing open questions in theoretical biology – it is also connected to the subject we will develop in the next section, artificial cells. In a different dimension, let us remark that computational properties (qualitatively not very different from the properties we have described here) are very often formulated when authors enter into the bewildering variety of informational processes performed by cells in colonies (Shapiro 1988), multispecies aggregates (Sonea 1990), tissues and organs (Paton *et al.* 1992), nervous systems and so on.

5.4 ARTIFICIAL CELLS

As Rosen (in press) comments, the idea of making 'artificial cells' was one of the earliest nineteenth-century scientific myths. D'Arcy Thompson's famous book (1917) gives some of the flavour of this biomimetic enterprise as it was pursued in that century: the emphasis was on producing artificial cells working with oil droplets in various ionic baths, mimicking more and more cellular behaviours, in the hope of crossing a certain 'threshold'. Some of these artificial systems were actually quite startling.

Nowadays, among the scientific fields we have mentioned, some of them are again not far away from the mythical enterprise of making an artificial cell. In molecular computing (or as Conrad (1992) has put it: molecular information science and technology), experimental works of Kuhn (1985) and others have addressed the problem of combining the properties of bilayer membranes with the processing of enzyme and protein networks (important advances in optical processing by means of chemical reactions and enzymic procedures have been reported too). As a thinking model, Conrad (1990) has proposed 'evolutionary factories' for the systematic development of functional proteins. In biotechnology, some authors have claimed the necessity of developing a new discipline of 'metabolic engineering', increasing the predictive power of its practitioners in order to restrict undesirable characteristics of microrganisms with industrial applications and to design more efficient transformed species (Stephanopoulus and Vallino 1991; Bailey 1991). Two other fields are actually converging towards the potential goal of artificial cells: nanotechnologies and molecular medicine. It is obvious that the present rush of micromachine builders towards the molecular scale converges, directly or indirectly, on the cellular system. The proton-pumped rotatory engines of bacterial cilia, with their sophisticated gradient detection system, and the eukaryotic contractile flagellae, are molecular realizations of impressive performances that can be tapped for artificial purposes. Stripping a motile bacterium of most of its genome, just leaving the carcass, the control of motion and the housekeeping system, and replacing instead other genes codifying for certain desired products, might constitute a powerful form of magic bullet (for molecular medicine) or might be applied to some other sort of molecular biomaterial production. The exact and punctual delivery of cargo on a molecular scale by means of artificial cellular robots would be a feat comparable to the proposed use of autonomous robots for sideral exploration or for the realization of domestic tasks.

Perhaps we are closer, in the cellular realm, to attaining that utopic goal; at least in a theoretical dimension. As far as the sequencing of genomes and the determination of the structure and function of the individual enzymes and proteins reach a reasonable completion for particular microorganisms, the next logical step should be the integration of the whole information obtained, looking for realistic computational models of the simplest cells (Marijuán 1991). Anecdotally, a recent editorial of the journal *Nature* ('Imaginative biology', 1992, p. 272) has pondered the plausibility of this goal too: 'Is it even time to embark on the construction of a mathematical model of a cell, no doubt the megaproject that will follow the human genome projects?' Such a multidisciplinary project would have numerous applied implications, but important theoretical ones too. As we can witness through reviews on the subject (Marijuán and Westley 1992; Paton 1993), the theoretical models of the cellular system have substantially increased their foundations and scope during recent decades and we are experiencing a real inflation of research. Actually, an ambitious international project to model the cell, and to seriously build the theoretical foundation for creating artificial cells, could contribute to open a firmer integrative process in the present-day multiplicity of theoretical developments.

The idea of artificial cells and the evolutionary factory proposed by Conrad, suggest a combined thought experiment. Let us imagine a medium with nutrients; we put in it a membrane 'bag' with enzymes, proteins and channels, and we connect it to a peptide-custom synthesizer. Let us suppose that the custom-peptides are long enough to replace any of the elements in the bag; we have to assume too that the connection between the bag and the synthesizer can be realized throughout an ideal pipe without any disturbances, and in both senses if necessary. (Actually the 'bag' consists in the simplest prokaryotic cell stripped away from its complete genome, RNAs and ribosomes.) Three related questions are: what information – or better still, what sensors – need we put into the bag (the cell) in order to keep control of the ongoing dynamics? What computer processes have to be performed on the signals coming from the cell, for example, to compensate realistically via the synthesizer for the protein turnover? And finally, what fluxes of peptides should we introduce into the system (for example, with a delay time of a couple of minutes) to develop the whole array of cellular functions? The thought experiment – just to separate a cell from its genome, and to artificially reconstruct the mutual flow of information – is a sort of materialization of a M/R system. It might involve any degree of complexity: growth, reproduction, intercellular communication, specialization, colonies, medium depletion, mutations and selection (with *ad hoc* development of proteins), and so on.

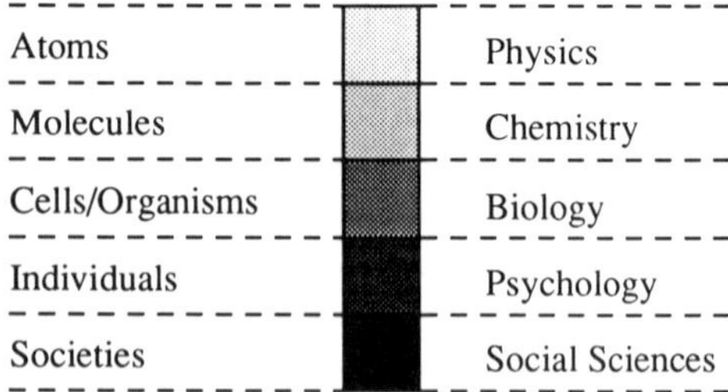

Figure 5.4 The horizontal hierarchical representation of the sciences.

5.5 PHILOSOPHICAL COROLLARY

Finally, let us sketch a brief philosophical corollary extracted from the multidisciplinary peculiarities that the informational study of the cell shows (Marijuán 1991; Marijuán and Westley 1992).

First of all, one of the thought obstacles to overcome in the development of any multidisciplinary study is the lack of philosophical sophistication concerning the relationship between disciplines. The spatial disposition usually attributed to the sciences (horizontal strata hierarchically arranged) gives a very simplified account of the ongoing disciplinary relationships. The logical-positivist emphasis on reduction between adjacent fields has strongly predisposed towards this hierarchical view, suggesting an order of disciplines that correlates with the corresponding 'material levels' of reality (Figure 5.4). Even antireductionist authors like von Bertalanffy, and with him most systems theory scholars, have not discussed this dubious correspondence between scientific and material layers:

> We cannot reduce the biological, behavioral, and social levels to the lowest level, that of the constructs and laws of physics. We can, however, find constructs and possibly laws within the individual levels. The world is, as Aldous Huxley once put it, like a Neapolitan ice cake where the levels, the physical, the biological, the social and the moral universe represent the chocolate, strawberry, and vanilla layers. We cannot reduce strawberry to chocolate . . .
>
> Bertalanffy (1956, p. 8)

In Boulding (1956, p. 13) the cake contains more layers: 'Every discipline studies some kind of "individual" – electron, atom, molecule, crystal, virus, cell, plant, animal, man, family, tribe, state . . .'

Instead of layers (reducible or irreducible ones), the plurality of disciplines needed for the general study of the cell suggests an alternative mapping – or at least, a countervailing view – where basic vertical sciences mutually overlap and combine their conceptual tools, as shown in Figure 5.5. Obviously we can apply physics outside the atom – we apply it to cells and organisms, and even to planets and stars! In general, knowledge about any

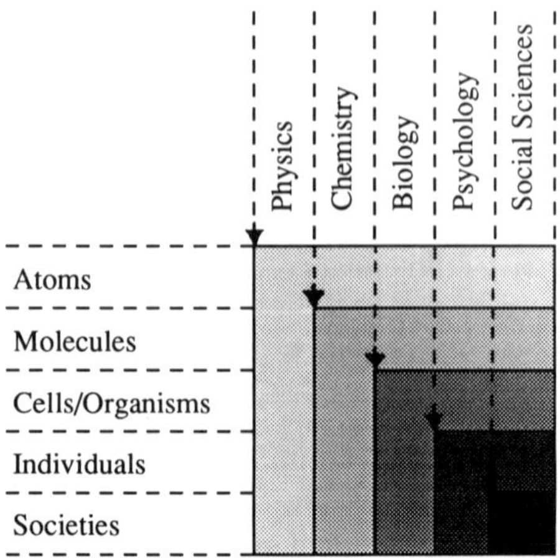

Figure 5.5 The overlapping of successive vertical disciplines; the study of objects in the lowest strata shows the highest levels of multidisciplinarity and complexity.

object forces us to use and combine a plurality of scientific approaches and, vice versa, every basic discipline can be applied to a plurality of objects; Figure 5.5 illustrates these two points. In fact any careful inspection of the numerous disciplines and subdisciplines that have recently emerged shows a variety of combinatory dynamics at work in the relationships between sciences (e.g., sociobiology, biopsychology, physical-chemistry, biochemistry, psychosociology; let us note too that we can obtain these mixed disciplines in the successive overlappings of Figure 5.5). Like specialized sensory modalities in nervous systems, the partial disciplinary contents of the external world are continuously mixed and rearranged – in this respect the system of the sciences appears as a sophisticated 'sensorium' for the whole social organism.

From this more representative 'map of the sciences', the emerging relationship between living beings, computer sciences and technologies, and other scientific and engineering fields, appears in a different light. To some extent, a new fundamental science looms – information science (see different suggestions in Conrad and Liberman 1982; Rosen 1985*b*; Scarrot 1986; Lima de Faria 1988; Stonier 1990; Welch 1992). Like physics and chemistry, information science would constitute a compositional 'texture' of the sensible universe. Its concepts could become as real and formalizable as the concepts of these two fundamental sciences. But we have not yet adumbrated them for a series of reasons (Stonier 1990).

Nowadays information is only a perspective (yet to be elaborated rigorously: yet to be advanced to scientific status). The successful development of this perspective would have to establish a consistent body of relationships between a group of tightly interrelated concepts. The conceptual puzzle might include: information, processor, symmetry, function, specialization, complexity, activation-inhibition, self-organization, integration, memories, knowledge accumulation and so on. As far as these terms could be elaborated, distinguished, quantified or related to laws, the qualitative perspective would dissolve and the new science would emerge.

A mature information science should include an 'ecology' of interrelated fields (like physics and chemistry), many of them nowadays existing in other scientific and engineering areas: quantum measurement and Maxwell's demon, information theory, automata theory, game theory, emergent computation, distributed systems, parallel processing, von Neumann systems, complexity theory, thermodynamics of information, enzyme networks, and so on. New interdisciplinary fields would emerge from the collaboration with other disciplines: 'information physics', 'information chemistry', 'socioinformation' (not far away from present-day interdisciplinary developments in molecular computing, artificial life, artificial intelligence) as shown in Figure 5.6.

Concerning the interdisciplinary relationship with biology, if the impact of physics and chemistry upon the life sciences produced biophysics and biochemistry respectively, it seems that information science should produce a bioinformation (bioinformatics?) scientific discipline. So, bioinformation seems to be a general conceptual framework which needs to be developed.

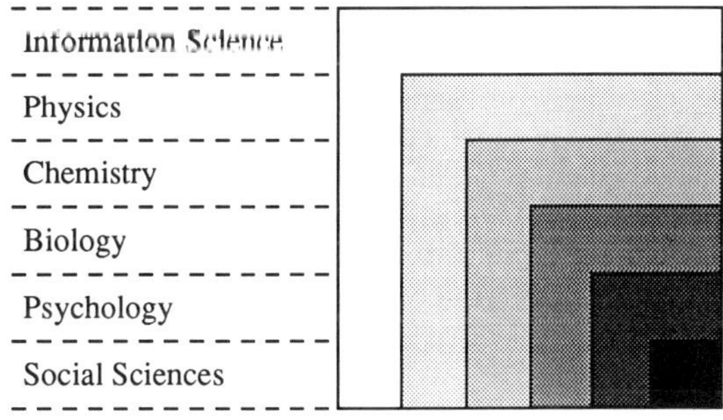

Figure 5.6 The proposed 'information science' and its relationships with the basic sciences.

REFERENCES

Alberts, B., Bray D., Lewis J., Roff, M., Roberts, K. and Watson, J. D. (1983). *The Cell*. Garland, New York.

Bailey, J. E. (1991). Toward a science of metabolic engineering. *Science* **252**, 1668–74.

Barrington, E. J. W. (1980). Hormones and evolution: after 15 years. In *Hormones, Adaptation, and Evolution* (ed. S. Ishii *et al.*). Japan Sci. Soc. Press, Tokyo/Springer, Berlin.

Beer, R. D. (1990). *Intelligence as adaptive behavior: An experiment in computational neuroethology*. Academic Press, San Diego, CA.

Beni, G. and Hackwood, S. (1991). The maximum entropy principle and sensing in swarm intelligence. *Proceedings of the European Conference on Artificial Life*, Paris.

Bertalanffy, L. von (1956). General systems theory. *Yearbook of the Society for the Advancement of GST* (ed. L. von Bertalanffy and A. Rapoport).

Boulding, K. (1956). General systems theory – The skeleton of science. *Yearbook of the Society for the Advancement of GST* (ed. L. von Bertalanffy and A. Rapoport).

Bray, D. (1990). Intracellular signalling as a parallel distributed process. *J. Theoret. Biol.* **143**, 215–31.

Brooks, R. A. (1989). A robot that walks: emergent behaviors from a carefully evolved network. *Neural Comput.* **1**, 253–62.

Brown, G. C. (1991). Total cell protein concentration as an evolutionary constraint on the metabolic control distribution in cells. *J. Theoret. Biol.* **153**, 195–203.

Brown, S. (1969). *Laws of Form*. Allen & Unwin, London.

Campbell, J. H. (1991). Transgenerational effects of drugs and their interpretation: the cybernin system. In *Functional Neuroteratology of Short Term Exposure to Drugs* (ed. T. Fujii and G. J. Boer). Teikyo University Press, Tokyo.

Capstick, M. H., Liam Marnane, W. P. and Pethig, R. (1992). Biologic computational building blocks. *Computer* **25**(11), 22–9.

Changeux, J. P. (1965). The control of biochemical reactions. *Scientific American*, April, 295–304.

Colombo, M. F., Rau, D. C. and Parsegian, V. A. (1992). Protein solvatation in allosteric regulation: a water effect on hemoglobin. *Science* **256**, 655.

Conrad, M. (1972). Information processing in molecular systems. *Curr. Modern Biol.* **5**, 1–14.

Conrad, M. (1984). Microscopic-macroscopic interface in biological information processing. *BioSystems* **16**, 345–63.

Conrad, M. (1985). On design principles for a molecular computer. *Commun. ACM* **28**(5), 464–80.

Conrad, M. (1990). Molecular computing. In *Advances in Computing* (ed. M. Yovits). Academic Press, New York.

Conrad, M. (1992). Molecular computing: the lock-key paradigm. *Computer* **25**(11), 11–20.

Conrad, M. and Liberman, E. A. (1982). Molecular computing as a link between biological and physical theory. *J. Theoret. Biol.* **98**, 239–52.

Crabtree, B. and Newsholme, E. A. (1987). A systematic approach to describing and analysing metabolic control systems. *TIBS, Trends Biochem. Sci* **12**, 4–12.

Discussion Forum (1987). Discussion forum on metabolic control systems. *TIBS, Trends Biochem. Sci.* **12**, 12–19.

Emmeche C. (1991). A semiotic reflection on biology, living signs, and artificial life. *Biology and Philosophy* **6**(3), 325–40.

Faber, D. S., Young, W. S., Legendre, P. and Korn, H. (1992). Intrinsic quantal variability due to stochastic properties of receptor-transmitter interactions. *Science* **258**, 1494–8.

Fujita, T. (1988). *Paraneurons*. Springer-Verlag, Tokyo.

Gatlin, L. (1972). *Information Theory and the Living Systems*. Columbia University Press, New York.

Hackwood, S. and Beni, G. (1992). Self-organization of sensors for swarm intelligence. *Proceedings of the International Conference on Robotics and Automation*, Nice (France).

Herrera, E. (1991). *Bioquímica: Aspectos estructurales y vías metabólicas*. McGraw-Hill Interamericana de España, Madrid.

Imaginative Biology (1992). Editorial comment. *Nature* **356**, 272.

Jones, S. W. (1991). Not an open-and-shut case (News and Views). *Nature* **353**, 603–4.

Kacser, H. (1983). The control of enzyme systems in vivo: elasticity analysis of the steady state. *Bioch. Soc. Trans.* **11**, 35–40.

Kauffman, S. A. (1974). The large scale structure and dynamics of gene control circuits: an ensemble approach. *J. Theoret. Biol.* **44**, 167–89.

Kauffman, S. A. (1985). Self-organization, selective adaptation and its limits. In *Evolution at a Crossroads* (ed. Depew and Weber). MIT Press, Cambridge, MA.

Koshland, D. E. (1986). Molecular mechanisms for

memory. In *Fast and Slow Chemical Signalling in the Nervous System* (ed. Iversen and Goodman). Oxford University Press.

Koshland, D. E., Goldbeter, A. and Stock, J. B. (1982). Amplification and adaptation in regulatory and sensory systems. *Science* **217**, 220–5.

Kraut, J. (1988). How do enzymes work? *Science* **242**, 533–9.

Kuhn, H. (1985). Molecular engineering – a begin and an endeavor. In *Bioelectronic and Molecular Electronic Devices*. Research and Development Association for Future Electronic Devices, Tokyo.

Kung, C., Yoshiro, S. and Martinac, B. (1990). Mechano-sensitive ion channels in microbes and the early evolutionary origin of solvent sensing. *Curr. Topics Membr. Transp.* **36**, 145–53.

Le Roith, D., Delahunty, G., Wilson, G. L., Roberts Jr, C. T., Shemer, J., Hart, C., Lesniak, M. A., Shiloac, J. and Roth, J. (1986). Evolutionary aspects of the endocrine and nervous system. *Rec. Prog. Hormone Res.* **42**, 549–87.

Liebovitch, L. S. and Toth, T. I. (1991). A model of ion channel kinetics using deterministic chaotic rather than stochastic processes. *J. Theoret. Biol.* **148**, 243–67.

Lima de Faria, A. (1988). *Evolution without Selection – Form and Function by Autoevolution*. Elsevier, Amsterdam.

Margalef, R. (1980). *La Biosfera: Entre la termodinámica y el juego*. Ediciones Omega, Barcelona.

Marijuán, P. C. (1991). Enzymes and theoretical biology: Sketch of an informational perspective of the cell. *BioSystems* **25**, 259–73.

Marijuán, P. C. and Westley, J. (1992). Enzymes as molecular automata: A reflection on some numerical and philosophical aspects of the hypothesis. *BioSystems* **27**, 97–113.

Miller, C. (1991). 1990: *annus mirabilis* of potassium channels. *Science* **252**, 1092–6.

Nicolis, G. and Prigogine, I. (1989). *Exploring Complexity*. Freeman, New York.

Okamoto, M., Katsurayama, A., Tsukiji, M. and Hayashi, K. (1980). Dynamic behavior of enzymatic system realizing two factor model. *J. Theoret. Biol.* **83**, 1–16.

Okamoto, M., Sakai, T. and Hayashi, K. (1989). Biochemical switching device: how to turn (off) the switch. *BioSystems* **22**, 155–62.

Ottaway, J. H. (1988). *Regulation of Enzyme Activity*. IRL Press, Oxford.

Paton, R. C. (1992). Understanding biosystem organization. Part 2: towards a theoretical framework. *International Journal of Science Education* **15**(6), 637–53.

Paton, R. C. (1993). Some computational models at the cellular level. *BioSystems* **29**, 63–75.

Paton, R. C., Nwana, H. S., Shave, M. J. R. and Bench-Capon, T. J. M. (1992). Computing at the tissue/organ level (with particular reference to the liver). In *Towards a Practice of Autonomous Systems* (ed. F. J. Varela and P. Bourgine). MIT Press, Cambridge, MA.

Rand, R. P. (1992). Raising Water to New Heights. *Science* **256**, 618.

Rashevsky, N. (1954). Topology and life: in search of general mathematical principles in biology and sociology. *Bull. Math. Biophys.* **16**, 317–48.

Rosen, R. (1967). Two-factor models, neural nets and biochemical automata. *J. Theoret. Biol.* **15**, 282–97.

Rosen, R. (1979). Some comments on activation and inhibition. *Bull. Math. Biol.* **41**, 427–45.

Rosen, R. (1985*a*). Effective processes and natural law. In *The Universal Turing Machine – A Half-Century Survey* (ed. R. Herken). Kammerer and Unverzagt, Hamburg-Berlin.

Rosen, R. (1985*b*). Organisms as causal systems which are not mechanisms. In *Theoretical Biology and Complexity: Three Essays on the Natural Philosophy of Complex Systems* (ed. R. Rosen). Academic Press, Orlando.

Rosen, R. (in press). On psychomimesis.

Sampson, J. R. (1976). *Adaptive information processing*. Springer, New York.

Scarrot, G. (1986). The need for a 'science' of information. *J. Inform. Technol.* **1**(2), 33–8.

Shapiro, J. A. (1988). Bacteria as multicellular organisms. *Scientific American* **256**(6), 82–9.

Sonea, S. (1990). Bacterial (prokaryotic) communication. In *The Semiotic Web* (ed. T. A. Sebeok and J. Umiker-Sebeok). Mouton de Gruyter, Berlin.

Srere, P. A. (1988). Complexes of sequential metabolic enzymes. *Ann. Rev. Biochem.* **56**, 89–124.

Stadtman, E. R. (1992). Protein Oxidation and Aging. *Science* **257**, 1220–4.

Stahl, W. R. (1966). *Natural Automata and other Useful Simulations*. Macmillan, New York.

Stephanopoulus, G. and Vallino, J. J. (1991). Network rigidity and metabolic engineering in metabolite overproduction. *Science* **252**, 1675–81.

Stock, J. (1987). Mechanisms of receptor function and the molecular biology of information processing in bacteria. *BioEssays* **6**(5), 199–203.

Stock, J. B., Stock, A. M. and Mottoner, J. M. (1990). Signal transduction in bacteria. *Nature*

344, 395–400.

Stonier, T. (1990). *Information and the Internal Structure of the Universe: An Exploration into Information Physics*. Springer, London.

Stonier, T. (1992). *Beyond Information: The Natural History of Intelligence*. Springer, London.

Sugita, M. (1961). Functional analysis of chemical systems *in vivo* using a logical circuit equivalent. *J. Theoret. Biol.* **1**, 415–30.

Sugita, M. (1963). Functional analysis of chemical systems *in vivo* using a logical circuit equivalent, II. The idea of a molecular automaton. *J. Theoret. Biol.* **4**, 179–92.

Szent Györgyi, A. (1957). *Bioenergetics*. Academic Press, New York.

Szent Györgyi, A. (1968). *Bioelectronics*. Academic Press, New York.

Tchuraev, R. N. (1991). A new method for the analysis of the dynamics of the molecular genetic control systems. I. Description of the method of generalized threshold models. *J. Theoret. Biol.* **151**, 71–87.

Thomas, R. (1991). Regulatory networks seen as asynchronous automata: a logical description. *J. Theoret. Biol.* **153**, 1–23.

Thompson, D. W. (1917). *On Growth and Form*. Cambridge University Press, London.

Tomkins, G. M. (1975). The metabolic code. Biological symbolism and the origin of intercellular communication is discussed. *Science* **189**, 760–3.

Weisbuch, G. (1986). Networks of automata and biological organization. *J. Theoret. Biol.* **121**, 255–67.

Welch, G. R. (1987). The living cell as an ecosystem: hierarchical analogy and symmetry. *TREE* **2**(10), 305–9.

Welch, G. R. (1992). An analogical 'field' construct in cellular biophysics: history and present status. *Prog. Biophys. Molec. Biol.* **57**, 71–128.

Welch, G. R. and Clegg, J. S. (eds.) (1987). *Organization of Cell Metabolism*. Plenum Press, New York.

Westley, J. (1969). *Enzymic Catalysis*. Harper & Row, New York.

THE MOLECULAR COMPUTER 6

Chris Winter

6.1 INTRODUCTION

The 'molecular computer' has become the alchemist's dream for many materials and systems designers. It is believed that it will extend computer systems beyond the perceived limitations of silicon. The advantages claimed for the molecular computer have included smaller size devices, three-dimensional systems, natural parallelism (Carter 1982; Barker 1987*a*) and even evolvability (Conrad 1983, 1990). However the materials progress towards this goal is slow and disjointed. Many of the developments and lines of research point to completely different architectures and applications. This means that it is difficult to judge what a biocomputer will be like or what tasks it will fulfil. Equally, assumptions about the future of silicon devices from supporters of the biocomputer often appear deliberately pessimistic. They hardly reflect the long history of silicon-based systems which, by constant improvement, have consistently outpaced competing materials technologies.

The purpose of this article is to examine the basic ideas behind the molecular computer, the nature of the materials and techniques that have been suggested as building blocks, some of the systems constraints that will limit the choice of materials and architectures and, finally, to describe one possible route towards the construction of such a computer. The intention is not to cover the field exhaustively – the diversity of approaches and possible relevant background material makes this impractical – rather it is to indicate the most advanced materials work and highlight some of the practical questions surrounding the biocomputer. To enable the reader to pursue the various lines further, the references selected have been chosen from the growing collection of conference proceedings and books that describe developments in the field, thus serving as pointers to more detailed sources.

In Section 6.2 the different systems covered by the loose term 'biocomputer' are classified, and the advantages and problems of each different approach described. This chapter will then concentrate on the 'non-evolvable molecular computer' defined there. Once the family type of biocomputer has been selected for study, it is possible to narrow the choice of architectures, fabrication technologies and materials. In the following sections these three areas will be addressed. In particular the interconnected nature of the choices made in each area will be stressed. Finally in Section 6.6 an attempt to bring together one particular biocomputer is presented. This will be used to illustrate some of the difficulties faced by the molecular designer in attempting to produce a functional system.

6.2 THE BIOCOMPUTER

The terms 'biocomputer', 'molecular computer' and 'molecular electronics' have come to mean different things to different people. The field covered by the latter term is particularly large (see, for instance, Aviram 1989).

Because of these different perceptions concerning the nature of molecular computing it is worth describing various different systems that can be grouped under the loose title 'biocomputer' and briefly discussing the merits of each class.

'Biocomputing' can be taken to mean designing computers based either on biological principles or made using biological molecules. Broadly such systems can be divided into five classes:

1. those made from biological or organic materials but imitating a typical silicon serial architecture;
2. those made from biological or organic materials and using a novel architecture normally abstracted from biology;
3. those made from biological or organic materials and designed as an 'evolving' system;
4. those made from silicon but with a novel architecture abstracted from biology;
5. those made from silicon but with an 'evolving' architecture.

The term 'architecture' is used loosely to cover aspects of the design of a system, namely, the way the overall computation is arranged: for example, whether it is a serial von Neumann computer or some form of parallel computer and what is the nature of the logic elements used to produce the computation, i.e., Boolean logic gates such as NAND or NOR gates. The phrase 'traditional architecture' refers to a silicon-based Von Neumann serial computer implemented with Boolean logic elements. Clearly there are plenty of non-traditional silicon architectures available (e.g., Hillis 1984) and others based on gallium arsenide, which implement parallel architectures normally using Boolean logic. In this article I will focus only on the types 1 and 2 'molecular computer' as the simplest, purely molecular implementation of the biocomputer. These may not prove to be either the most useful or the easiest to construct of the five types described. However, a range of techniques now exists (described in Section 6.5) for manipulating certain molecules. These techniques can be combined with the detailed molecular knowledge available on some organic materials, such as phthalocyanines, to produce the outline structure of a molecular computer (Section 6.6). This should not be taken to mean that the other forms of biocomputer are less interesting. Indeed developments in molecular biology and genetics may soon give us more powerful tools for manipulating proteins and similar biological molecules.

The basic material of the biocomputer could be either molecules (types 1–3) or silicon (types 4 and 5). The dimensional advantages cited in favour of using molecules over silicon include device size and functionality and the ability to build three-dimensional systems (Barker 1987*a*; Carter 1982; Conrad 1985). Biocomputers made from organic or biological materials may well have smaller device sizes than those made from silicon. However if each data bit is represented by a pulse of carriers (e.g., electrons) rather than one bit per carrier, or if the functional logic elements (e.g., logic gates) cannot be produced from a minimum number of molecules, then much of the size advantage could be lost. The advantage offered by access to the third dimension depends critically on the assembly techniques and their relationship to the desired function of the system. As will be described below, we do not possess the ability to locate molecules precisely in a three-dimensional array, and the structures chosen need to reflect the assembly techniques available. A further advantage with molecular systems is that they generally operate with lower switching energies and thus overcome some of the heat dissipation problems associated with a high density of logical switching elements. In comparison with the dimensional advantages of molecules, silicon systems possess advantages of well-characterized materials with well-defined manufacturing techniques. The smallest current line widths

(<100 nm) are not far removed from the diameter of a typical protein molecule (~50 nm), which places a stringent limit on the justification of using proteins as, say, molecular transistors. If a powerful architecture can be abstracted from biology that maps in a simple way onto transistors, then silicon would remain the first choice for the biocomputer.

The architectures outlined in the classes above fall into three groups. The first of these involves discrete components, each defined precisely in function and position. This is the traditional silicon architecture for which many molecular systems are being designed (Metzger and Panetta 1987; Aviram 1993). The second group of architectures is typically abstracted from biology. They include cellular automata and neural networks, in some of which a simple repeating pattern of elements is required. The general thesis proposed here is that attempts to mimic traditional architectures force the designer to use molecules in an unnatural manner. In comparison the careful selection of an appropriate architecture based on a simple repeating pattern makes the potential materials designers' task considerably easier. The final class of architectures – those that can alter their components in real time to adapt to changes – are not discussed here. They form a fascinating group of systems that have been discussed extensively by Conrad (1983, 1990). Their practical realization as molecular systems is probably further away than any of the other possibilities, although recent developments in evolving RNA systems may reverse this situation (Joyce 1989*a*, *b*). Silicon systems based on field programmable gate arrays (FPGAs) could be the first example of 'evolving' hardware. Conrad (1990) argues that pattern recognition is the key to molecular-based evolving systems, in which case FPGAs may not possess enough processing power to produce a suitable system.

There is an alternative way to view molecular functions which is markedly different in conceptual approach to the other techniques; it is to consider the molecule not as a logic or circuit element, but rather as a mechanical machine (Drexler 1987; Pollard 1992; Welch 1987). This approach is not considered here because the ideas from the field do not appear to cross-fertilize with ideas from the molecular electronic approach described here.

In the succeeding sections the discussion will proceed from architectures to system considerations to molecules. This is because there are fewer systems sympathetic to be being built from simple molecular elements than molecules in which switching has been observed. Indeed almost any molecule can be made to change state, and thus be proclaimed a 'switch'. Selection of the appropriate molecules from the myriad of possibilities requires careful analysis of their system implications.

6.3 ARCHITECTURES

Two main architectures have been proposed for making a biocomputer. They differ both in the degree they mimic biology and the extent to which they overcome the problems and exploit the advantages of using a molecular technology.

6.3.1 SILICON HYBRIDS

The earliest molecular systems proposed were based on conjugated polymers where the data was transported as solitons and switched by elements bearing a marked resemblance to VLSI design concepts; a soliton being a structure that propagates without dispersion or dissipation. In this case the soliton consisted of a deformation of the alternate double-single bond alternation of a long chain polyacetylene (Carter 1982). This choice of material and architecture leads to the same acute problems of manufacture, design and testing that have caused silicon systems designers to look to alternative architectures (Barker 1987*a*, *b*). None of these problems are dealt with by mimicking the silicon architecture with molecules. However, copying silicon does reduce the design problem to one solely of miniaturization. The speed of implementation would be greater still if much of the material of a biocomputer could be made from a well-characterized substrate such as silicon. Aviram (1993) has proposed just such a hybrid device. In this silicon-organic hybrid the conducting paths and, presumably, some of the supporting circuitry such as the interface are made using 'standard' silicon processing techniques. However, where a transistor is called for, the tracks are etched to leave a gap of the order of 10–50 nm where an appropriate molecule would be adsorbed. By selecting the material of the electrodes, and making the gap the appropriate shape, the molecule could be induced to react onto the surfaces in the desired configuration. In this way a controlled substrate could simply be dipped into a solution containing the 'transistor' molecules which would then be adsorbed in the appropriate places in a period of 1–2 minutes. This is an elegant production technique but appears to leave a range of questions unanswered: the miniaturization is limited to treating molecules as transistors, the size of the device is limited by lithographic technology, the design difficulties of VLSI are carried through to these ULSI circuits, and it is not clear what the relationship between pulse size and bit logic will be. Perhaps an intermediate stage to the biocomputer may lie in such hybrid devices but much of the potential power of the biological system will have been dispensed with.

6.3.2 BIOLOGICAL ARCHITECTURES

There are two widely used abstractions from biology – neural networks and cellular automata (von Neumann 1951, 1966). The neural network approach has become the technique of choice amongst those studying alternatives to traditional architectures in silicon; it has been less studied in organics despite the closeness of the biological analogue (Samsonovich 1991). Being a cell-based abstraction it requires long-range interactions between elements and a high degree of fan-in, fan-out. In contrast, the cellular automata approach (Carter 1984; Barker 1987*a*, 1990) offers many advantages to the would-be system designer using molecules. It consists of simple elements such as finite state machines that can be formed from a minimum number of molecules (Section 6.6), it has a repetitive structure that might be produced using a simple mix of two to four molecules self-organizing in the appropriate pattern, it uses short range interactions with no large fan-in/fan-out, and it is claimed to be fault-tolerant (Barker 1987*b*).

Cellular automata consist of a one-, two- or three-dimensional grid of cells. Each cell may exist in any one of a number of states. The state of any given cell depends on that of the cells around it. Different cellular automata models vary in the number of states per cell, the type of cells that influence the state (e.g., nearest four or eight neighbours, or even next nearest neighbours as well) and the rules governing how the transitions map one cell state to the next. It can be shown that the long-range interactions of neural networks can be mapped onto the cellular space (Barker 1987*b*) and that cellular automata are 'computationally universal' – that is they can be used to solve any computational problem. The standard proof that cellular automata are computational universal relies on showing that the cells can be connected together to form a universal computing element such as a NAND or a NOR gate. It is then necessary to show that these gates can be connected to form any combination logic circuit and that, in conjunction with a pulsed clock, they can be made into any sequential logic circuit. It then follows that the computer resulting is equivalent to a Turing machine. Such a computer is described as being 'uniform' which implies that we only require a single finite program to deal with all input sizes of a given problem.

The implementation and difficulties with

cellular automata will be discussed in subsequent sections. However some of them are noted here. First, Boolean logic is 'non-conservative', which means that data is both created and destroyed. For instance in a NAND gate, if both inputs are '1' the output is '0' and if both inputs are '0' the output is '1'. Data bits are effectively created or destroyed in these operations, thus logic elements require the equivalent of a high-energy feed and a low-energy sink to produce/remove the data carriers. As will be discussed later, conservative logic, where data is not so altered, is much easier to implement in a molecular structure (Barker 1987*a*, *b*). Secondly the requirement for a clocking pulse is very restrictive. The molecular designer would like to see cellular automata models which used conservative logic, asynchronous stochastic processing and were fault-tolerant! The challenge of designing a system that meets these parameters and can still be shown to be 'computationally universal' complements the challenge presented to molecular designers in producing and manipulating suitable molecules.

6.4 SYSTEM CONSIDERATIONS

In this section systems limitations are described that restrict the possible molecular options to a relatively limited set. Each subsection examines one specific issue although the various issues are interrelated. The balance of the solutions to all the issues addressed may well change with time, as more results become available from the diffuse research in this worldwide field. However the nature of the questions will probably remain the same.

6.4.1 DATA TRANSPORT

Information can be transported around molecular systems in a variety of ways – by propagating conformational changes (such as solitons), or by moving electrons, ions or small chemical molecules. Two major issues need to be addressed – the type of information carrier and the number of carriers used to code for a bit of information. In a conventional silicon device the number of electrons in the pulse is sufficient for stochastic effects to be ignored. This requires many hundreds to thousands of electrons per bit. In this kind of system such a pulse would not be localized on a single molecule. To treat the pulse as a single entity would require each conducting pathway or switch to consist of many parallel molecular-level paths. The simple identity of one molecule with one logic step would be lost. Furthermore, as Barker (1987*b*) has pointed out, in silicon the pulse is constantly destroyed and regenerated; implementation of a similar system using a sink to remove unwanted carriers and a high-energy line to provide new carriers to regenerate the pulse does not appear possible at the molecular level. His view is that the future of molecular computing lies in a regime where one logic bit is represented by just one carrier. Indeed, if the future lies with molecular scale circuits, rather than circuits made of molecules operating as though they are bulk materials, then we will require this level of information coding to exploit the molecular power. Silicon nanotechnology is moving towards single electron logic with various coulomb-blocking devices (Likharev 1990; Averin and Korotkov 1991). The possible data carriers are described in the following sections.

Electrons

Transport of 'free' electrons is discussed here rather than the soliton transport observed in polyacetylene and similar polymers (Carter 1987). Early attempts at designing molecular computers concentrated on electronic data transfer as can be seen in the work on molecular rectification (Aviram *et al*. 1982; Metzger and Panetta 1987). Many papers have been published on electronic transport in organics, particularly crystals which show a range of conduction from insulator to superconductor, and on transport in conjugated polymers (Aviram 1989; Lazarev 1991). Although electron transport in proteins has been studied and conjectured (Isied 1991), the results need to be carefully interpreted.

Proteins are generally dielectric materials, displaying characteristics more correctly described as 'semi-insulating'. There is a clear distinction between a conductor (where free electrons exist) and a conductive pathway (for instance the conduction band of a semiconductor or even an insulator) into which charge may be injected. The problem for the materials designer lies not in producing suitable macroscopic conductivity, but in producing the appropriate effects in the isolated molecules required for a molecular circuit. At the later level purity is a major problem. The behaviour of single electrons and molecules will be very difficult to predict and control if the material is ill-defined (less than 99.9999% pure) or amorphous or polycrystalline in physical form. Yet these two criteria are seldom met for an organic sample. Can we produce materials suitable for one bit per carrier logic using electrons, or will the reproducibility of such devices be too limiting?

Ions and protons

The concentration on molecular electronics has tended to lead to less interest in the other forms of data processing common to biological systems – that of ions and chemical messengers. This arises because ion transport requires a liquid or gel matrix to proceed at a reasonable rate and thus ion-based devices and systems are radically different from the solid-state systems studied by electronics engineers. However, ion transport and switching is at the core of neural data processing and cell control, with Ca^{2+}, Na and K^+ all being critically involved (Hall 1992; Levitan and Kaczmarek 1991). Hydrated protons in mitochondria are a key element in energy production and are pumped across cell and mitochondria membranes by various proteins. All of these offer some interesting advantages over electron-based systems: localization, the ability to control side reactions and the possibility of 'ion-multiplexing' (using different ions to carry different signals) could all be exploited. Much less is known about designing ion switching materials, with most of the work being done on proteins extracted from biological membranes.

Genetic engineering techniques could be used to produce tailor-made complex molecules. Unfortunately such complex molecules are difficult to study. Thus it is unclear what part of the molecule needs to be modified to produce a new or desired action. Simple synthetic peptides showing ion switching effects would be a more useful starting point than naturally occurring proteins. Research on the nature of ion logic is needed to address the questions of whether we can produce ion-switched ion gates, whether we can synthesize/extract an ion integrator and what other ion-based logic elements exist naturally. Developments of an ion–electron or ion–photon interface are, however, taking place as part of the work on biosensors and some of these developments are discussed below. Ions could yet prove a more versatile and easier carrier than electrons to work with in the one bit per carrier regime.

Small molecules

Very few studies have been directed towards carrying data on small molecules (Miller and Lau 1987). Yet, in some ways, small molecules are the most biological approach to data transport. They offer considerable advantages: just as a large molecule involved in the processing of data can be considered as representing a large number of logic elements (see below) so a small molecule can represent considerably more than one bit of data. This point is made by Conrad (1985, 1990), who emphasizes the fact that molecular pattern recognition is at the heart of biological processing. The 'shape' of a molecule can convey a great deal of information. Thus many different 'shapes' or molecules are needed to exploit molecular data flow. 'Molecular multiplexing' would permit data processing where the density of bits processed per unit volume would probably exceed any other conceivable system. One of the best goals for research in 'biocomputing' must be to develop protein systems that carry out computational processes by operating on small molecules.

Conformational changes

Carter (1982), in one of the earliest attempts at linking molecular properties to system architecture, first proposed the use of 'solitons' in polyacetylene to transmit and process data. A number of elegant devices were proposed by Carter (1982, 1987) and Groves (1987). Although these would be of considerable interest if they could be synthesized, the gap between our current synthetic abilities and those needed to produce such enormous, highly defined polymers is still too great. However, solitons and other conformational states of molecules, could be used for molecular-level data processing. Despite the elegance of Carter's work all the spin-off studies have led to bulk polymer devices, some of which are very interesting but hardly pointers to molecular-level computing.

6.4.2 MOLECULAR LOGIC ELEMENTS

It is possible to imitate silicon directly and treat molecules as transistors (or even resistors/capacitors), replacing on a one-for-one basis those already used in silicon with the smaller molecular version. The early days of molecular computing were dominated by the attempt to produce a molecular transistor (Aviram *et al.* 1982; Carter 1982) and thus produce a standard Boolean logic element. Some current approaches are still targeted towards such goals (Aviram 1993). However, if we examine biological systems we do not see molecules behaving in a fashion that resembles discrete transistors. A logic diagram for, say, a protein looks more like a complex set of logic gates (Barker 1987*b*; Capstick *et al.* 1990). This is to the advantage of the designer of the biocomputer. It means that the effective size of a molecular logic element is that much smaller and there is no need to connect the transistors to produce logic elements. The task is simplified to connecting the logic elements together to produce computational elements. A simple repeating pattern of logic elements can easily be used to produce an interesting structure such as a cellular automaton. A simple repeating pattern of transistors would not be so easy to exploit. As will be discussed below, repetitive structures are essential if we are to use our current molecular handling techniques.

Cellular automata are attractive because it should be possible to design molecules that function as finite state machines. Consider the phthalocyanine molecule shown in Figure 6.1. With the correct selection of R groups this molecule can store up to four holes. It can thus act as an electron summation or integration unit. Such a logic element forms the heart of a finite state machine. By varying the nature of each R group and its linkage to the central ring the redox potential of the molecule may change or remain the same as each hole is added, the choice between the two behaviours being made by the designer of the molecule. This could be used to control

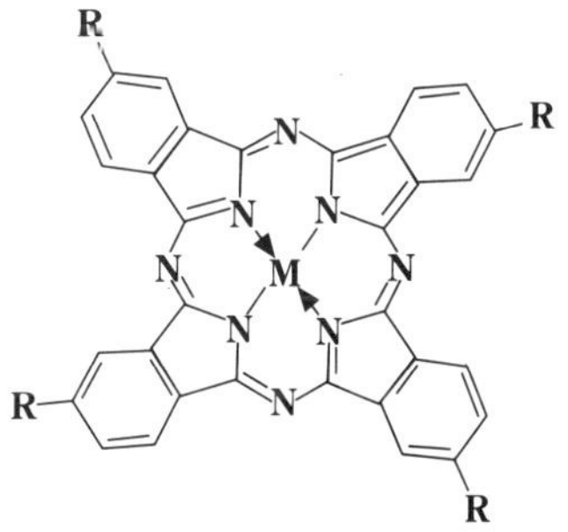

Figure 6.1 Schematic structure of a metal phthalocyanine.

some of the transition rules determining its behaviour if it were treated as a finite state machine (Carter 1984; Barker 1987*a*, *b*). Thus by considering the molecule not as an analogy to its silicon basic functional unit but rather as a separate logic element in its own right, we can exploit a much greater miniaturization of the size of the logic elements. This is similar to the argument that pattern recognition, rather than bit switching, should be at the heart of molecular computing because of the 'fit' to molecular behaviour.

The nature of the molecular components places the following two constraints on the design of a system. There should be no destruction or creation of data as it propagates through the system – the logic must be 'bit' conservative. In a traditional NAND gate two input bits give rise to no output bits (thus two bits have been removed). In comparison Fredkin gates are an example of conservative logic where data is not destroyed (the advantages and disadvantages of various forms of conservative logic are discussed by Barker 1987*a*, *b*). The second constraint is due to the difficulty of process synchronization, there being no easy equivalent of a clock pulse in a molecular system. Indeed central timing could be viewed as a retrograde step in a locally determined logic system such as cellular automata. However no proof exists that an asynchronous, conservative cellular automaton would be computational universal.

6.4.3 STOCHASTIC VS. DETERMINISTIC

Clearly, if the intention in making a molecular computer is simply to mimic silicon on a smaller scale or in three dimensions then pulses of carrier will be used and the architecture will be such that there are many parallel strands making a single logic pathway. In this case the failure of a single molecule will not be significant. However, if the one bit per carrier scheme is carried out then the situation will be very different (Barker 1987*a*, *b*). The purity of organic materials seldom exceeds 99.9%, which implies that there are likely to be carrier traps that unpredictably remove pieces of information from the system. Also it will be difficult to arrange for all the logic steps to be sufficiently far from equilibrium that the error rate is sufficiently low. Thus there will always be bit-level noise. It is difficult to see how to clock a one bit–one carrier–one molecule process, which means that the temporal sequence of operations for a given data bit will also be probabilistic. It is likely that a molecular computer will not operate by deterministic logic like a silicon computer which dissipates considerable amounts of energy at each processing step to ensure that the outcome is within well-characterized bounds; rather the individual logic steps will occur through near-equilibrium processes, dissipating minimal amounts of energy and occurring in a stochastic manner. This difference may eventually prove the heart of the difference between current and future systems. The question remains then to identify those problems best suited to non-deterministic computing and the architectures which maximize the computing power available through stochastic processing.

6.4.4 ASSEMBLY CONSTRAINTS AND THE THIRD DIMENSION

Silicon technology is (currently) planar. The tracks often overlay and some devices can be made to overlay as well; however, the stacking in three dimensions is unlikely to exceed ten devices. If the smallest transistors in a molecular computer had a line width of 1 μm, access to the third dimension, if it could be realized, could give many orders of magnitude more devices per mm^2 of substrate. Utilization of the third dimension is not easy, particularly if a unique wired structure is required rather than a repeating one. Current techniques produce a layered architecture. They give some control in the *z*-direction but no structure in the *xy* plane (beyond that present in the substrate if epitaxial growth is achieved). It is better to justify biocomputers on the grounds of miniaturization and novel architectures than the claimed access to the third dimension. We do not really possess the tools to enable us to manipulate molecules in 3D except in the crudest possible way. The only construction technique that currently lets us manipulate single molecules is the scanning tunnelling microscope (STM), but it would be impracticable to assemble the millions of molecules needed for a real device by moving one at a time using an STM.

6.4.5 THE MACROSCOPIC–MOLECULAR INTERFACE

Converting data from a macroscopic level down to the level of one bit per carrier and locating it on the desired molecular elements represents a major problem. A wide variety of approaches have been suggested as possibilities (Nagel 1987; Lawrence 1987; Barker *et al.* 1990). However, unless the details of the system are known – the nature of the carrier, the architecture, the dimensionality of the system – it is difficult to select an appropriate interface technology. Conrad (1985) has suggested some very novel macroscopic–microscopic conversions suitable for pattern recognizing systems. These pattern converters might be implemented in ionic systems using a photon-induced release of ions from an ionophore.

The STM (scanning tunnelling microscope) is the only technique available for addressing individual molecules in a dense array. The most elegant work with this technique has recently been reported by Rabe *et al.* (1993) which shows individual molecules adsorbed onto a surface. There are many limitations to the use of an STM since it looks at tunnelling currents. If multilayers or thick layers (>10 nm) of a dielectric material are examined using an STM then it is unlikely that any useful molecular property will be observed since the tunnelling currents would almost certainly be associated with defects. Considerable caution in the interpretation of some of the published results and extension of the idea to universal molecular addressing is required. Rectification by single molecules, reported in some early studies, is now believed to have arisen from artefacts. For addressing a whole molecular chip a mechanical scanning technique would be unsuitable, and something more akin to address lines would be required. It is likely that the STM will remain a visualization tool rather than an addressing tool.

Localized charge injection from multiple quantum well (MQW) nanodots into single molecules bound to the top of such a dot might be used for a semiconductor–molecular interface. MQW nanodots are small islands (diameter <10 nm) of a MQW superlattice made by etching through the superlattice to leave islands of the superlattice on the semiconductor substrate. The islands are at larger distances from each other than their diameter.

6.5 MOLECULAR SYSTEMS

In this section some molecular handling techniques are described and a few brief comments given concerning the multiplicity of molecules that could be used to produce a molecular logic element.

6.5.1 ASSEMBLY TECHNIQUES

A variety of techniques exist for manipulating small numbers of molecules or individual molecules to produce thin or partially ordered films. These are summarized below.

Langmuir–Blodgett films

The Langmuir–Blodgett (LB) method of producing monomolecular layers of amphiphilic molecules was first described by Blodgett (1935). The technique works with molecules possessing a hydrophilic surface and a hydrophobic surface (typically a hydrophilic head group and a hydrocarbon tail). A film of the material is spread on an aqueous subphase to give a submonolayer coverage of the water surface. The film is then compressed by PTFE barriers or tapes until the molecules pack closely together. The molecules align with the hydrophobic head groups at the air–water interface and the hydrophilic tails point away from the subphase. Under appropriate subphase and surface pressure conditions the molecules can be transferred to a solid substrate which is passed slowly through the air–liquid interface. The transfer generally occurs on both the up and down strokes, although some materials and conditions give rise to deposition on only one stroke. Not all films remain in the layered structure following deposition; X-ray and other characterization tools are needed to demonstrate the nature and quality of the films produced. A variety of very elegant experiments have been done on a wide range of materials (Barlow 1980). The four principal classes that have been studied are the simple aliphatic molecules (Blodgett 1935), aromatics such as phthalocyanines (Hahn *et al.* 1987) and porphyrins (Miller *et al.* 1985), polymerizable materials that are deposited as monomers then polymerized *in situ* such as ω-tricosenoic acid (Peterson and Russell 1985), and preformed polymers including polypeptides (Winter and Tredgold 1985; Hodge *et al.* 1985). The polypeptides form some of the best ordered and most stable materials (Wegner 1993).

The LB technique can be used to produce a layered structure with a great deal of control in the *z*-direction. In comparison control of the structure in the *xy* plane is more difficult, even with a simple material like stearic acid or ω-tricosenoic acid. The precise deposition details can strongly affect the film structure (Peterson and Russell 1985). Mixtures of two components have been tried – typically to improve the deposition of rigid aromatics – but no great success has been achieved in producing a controlled two-dimensional structure. Thus as a building block for the molecular computer the LB technique has many limitations, even though it offers some access to the 'third dimension'.

A molecular computer fabricated using Langmuir–Blodgett films would require a simple arrangement of molecules in the *xy* plane and any complex order in the *z*-direction. Order in the *xy* plane of the form desirable for cellular automata might be possible if good epitaxial deposition on an ordered substrate could be demonstrated. Such epitaxial growth is unlikely to occur on inorganic substrates for the following reasons. An epitaxially grown film would be limited, probably, to a mix of two materials compounds; thus each molecule would have to act as the 'cell' of a cellular automaton or other repeating pattern system. A large molecule (that is at least the size of a phthalocyanine molecule) would be required to contain all the logic required of the finite state machine central to the cellular automata. Such a molecule would not easily lattice match to an inorganic substrate. The solution could be to exploit the possibilities that arise from the two-dimensional networks described below. These networks have already been deposited as Langmuir–Blodgett films (Morgan *et al.* 1992, 1993) and would make ideal candidates for substrates onto which arrays of proteins could be deposited epitaxially.

Adsorbed films

Adsorbed films are one alternative to Langmuir–Blodgett films. They can be produced by physiochemical adsorption from a hydrocarbon solvent (Tredgold and Smith 1982), but are limited to a single monolayer in thickness. A development of the adsorption technique using adsorption from alternate aqueous and hydrocarbon phases has been demonstrated to produce multilayer films (Tredgold *et al.* 1985). Although multilayer

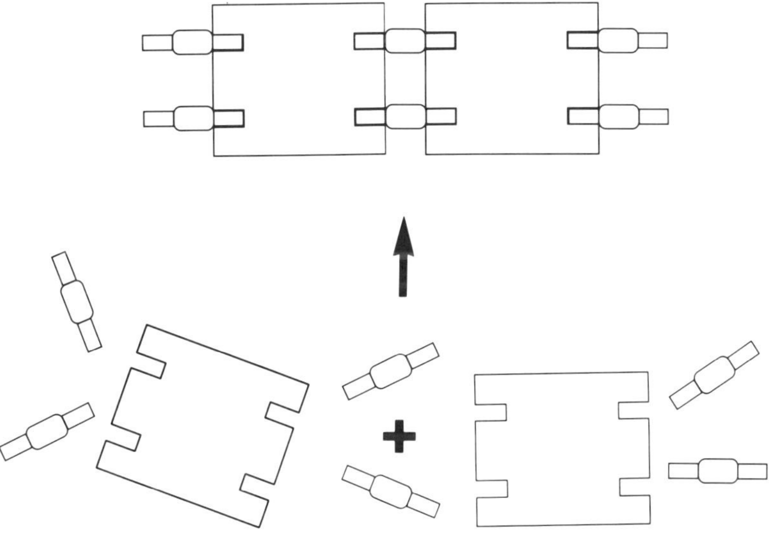

Figure 6.2 Schematic diagram of the assembly of Streptavidin (large squares) and *bis*biotin (small shapes) into a linear chain.

adsorbed films offer similar structural advantages and limitations to Langmuir–Blodgett films the range of materials studied is limited and less interesting. Netzer *et al.* (1984) extend the possibilities of using adsorbed films considerably by using chemical modification to produce a fresh surface for continuing deposition from a hydrocarbon phase. Netzer's technique is more flexible but has yet to produce films of hundreds of layers or be used with molecules that could be used as active computational elements.

Ordered synthesis

In LB and adsorbed films order in the *xy* plane is largely uncontrolled. Both techniques can only be used to control the order in the *z*-direction. Morgan *et al.* (1992, 1993) have published results on a different system which concentrates on producing order in the *xy* plane. In their approach a matrix of protein molecules and ligands are assembled into a two-dimensional network or a one-dimensional chain as shown in Figure 6.2. It is possible to form such self-assembled structures if proteins with multiple ligand binding sites are exposed to a solution containing a ligand with two binding groups joined by a spacer chain. This was demonstrated using Avidin or Streptavidin as the protein and *bis*biotin as the ligand (Morgan *et al.* 1993).

If Morgan's ordered networks can be made, then the proposals put forward by Ulmer (1982, 1987), Carter (1983) and others to order organic materials by deposition on a two-dimensional matrix could be realized. Ulmer's (1982, 1987) suggestion of using proteins as attachment sites for active molecules would become feasible. Ulmer argues that the use of proteins as the active ingredient in such films is limited, but that they should be used for the ordering of other groups. Combined with the Avidin-*bis*biotin-type network shown in Figure 6.2 Ulmer's approach might lead to a regular two-dimensional array where the lattice spacing matches the requirements suggested above. Morgan's films could also be used as substrates onto which Langmuir–Blodgett films could be epitaxially deposited.

Artificial membranes

The use of ions, or hydrated protons, as carriers of information rather than electrons presents a completely different set of molecular options. In particular, switchable ion channels – either biological or synthetic – could form the basis of molecular logic elements. The design of ion-switching materials requires a very different approach to the use of electronic switching: such materials are only likely to work in aqueous or gel phases, especially if the ion channels are of biological extraction. A number of techniques exist to study artificial and biological membranes as model systems for ion switching, for example, the Montal–Mueller technique, black lipid membranes and the formation of lipid vesicles. Studies of ion channels and materials like bacteriorhodopsin in artificial membranes (Hong 1986) may point to whole new areas of potential for molecular computing.

6.5.2 ACTIVE MOLECULES

A wide range of molecules have been proposed as molecular switches. Potentially any molecule that changes state could be proposed for storage or switching. (Some examples of the possibilities are included in, among others, Aviram *et al.* (1982), Carter (1982), Haddon and Stillenger (1982) and Hong (1986)). However the potential or limitations of a given molecule are only apparent when system considerations are taken into account. Therefore the following list of questions should be asked when designing a molecule. Some of these questions are obvious to chemists, others to systems designers; however the literature contains many examples that fail several criteria.

1. Organic materials
   (a) Can it be synthesized easily?
   (b) Can the state transition be manipulated at a molecular level or is it a bulk property?
   (c) Can the molecule be manipulated in a usable physical form?
   (d) Can the molecule perform a logical function that fits into the system architecture?
2. Biological molecules
   (a) Is the molecule stable under reasonable operating conditions (0–60°C is considered just acceptable)?
   (b) Are techniques such as genetic engineering realistically available for the proposed molecule?
   (c) Can the molecule be manipulated easily to produce suitable architectures?
   (d) Is the logic you are proposing to do already at the same scale as the projected limit for silicon (i.e., one transistor per protein)?

As past rivals to silicon have found, it can be very difficult to gain the acceptance of a new material by industry unless its advantages are outstanding!

6.6 ONE POSSIBLE SYSTEM

The following represents one possible solution to the biocomputer. It is described here to illustrate the interweaving of the challenges presented to the would-be system builder and is purely conjectural in nature. However many of the steps described in building such a biocomputer have been demonstrated individually and are now being worked on as part of our program aimed at building a molecular computer. What remains is to bring the various components together and to test whether such a biocomputer can perform a computationally useful task.

First the architecture: in this scheme this would be a cellular automaton where the transition rules were selected to ensure reasonable computational powers under the following constraints:

1. conservative logic;
2. asynchronous timing;

3. transition rules consistent with molecular states;
4. tolerance of faulty cells.

These subjects have each been discussed above, but a summary of the reasoning is given here. Conservative logic is necessary because in Boolean logic bits are not conserved and thus a power line has to be brought to each logic element to enable carriers to be removed or added. It is impractical to produce a power line at the molecular level. Asynchronous timing is required, otherwise a clock pulse must be provided at the molecular level, which would appear difficult if using one bit per carrier logic. The stochastic nature of molecular-level processing and the impurity of materials also means that we cannot rely on guaranteed timing loops or indeed that all the cells will function, which points to both asynchronous timing and the need for fault tolerance. Finally, whatever cellular automata we wish to use, the transition table for the state changes must be able to be implemented in a simple manner with the molecules available. Complex rules that require a many-molecule size logic element are beyond the fabrication and handling technologies described in Section 6.5. The intention in this article is to focus on whether a biocomputer might be made using today's molecular handling technology and thus to point to further studies on whether there are cellular automata-like architectures that are consistent with these handling technologies and are computationally useful.

The assembly technique to produce the cellular structure would be the first fabrication step. A modified version of the Streptavidin-*bis*biotin technique would be used, adapted to give a two-dimensional network deposited by adsorption as a monolayer on a solid substrate. The main purpose of this monolayer is to produce a regular array of binding sites with a repeating cell spacing of 50–100 nm, where a cell is defined as each protein molecule and its attachments. The elements of the network might also be able to contribute towards the active functions of the biocomputer. The size of the cells would be less than the likely minimum size of a silicon transistor.

The main element of the cell would be a phthalocyanine of the form shown in Figure 6.1. It would be attached directly to the protein molecules in the protein-ligand lattice and act as the summation element of the finite state machine. The redox potentials of the five states of the molecule would be crucial in determining the nature of many of the transition rules. Fortunately, the redox behaviour is precisely the part of the system most under the control of the molecule designer. The difficulty comes in linking these rings together to form the cellular network. The ideal arrangement would have the electronic state of each phthalocyanine coupled to the central protein, so that the information passed through the coupling ligand. However, it is likely that the phthalocyanines would need to be linked in a secondary lattice. Phthalocyanines can be made to form linear chains by linking them through the central metal ion using ligands (the so-called 'shish-kebab' structure). What is needed is a derivation of this structure. If a cross-shaped ligand which bound four phthalocyanines could be made then we might envisage a manufacturing pathway that went as follows (Figure 6.3):

1. Deposit a protein/ligand film consisting of a tetrameric protein which is needed to produce the 2-D network (Figure 6.3).
2. Bind four phthalocyanines onto the equivalent sites on the protien.
3. Add cross-shaped ligand (Figure 6.4, 2-D network substrate omitted for clarity).
4. Add 'shish-kebab' ligand to link centres.

The separation of the phthalocyanines should be able to be controlled so that the inter- and intra-protein distances are different and thus

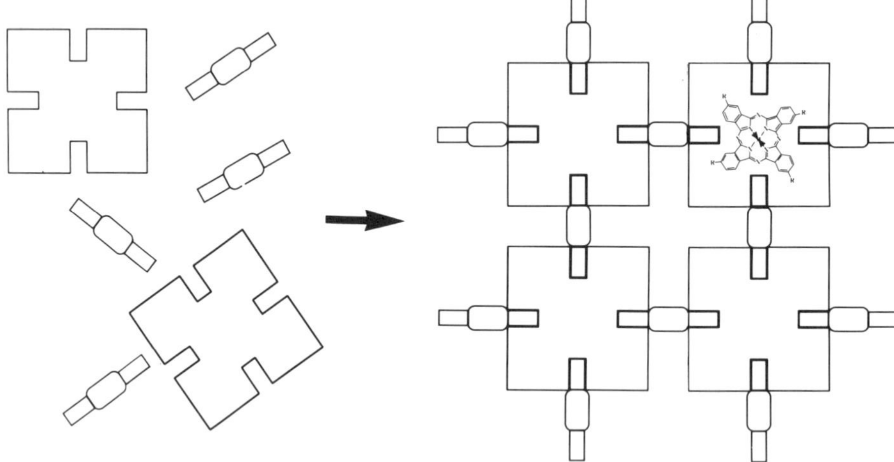

Figure 6.3 Construction of a two-dimensional network from a tetrameric protein and a suitable ligand. Phthalocyanine is bound to each protein subunit.

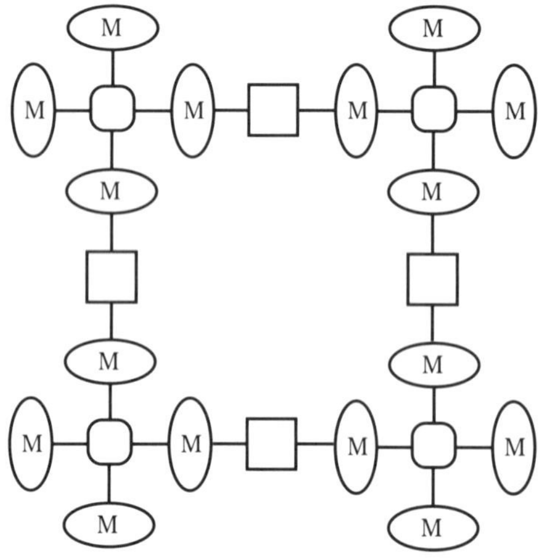

Figure 6.4 The phthalocyanines shown in Figure 6.3 can be linked using 'shish-kebab' type ligands to form a network. Two ligands are used, linking the phthalocyanines on a given protein and between different protein centres.

the cross-shaped and normal ligands should have a higher affinity for their particular sites. Work on suitable ligands to span the distances involved (perhaps 20–30 nm) would be required. However, each of these steps appears to be feasible. Each cell would now have four phthalocyanine rings and, according to the selected phthalocyanines, anything upwards of 20 states (depending on the degeneracy of various ways of distributing the electrons over the four rings). Quite complex transition rules should prove possible.

Finally, how would we address such a system? It is not necessary to address all the cells of an automaton at once. We might imagine that the substrate has a series of interface nanodots, to which some of the proteins have bound, and which influence the state of the attached phthalocyanine. These dots might be spaced a few cells apart so that they can be addressed by macroscopic electron waveguides or similar single electron systems (Barker 1987*b*; Likharev 1990; Averin and Korotkov 1991). The data processing would then occur in the molecular structure and be output in a similar manner.

The above represents an attempt to weave together the molecular and architectural considerations into a system that could perhaps produce a viable biocomputer. The abstraction from biology is considerable, but it lies close to the bounds of what might be done with today's molecular technology. If such a biocomputer could be fabricated, and if the limitations of a molecular structure could be shown to be acceptable for producing a computationally useful cellular automaton, then the miniaturization that would be achieved by reducing the size of a single cell of a cellular automaton to that of a protein molecule would be potentially immense.

6.7 CONCLUSIONS AND PROSPECTS

In this chapter, a brief summary of possible lines of research leading to a molecular computer has been given. It is easy to observe a state change in a molecule and claim it to be a functional unit in a biocomputer; however, it is necessary to keep in mind a wider range of issues than just the demonstration of simple switching. These include: how the system is to be assembled, the nature of the molecular logic element, the nature of the architecture and the benefit the assembled system offers over a silicon system. An attempt to synthesize all these lines of work has been made in Section 6.6. Progress on molecular handling techniques and molecules as logic elements rather than as switches or transistors is enabling us to see some realistic routes toward the dream of a biocomputer. However, what is really needed is more discourse between the system architects and the molecular designers so that a constant feedback of ideas from one to another will stimulate mutually compatible developments. Are there any alternatives to cellular automata that are simple and molecule-friendly? Can cellular automata meet the criteria listed above and still be computationally useful?

In the short term it is likely that the main result of these studies may be ever more powerful ways of employing silicon in novel parallel architectures. In the longer term the evolving systems of Conrad (1985) offer still more unique computing, but for the molecular computer a combination of current technologies, molecules and architectures could yet prove to be a viable combination.

REFERENCES

Averin, D. V. and Korotkov, A. N. (1991). Correlated single-electron tunnelling via ultra-small metal particles. In *Molecular Electronics: Materials and Methods* (ed. P. I. Lazarev). Kluwer Academic, Dordrecht.

Aviram, A. (ed.) (1989). *Molecular Electronics – Science and Technology*. Engineering Foundation, New York.

Aviram, A. (1993). A view of the future of molecular electronics. *Molecular Crystals and Liquid Crystals* **234**, 13–28.

Aviram, A., Seiden, P. E. and Ratner, M. A. (1982). Theoretical and experimental studies of hemiquinones and comments on their suitability for molecular information storage elements. In *Molecular Electronic Devices* (ed. F. L. Carter). Marcel Dekker, New York, pp. 5–18.

Barker, J. (1987*a*) Prospects for molecular electronics. *Proceedings of the Sixth European Microelectronics Conference*, pp. 13–26.

Barker, J. (1987*b*). Complex networks in molecular electronics and semiconductor systems. In *Molecular Electronic Devices II* (ed. F. L. Carter). Marcel Dekker, New York, pp. 640–74.

Barker, J. (1990). Granular nanoelectronics. In *Granular Nanoelectronics*, NATO ASI Series B, Vol. 251 (ed. D. K. Ferry, J. R. Barker and C. Jacoboni). Plenum Press, New York, pp. 327–42.

Barker, J., Connolly, P. C. and Moores, G. (1990). Interfacing to biological and molecular structures. In *Granular Nanoelectronics*, NATO ASI Series B, Vol. 251 (ed. D. K. Ferry, J. R. Barker and C. Jacoboni) Plenum Press, New York, pp. 425–40.

Barlow, W. A. (1980). *Langmuir–Blodgett films*. Elsevier, Amsterdam.

Blodgett, K. B. (1935). Films built up by depositing successive monomolecular layers on a solid surface. *Journal of the American Chemical Society* **57**, 1007–22.

Capstick, M. H., Pethig, R., Gascoyne, P. R. C. and Becker, F. F. (1990). Studies of the logic of electron transfer in oxidoreductase enzymes. *International Conference IEEE Engineering in Medicine and Biology* **12**(4).

Carter, F. L. (1982). Conformational switching at the molecular level. In *Molecular Electronic Devices* (ed. F. L. Carter). Marcel Dekker, New York, pp. 51–72.

Carter, F. L. (1983). Molecular level fabrication techniques and molecular electronic devices. *Journal of Vacuum Science Technology* **1**(4), 959–68.

Carter, F. L. (1984). The molecular device computer: point of departure for large scale cellular automata. *Physica* **10D**, 175–94.

Carter, F. L. (1987). Soliton switching and its implications for molecular electronics. In *Molecular Electronic Devices II* (ed. F. L. Carter). Marcel Dekker, New York, pp. 149–82.

Conrad, M. (1983). *Adaptability: The Significance of Variability from Molecule to Ecosystem*. Plenum Press, New York.

Conrad, M. (1985). On design principles for a molecular computer. *Communications of the ACM* **28**(5), 464–80.

Conrad, M. (1990). Molecular computing. In *Advances in Computers*, Vol. 31 (ed. M. C. Yovits). Academic Press, Boston, pp. 236–324.

Drexler, K. E. (1987). Molecular machinery and molecular electronic devices. In *Molecular Electronic Devices II* (ed. F. L. Carter). Marcel Dekker, New York, pp. 549–72.

Groves, M. P. (1987). Dynamic circuit diagrams for some soliton switching devices. In *Molecular Electronic Devices II* (ed. F. L. Carter). Marcel Dekker, New York, pp. 183–204.

Haddon, R. C. and Stillenger, F. H. (1982). Molecular memory and hydrogen bonding. In *Molecular Electronic Devices* (ed. F. L. Carter). Marcel Dekker, New York, pp. 19–30.

Hahn, R. A., Barlow, W. A., Steven, H. S., Eyres, L. B., Twigg, M. V. and Roberts, G. G. (1987). Langmuir–Blodgett films of substituted aromatic hydrocarbons and phthalocyanines. In *Molecular Electronic Devices II* (ed. F. L. Carter). Marcel Dekker, New York, pp. 459–74.

Hall, Z. W. (ed.) (1992). *An Introduction to Neurobiology*. Sinauer Associates, Sunderland, UK.

Hillis, W. D. (1984). The connection machine: a computer architecture based on cellular automata. *Physica* **10D**, 213–28.

Hodge, P., Khoshdel, E., Tredgold, R. H., Vickers, A. J. and Winter, C. S. (1985). Langmuir–Blodgett films from preformed polymers. *British Polymer Journal* **17**(4), 368–70.

Hong, F. T. (1986). The bacteriorhodopsin model membrane system as a prototype molecular computing element. *Biosystems* **19**, 223–36.

Isied, S. S. (1991). Metal to metal intramolecular electron transfer across peptide and protein bridges. In *Molecular Electronics: Materials and Methods* (ed. P. I. Lazarev). Kluwer Academic, Dordrecht.

Joyce, G. (1989*a*). A novel technique for the preparation of mutant RNA. *Nucleic Acids Research* **17**(2), 711–22.

Joyce, G. (1989*b*). Building the RNA world: evolution of catalytic RNA in the laboratory. *Molecular Biology* **94**, 361–71.

Lawrence, A. F. (1987). How do we talk to molecular level circuitry? In *Molecular Electronic Devices II* (ed. F. L. Carter). Marcel Dekker, New York, pp. 253–68.

Lazarev, P. I. (ed.) (1991). *Molecular Electronics: Materials and Methods*. Kluwer Academic, Dordrecht.

Levitan, I. R. and Kaczmarek, L. K. (1991). *The Neuron*, Oxford University Press.

Likharev, K. K. (1990). Single electronics: correlated transfer of single electrons in ultra-small junctions, arrays and systems. In *Granular Nanoelectronics*, NATO ASI Series B, Vol. 251 (ed. D. K. Ferry, J. R. Barker and C. Jacoboni). Plenum Press, New York, pp. 371–92.

Metzger, R. M. and Panetta, C. A. (1987). Towards organic rectifiers. In *Molecular Electronic Devices II* (ed. F. L. Carter). Marcel Dekker, New York, pp. 5–26.

Miller, A., Knoll, W., Mohwald, H. and Ruadel-Teixier, A. (1985). Langmuir–Blodgett films containing porphyrins in a well-defined environment. *Thin Solid Films* **133**, 83–92.

Miller, L. L. and Lau, A. N. K. (1987). Chemical communications involving electrically simulated release of chemicals from a surface. In *Molecular Electronic Devices II* (ed. F. L. Carter). Marcel Dekker, New York, pp. 5–26.

Morgan, H., Taylor, D. M., D'Silva, C. and Fukushima, H. (1992). Self-assembly of Streptavidin/bisbiotin monolayers and multilayers. *Thin Solid Films* **210**, 773–5.

Morgan, H., Taylor, D. M., Fukushima, H. and D'Silva, C. (1993) Self-assembly of Streptavidin/bisbiotin monolayers and multilayers. *Molecular Crystals and Liquid Crystals* **235**, 121–6.

Nagel, D. J. (1987). Macro to molecular connections. In *Molecular Electronic Devices II* (ed. F. L. Carter). Marcel Dekker, New York, pp. 381–404.

Netzer, L., Iscovici, R. and Sagiv, S. (1984). Adsorbed monolayers versus Langmuir–

Blodgett monolayers – why and how? *Thin Solid Films* **99**, 235–42.

Peterson, I. R. and Russell, G. J. (1985). The deposition and structure of Langmuir–Blodgett films of long chain fatty acids. *Thin Solid Films* **134**, 143–52.

Pollard, T. D. (1992). Proteins as machines. *Nature* **355**, 17–18.

Rabe, J. P., Lambin, G., Delvaux, M. H., Calderone, A., Lazzaroni, R., Brédas, J. L. and Clarke, T. C. (1993). A quantum chemical approach to the STM imaging of organic molecules on graphite. *Molecular Crystals and Liquid Crystals* **235**, 75–82.

Samsonovich, A. V. (1991). Molecular level neuroelectronics. In *Molecular Electronics: Materials and Methods* (ed. P. I. Lazarev). Kluwer Academic, Dordrecht.

Tredgold, R. H. and Smith, G. W. (1982) Schottky photodiodes incorporating monolayers formed by adsorption and the Langmuir–Blodgett technique. *IEE Proceedings I, Solid State and Electron Devices* **129**(4), 137–40.

Tredgold, R. H., Winter, C. S. and El-Badawy, Z. I. (1985). Multiple monolayer adsorption – a new technique for the production of noncentrosymmetric films. *Electronics Letters* **21**(13), 554–5.

Ulmer, K. M. (1982) Biological assembly of molecular ultracircuits. In *Molecular Electronic Devices* (ed. F. L. Carter). Marcel Dekker, New York, pp. 213–22.

Ulmer, K. M. (1987) Self-organising protein monolayers as substrates for molecular device fabrication. In *Molecular Electronic Devices II* (ed. F. L. Carter). Marcel Dekker, New York, pp. 573–90.

von Neumann, J. (1951). The general and logical theory of automata. In *Cerebral Mechanisms in Behaviour, Proceedings of the Hixon Symposium* (ed. L. A. Jeffress). Wiley, New York, pp. 1–31.

von Neumann, J. (1966). *The Theory of Automata: Construction, Reproduction and Homogeneity.* University of Illinois Press, Urbana, IL.

Wegner, G. (1993). Control of molecular and supramolecular architecture of polymers, polymer systems and nanocomposites. *Molecular Crystals and Liquid Crystals* **235**, 1–34.

Welch, G. R. (1987). The living cell as an ecosystem: hierarchial analogy and symmetry. *Trends in Ecology and Evolution* **2**, 305–9.

Winter, C. S. and Tredgold, R. H. (1985). Langmuir–Blodgett multilayers of polypeptides. *Thin Solid Films* **123**, L1–3.

COMPUTING DENDRITIC GROWTH 7

Patrick Hamilton

7.1 INTRODUCTION

Classical brain research usually works at the level of microscopic analysis as well as at the level of macroscopic research on various levels of the cortical matter. From that, structural models are extracted (for example, the definition of functional blocks within the brain). In contrast, cell biology examines the habits of single cells with their physiological abilities to extract from this their own models (Brown 1991); for example, the cellular processes which accomplish growth and interaction. Both methods of research operate on the level of spatial and temporal spot checks. A direct and ongoing observation of nerve growth processes in the brain cannot take place.

To understand more about the mechanisms underlying neuronal tree building and, by that, the whole realm of neuronal plasticity, it is necessary to study these effects continuously. In the study of plasticity of neurons, where, for example, informationally relevant parts grow or shrink, we have put our emphasis on studies concerning temporal morphological effects, especially during the period of embryonic growth (Brandt 1981).

At this point, we recognize the application of computer science to this type of research in a way that will allow the researcher to experiment freely with parameters and add given knowledge to computer design. Simulations, especially if they are based on graphic interfaces, allow a much quicker evaluation of given hypotheses, as an ongoing observation is possible, and developmental times and growth times can be considerably shortened. Within this context computer graphics give excellent possibilities to visualize complex objects and phenomena, not only in the area of mathematics (e.g., fractals), but in the area of biosciences as well.

Our experimental setting simulates the interactive 'growth' of various types of nerve cells, deriving from different 'genomes' and producing a set of 'phenotypes', which depend not only on a variety of 'genomic' information, but also on external factors, such as general and local age as well as on influences deriving from other cells. The researcher is creating code sequences, which can be compared with the DNA to describe a specific cell type. These sequences can be understood as 'genomes' and contain the entire layout of the forthcoming object. By a type of compiling process this information is taken to produce a currently two-dimensional pattern of the object, which we call the 'phenotype'. In cases where the growth was only affected by the genome, a 'genotype' form is received. (These definitions do not exactly match the biological usage of these words, but illustrate similar effects.)

7.2 THE FRACTAL DIMENSION

We had to recognize in our own approach to simulating neuronal growth that there is no adequate starting point by using Euclidean geometry, which only speaks of basic forms like cycles, rectangles, and so on. The objects needed for our simulations are much more of a fractal nature, which allows us to take

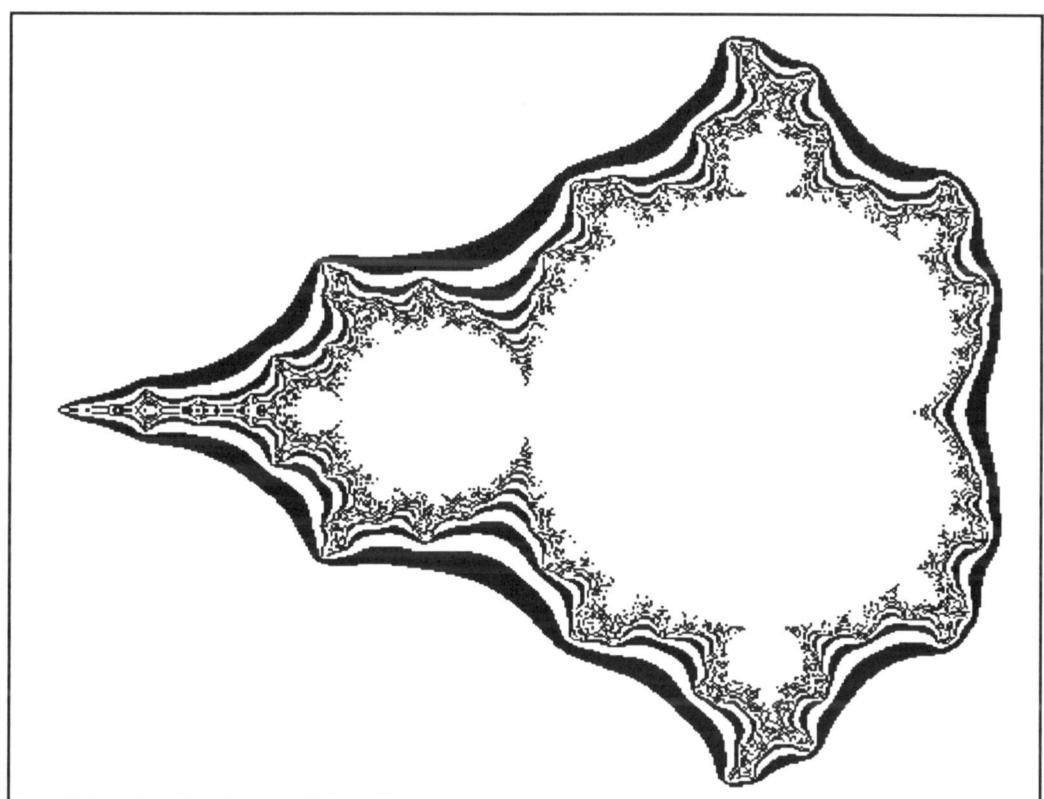

Figure 7.1 Mandelbrot's 'appleman', a non-Euclidean object.

into consideration the irregularity of living objects to some extent (Barnsley, Devaney, Mandelbrot, Peitgen, Saupe and Voss 1989). Figure 7.1 shows a well-known non-Euclidean object. As an appropriate fractal subquantity we chose the so-called 'rewriting string-systems' (Prusinkiewicz 1989).

In our own research, we started to use this method to generate the primary dendritic tree and to take the resulting information to add further details. Discrete interaction of specific cell areas with other totally different structured cell types was included, for example, growth-cones (Davies 1986; Purves 1988) with developmental physiology. The aim of our computer-based models is a deeper understanding of neuronal interaction, by studying the development and growth of these nerve cells. Our main emphasis is on spatial development and the interaction of several nerve cells. By using this approach, we want to contribute to the discussion about artificial neural nets by following Carver Mead (Schrade 1989), who talks about neuronal architecture instead of pure random modelling. In this way, we want to contribute to connectionism and neuroinformatics in terms of an interdisciplinary work, in which we want to understand more about the unfolding of the neuronal structure of the brain.

7.3 THE CONCEPT OF REWRITING GRAPH GRAMMAR

As shown in former work (Hamilton, Kurthen and Linke 1991; Hamilton 1992), we have chosen a grammatical concept to create dendritic trees of nerve cells. For this purpose a three-step procedure has to take place:

1. The user has to define distinct sequences of characters and global variables.
2. By using a special rewriting process these chains are joined.
3. The outcoming chain is used to run a painting tool, which draws the object.

Coming from the idea that the setup of a nerve cell is somehow coded within DNA, we are able to transform given codes into such trees, which yield an extremely high information density coded into the initial strings. Proceeding this way implies that a high measure of well-defined information must be coded in advance by giving these distinct strings. The initial settings within our concept must be handled like a computer program, which cannot be set up by randomly chosen words, but has to follow a very specific syntax, before a type of compiling process creates the resulting code.

Our work was developed by using the 'Lindenmayer systems'. They were introduced by the biologist Lindenmayer in the late 1960s (Lindenmayer 1968*a*, 1968*b*). He used 'rewriting string systems' to simulate the growth of plants. Since that time, the L-system has found an important place in various areas, such as the theory of formal languages or biomathematics (Rozenberg 1980). This technique was used to describe the growth of plant cells, but not to describe dendritic growth.

In this deterministic L-system (DOL-system) Lindenmayer (1974) showed that the way the branching of simple filamentous organisms takes place can be formalized. The L-system in itself has no drawing instructions of its own; instead it contains a recursive mechanism, by which the complete information is built from a set of given rules. Once the final instructions are set up by a rewriting process, they can be used for further purposes.

The basic idea of graph grammar contains:

1. an initial word put together from letters of a defined alphabet;
2. exchange rules, where a specific letter is replaced by an attached word;
3. a rewriting process where each letter of the initial word is replaced sequentially by the appropriate exchange word if there is one;
4. a given number of iterations to rewrite the resulting string.

Following this concept, an alphabet $\mathfrak{Z}_L$ can be defined as a set of characters $\mathfrak{Z}_L = \{a, b, c, d, \ldots\}$. Following Prusinkiewicz (Prusinkiewicz 1989) a string OL-system is defined as an ordered triplet, put together out of a single letter w of the given alphabet $\mathfrak{Z}_L$ and a word $\wp$.

While generating an object, each letter w_i is sequentially replaced by its production $\langle \mathfrak{Z}_L, w_i, \wp_i \rangle$. In case no production is given, the letter is exchanged for itself (Figure 7.2(b)).

To run the L-system, the user has to define the initial word $\Im$ from the given alphabet $\mathfrak{Z}_L$, as well as a set of productions $K_n = \langle \mathfrak{Z}_L,$

(A) Definitions:

alphabet $\mathfrak{Z} = \{+, -, [,], F, X, Y, Z\}$
initial word: X + F − [X]
productions:
$\mathfrak{R}_1 = <\mathfrak{Z}, X, Y [F-F+]>$
$\mathfrak{R}_2 = <\mathfrak{Z}, Y, FZF>$
$\mathfrak{R}_3 = <\mathfrak{Z}, Z, +[X]->$
number of iterations:4

(B) The 'rewriting' process:

X+F−[X]
in the beginning: Y[F−F+]+F−[Y[F−F+]]
1. generation: FZF[F−F+]+F−[FZF[F−F+]]
2. generation: F+[X]−F[F−F+]+F−[F+[X]−F[F−
3. generation: F+]]
F+[Y[F−F+]]−F[F−F+]+F−[F+[Y[F−
4. generation: F+]]−F[F−F+]]

(C) Extraction of unidentified Turing symbol:

Turing alphabet: $\mathfrak{Z}_r = <F, +, -, [,]>$
original form: F+[Y[F−F+]]−F[F−F+]+F−[F+[Y[F−F+]]−F[F−F+]]
usable code: F+[F−F+]−F[F−F+]+F−[F+[F−F+]−F[F−F+]]

Figure 7.2(a) An example of the rewriting process.

F Move pen forward one step in current direction and draw line

f Move pen forward one step in current direction without drawing

+ adds α to current angle

− subtracts α from current angle

[put current plot information onto the computer stack

] reset plot information to last stack information

Figure 7.2(b) The meanings of various operators.

w_n, $\wp_n\rangle$, where w_n characterizes the starting letter and $\wp_n$ the exchange rule. Finally the user has to define the number of iterations i of the whole process.

Once the rewriting process is finished, those letters which belong to a defined Turing alphabet $\mathfrak{Z}_T = \{F, f, +, -, , \{, \}, \ldots\}$ are used to run a 'LOGO'-fashioned drawing unit (Turing 1950) (Figure 7.2(a)). 'LOGO' is a simple programming language, where a 'turtle' can be moved on a graphic surface by using simple commands like 'forward', 'turn left', 'turn right', and so on. The underlying concept derives from the classical Turing machine. To accomplish this, each letter of the Turing alphabet $\mathfrak{Z}_T$ must correlate to a special action like 'move forward', 'turn right', 'turn left', 'save current information', 'restore current information and so on (Figure 7.2(b)). Only those letters which are part of the Turing alphabet $\mathfrak{Z}_T$ will be taken to run the drawing unit. All other letters are skipped.

This interaction is illustrated in Figure 7.3. We have taken an extreme example, which usually would not appear in nature, but gives an impression of the entire process. For better understanding, all hidden characters are included in a type of pseudosyntax. This can be seen at the right-hand side of Figure 7.3: on top the initial chain $\mathfrak{I}$ can be found, followed by the productions $\langle \mathfrak{Z}_L$, X, FY$\rangle$, $\langle \mathfrak{Z}_L$, Y, [+Z] [−Z]$\rangle$ and $\langle \mathfrak{Z}_L$, Z, [+F+F+F+F] [−F−F−F−F]FF$\rangle$ written down in a semi-graphical fashion.

In observing the ongoing process of rewriting, the development of the given object can be studied. At the end of the first rewriting run, which is equal to the object after the first generation, a visible line with an invisible $\underline{\underline{Y}}$ on top can be found. After the process has been repeated once, $\underline{\underline{Y}}$ is exchanged for two visible branches with invisible $\underline{\underline{Z}}$ operators at their ends. Once a further rewriting iteration has taken place, those $\underline{\underline{Z}}$s are replaced by visible semicircles and invisible $\underline{\underline{X}}$ operators.

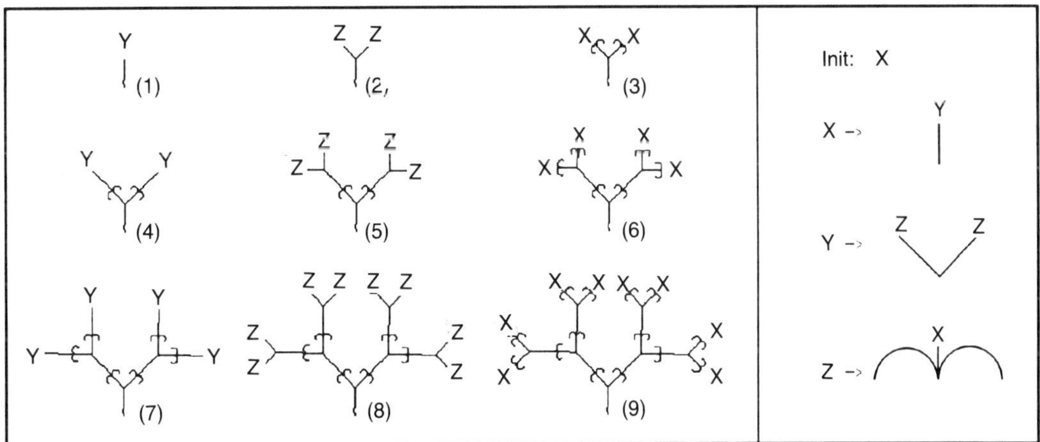

Figure 7.3 An example of rewriting a string system.

This operation goes on until the required number of iterations has been accomplished. Any given syntax will operate in the same way, although further basic elements have been added in our system.

7.4 BUILDING A NEURON MODEL

Looking at neurophysiology, each cell takes an unique direction, position and function within a body of cells. This gives the possibility of defining each neuron individually, without having much knowledge about its specific form or configuration. We have done this by using the following model. A single neuron n_i is part of a set of neurons N (for example, the brain or a specified part of it). Each neuron n_i can initially be described by the following parameters:

$$n_i = n[\zeta[\Psi], F_a, (X_s/Y_s), \alpha_s, S_f[\ldots]] \in \mathrm{N} \quad (7.1)$$

where:

1. ζ [Ψ] defines the specified type of the neuron, Ψ describes the preset form and growth characteristics;
2. F_a indicates an aging factor of the current neuron n_i;
3. (X_s/Y_s) defines the starting position of growth within a target area;
4. α_s defines the initial direction of growth;
5. $S_f[\ldots]$ defines the probabilities of various side effects, influencing the construction.

To transform this model to graph rewriting systems, we postulate that each neuron contains its DNA, which is somehow organized in a rewriting string fashion. $\zeta[\Psi]$ can be understood as the 'phenotype', which derives from the genome Ψ. The information about growth is taken from Ψ, which is similar to the inital setting of the rewriting input (Figure 7.2(b)). Through the rewriting process which uses Ψ, the growth process of the cell is simulated until at the end of the rewriting process the entire 'phenotype' ζ can be found. The growth of the given neuron will start at (X_s/Y_s) and its initial direction is given by α_s. Those factors, summarized by $S_f[\ldots]$ may influence this growth, so that the resulting 'phenotype' differs from the genuine 'genotype'. The meaning of $S_f[\ldots]$ will be discussed in detail in the next section.

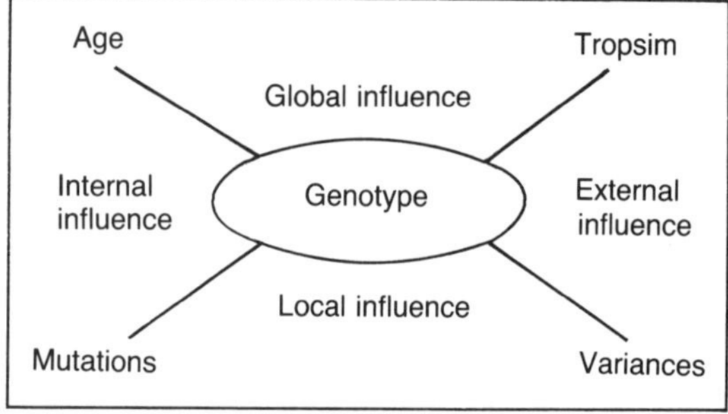

Figure 7.4 Growth influencing factors.

7.5 INFLUENCING THE NEURONIC SHAPE

Once a neuron is growing in nature, various side effects influence and attract the direction of growth. They act in part globally and in part locally. Global effects are those which influence a whole set of neurons equally, while local effects will only happen to a specific neuron. Besides this, there are external and internal effects. Both influence the appearance, but internal effects will change the genomic information irreversibly, while external effects cause no genomic alteration. Figure 7.4 summarizes this situation.

7.5.1 MUTATIONS

In biology, mutations can be understood as changing the genotype of a cell, randomly triggered by different processes. We have taken this idea and included it in the rewriting process. This means that once a specific mutation takes place all following rewriting processes will include this change. Figure 7.5 shows the results of a mutation simulation.

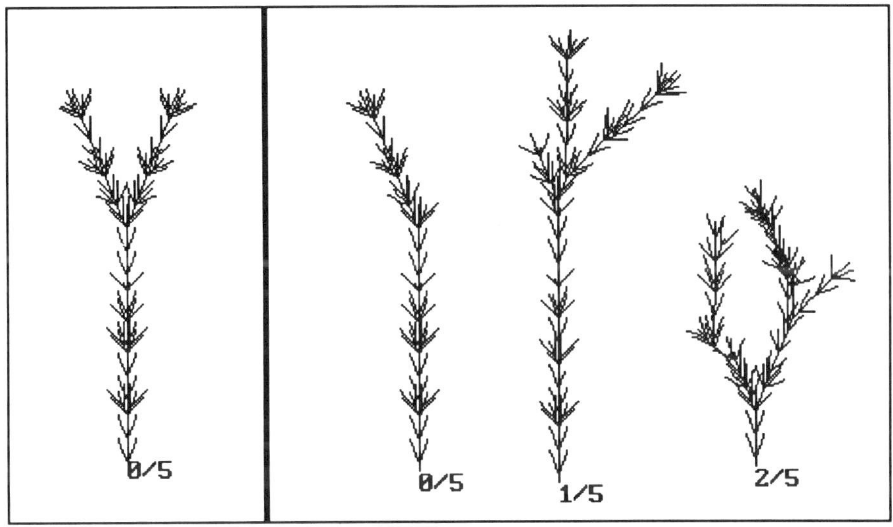

Figure 7.5 Changing the phenotype by mutation. The original is on the left, the mutated objects on the right.

Erase mutation

This first type of mutation erases a letter during the process of copying, so that a deletion (in terms of molecular genetics) within our model world takes place.

ABC D EF → ABC_EF

Change mutation

Run by a random process one incoming letter is taken off during the process of copying and exchanged for another randomly selected letter.

ABC D EFG → ABC A EFG

Inversion mutation

A subchain of randomly defined length is taken from the incoming chain. This sequence is inverted and put back into the new chain.

ABC DEFG HIJ → ABC GFED HIJ

7.5.2 AGING

Looking at age-dependent growth, it can be observed that specific factors change their habits while system time passes. Usually the whole system is affected by that. We have implemented this global factor in a way that internal values (for example, stepwidth, angle, sprouting) of a cell are changed by an age-dependent random process. In nature there are times when sprouting of axons is tremendously high (for example, at a very young age), while the same process will happen less often when the whole set of neurons has become old. Figure 7.6 summarizes what this kind of simulation produced.

7.5.3 VARIANCES

Another important factor, overlaid on the drawing process, is a defined statistical variance of angles and stepwidth through the whole process. This factor produces a very important difference between a dead and a vivid looking object. The second cannot be

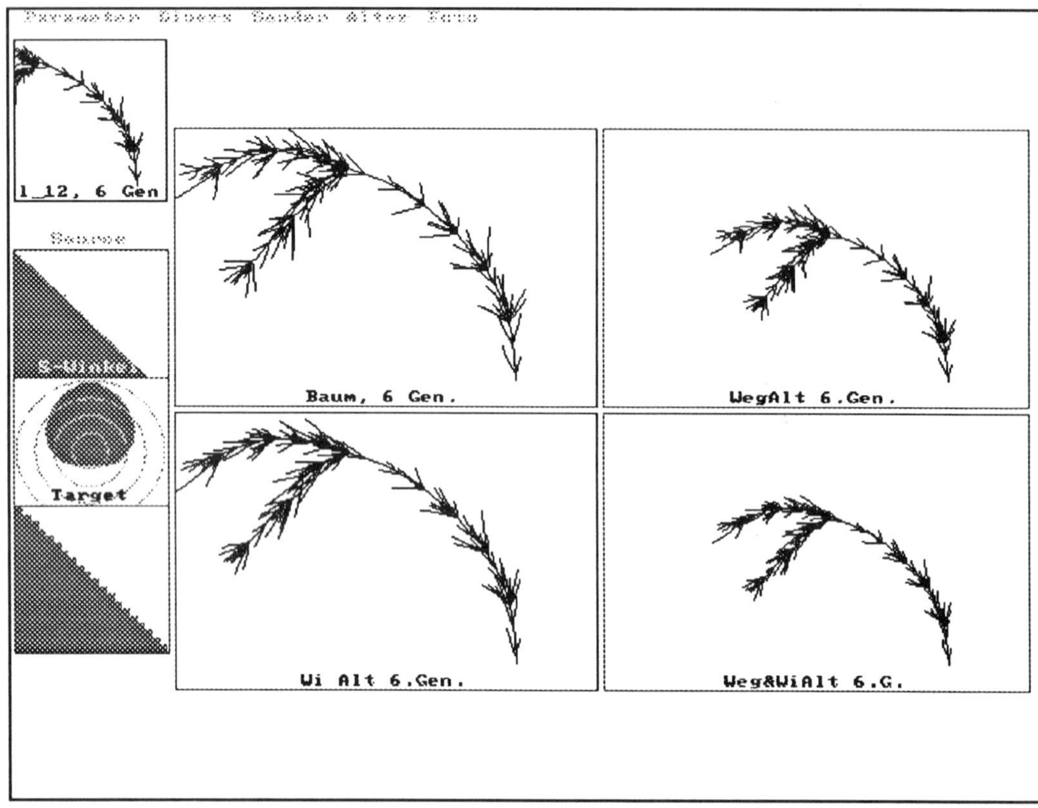

Figure 7.6 The results of the aging factor, influencing the direction (top left), the angle (bottom right) and influencing both (bottom left).

reconstructed exactly like the former one, although the same genotype is used.

7.5.4 TROPISM

As shown in various enquiries, dendrites grow in the realm of an inhibitory or excitatory gradient field. Our first studies used a rotation symmetric potential field globally affecting all those dendrites growing in this field. A force slightly changing the direction of growth towards the centre of the tropistic field is superimposed. The deviation depends on the strength of the gradient field and the relative angle towards the centre (Figure 7.7).

7.6 EXPANDING THE MODEL

7.6.1 MODIFYING THE REWRITING STRING PROCEDURE

In Figure 7.8 we have shown the general principle of a rewriting procedure. As seen here, the entire process can be understood as a repetitive sequential exchange, which happens n times in breadth-first progression. One of the disadvantages of working this way is that the length of the chain is growing very fast and many void operations must be included. To solve this problem a depth-first approach would be much more appropriate (Figure 7.9). In this case each letter of the initial chain $\Im$ is exchanged immediately by

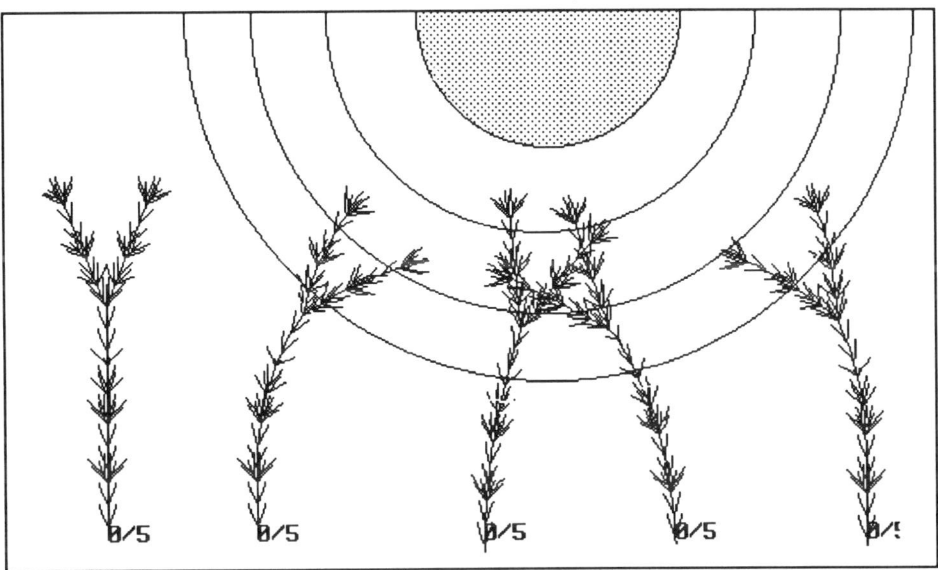

Figure 7.7 Growth influenced by a tropic field.

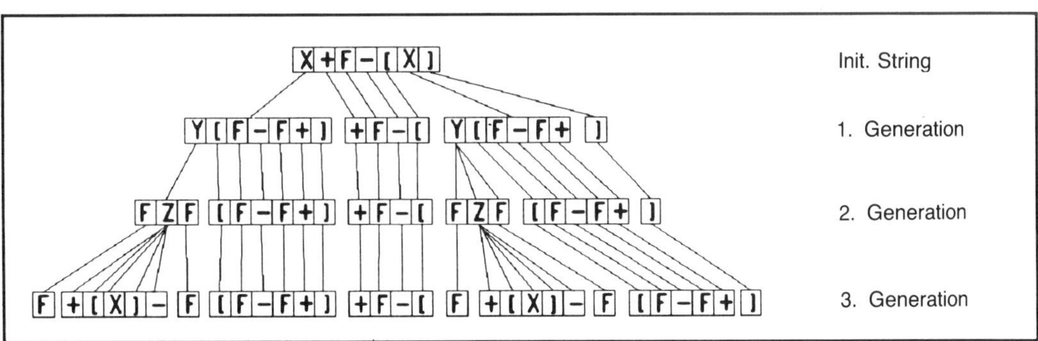

Figure 7.8 Showing the breadth-first approach during the rewriting process.

the given production. After that each letter from the newly given subchain is sequentially exchanged by the given productions. If no production exists or if the maximum depth of iterations is reached, the next letter of the current subchain is taken. Once the end of a subchain is reached, the prior subchain, which is one level higher, will be continued. The entire process goes on until the end of the initial chain is reached.

In using the depth-first approach, a 'naturalization' process can be added to the resulting chain much more easily.

7.6.2 NATURALIZING OBJECTS

What makes the difference between deterministic and chaotic forms on one side and biological looking forms on the other side can be described by the term 'naturalness'. In using this word, a type of growth is characterized, which follows clearly given guidelines

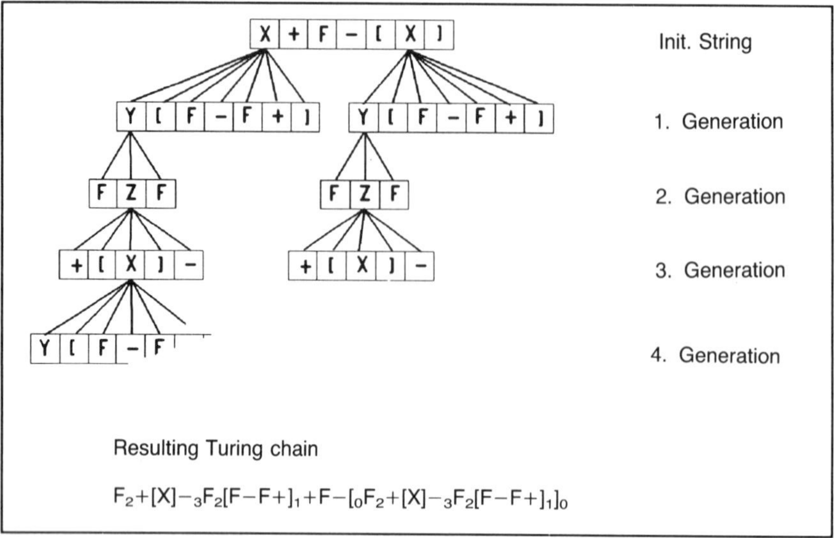

Figure 7.9 Showing the depth-first approach during the rewriting process. The chain at the bottom shows the resulting Turing sequence. The indices mark the deriving generation of the letter.

but includes a given measure of unique elements as well. Looking into any area of biology makes this term evident, for all living objects show similar patterns. The effect can be accomplished by superimposing a deterministic genotype on any type of stochastic variance. In so doing several types of representation can be defined:

1. Deterministic representation: no variances except those deriving from the 'genome' have taken place. Whenever the same 'genome' is used again, it will always bring forth exactly the same 'phenotype'. This representation can only occur if no stochastic component is superimposed.
2. Natural representation: within a small bandwidth using a minimal stochastic component, figures can be found which get close to a biological outlook.
3. Stochastic representation: once the influence of the stochastic component is growing, the given genotype becomes very quickly distorted or even destroyed so that it cannot be recognized any longer. The stronger the influence becomes, the more the drawing gets a purely random form.

To include a stochastic naturalization factor of a defined range, we have added a random angle β from a given stochastic distribution Γ_β, which follows each given 'F'-operator. The band width of the given distribution Γ_β is defined in advance as a global constant for each object, for example:

$$F + F - FF \rightarrow F\beta + F\beta - F\beta F\beta.$$

Looking at Figure 7.10, we can see the effects of this angle β. In all cases we have chosen exactly the same initial data and productions. The only difference was made in changing the distribution width Γ_β. The picture shows clearly the move from an absolutely symmetric object (e.g. a snowflake), passing through various degrees of natural looking objects and ending in complete chaos. Taking the idea of this picture, it is obvious that dendritic structures need a small β-distribution.

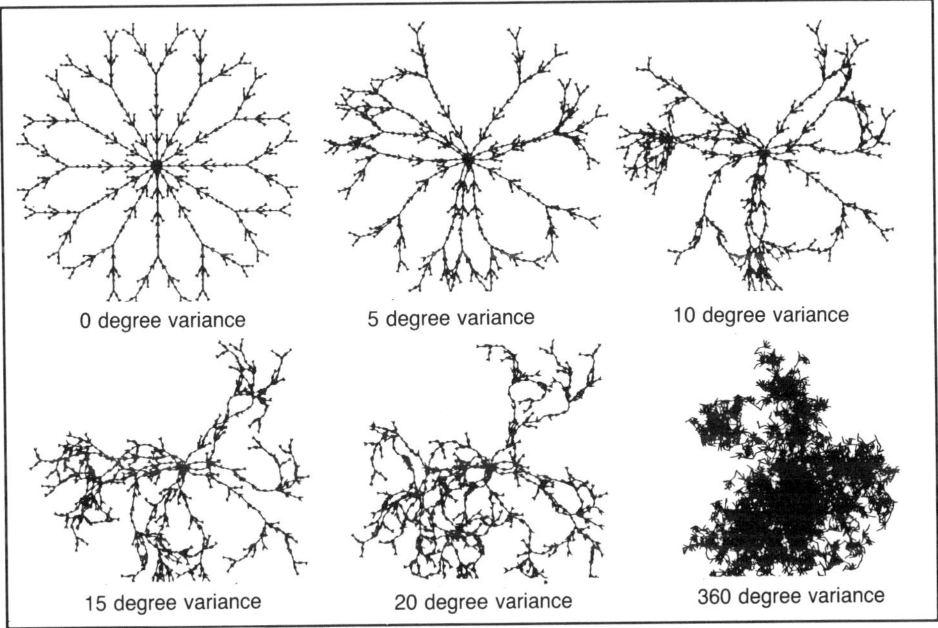

Figure 7.10 Effects of naturalization.

7.7 DISCUSSING SOME ASPECTS

7.7.1 SYNTACTICAL INFLUENCES

In this section, we discuss the effects of changing syntactical elements. It is our aim to give some understanding of the great importance of the proper setting for the given productions and the initial string. Only by understanding the effects derived from a distinct combination of letters from the given 'rewriting' alphabet $\mathfrak{Z}_L$ is it possible to use this method effectively in simulating given dendritic structures.

For this purpose, we have plotted the permutations of a simple set of productions in Figure 7.11. First we fixed the initial string $\mathfrak{I}$: X, the production $\langle \mathfrak{Z}_L$, Y, [+F][−F]$\rangle$ and the number of iterations n, while $\langle \mathfrak{Z}_L$, X, . . . $\rangle$ containing the operators $\underline{\underline{F}}$, $\underline{\underline{X}}$ and $\underline{\underline{Y}}$ was permutated. In Figure 7.11, drawings (B) and (F) are different. The question is why. A closer look at (B) shows that during each iteration a further operator $\underline{\underline{F}}$ is added before any other operator type. This means that a stem will come up first which will grow continually. At the apex of this stem all other operators will follow. Because of the given productions they will include branches but no further stretching of the stem. In looking at (F), we find the opposite result: this time during each iteration a further operation $\underline{\underline{F}}$ is added after any other operator. Out of this, the branching will take place at the initial geometric point, while a stem will grow separately out of this very location. The remaining four drawings show a high measure of similarity. The reason for that can be found in the medial structure of the final string, which is identical for all four of them and has the structure '. . . FYFY . . .'. Its meaning is: take one step of growth, build two branches as defined by operator $\underline{\underline{Y}}$, take another step of growth, build branches, and so on. The slight variations between (A), (C), (D) and (F) derive from the differing first part and last

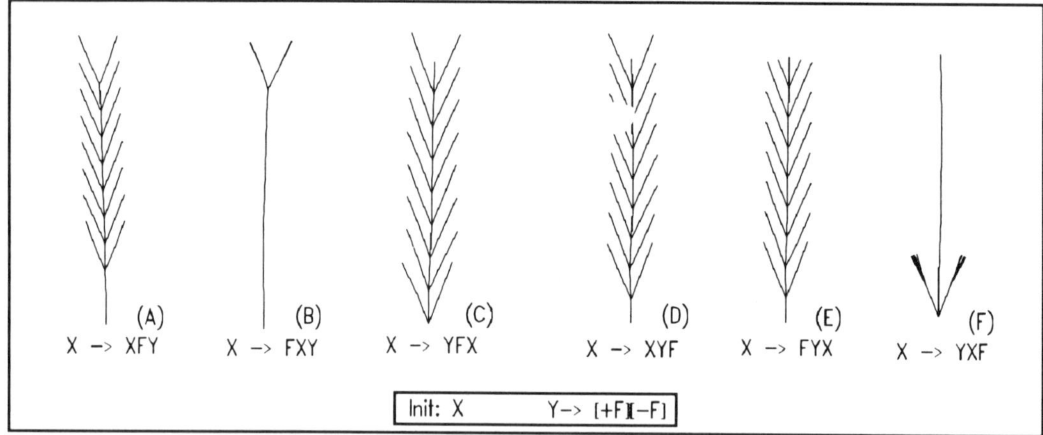

Figure 7.11 Results of permutating the production $\langle \mathfrak{Z}_L, X, \ldots \rangle$. The sequence of the letters is permutated.

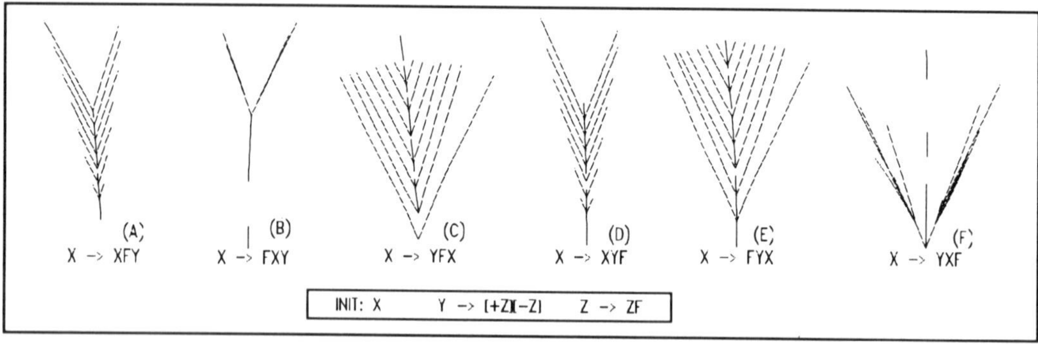

Figure 7.12 Results of permutating the sequence of $\langle \mathfrak{Z}_L, Y, FXY \rangle$.

part of each of the final strings, where a varying number of $\underline{\underline{F}}$ operators can be found.

In the next step we have taken the same conditions, but have replaced the operator $\underline{\underline{F}}$ by a production $\langle \mathfrak{Z}_L, Z, ZF \rangle$ within the given production $\underline{\underline{Y}}$. $\underline{\underline{Z}}$ can be understood as a dynamic operator, since its length will be changed by each iteration. In coutrast to that $\underline{\underline{F}}$ was static, as it did not show any growth over a period of time.

In using $\mathfrak{I}$: X, $\langle \mathfrak{Z}_L$, Y, [+Z] [−Z] $\rangle$ and a constant number of iterations n, we received the permutations shown in Figure 7.12. We must expect changes in the branching pattern, although the stem will be similar to the previous example. If the drawings (A), (C), (D) and (E) are compared, two subclasses can be found now. In (C) and (E) a decrease in length from bottom to top can be found in the branches. This means that the eldest branch will be situated at the bottom, the youngest at the top.

The syntax necessary for that shows that new $\underline{\underline{Y}}$ operators will be placed into the growing chain behind the given $\underline{\underline{Y}}$ operators. In (A) and (D) the opposite happens. New $\underline{\underline{Y}}$ segments are added to the chain before the given $\underline{\underline{Y}}$ operators. In consequence the older parts are shifted more and more to the top. A simple observation of living nature shows

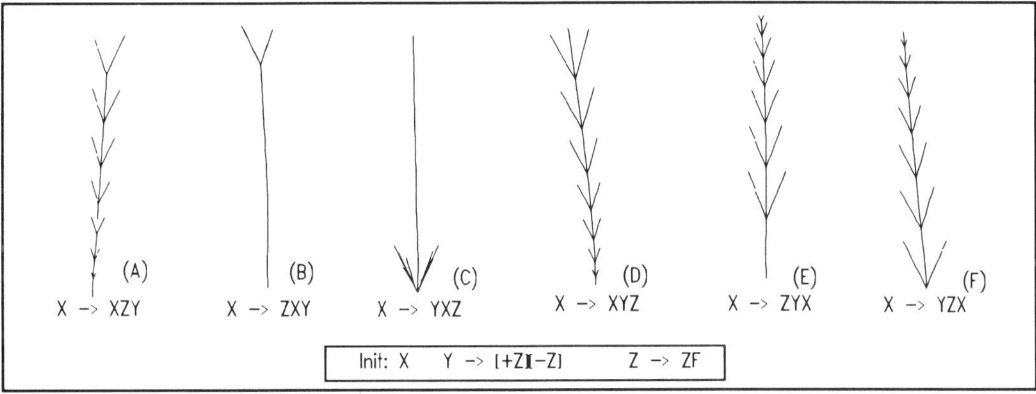

Figure 7.13 Results of permutating the sequence of $\langle \mathfrak{Z}_L, Y, XYZ \rangle$ with $\langle \mathfrak{Z}_L, Z, ZF \rangle$.

that this type of growth would usually not appear. Through this we receive a very elementary formal restriction in setting up biologically motivated objects.

If production $\langle \mathfrak{Z}_L, Z, ZF \rangle$ is a replacement for all $\underline{\underline{F}}$ operators and everything else is kept similar to Figure 7.11, the results can be found in Figure 7.13. Here again the length of a given segment can be used as the measure for its age. As in Figure 7.12 'unbiological' forms like (A) and (D) can be found.

7.7.2 AN AGE-DEPENDENT GROWTH OPERATOR

Although the results in using $\langle \mathfrak{Z}_L, Z, ZF \rangle$ are impressive, they do not properly match the biological situation, because the growth is constant. In reality growth underlies a declining function, where the growth characteristics are similar to $\underline{\underline{Z}}$ at the beginning, but shift slowly to the characteristics after a distinct number of iterations. We have included this type of $\underline{\underline{G}}$ operator in our simulations. In Figure 7.14, we have set up a comparison of these operators for a similar set of rules, defined by:

$\mathfrak{I}$: X
$\langle \mathfrak{Z}_L, X, \mathscr{L}_1 YX \rangle$
$\langle \mathfrak{Z}_L, Y, [+\mathscr{L}_2][-\mathscr{L}_2] \rangle$
$\langle \mathfrak{Z}_L, Z, ZF \rangle$
$n_{\text{Iter}} = 8$

with:

$\mathscr{L}_1 \in F, Z, G.$

This setting represents a 'tree', which grows over a period of n_{Iter} generations. In the left-hand column the static $\underline{\underline{F}}$ operator is used in production $\underline{\underline{Y}}$, whereby no dynamic growth will happen in the twigs. The right-hand column uses the constant growth operator $\underline{\underline{Z}}$ in production $\underline{\underline{Y}}$, so that a continuous growth takes place in the branches during the entire time of rewriting. In comparison with that, the $\underline{\underline{G}}$ operator is used in the middle column. The expected results are found, where at the beginning a strong growth takes place, which is reduced to zero as time progresses. Looking at the rows, we have done the same in each case concerning the $\underline{\underline{X}}$ operator. The top row uses the static $\underline{\underline{F}}$ operator in the stem, while the bottom row uses the constant growth operator $\underline{\underline{Z}}$. The middle row shows the results if $\underline{\underline{G}}$ is in use.

7.7.3 GENERAL GROWTH

We found in our simulations that the time marks given by the rewriting string procedure do not represent a continuous clock.

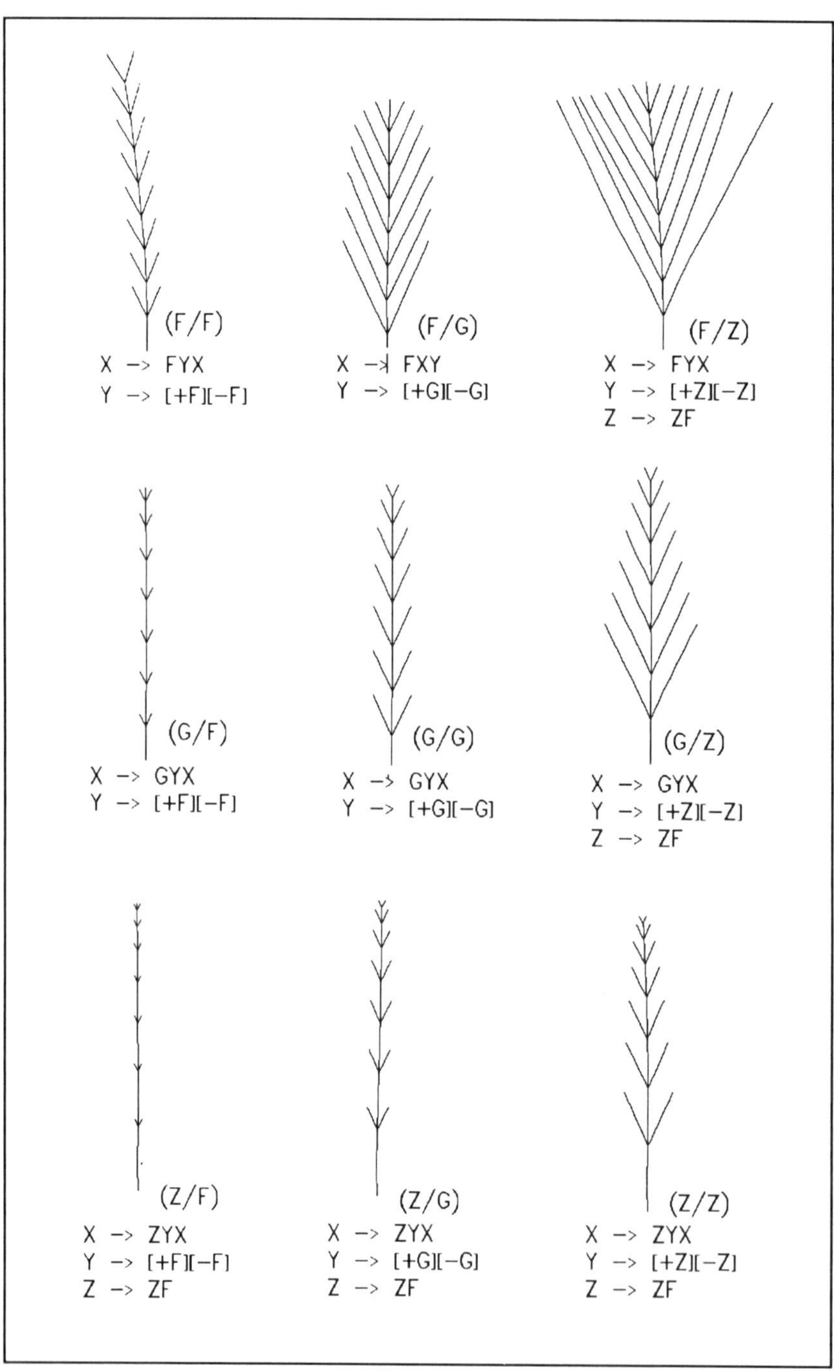

Figure 7.14 Comparing the influence of the operators 'F', ⟨3,Z,ZF⟩ and 'G'.

Furthermore, the given results can be understood as intermediate stepping stones which the growing system has to pass. Therefore a method had to be found to use the given information in such a way that a continuous timescale of growth can be accomplished. If we speak of 'continuous' processes, we are aware, that a 'von Neumann computer' would never really allow such a process in the genuine meaning of the word. Because of that, we use this term as a description for incremental steps of growth. One approach to this rescaling can be done by using the actual number of the iteration where a distinct letter $\mathcal{L}_i(\Re_T, \Delta_T)$ was finally replaced by the rewriting process $\Re_T$ plus the number of elements being placed between the current letter and its origin Δ_T. If all letters $\mathcal{L}_i(\Re_T, \Delta_T)$ are numbered in such a way that their first priority of order derives from $\Re_T$ and their second priority from Δ_T the resulting numbers can be used for the description of a continuous growth process. For each time T an object $\mathfrak{C}_T$ can be set up, which uses only those letter $\mathcal{L}_i(\Re_T, \Delta_T)$ whose resulting index number is smaller than T. Only those letters affected by this counting and skipping process will run the drawing process. Figure 7.15 shows at the bottom the nonequal timesteps deriving from the pure rewriting process, while the upper half of the figure shows the continuous growth process as discussed.

7.7.4 TROPIC INFLUENCES

Using the idea of a local tropic field, we can find within the realm of neurophysiological research that dendrites are able to find their way to their targets with the help of chemical markers and gradient fields (Brown 1991) (Plate 1). We have used this concept for another working hypothesis, which says that each dendritic tree builds up a specific far-reaching chemical 'attractor' field (Figure 7.16 left columns). Please note that the term 'attractor' used here stands for biochemical substances, which are distributed at specific locations by nerve cells and which can be detected by other cells. The complete field is superimposed by local chemical 'attractor' sources, put in defined places within the tree. Besides this, each neuron has sensors in each tip of its dendritic tree, which sense the gradient field of a special type of 'attractor'. This corresponds to biochemical attractors in real neurons (Shatz 1992).

Similar to the effects of tropism, these effectors force the whole object to grow towards the maximum gradient of the attractor field. Our working hypothesis is that each object owns attractor and effector elements. The effectors are defined in a way that they pour out a quantity of chemical transmitter, once the desired letter in the given chain is activated. Because of its position and history,

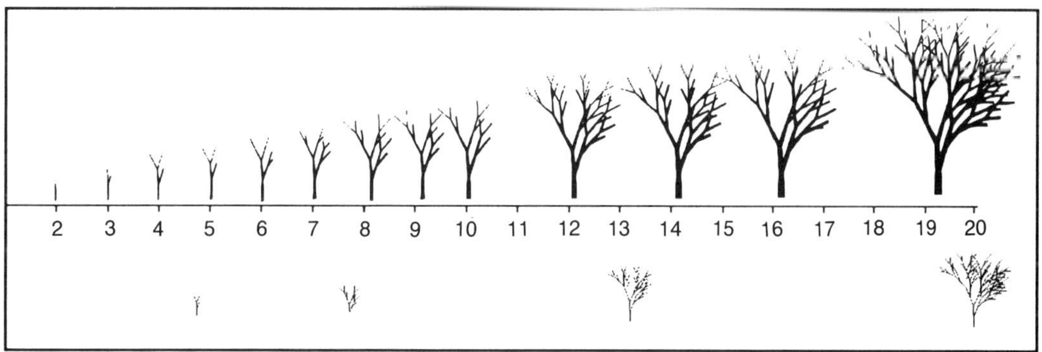

Figure 7.15 Comparing the timescale given by the rewriting process with the continuous timescale.

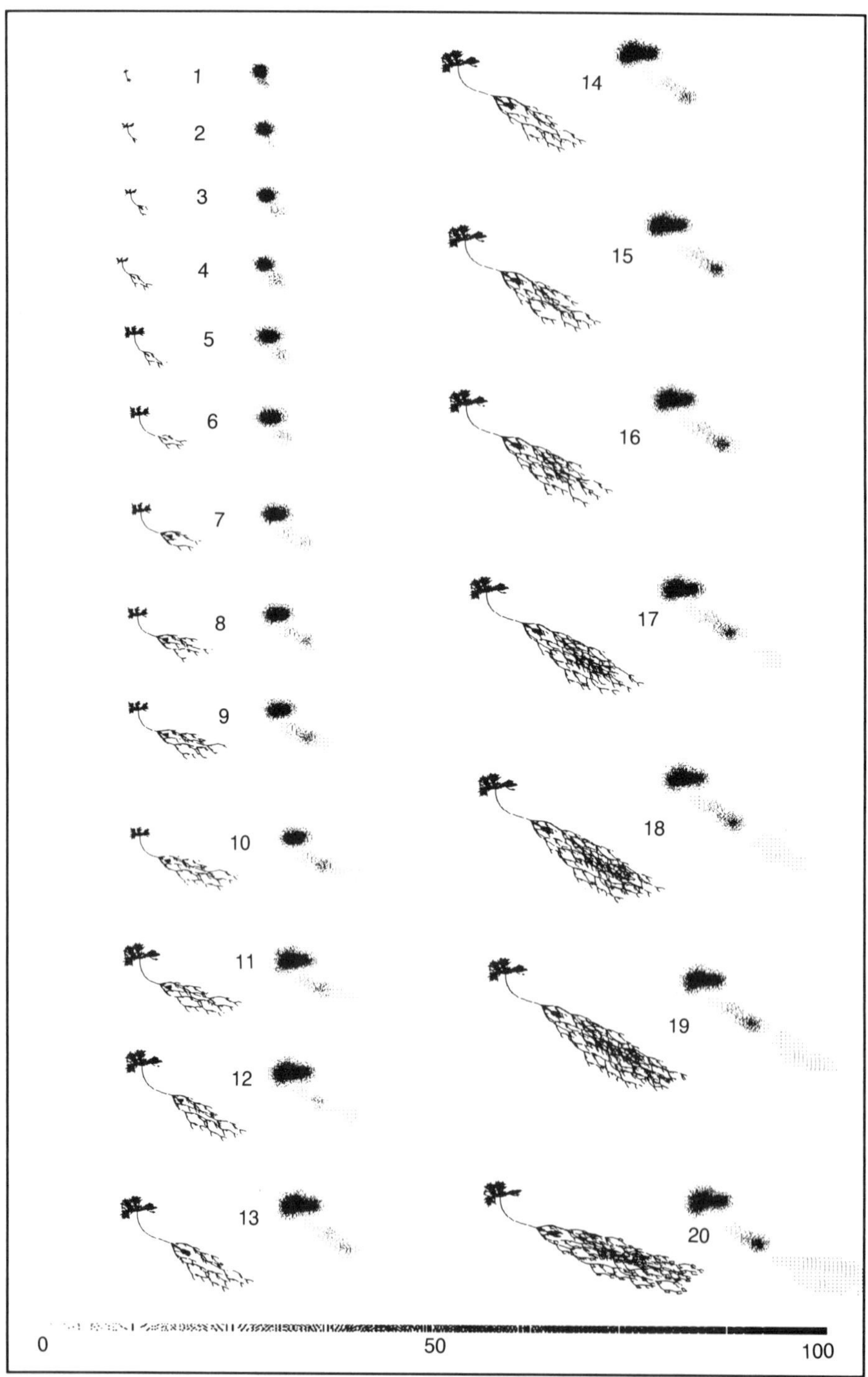

Figure 7.16 Steps of simulated growth (left); growth of an affliated tropic field (right).

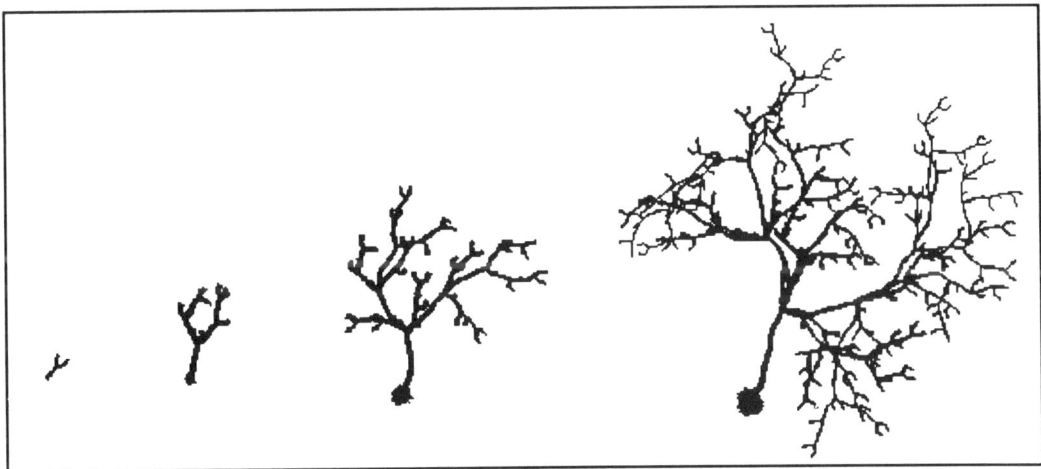

Figure 7.17 Simulated growth of a simulated Purkinje-type cell.

this happens only once at a distinct time, which correlates to the general clock of growth. The attractors on the other hand sense the strength of the given chemical field, as well as its field gradient. Comparable to what happened within the tropistic field, they try to shift along the field vector. In this way, the genotype of the dendritic tree is forced to change its direction incrementally.

7.7.5 CURRENT RESULTS

Currently, we are able to generate various dendritic trees by changing the genomes and the influencing local and global factors (Figure 7.17). The phenotype can be influenced by a variety of mutational types while the dendritic tree is growing. A single tropistic field can be placed in the region of growth, so that the phenotype is changed by that. A nonlinear aging factor is included, to vary the rate of sprouting, angles and stepwidth.

Further steps have been taken to observe the build-up of specific asymmetric potential fields as mentioned previously. These are needed in the realm of growth-cone theory, to understand the processes of interneuronal attraction and building of synapses. With the described tool we are able to set up genome types of various artificial nerve cells. We have worked with the described external and internal factors, for example mutation, to see the change in the phenotype in comparison with the genotype. First experiments with an external static tropistic field have been made, where the phenotype was influenced in its growth direction. Aging factors have been included. We were able to force lesions to be put into the tree structure, so that parts of the tree could be 'transplanted' to other locations within the tree and further studies of regrowth could be made.

Currently we are aiming towards studies of the temporal interaction of several nerve cells, as they are found during embryogenesis. Our global aim herein is to simulate and verify the very growth and interactive behaviour of multiple cells and their outcome in terms of connections (Abeles 1982; Braitenberg 1991).

REFERENCES

Abeles, M. (1982). *Local Cortical Circuits.* Springer, Heidelberg.

Barnsley, M. F., Devaney, R. L., Mandelbrot, B.

B., Peitgen, H. O., Saupe, D. and Voss, R. F. (1989). *The Science of Fractal Images*. Springer, Heidelberg.

Braitenberg, V. (1991). *Anatomy of the Cortex*. Springer, Heidelberg.

Brandt, I. (1981). Brain growth, fetal malnutrition. *J. Perinat. Med.* **9**(3), 3–26.

Brodal, A. (1981). *Neurological Anatomy in Relation to Clinical Medicine*, 3rd edn. Oxford University Press.

Brown, M. C. (1991). *Essentials of Neural Development*. Cambridge University Press.

Cotterill, R. (1988). *Computer Simulations in Brain Science*. Cambridge University Press.

Davies, A. M. (1986). Different factors from the central nervous system and periphery regulate the survival of sensory neurons. *Nature* **4**, 11–20.

Hamilton, P. (1991). Computersimulationen zum Nervenwachstum auf der Basis von Lindenmayer-Systemen. *Informatikberichte Univ. Bonn.* **85**, 19–28.

Hamilton, P. (1992). *Einführung in die neuronalen Netzwerke*, 1st edn. VDE-Verlag, Berlin.

Hamilton, P., Kurthen, M. and Linke, D. B. (1991). An approach to cerebral neuronal plasticity on the basis of Lindenmeyer L-systems. *Cognitive Systems* **3–2**(November), 162–78.

Lindenmayer, A. (1968*a*). Mathematical models for cellular interactions in development, I: filaments with one-sided inputs. *J. Theo. Biol.* **18**, 280–99.

Lindenmayer, A. (1968*b*). Mathematical models for cellular interactions in development, II: simple and branching filaments with two-sided inputs. *J. Theo. Biol.* **18**, 300–15.

Lindenmayer, A. (1974). Adding continues components to L-systems. In *L-Systems* (ed. G. Rozenberg). Springer, New York.

Prusinkiewicz, P. (1989). *Lindenmayer Systems, Fractals and Plants*. Springer, New York.

Purves, D. (1988). *Body and Brain*. Harvard University Press, Cambridge, MA.

Rozenberg, G. (1980). *The Mathematical Theory of L-Systems*. Academic Press, New York.

Schrade, M. (1989). The guru of silicon chips, Carver Mead is shaping an entire industry with his approach to the gray matter of computer brains. *Los Angeles Times*, p. 3, 21 May 1989.

Shatz, C. J. (1982). Das sich entwichelnde Gehirn. *Spektrum der Wissenschaft* **11**, 44–52.

Turing, A. M. (1950) Computing machinery and intelligence. *Mind* **60**, 433–60.

PART TWO

TISSUES

THE BRAIN AS A METAPHOR FOR SIXTH GENERATION COMPUTING 8

Michael Arbib

8.1 INTRODUCTION

This chapter is designed to expand our concepts of computation to embrace the style of the brain. This style depends on the constant interaction of concurrently active systems, many of which express their activity in the interplay of spatiotemporal patterns in manifold layers of neurons. In order to develop our appreciation of developments in this area, we distinguish two methods of approaching problems, namely computational neuroscience and neural computing. The two subjects have different goals. Nonetheless, they overlap and are mutually stimulating.

In **computational neuroscience**, we use computational concepts and methodologies to study the structure and function of real brains. Here the guiding criterion is explanation of biological data. Questions addressed by computational neurobiology include the following. How do the many regions of the human cerebral cortex cooperate? How can the brain of a frog extract visual data which enable it to detour around a barrier to reach its prey? What neural networks are involved in the control of hand movements? What changes in neural networks account for the memory of specific events and the acquisition of novel skills?

In **neural computing**, the aim is to apply ideas about neural networks in the development of the next generation of computers and intelligent systems. In my own group, we take the neural networks modeled above and study how best to control a mobile robot, or develop the next generation of machine vision systems. Other workers take various technological models of adaptive neural networks and apply them, for example, in building pattern recognition devices or adaptive controllers for intelligent machines, with little attention to biological inspiration.

I argue that the study of the brain defines a sixth generation of computers (Arbib 1987a) which will be characterized by melding into computer design the insights of brain research and cognitive science as catalyzed by developments in computer networking and parallel computation and such new implementation technologies as optical computing. In analyzing the history and current status of computers, commentators have distinguished five generations of computers. The first four generations are characterized by the underlying hardware:

1. vacuum tube;
2. transistor;
3. integrated circuit;
4. very large scale integration.

Where the fifth generation might be seen as marrying the 'state of the art' in AI software (e.g., expert systems) to appropriately engineered hardware (e.g., MIMD hypercubes), the development of sixth generation systems, the computer systems for 1995–2005, will be heavily influenced by discoveries concerning information processing in the brain. Since billions of human brains already exist, the goals of such computer design must be to develop computers that complement brain function, rather than simply emulate it, looking for promising contributions from neural nets, physics, and 'classic' computer science. There is no ready made body of wisdom in neuroscience waiting to be applied. Thus we cannot simply follow the 'neural blueprint' to build a better computer. Rather, fundamental research in neuroscience must continue, but now structured to extract key principles of information processing that can enter into the design of the next generation of computers, building, for example, on the interaction between neuroscientists and AI workers in the study of vision (Arbib and Hanson 1987).

Inspiration from the brain leads away from emphasis on a single universal machine toward a device composed of different structures, just as the brain may be divided into cerebellum, hippocampus, motor cortex, and so on. Thus we can expect to contribute to neural computing as we come to chart the special power of each structure better. Subsystems will include general-purpose engines and special-purpose machines some of which (such as the front ends for perceptual processors, and devices for matrix manipulation) will be highly parallel. Sixth generation systems may be characterized by:

1. cooperative computation: the computer will be a heterogeneous network of special-purpose and general-purpose subsystems;
2. perceptual robotics: increasingly, computers will have intelligent perceptual and motor interfaces with the surrounding world;
3. learning: many of the subsystems will be implemented as adaptive 'neural style' networks.

Just as the interface to computers has progressed from bits to symbols to interactive graphics, the next generation will be more 'action oriented' with computers including robotic actuators and multimodal intelligent interfaces among their subsystems. Such computers will 'grow' with the design of new subsystems using, for example, silicon compilers to specify new chips, and mechotronics for integrated design of robotic subsystems and their controllers. The design of these perceptual robotic systems will be heavily influenced by the analysis of neural mechanisms for perception and the control of movement. Moreover, sixth generation computers will be machines that learn, with principles of adaptive programming that incorporate lessons from brain theory and AI. Much of the resurgence of enthusiasm for neural network technology in the 1980s was based on the excitement over a variety of learning rules for neural networks, especially those that address the problem of training hidden units. However, learning is not the sole key to neural computing, and much is to be learned from the explicit design of networks for cooperative computation, perceptual robotics and other intelligent functions (Arbib and Buhmann 1992).

Note the blurring of the line between hardware and software. For many applications, we may take an existing machine and program it or, appealing to the power of neural networks, adapt it to a new application. However, for other applications, the design of a new architecture, the specification of physical systems and their relationships, will be an integral part of tackling that application. The computer scientist of the future will see hardware and software issues, as well as learning components and robotic interfaces, as integral parts of the overall design process. Such a view leads to the apparently paradoxical claim that neural computing is not restricted to neural networks. This is because the design of sixth generation systems must embrace at least two grains of parallel and distributed computing: 'coarse-grain' analysis of the overall system as a network of interacting subsystems (whether physically distinct, or processes played out over a set of processors) and 'fine-grain' neural analysis using artificial neural networks to build many of these subsystems.

8.2 SCHEMA THEORY

Schema theory (see Arbib (1992) for a review) provides the beginnings of a methodology for parallel and distributed computation. A top-down analysis starts with the choice of some overall behavior for study. Proceeding with a functional analysis, we explain behaviors in terms of the interaction of functional units called **schemas**. Then, as we seek to bridge from large unit to neuron, we may elaborate our schemas into networks of interacting subschemas until finally we may realize our constructs in terms of arrays, columns or modules implemented as neural networks, or in terms of other appropriate circuitry. A sixth generation system may well be hybrid, using, for example, neural networks for low-level sensory processing and adaptive subsystems and conventional digital circuitry for arithmetic processing. In the brain, everything is a neural network. In designing a sixth generation computer, technological efficacy is the touchstone, and only certain subsystems are best realized neurally.

The style of cooperative computation is far removed from serial computation and the symbol-based ideas that have dominated conventional AI. The representation of the world is the pattern of relationships between all its partial representations. Much work in AI contributes to schema theory, even when it does not use this term. For example, Brooks (1986) builds robot controllers using layers made up of asynchronous modules that can be considered as a version of schemas. This work shares with schema theory, with its mediation of action through a network of schemas, the point that no single, central logical representation of the world need link perception and action. A **perceptual schema** embodies the process that allows the computer or organism to recognize a given domain of interaction. Various schema parameters represent properties such as size, location and motion. Given a schema, we may need several schema instances, each suitably tuned, to subserve our perception of several instances of its domain. A schema assemblage – an assemblage of perceptual schema instances – provides an estimate of environmental state with a representation of goals and needs. New sensory input as well as internal processes update the schema assemblage. The internal state is also updated by knowledge of the state of execution of current plans made up of **motor schemas** which are akin to control systems but distinguished in that they can be combined to form coordinated control programs which control the phasing in and out of patterns of coactivation, with mechanisms for the passing of control parameters from perceptual to motor schemas. Draper *et al.* (1989) discuss schemas for visual interpretation which do not involve motor considerations. By contrast, Arbib (1990a) relates perceptual and motor schemas and neural networks in the control of hand movements (of both human hands and dextrous robot hands); while Lyons and Arbib (1989) describe the RS language specifically developed for specification of robot schemas.

The RS (robot schema) language (Lyons and Arbib 1989) formalizes each schema as a port automaton with a set of input and output ports through which it can communicate with other instances. As action and perception progress, certain schema instances need no longer be active (they are deinstantiated), while new ones are added (instantiated) as new objects are perceived and new plans of action are elaborated. A basic schema definition includes a C-like behavior specification making explicit how an instance of the schema behaves. Schemas may be formed hierarchically as assemblages – which also have ports but have behavior defined through the interactions of instances.

An assemblage may itself be considered a schema for further processes of assemblage formation; and the network itself will be dynamic, growing and shrinking as various instantiations and deinstantiations occur. The RS syntax for an assemblage tells us how to put schemas together in a way which does not depend on how the behavior specification is given. It is thus simple to extend RS to make it possible, in the basic schemas, to define the behavior directly in terms of a neural network as well as by a C-like program.

While work on schemas has to date yielded no single formalism, we do see the evolution of a theory of schemas as programs for a system which has continuing perception of, and interaction with, its environment, with concurrent activity of many different schemas passing messages back and forth for the overall achievement of some goal. In current computer science, we see a convergence between schema theory and object oriented concurrent programming (e.g., Yonezawa and Tokoro 1987). When we turn to computational neuroscience, we further require that the schemas be implemented in specific neural networks (Arbib 1989). In this spirit, Weitzenfeld (1992) has developed the object oriented abstract schema language, ASL, and charted how it can provide the framework for a variety of domain-specific schema languages including a neural/schema language based on the earlier neural simulation language, NSL (Weitzenfeld 1991).

It is also useful to view cooperative computation as a social phenomenon. A schema is a self-contained computing agent (object) with the ability to communicate with other agents, and whose functionality is specified by some behavior. Whereas schema theory was motivated by analogies with the functions of interacting brain regions, much work in distributed artificial intelligence (DAI, cf. Bond and Gasser 1988; Gasser and Huhns 1989) was motivated by a social analogy in which the schemas are thought of as 'agents' analogous to people interacting in a social setting to compete or cooperate in solving some overall problem (Kornfeld and Hewitt 1981; Minsky 1985). The full development of sixth generation computing will see the transition from social analogy to social actuality. We will not only need to integrate schemas within the processors of the distributed network that constitutes a single machine, we shall also have to understand how vast problems (such as ecological control) can only be handled by human–machine networks where agents may be schemas, machines, people or groups of these.

8.3 THE MANY CONCEPTS OF THE NEURON

For many behaviors, analysis at the level of single neurons may be superfluous. It is certainly premature in the study of language. However, even when this further level of analysis is appropriate, as in a number of the problems of visuomotor coordination, the question of how detailed the analysis of each neuron must be remains.

At the basis of much current work in neural computing is the McCulloch–Pitts neuron, often embellished by a learning rule of some kind, or the leaky integrator model (often referred to anachronistically as the Hopfield neuron if it is connected to other neurons in a symmetric way).

In computational neuroscience, real neurons may be represented somewhat grossly by the leaky integrator model, or at a finer level in terms of a connected network of compartments described, for example, by the Hodgkin–Huxley equation for the dynamics of membrane potential. Hodgkin and Huxley (1952) showed how much can be learned from analysis of membrane properties in understanding the propagation of electrical activity along the axon; and Rall (1970) and Shepherd (1979) are among those showing that study of membrane properties in dendrite, soma and axon can help us understand small neural circuits, as in the olfactory lobe or for rhythm generation (McKenna, Davis and Zornetzer 1992). Nonetheless, the complexity of such analysis makes it in many cases more insightful to use a lumped representation of the individual neurons to understand the properties of large networks.

Briefly, the McCulloch–Pitts model is a synchronous discrete time model. The output of each neuron can have the value of either 0 ('not firing') or 1 ('firing'). The output line ('axon') branches repeatedly, finally ending in junction points, called synapses, with other neurons or with effectors. The neuron itself may receive synapses from other neurons or from receptors. We denote the output of the neuron by y; it also has n inputs $x_1(t), \ldots, x_n(t)$, corresponding to synapses with weights $w_1, \ldots, w_n$ and a threshold (θ). We say a synapse is excitatory if its weight is positive, inhibitory if negative. The McCulloch–Pitts neuron has no memory for inputs as such – most learning rules in current models take the form of schemes for adjusting the ws. With the current weights, the input at time t determines the output at time $t + 1$ (the time unit is called the synaptic delay) according to the rule

$$y(t + 1) = 1 \quad \text{if and only if} \quad \Sigma w_i x_i > \theta.$$

At an intermediate level of detail, the neuron may be modeled as a leaky integrator. Here time is continuous, and two quantities are used to describe the behavior of the neuron, a membrane potential $m(t)$ and a firing rate $M(t)$. A few classes of real neurons communicate by the passive propagation (cable equation) of membrane potential down their (necessarily short) axons. However, most communicate by the active propagation of 'spikes' (temporarily localized but large changes of membrane potential). The generation and propagation of such spikes has been described in detail by the Hodgkin–Huxley equations. However, the leaky integrator model omits such details, simply approximating $M(t)$, the average rate of spike generation over some recent time interval, as a function f of $m(t)$, $M(t) = f(m(t))$. For example, f might be the sigmoidal function $k/[1 + \exp(-m/\theta)]$, increasing from 0 to its maximum k as m increases from $-\infty$ to $+\infty$. This is the neuron model used in the studies reported for *Rana computatrix*. A review of the full range of neuron models is beyond the scope of this chapter. Instead we shall concentrate on one example, the neural mechanisms revealed in biological study of networks which control rhythmic movements; specifically, the neural network of the lobster stomatogastric ganglion which controls rhythmic chewing (Selverston *et al.* 1976). This will show how synaptic performance can be dramatically modulated by chemical means.

Lobsters have three teeth in their stomachs with attached muscles: two lateral teeth that move in and out; and a medial tooth that moves forward and backward. A major component of the control circuit is shown in Figure 8.1(a). DG (together with another neuron, GM, not discussed here) control the medial tooth, while LG, MG, and LPG control the lateral teeth. But since almost all the connections are inhibitory (the open circles), how is an oscillation initiated, let alone maintained? It has been hypothesized that INT 1 (interneuron 1) and the electronically coupled pair of cells LG/MG (the lateral gastric and medial gastric motoneurons) form an oscillator that works as follows: INT 1 and LG/MG are tonically active, but connected by reciprocal inhibition. The key is that each has fatigue and postinhibitory rebound. After one half fires, for a while inhibiting the other half of the pair, its level of inhibition decays to the point where the other cell becomes active. After inhibition is removed, the cell fires at above its tonic level. The tonic level is not enough to inhibit the other cell; the rebound level is. Thus firing of the cells alternates.

The interaction between LPG (the two lateral posterior gastric motoneurons) and LG/MG is probably too weak for them to function as such an oscillator. However,

INT 1 excites DG (the dorsal gastric motoneuron), and since INT 1 inhibits LG/MG which in turn inhibits LPG the circuit may suffice to entrain the endogenous bursting of LPG. Thus INT 1, DG and LPG burst (almost) in phase. This ensures a phase coupling between the control of the medial tooth (by DG) and the control of the lateral teeth (by LG, MG and LPG) which yields the 'squeeze mode' of chewing, in which the cusps of the three teeth move together simultaneously, with forward movement of the medial tooth and a small (or zero amplitude) backward movement of the two lateral teeth.

This discussion has already shown that biology introduces us to mechanisms of synaptic action beyond the McCulloch–Pitts neuron or leaky integrator. To carry this class of biological 'innovations' further, we must note that, to the class of 'usual' synaptic transmitters which have an immediate excitatory or inhibitory effect on the postsynaptic membrane, there has been added another class, the peptides, which act to change cellular properties on a timescale of seconds or more. Recent work (Heinzel and Selverston 1988) has shown that the properties of the stomatogastric oscillator can be dramatically changed by bathing it in a peptide called proctolin. They showed that proctolin can change the gastric oscillator by targeting on LG and DG as follows (Figure 8.1(b)): While DG is normally driven by strong excitatory input from interneuron 1 (INT 1), with proctolin it becomes a pacemaker, or endogenous burster – that is, it needs no input for its output to take the form of cyclical bursts of activity. LG shows plateauing (i.e., it exhibits a new stable resting potential) and the DG to LPG plateauing becomes stronger, leading to disinhibition of LG from LPG. The plateauing LG inhibits its synergistic motoneuron MG more strongly than before, causing a delay of its burst. As a result of these proctolin-induced changes, the medial tooth subsystem (see the connections of DG in Figure 8.1(b)) runs with a different phase in relation to the lateral tooth subsystem (LG, MG, LPG) from that in the normal situation described above. This corresponds to the 'cut and grind' mode of chewing in which the serrated edges of the lateral teeth meet (cut) and the cusps of the lateral teeth rasp backward along the file of the medial tooth (grind), whereas in the 'normal' squeeze mode only the three cusps are pressed against each other. We thus see the many dimensions of real neurons still unexplored in neural computing.

While I have used the lobster stomatogastric rhythm here, the lobster also exhibits a pyloric rhythm. The mechanisms controlling it include pacemakers, plateauing (which, as

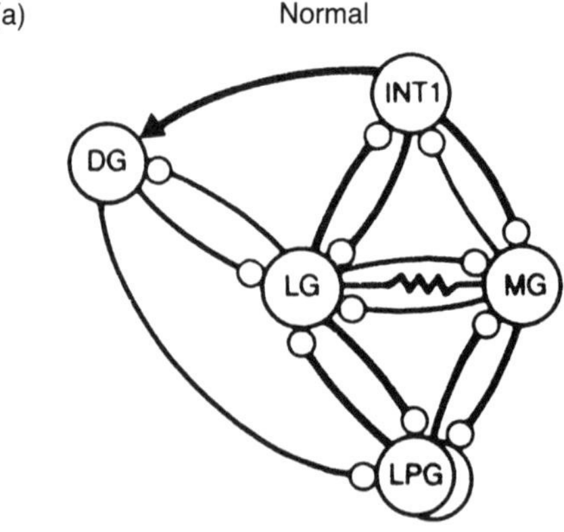

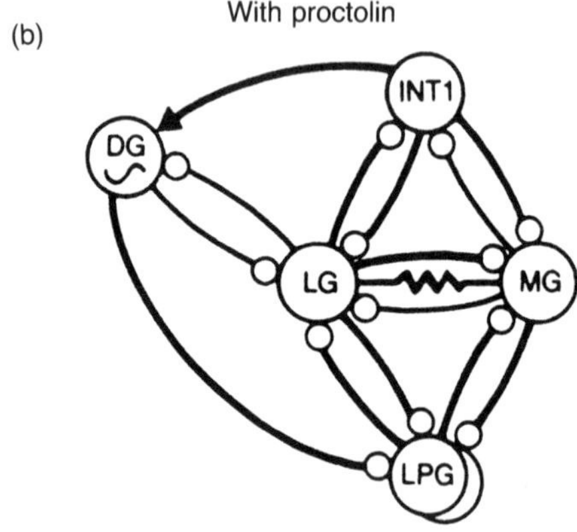

Figure 8.1 Proctolin modulation of the gastric oscillator in the lobster stomatogastric ganglion turns the effective connectivity from that shown in (a) to that shown in (b).

we have seen, is a membrane change which implies bistability of membrane potential) and network interactions. These have been elucidated by Eve Marder and her colleagues, who also work on neuromodulation by a variety of new neurotransmitters (see the review by Harris-Warwick and Marder (1991)).

8.4 LESSONS FROM BRAIN ORGANIZATION

The principles of brain theory that theory must be action-oriented and that the brain is an action-oriented layered somatotopically organized computer (Arbib 1972) also have use for neural computing. The first principle, which stresses the need to place brain function in the context of the organism's ongoing interactions with its environment, also points to the increasing integration of computer systems with elaborate perceptual and motor devices, that is, to a blurring of the line between 'mainstream' computer science and perceptual robotics.

Central to the action-oriented view is that the system (human, animal or computer/robot) must be able to correlate action and the results of interaction in such a way as to build up internal 'models' of the world (Craik 1943). The stress on interaction with the real world forces us to look at the spatial dimensions of behavior. The brain evolved to solve problems in the physical world – biological systems try to solve the whole problem: sensory and motor. This involves data acquisition and simple control, coordination and organization: planning, long-term memory, and so on. All these will be important in future computer designs. This trend is now explicitly recognized in a conference series which encourages the interchange between biological and robotics studies (see Meyer and Wilson (1991) for the proceedings of the first meeting). The brain is a network of neurons and so has components which are not only susceptible to drugs and hormonal influences but also have their own style of unit information processing. It is organized heterarchically and hierarchically, both structurally and functionally. The crucial point here is that no single-level model can encompass any sizable portion of brain function. An analysis of different brain regions for their own 'style' is important. In many such regions (especially near the sensory periphery), the coordinates of the neural layers map the sensory input in a way which reflects (with distortions that reflect information richness) the coordinates of the sensory surface: **retinotopy** (retina place) in the visual system, **somatotopy** (body place) in the somatosensory system, and **tonotopy** (place codes frequency) in the auditory system. Plasticity then adapts the maps to the needs of the systems.

Finally, I stress that, in view of the layered structure of the brain, even sequential tasks will involve highly distributed computation – the brain is not a serial computer. For example, the most satisfactory explanation of the series of eye movements in mammalian visual perception involves parallel interaction of many neurons in frontal and parietal cortices, thalamus, basal ganglia and midbrain to determine each movement and to update the internal representation with the input from the new fixation (Dominey and Arbib 1992).

Biology offers ideas for vision, face recognition, motor control and so on, although we look to biological systems for *a* solution to a problem, not the one that is optimal for given technological components. The brain uses slow neurons with rich connectivity, so if we use very fast, very precise components in building a computer, we may (but may not) see computer architectures quite different from those of the brain. Thus the hardware of the brain offers nothing to technology in speed, but may offer ideas on connectivity and adaptability. For example, the cerebellum plays a key role in the coordination of movement (Ito 1984). A Purkinje cell in the cerebellar cortex has of the order of 10^5

parallel fibre inputs. A key question for brain theory is to understand the role of Purkinje cells in motor coordination. Neural computing might then ask: given much faster components can we achieve the same function with a fan-in of only 10 or 10^2? Distributed coding in neural systems: the way in which a pattern is coded not as a string of arbitrary symbols as in most current computers but rather as activity in a network. Ideas on distributed encoding go back at least to Pitts and McCulloch (1947) and were reviewed by Erickson (1974), while Hinton *et al.* (1986) developed a theoretical perspective on the relative merits of distributed and localized encoding of information. Enriching this framework, neuroscientists have found distributed encoding in different regions of the mammalian brain: Georgopoulos *et al.* (1983) developed the 'vector model' for the encoding of reaching direction by the averaged 'vote' of a population of neurons in motor cortex; McIlwain (1982) demonstrated the widespread effects of even microstimulation of superior colliculus; and Andersen *et al.* (1985) suggested how distributed encoding in parietal cortex can combine precision with economy. Touretzky and Hinton (1985) and Shastri (1989) have placed the issue of distributed encoding in the context of cognitive

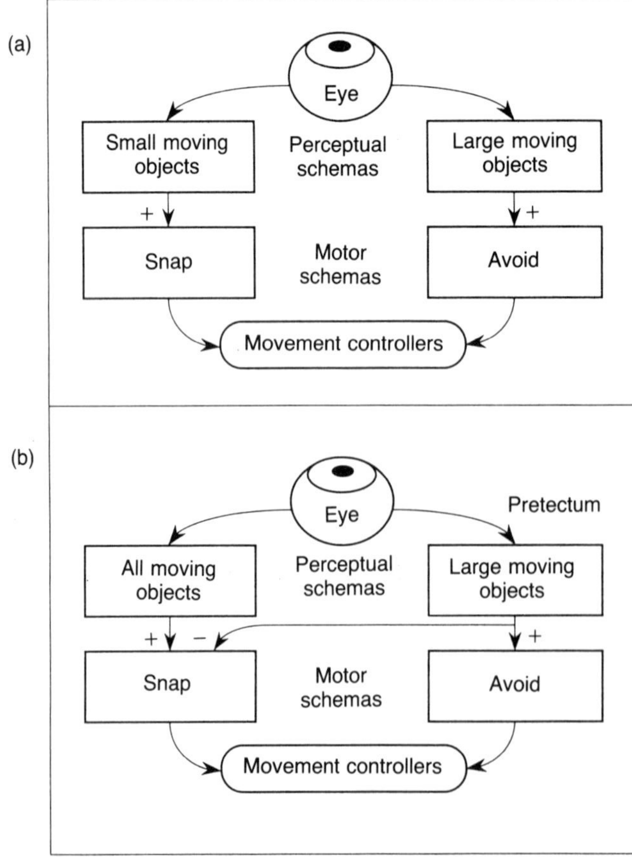

Figure 8.2 (a) The 'naive' schema program for the toad's snapping and avoidance behavior; (b) the schema program revised in light of data on the effect of lesioning the pretectum.

modeling by relating 'symbolic' and 'distributed' representations. Many brain theorists instead see the redundancy of the brain as resulting in part from its layered structures in which nearby elements have overlapping receptive fields.

The study of specific neural systems has many implications for the development of novel computer architectures. The next section introduces the study of visuomotor coordination in frog and toad as a window on a wide variety of neural mechanisms of interest from both a biological and a technological perspective.

8.5 *RANA COMPUTATRIX* – THE FROG THAT COMPUTES

We must aim not so much to understand responses in a single region of the brain but to try to understand how many different parts cooperate to subserve a task. A vocabulary is required in which to express the interactions between many parts of the brain at a high level, so that we may then formulate the more detailed hypotheses about how the functions of individual neurons fit into those constituent activities. We thus return to the two-level modeling methodology which complements neural networks with schemas which describe the basic subfunctions of some overall computation. We emphasize the role of patterns of activity in arrays of neurons, but also make explicit the 'cooperative computation' between brain regions. Schema theory provides a functional level of analysis between behavioral and lesion studies and the analysis of neural circuitry.

The emphasis in studying *Rana computatrix*, our computational model of the frog, is on visuomotor coordination and two constraints will be applied: hypotheses about how schemas are played out over one or more regions of the brain are constrained by lesion data showing how the behavior of animals changes when one or more brain regions are missing. Similarly, neural network models are constrained by detailed studies on anatomy and physiology.

We start with an example of how lesion experiments can test a schema model of a simple behavior. Frogs and toads snap at small moving objects and jump away from large moving objects. A simple schema model of such behavior is summarized diagrammatically in Figure 8.2(a). One perceptual schema recognizes small moving objects, another recognizes large moving objects. The first schema activates a motor schema for approaching the prey; the latter activates a motor schema for avoiding the enemy. These in turn feed the motor apparatus.

Lesion experiments can put such a model to the test. It was hypothesized that recognition of small moving objects occurred in the tectum (the primary target of retinal fibres in the mid-brain) while the pretectum (a region in front of the tectum) was perhaps the locus for recognizing large moving objects. Peter Ewert (1976) lesioned the pretectum and discovered that while the claim about the pretectum may be true, the model in Figure 8.2(a) is wrong. That model would predict that a toad with lesioned pretectum would be unresponsive to large objects but would respond in a normal manner to small objects. However, the facts are quite different. The pretectum-lesioned toad will approach any moving object and snap at it. This leads to the new schema model of Figure 8.2(b). The perceptual schema in the pretectum recognizes any large moving object, but the other perceptual schema now operates not only with small moving objects (initial model) but with any moving object. The model is completed by the inclusion of the inhibitory link between the pretectal perceptual schema and the tectal motor schema (Figure 8.2(b)). Note, then, that we may still talk of a schema for recognition of small moving objects (i.e., the perceptual schema that controls approach behavior), but that it is not localized in any one region. Rather, it involves the cooperative computation of both tectum and pretectum.

Similarly, the current model posits that the pretectum plays a crucial role in both approach and avoidance behavior.

The strategy presented here is to maintain models at a level which aids us in understanding behavior. As such, much of the beautiful geometry of the brain is not included (not without regret, not without a whispered promise to return to it when the time seems ripe to extend the details of our model) and we set up differential equations to describe the cellular dynamics. Sometimes these models are simply in terms of interacting schemas; in other cases those schemas will be explicated in terms of relatively simple neural networks. As computing power becomes greater, we can expand the number of schemas to explain a richer range of behavior, and expand the detail of the neurons to catch more subtleties of the schemas' internal processing.

With this background, we briefly examine an early two-part example of how we capture some interesting behavioral properties in simple neural networks (see Arbib 1989, Sec. 7.3 for a textbook exposition). Figure 8.3 shows our model of the tectal column (see Lara, Arbib and Cromarty 1982): a simplified network based on a 'vertical sample' of the actual anatomy of frog tectum. The glomerulus (GL) is a twining together of neural fibres. Both the large pear cell (LP) and the small pear cell (SP) send excitatory signals back to the glomerulus, to engage in a pattern of self-excitation where their own activity is fed back to increase that activity even further. Left to itself, this recurrent circuit would build up excitation to the point that eventually the conjoint activity of LP and SP can trigger an output from the pyramidal cell (PY). However, this can be blocked by the stellate neuron (SN). This inhibitory neuron can gain sustenance from the excitatory loop and try to inhibit the recurrent activity. With parametric analysis, we were able to set the time constants and connection strengths to obtain a network facilitation behavior observed in real

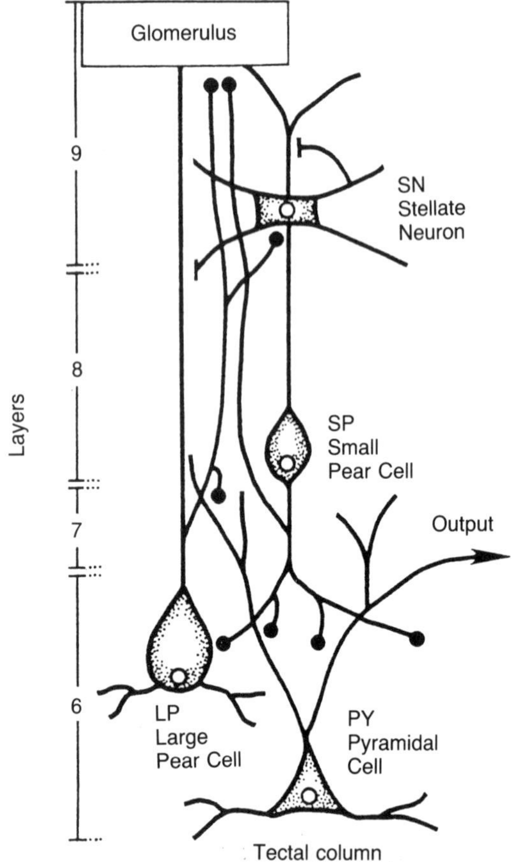

Figure 8.3 Neurons and synaptology of the model of the tectal column. The numbers at the left indicate the different tectal layers. Cell types: LP, the large pear-shaped neuron with dendritic appendages and ascending axon; PY: the large pyramidal neuron with efferent axon; SP: small pear-shaped neurons with ascending axons; and SN: stellate neurons. The glomerulus comprises the LP and SP dendrites and recurrent axons as well as optic and diencephalic terminals. The LP excites the PY, the SN and the GL, and is inhibited by the SN. The SP excites the LP and PY cells, and it sends recurrent axons to the glomerulus; it is inhibited by the SN. The SN is excited by LP neurons and diencephalic fibres and it inhibits the LP and SP cells. The PY is activated by the LP, SP, and optic fibres, and is the efferent neuron of the tectum (after Lara, Arbib and Cromarty (1982)).

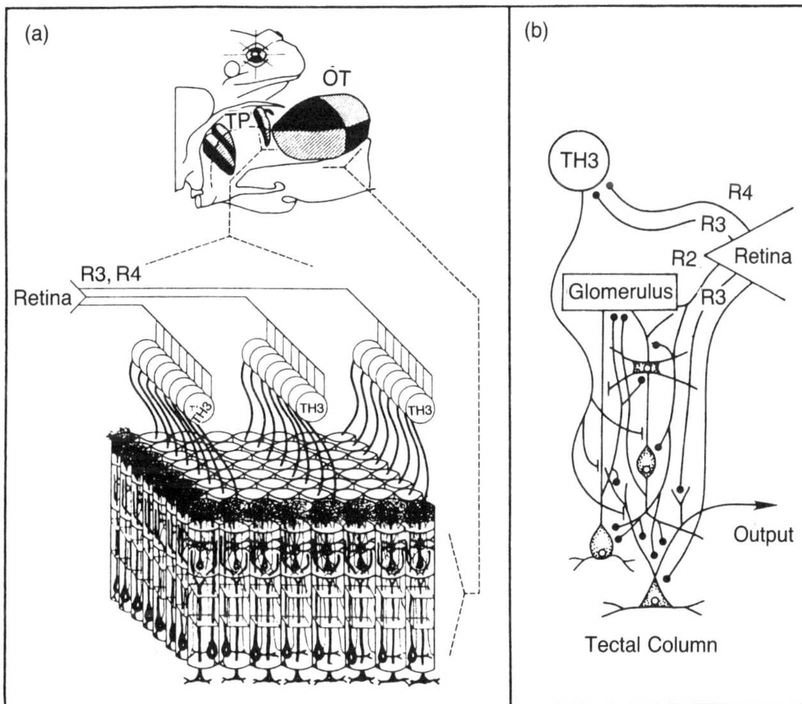

Figure 8.4 Interactions among retina, optic tectum and pretectum in the model of Cervantes-Perez, Lara and Arbib (1985). The retina sends fibres in a retinotopical fashion to both optic tectum (class R2, R3 and R4), and pretectum (class R3 and R4). (a) TH3 neurons also project retinotopically to the optic tectum. For simplicity we only show the projection of three rows of TH3 cells projecting upon the tectal columns. (b) A closer look of the interactions among retinal, tectal and pretectal cells. The TH3 cell of the pretectal column inhibits LP, SP and PY of the tectal column corresponding to its retinotopic projection.

frogs by Ingle (1975). If you show a worm to a frog for a third of a second, there is no response, but presentation for two thirds of a second would be enough to yield a response. However, if you present the worm for a third of a second, take it away for two or three seconds, then show it again for another third of a second, the animal will respond the second time. The first subthreshold stimulation facilitates the response of the second.

To proceed further we develop the idea that the tectum (like many other parts of the brain) can, at least to a first approximation, be presented as an array or tiling of local units. To model such an array, we complement the description of the unit by specifying the interactions between the units, and then try to understand how processes are spread out over the architecture. The second network of *Rana computatrix* models the tectum as an 8 × 8 array of the individual tectal columns of Figure 8.4(b) augmented by a very simple model of the pretectum (Figure 3; Cervantes-Perez, Lara and Arbib 1985). Ewert (1976) exposed toads to three types of stimulus: 'worms' (rectangles moving along their long axis), 'antiworms' (rectangles moving orthogonal to their long axis), and squares. The result was a parametric description of the animal's behavior. It responds better and better to longer and longer worms, it does not snap at any but the shortest antiworms, and there is a 'tug of war' between the two dimensions in determining the response to the square. We were able to show that the model of Figure 8.4 could be so tuned that the tectum could mediate response to movement extended in either direction, while the pretectal cells would respond best to objects elongated in the antiworm direction (antiworms or squares) and inhibit the tectum accordingly to exhibit pyramidal cell output whose firing rate correlated well with the rate of behavioral response shown in Ewert's data. Note that this study extends the model of Figure 8.3 in two ways. Firstly, it uses Ewert's data to refine the description of the schemas by going from a simple dichotomy (small vs. large moving objects) to a parametric analysis of response to moving configurations. Second, it provides a neural network consistent with the schema model, with inhibition from pretectal neurons 'sculpting' the relatively undifferentiated response of tectal neurons to retinal stimulation. Much work in the last ten years has gone into refining the neurophysiological validity of the models, and in extending their applicability to include response to predators, detour behavior and habituation phenomena, as well as basic prey-catching behavior (see Arbib and Ewert (1991) for a selection of more recent data and models).

8.6 COOPERATIVITY AND SELF-ORGANIZATION

A key issue for cooperative computation or, better still, computation based on the competition and cooperation of concurrently active agents, is how local interaction of systems can be integrated to yield some overall result without explicit executive control. Our study of *Rana computatrix* gives us two grains of cooperative computation: the fine-grain style in which many processes work in parallel to process some array of information; and a coarse-grain style, where relatively diverse systems – our schemas – interact. What gives our approach its special flavor is that when we think about the brain, we want to understand to what extent the schemas can be localized in the brain or implemented in neural networks. With this we turn to a crucial analogy between cooperative computation in neural networks and 'cooperative phenomena' as studied by physicists, in which the statistical properties of atoms or molecules 'cooperate' to yield properties of bulk matter which appear quite novel in terms of the properties of a few atoms considered in isolation.

Cragg and Temperley (1954), a neuroanatomist and a physicist, observed that the formation of patterns across the cerebral cortex might well be compared to the formation of domains of magnetization in a magnet – regions in which the atomic magnets line up with one predominant direction of the magnetization. Perhaps the first person to really develop this analogy was Bela Julesz (1971), who offered a cooperative computation model of depth perception, with two arrays of spring-coupled magnets, one driven by the left eye and one by the right. The orientation of each magnet encodes the depth of the corresponding point in the visual field. Julesz showed how to couple the magnets both within and across the two arrays to have the resultant orientation constitute a coherent hypothesis for the actual depths of the objects which stimulated the two retinas. This provides us with a concrete attempt to take some of the notions of statistical mechanics and carry them into the modeling of an interesting perceptual phenomenon. It was carried even further by the effects of hysteresis (Figure 8.5). If you are looking at two points, one projecting to each eye, they have to be very close in where they project before you can fuse them. If you now move them apart, you will see them as one for a much longer

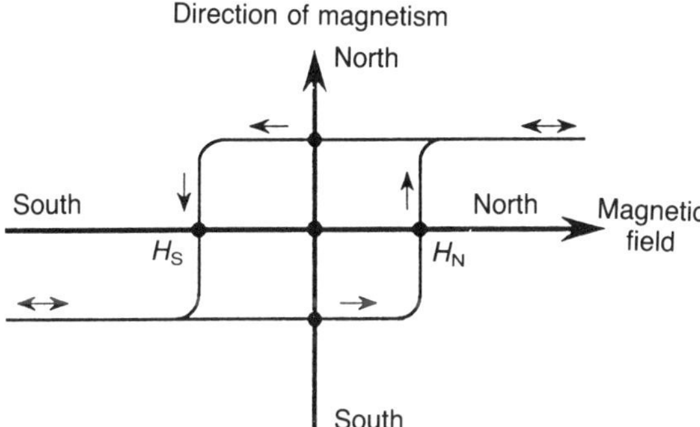

Figure 8.5 An example of a hysteresis cycle, plotting the direction of magnetization of a piece of iron against the magnitude of an external magnetic field. Because the magnetic field must flip each atomic magnet against the cooperative effect of interaction with those that remain unflipped, the field which induces a south–north transition, H_N, is much larger than the reversal field H_S.

range. This is hysteresis: there is no single-valued function from separation of stimuli on the two eyes to whether or not fusion is achieved. Rather, there is a history-dependent curve. In the ensuing years, a number of people such as Dev (1975) and Marr and Poggio (1977) began to build neural models that rationalized Julesz's magnetic analogy, while Harth *et al.* (1970) developed neural net models that explicitly exhibited hysteresis.

It has been well demonstrated that synaptic competition plays a crucial part in the plastic behavior exhibited by the brain (Merzenich *et al.* 1990). Indeed a good rule of thumb seems to be that if genetics gives a rough sketch of synaptic connectivity then experience tunes it, and in some cases greatly modifies it. One of the first people to model this was von der Marlsburg (1973).

The classic rules for adjusting the synaptic weights are: (1) the Hebb rule for correlation learning (Hebb 1949) but see Milner (1957) for the use of inhibitory synapses, von der Marlsburg for the use of 'normalization' to prevent saturation of synaptic weights and the useful perspective on such competitive learning offered by Rumelhart and Zipser (1985) and Grossberg (1987); and (2) the Perceptron learning rule (Rosenblatt 1962). Arbib (1987) provides an exposition of these models and an introduction to the mathematical theory of the Perceptron's capabilities (see Chapters 4 and 5).

Hebb's idea was that learning would occur if synapses could automatically strengthen themselves to record associations, strengthening a connection onto a cell whenever that input helps fire the cell. In other words, if input x_i is positive (there is an active input on the ith line), and output y is positive (the cell is responding), Hebb's rule is to strengthen the corresponding connection, that is, increase the synaptic weight s_i. The trouble with this original Hebb model is that every synapse will eventually get stronger and stronger till they all saturate, thus destroying any selectivity of association. Von der Marlsburg's (1973) solution was to normalize the synapses impinging on a given neuron. First compute the Hebbian 'update', and then divide this by the total putative synaptic weights to get the final result, that is, if Hebb would say Δs_i should equal kx_iy, then the normalization rule replaces s_i by

$$\frac{s_i + \Delta s_i}{\sum_j (s_j + \Delta s_j)}$$

where the summation j extends over all inputs to the neuron. Thus this new rule not only increases the strengths of those synapses whose inputs were most strongly correlated with the cell's activity, it also decreases the synaptic strengths of other connections to these cells in which such correlations did not arise.

However, another problem is that a lot of nearby cells may, just by chance, all have initial random connectivity which makes them easily persuadable by the same stimulus. Or the same pattern might occur many

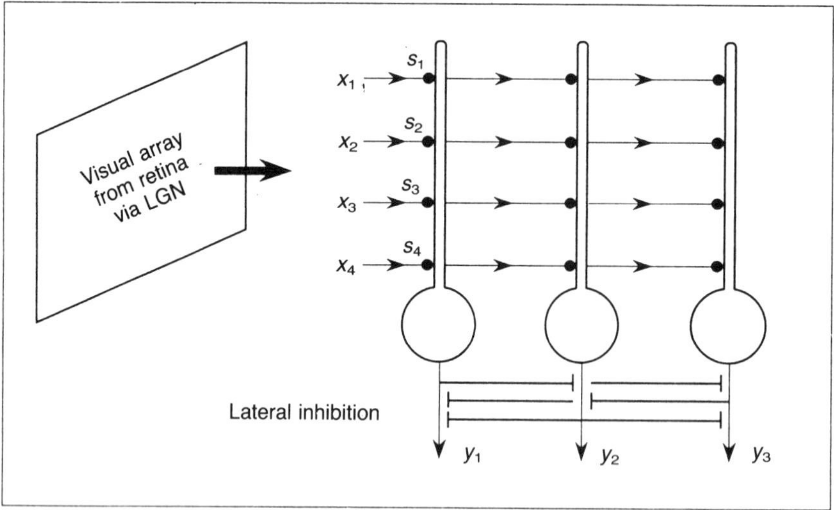

Figure 8.6 Hebbian synapses are augmented by normalization and lateral inhibition in the adaptive feature detectors of von der Malsburg (1973).

times before a new pattern was experienced by the network. In either case, many cells would become tuned to the same pattern, with not enough cells left to learn important and distinctive patterns. To solve this, von der Malsburg introduced lateral inhibition into his model of cortical plasticity (Figure 8.6). This is the connectivity pattern in which activity in any one cell is distributed laterally to reduce (partially inhibit) the activity of nearby cells. This ensures that if one cell – call it A – were active, its connections to nearby cells would make them less active, and so make them less likely to learn, by Hebbian synaptic adjustment, those patterns that most excite A.

In summary: initially, the inputs are pretty much randomly connected to the cells of the processing layer. As a result, none of these cells is particularly good at pattern recognition. However, by sheer statistical fluctuation of the synaptic connections, one will be slightly better at responding to a pattern than others were; it will thus slightly strengthen those synapses which allow it to fire for that pattern; through lateral inhibition, this will make it easy for cells initially less well tuned for that pattern to become tuned to it. Thus, without any teacher, this network automatically organizes itself so that each cell becomes tuned for an important cluster of information in the sensory inflow.

Von der Malsburg's paper exhibits the ideas of 'neural Darwinism' (Edelman 1987) in miniature. The cortical cells form a repertoire of feature detectors. The one that is initially the best (but still a poor) detector for a cluster of patterns has its synapses tuned, by a process of Hebbian learning plus normalization, by repeated exposure to these patterns so that it becomes a finely tuned feature detector responding vigorously to patterns in or near that cluster. Edelman would emphasize the initial predisposition and so speak of the learning process as one of selection; others would note that the process of synaptic adjustment changes the cell so much that we cannot think of a detector that is already adequate being simply selected from a preexisting repertoire. Whatever the terminological choice, these ideas play an important role in our understanding of development and learning in natural and artificial networks.

8.7 SCHEMAS AND PROGRAMMING METHODOLOGY FOR THE SIXTH GENERATION

It has become a truism in the computer industry that software is at least as important as hardware in the advance of computer technology. But some have argued that, with the use of the learning and self-organization principles of artificial neural networks, programming will no longer be necessary. However, if we consider that the human brain is a giga-gigaflop machine (10^{15} synapses running at a millisecond 'clock rate'), and that it takes 25 years for such a system to learn the skills for which we grant a Ph.D., it can be seen that more efficient techniques will be required for even the fastest and most adaptive of neurooptical computers. In short, we will still require sophisticated programming to reap the full benefits of sixth generation computers. This requires a methodology which allows an overall task to be decomposed into subtasks, with further stepwise refinement (and various processes of iteration up and down the levels of complexity) being required before subtasks are derived for which implementation is a straightforward process. Schema theory, as introduced above, provides the beginnings of such a methodology for parallel and distributed computation in the context of brain theory and perceptual robotics.

Classically, something is computable if it can be mapped into a recursive function f: $\mathbf{N} \rightarrow \mathbf{N}$. Such processes can, more generally, be captured by a finite (but expandable) graph of finite connectivity with a finite set of neighborhood functions. This includes the 'two von Neumanns' of stored programs and cellular automata (von Neumann, Burks and Goldstine 1947–8; von Neumann 1966). We get a starker generalization if we embrace the 'two Turings' of the classic model of serial computation (Turing 1936) and the 'stripe-machine' (Turing 1952) which explains pattern formation through processes of reaction and diffusion of morphogens in a discrete set of cells. The suggestion, then, is that our new generation of computers will still need a programming language for the formation of discrete assemblages (cf. the above remarks on schema theory) but that we will now allow building blocks to be continuous, generalizing a discrete finite basis to a continuous basis from which overall programs may be built. Ingredients for such building blocks include:

1. leaky integrator neurons;
2. dynamic systems (attractors/limit cycles);
3. optimization in the style of Hamilton's principle: stereo, motor control, etc.;

with modulation between the subsystems in the classic style of competition and cooperation in neural nets, and schemas tying it all together.

The bit no longer has to be the building block. It was not anyway. AI uses bits, not symbols, as the grounding level. But now we turn to analog VLSI inspired by the design of neural networks (Mead 1989) and optical computing as implementation technologies, and analog computers are once again part of the computer 'mainstream'. Filter operations can become unitary, rather than implemented by complex algorithms – for example, a lens is the best way to build a Fourier transform.

In the process, the line between computers and control disappears. Biological control theory usually studies neural circuitry specialized for the control of a specific function, be it the stretch reflex or the vestibulo-ocular reflex. Yet most behavior involves complex sequences of coordinated activity of a number of control systems. The notion of a coordinated control program (Arbib 1981) thus combines control theory and the computer scientist's notion of a program. These coordinated control programs can control the time-varying interaction of a number of control systems. In the diagrams representing such a program, there are lines representing both transfer of activation (as in computer flow diagrams) and transfer of data (as in control diagrams). However, just as we emphasized that sixth generation computers demand a programming methodology, so is it clear that symbolic coding remains a key issue. We will need to develop a whole new theory of computation to understand, for example, when analog 'searches' (e.g., simulated annealing, Kirkpatrick *et al.* (1983)) are more efficient than symbolic searches. The proper treatment of decision making under uncertainty will demand judicious blends of continuous processes and discrete graph structures, as we are beginning to see in the attempts to build large but tractable expert systems (Heckermann 1990). Von der Marlsburg (1988) has built on his earlier work on how synaptic changes may underlie the self-organization of retino-tectal connections to show how fast synaptic changes might offer a new approach to the problem of graph matching. This provides a new paradigm, the dynamic link architecture, for solving the binding problem of linking syntactically related items in a neurally based architecture which cannot use the explicit addressing of conventional computers to build pointers as in symbol-based AI. The method has already been used to build a promising prototype system for face recognition.

A classic form of program specification uses the notation $\{P\}S\{Q\}$ to specify that S be implemented as a function $f: X \rightarrow Y$ such that if $P(x)$ is true and $f(x)$ is defined, then $Q(f(x))$ is also true (Floyd 1967; Hoare 1969; Alagíc and Arbib 1977). To extend this to our proposed methodology for schemas, we may note with Arbib and Ehrig (1990) that module composition must be augmented with asynchrony and tunability. We thus have to relate the predicate logic style of program specification to the learning approach to neural net specification (which is in turn closely related to system identification). In training a neural network with a finite training set, we may write $\{x\}N\{y\}$ to require that the function of the net, f, satisfies $f(x_i) \approx y_i$ for each i, and some suitable sense of approximation. Here we have a finite set of requirements, rather than the global specification provided by the predicates P and Q of $\{P\}S\{Q\}$. However, we may achieve global specification of such systems by providing optimization rules $F(N)$ on the system N, with $F(N) = |f(x_i) - y_i|^2$ being a special case. More generally, a net defines a function in space and time, and we seek near-optimality.

Where computability theory has until now stressed the Turing machine halting problem this new methodology stresses the convergence problem for optimization techniques. Harking back to an issue raised in the previous section, we need to understand how symbolic structures may be seen as a limiting case of continuous activity so that recursive definition of program correctness may be related to recursive proofs of near optimality. In this regard, it may be encouraging that Manes and Arbib (1986) developed a general canonical fixpoint theory that applies to the recursive definition of functions in programming languages when taken in a partially additive category, yet also applies to more analytic problems when applied in a category of metric spaces. Many studies of adaptive neural networks have been in the context of learning strategies for pattern recognition and/or motor control. This ties in with the fact that for sixth generation computers, with their integral role for perceptual robotics, the specification of their performance will be neither in terms of symbolic predicates or numerical optimization, but will rather rest on relating schemas to 'world semantics', with success based on sensory and motor criteria for interaction with the rest of the world.

In high-performance computers, communication is a key issue. There is pressure for language as a high-level abstraction to reduce communication demands (cf. the idea of distributed artificial intelligence), as well as for new (e.g., optical) media to increase the bandwidth. So computers will have language, but it will have a far greater bandwidth for interprocessor communication than it will for human communication (though visualization seeks to make complex results immediately available without word-by-word encoding).

Finally, let me briefly note that, whatever our success in developing general principles of learning, schema methodology and optical implementation, the full success of sixth generation systems will rest in large part on the ability to store and retrieve much specific knowledge. In particular, the computer will have to be able to adapt itself to the needs of individual users. More to the point, relevant data to answer a user's query may be scattered over many databases, and ingenious new AI techniques will be needed to seek out and assemble the (possibly inconsistent) pieces of relevant data in a form that the user willl find intelligible. We may also note that expert systems need a great body of general knowledge to supplement their domain-specific knowledge if they are to avoid 'traps' that a human will readily see to be absurd. The Cyc project (Lenat and Guha 1990) is a valuable start in this direction, but needs to be dramatically augmented to handle sensorimotor databases and to take account of the learning techniques afforded by the new technology.

REFERENCES

Alagić, S. and Arbib, M. A. (1978). *The Design of Well-Structured and Correct Programs*. Springer-Verlag, New York.

Andersen, R. A., Essick, G. K. and Siegel, R. M. (1985). Encoding of spatial location by posterior parietal neurons. *Science* **230**, 456–8.

Arbib, M. A. (1972). *The Metaphorical Brain: An Introduction to Cybernetics as Artificial Intelligence and Brain Theory*. Wiley-Interscience, New York.

Arbib, M. A. (1975). Artificial intelligence and brain theory: unities and diversities. *Ann. Biomed. Eng.* **3**, 238–74.

Arbib, M. A. (1981). Perceptual structures and distributed motor control. In *Handbook of Physiology – The Nervous System, Vol. II, Motor Control* (ed. V. Brooks). Amer. Physiological Soc., Bethesda, MD, pp. 1449–80.

Arbib, M. A. (1982). *Rana computatrix*, an evolving model of visuomotor coordination in frog and toad. In *Machine Intelligence 10* (ed. J. Hayes and D. Michie). Ellis Horwood, Chichester, UK, pp. 501–17.

Arbib, M. A. (1987*a*). Neural computing: the challenge of the sixth generation. *EDUCOM Bull.* **23**(1), 2–12.

Arbib, M. A. (1987*b*). *Brains, Machines, and Mathematics*, 2nd edn. Springer-Verlag, Berlin.

Arbib, M. A. (1989). *The Metaphorical Brain 2, Neural Networks and Beyond*. Wiley-Interscience, New York.

Arbib, M. A. (1990*a*). Programs, schemas, and neural networks for control of hand movements: beyond the RS framework. In *Attention and Performance XIII* (ed. M. Ieannerod).

Arbib, M. A. (1990*b*). Visuomotor coordination: neural models and perceptual robotics. In *Visuomotor Coordination. Amphibians, Comparisons, Models, and Robots* (ed. J.-P. Ewart and M. A. Arbib). Plenum, New York.

Arbib, M. A. (1992). Schema Theory. In *The Encyclopedia of Artificial Intelligence*, 2nd edn. (ed. S. Shapiro). Wiley-Interscience, New York, pp. 1427–43.

Arbib, M. A. and Buhmann, J. (1992). Neural Networks. In *The Encyclopedia of Artificial Intelligence*, 2nd edn. (ed. S. Shapiro). Wiley-Interscience, pp. 1016–60.

Arbib, M. A. and Ehrig, H. (1990). Linking schemas and module specifications for distributed systems. *Proceedings of the Workshop on Distributed Computing Systems*, Cairo.

Arbib, M. A. and Ewert, J.-P. (eds.) (1991). *Visual Structures and Integrated Functions*. Research Notes in Neural Computing **3**, Springer-Verlag.

Arbib, M. A. and Hanson, A.R. (eds.) (1987). *Vision, Brain, and Cooperative Computation*. A Bradford Book/MIT Press, Cambridge, MA.

Arbib, M. A. and Robinson, J. A. (eds.) (1990). *Natural and Artificial Parallel Computation*. MIT Press, Cambridge, MA.

Bond, A. H. and Gasser, L. (eds.) (1988). *Readings in Distributed Artificial Intelligence*. Morgan Kaufmann.

Brooks, R. A. (1986). A robust layered control system for a mobile robot. *IEEE Journal of Robotics and Automation* **RA-2**, 14–23.

Cervantes-Perez, F., Lara, R. and Arbib, M. A. (1985). A neural model of interactions subserving prey–predator discrimination and size preference in anuran amphibia. *J. Theoret. Biol.* **113**, 117–52.

Cragg, B. G. and Temperley, H. N. V. (1954). The organisation of neurones: a cooperative analogy. *EEG Clin. Neurophysiol.* **6**, 85–92.

Craik, K. J. W. (1943). *The Nature of Explanation*. Cambridge University Press.

Dev, P. (1975) Perception of depth surfaces in random-dot stereograms: a neural model. *Int. J. Man-Machine Studies* **7**, 511–28.

Didday, R. L. and Arbib, M. A. (1975). Eye movements and visual perception: A 'two visual system' model. *Internat. J. Man-Machine Stud.* **7**, 547–69.

Dominey, P. F. and Arbib, M. A. (1992). A cortico-subcortical model for generation of spatially accurate sequential saccades. *Cerebral Cortex.* **2**, 153–75.

Draper, B. A., Collins, R. T., Brolio, J., Hanson, A. R. and Riseman, E. M. (1989). The schema system. *International Journal of Computer Vision.* **2**, 209–50.

Edelman, G. M. (1987). *Neural Darwinism. The Theory of Neuronal Group Selection*. Basic Books, New York.

Erickson, R. P. (1974). Parallel 'population' neural coding in feature extraction. In *The Neurosciences. Third Study Program* (ed. F. O. Schmitt and F. G. Worden). MIT Press, Cambridge, MA, pp. 155–69.

Ewert, J.-P. (1976). The visual system of the toad: behavioural and physiological studies on a pattern recognition system. In *The Amphibian Visual System* (ed. K. Fite). Academic Press, New York, pp. 142–202.

Floyd, R. (1967). Assigning meanings to programs. In *Mathematical Aspects of Computer Science*. Amer. Math. Soc., pp. 19–32.

Gasser, L. and Huhns, M. N. (eds.) (1989). *Distributed Artificial Intelligence*, Volume II. Pitman Publishing/Morgan Kaufmann Publishers.

Georgopoulos, A. P., Caminiti, R., Kalaska, J. F. and Massey, J. T. (1983). Spatial coding of movement: a hypothesis concerning the coding of movement direction by motor cortical populations. In *Neural Coding of Motor Performance* (ed. J. Massion *et al.*). Springer-Verlag, Berlin, pp. 327–36.

Grossberg, S. (1987). Competitive learning: From interactive activation to adaptive resonance. *Cognitive Sci.* **11**, 23–63.

Harris-Warrick, R. M. and Marder, E. (1991). Modulation of neural networks for behaviour. *Annu. Rev. Neurosci.* **14**, 39–57.

Harth, E. M., Csermely, T. J., Beek, B. and Lindsay, R. D. (1970). Brain functions and neural dynamics. *J. Theor. Biol.* **26**, 93–120.

Hebb, D. O. (1949). *The Organization of Behavior*. Wiley, New York.

Heckermann, D. E. (1990). Probabilistic Similarity Networks, Ph.D. Thesis in Medical Information Sciences, Report No. STAN-CS-90-1316, Stanford University, CA.

Heinzel, H.-G. and Selverston, A. I. (1988). Gastric mill activity in the lobster. III. Effects of proctolin on the isolated central pattern generator. *J. Neurophysiol.* **59**.

Hinton, G. E., McClelland, J. L. and Rumelhart, D. E. (1987). Distributed representations. In *Parallel Distributed Processing: Explorations in the Microstructure of Cognition*, (eds. D. Rumelhart and J. McClelland) The MIT Press/Bradford Books, Volume I, 77–109.

Hoare, C. A. R. (1969). An axiomatic basis for computer programming. *Comm. ACM* **12**, 576–80, 583.

Hodgkin, A. L. and Huxley, A. F. (1952). A quantitative description of membrane curtent and its application to conduction and excitation in nerve. *J. Physiol.* **117**, 500–44.

Hopfield, J. and Tank, D. W. (1986). Computing with neural circuits: A model. *Science* **233**, 625–33.

Ingle, D. (1975). Focal attention in the frog: Behavioural and physiological cortelates. *Science* **188**, 1033–5.

Ito, M. (1984). *The Cerebellum and Neuronal Control*. Raven Press, New York.

Julesz, B. (1971). *Foundations of Cyclopean Perception*. University of Chicago Press, IL.

Kirkpatrick, S., Gelatt, C. D. Jr. and Vecchi, M. P. (1983). Optimization by simulated annealing. *Science* **220**, 671–80.

Kornfeld, W. A. and Hewitt, C. (1981). The scientific community metaphor. *IEEE Trans. on Systems, Man and Cybernetics* **11**, 23–33.

Lara, R., Arbib, M. A. and Cromarty, A. S. (1982). The role of the tectal column in facilitation of amphibian prey-catching behaviour: a neural model. *J. Neurosci.* **2**, 521–30.

Lenat, D. B. and Guha, R. V. (1990). *Building Large Knowledge-Based Systems: Representation and Inference in the Cyc Project*. Addison-Wesley.

Lyons, D. M. and Arbib, M. A. (1989). A formal model of distributed computation for schema-based robot control. *IEEE J. Robotics Automation.*

Manes, E. G. and Arbib, M. A. (1986). *Algebraic Approaches to Program Semantics.* Springer-Verlag.

Marlsburg, C. von der (1973). Self-organization of orientation sensitive cells in the striate cortex. *Kybernetik* **14**, 85–100.

Marlsburg, C. von der (1988). Pattern recognition by labeled graph matching. *Neural Networks.* **141**, 8.

Marr, D. and Poggio, T. (1977). Cooperative computation of stereo disparity. *Science* **194**, 283–7.

McCulloch, W. S. and Pitts, W. H. (1943). A logical calculus of the ideas immanent in nervous activity. *Bull. Math. Biophys.* **5**, 115–33.

McKenna, T., Davis, J. and Zornetzer, S. F. (1992). *Single Neuron Computation.* Academic Press, San Diego, CA.

Mcllwain, I. T. (1982). Lateral spread of neural excitation during microstimulation in intermediate grey layer of cat's superior colliculus. *J. Neurophysiol.* **47**, 167–78.

Mead, C. (1989). *Analog VLSI and Neural Systems.* Addison-Wesley.

Merzenich, M. M., Recanzone, G. H., Jenkins, W. M. and Nudo, R. J. (1990). How the brain functionally rewires itself. In *Natural and Artificial Parallel Computation* (ed. M. A. Arbib and J. A. Robinson). MIT Press, Cambridge, MA, pp. 177–210.

Meyer, J.-A. and Wilson, S. W. (eds.) (1991). *From Animals to Animats: Proceedings of The First International Conference on Simulation of Adaptive Behavior.* MIT Press/Bradford Books.

Milner, P. M. (1957). The cell assembly: Mark II. *Psychol. Rev.* **64**, 242–52.

Minsky, M. L. (1985). *The Society of Mind*, Simon and Schuster, New York.

Pitts, W. H. and McCulloch, W. S. (1947). How we know universals, the perception of auditory and visual forms. *Bull. Math. Biophys.* **9**, 127–47.

Rakic, P. and Singer, W. (eds.) (1988). *Neurobiology of Cortex*, Wiley, Chichester/New York.

Rall, W. (1970). Dendritic neuron theory and dendrodendritic synapses in a simple cortical system. In *The Neurosciences. Second Study Program* (ed. F. O. Schmitt). Rockefeller University Press, New York, pp. 552–65.

Rosenblatt, F. (1962). *Principles of Neurodynamics.* Spartan, Washington, DC.

Rumelhart, D. E. and Zipser, D. (1985). Feature discovery by competitive learning. *Cognitive Sci.* **9**, 75–112.

Selverston, A. I., Russell, D. F. and Miller, J. P. (1976). The stomatogastric nervous system: structure and function of a small neural network. *Prog. Neurobiol.* **7**, 215–90.

Shastri, L. (1989). A connectionist encoding of semantic networks. In *Distributed Artificial Intelligence* (ed. M. N. Huhns). Pitman/Morgan Kaufmann, pp. 177–202.

Shepherd, G. (1979). *The Synaptic Organization of the Brain*, 2nd edn. Oxford University Press.

Touretzky, D. and Hinton, G. (1985). Symbols among the neurons: details of a connectionist inference architecture. *IJCAI* **85**, 238–43.

Turing, A. M. (1936). On computable numbers with an application to the entscheidungsproblem. *Proc. London Math. Soc.* Ser. 2, **42**, 230–65.

Turing, A. M. (1952). The chemical basis of morphogenesis. *Phil. Trans. Roy. Soc. Lond.* **B237**, 37–72.

von Neumann, J. (1966). *Theory of Self-Reproducing Automata* (edited and completed by A. W. Burks). University of Illinois Press, IL.

Weitzenfeld, A. (1991). *NSL: Neural Simulation Language, Version 2.1.* CNE-TR 91-05, University of Southern California, Center for Neural Engineering, Los Angeles.

Weitzenfeld, A. (1992). A unified computational model for schemas and neural networks in concurrent object-oriented programming. Ph.D. Dissertation, Department of Computer Science, University of Southern California, Los Angeles.

Yonezawa, A. and Tokoro, M. (eds.) (1987). *Object-Oriented Concurrent Programming.* The MIT Press.

ART 3: HIERARCHICAL SEARCH USING CHEMICAL TRANSMITTERS IN SELF-ORGANIZING PATTERN RECOGNITION ARCHITECTURES*

9

Gail A. Carpenter and Stephen Grossberg

9.1 INTRODUCTION: DISTRIBUTED SEARCH OF ART NETWORK HIERARCHIES

This paper incorporates a model of the chemical synapse into a new Adaptive Resonance Theory neural network architecture called ART 3. ART 3 system dynamics model a simple, robust mechanism for parallel search of a learned pattern recognition code. This search mechanism was designed to implement the computational needs of ART systems embedded in network hierarchies, where there can, in general, be either fast or slow learning and distributed or compressed code representations. The search mechanism incorporates a code reset property that serves at least three distinct functions: to correct erroneous category choices, to learn from reinforcement feedback and to respond to changing input patterns. The three types of reset are illustrated, by computer simulation, for both maximally compressed and partially compressed pattern recognition codes (Sections 9.20–9.26).

Let us first review the main elements of adaptive resonance theory. ART architectures are neural networks that carry out stable self-organization of recognition codes for arbitrary sequences of input patterns. Adaptive resonance theory first emerged from an analysis of the instabilities inherent in feedforward adaptive coding structures (Grossberg 1976*a*). More recent work has led to the development of two classes of ART neural network architectures, specified as systems of differential equations. The first class, ART 1, self-organizes recognition categories for arbitrary sequences of binary input patterns (Carpenter and Grossberg 1987*a*). A second class, ART 2, does the same for either binary or analog inputs (Carpenter and Grossberg 1987*b*).

Both ART 1 and ART 2 use a maximally compressed, or choice, pattern recognition code. Such a code is a limiting case of the partially compressed recognition codes that are typically used in explanations by ART of biological data (Grossberg 1982*a*, 1987*a*, 1987*b*). Partially compressed recognition codes have been mathematically analysed in models for competitive learning, also called self-organizing feature maps, which are incorporated into ART models as part of their bottom-up dynamics (Grossberg 1976*a*, 1982*a*; Kohonen, 1984). Maximally compressed codes were used in ART 1 and ART 2 to enable a rigorous analysis to be made of how the bottom-up and top-down dynamics of ART systems can be joined together in a real-time

* Editor's note: this chapter was first published in 1990 (Carpenter and Grossberg 1990). For more recent work, see Further Reading at the end of the chapter.

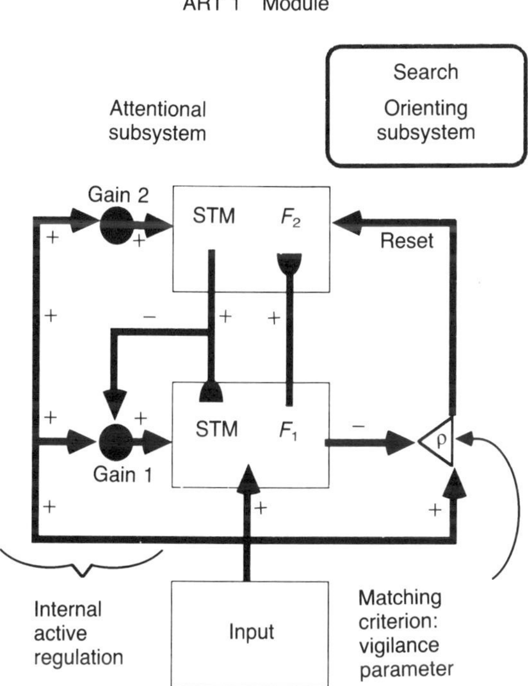

Figure 9.1 Typical ART 1 neural network module (Carpenter and Grossberg 1987*a*).

self-organizing system capable of learning a stable pattern recognition code in response to an arbitrary sequence of input patterns. These results provide a computational foundation for designing ART systems capable of stably learning partially compressed recognition codes. The present results contribute to such a design.

The main elements of a typical ART 1 module are illustrated in Figure 9.1. F_1 and F_2 are fields of network nodes. An input is initially represented as a pattern of activity across the nodes, or feature detectors, of field F_1. The pattern of activity across F_2 corresponds to the category representation. Because patterns of activity in both fields may persist after input offset yet may also be quickly inhibited, these patterns are called short-term memory, or STM, representations. The two fields, linked both bottom-up and top-down by adaptive filters, constitute the attentional subsystem. Because the connection weights defining the adaptive filters may be modified by inputs and may persist for very long times after input offset, these connection weights are called long-term memory, or LTM, variables.

An auxiliary orienting subsystem becomes active during search. This search process is the subject of the present article.

9.2 AN ART SEARCH CYCLE

Figure 9.2 illustrates a typical ART search cycle. An input pattern $\boldsymbol{I}$ registers itself as a pattern $\boldsymbol{X}$ of activity across F_1 (Figure 9.2(a)). The F_1 output signal vector $\boldsymbol{S}$ is then transmitted through the multiple converging and diverging weighted adaptive filter pathways emanating from F_1, sending a net input signal vector $\boldsymbol{T}$ to F_2. The internal competitive dynamics of F_2 contrast-enhance $\boldsymbol{T}$. The F_2 activity vector $\boldsymbol{Y}$ therefore registers a compressed representation of the filtered $F_1 \rightarrow F_2$ input and corresponds to a category representation for the input active at F_1. Vector $\boldsymbol{Y}$ generates a signal vector $\boldsymbol{U}$ that is sent top-down through the second adaptive filter, giving rise to a net top-down signal vector $\boldsymbol{V}$ to F_1 (Figure 9.2(b)). F_1 now receives two input vectors, $\boldsymbol{I}$ and $\boldsymbol{V}$. An ART system is designed to carry out a matching process whereby the original activity pattern $\boldsymbol{X}$ due to input pattern $\boldsymbol{I}$ may be modified by the template pattern $\boldsymbol{V}$ that is associated with the current active category. If $\boldsymbol{I}$ and $\boldsymbol{V}$ are not sufficiently similar according to a matching criterion established by a dimensionless vigilance parameter ρ, a reset signal quickly and enduringly shuts off the active category representation (Figure 9.2(c)), allowing a new category to become active. Search ensues (Figure 9.2(d)) until either an adequate match is made or a new category is established.

In earlier treatments (e.g., Carpenter and Grossberg 1987*a*), we proposed that the enduring shut-off of erroneous category representations by a nonspecific reset signal

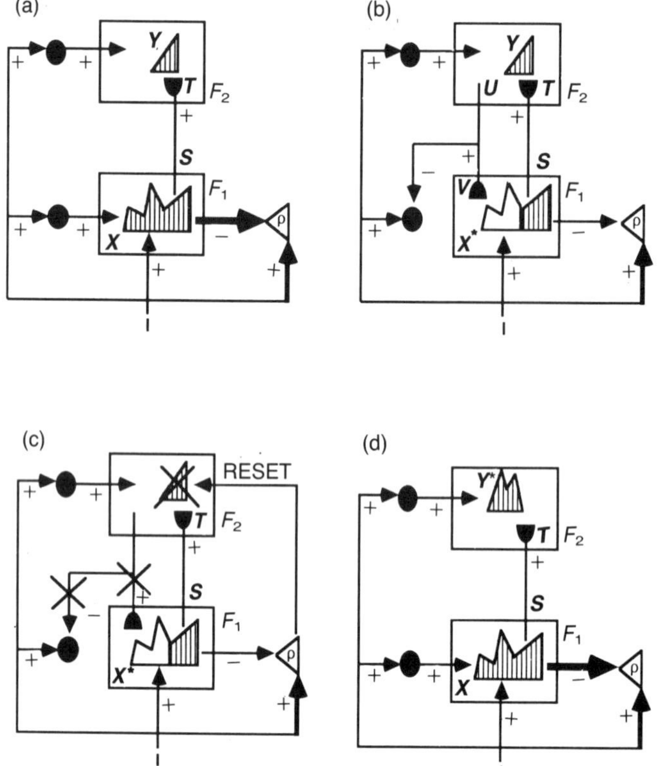

Figure 9.2 ART search cycle (Carpenter and Grossberg 1987*a*).

could occur at F_2 if F_2 were organized as a gated dipole field, whose dynamics depend on depletable transmitter gates. Though the new search process does not here use a gated dipole field, it does retain and extend the core idea that transmitter dynamics can enable a robust search process when appropriately embedded in an ART system.

9.3 ART 2: THREE-LAYER COMPETITIVE FIELDS

Figure 9.3 shows the principal elements of a typical ART 2 module. It shares many characteristics of the ART 1 module, having both an input representation field F_1 and a category representation field F_2, as well as attentional and orienting subsystems. Figure 9.3 also illustrates one of the main differences between the examples of ART 1 and ART 2 modules so far explicitly developed; namely, the ART 2 examples all have three processing layers within the F_1 field. These three processing layers allow the ART 2 system to stably categorize sequences of analog input patterns that can, in general, be arbitrarily close to one another. Unlike in models such as back propagation, this category learning process is stable even in the fast learning situation, in which the LTM variables are allowed to go to equilibrium on each learning trial. In Figure 9.3, one F_1 layer reads in the bottom-up input, one layer reads in the top-down filtered input from F_2, and a middle layer matches patterns from the top and bottom layers before sending a composite pattern back through the F_1 feed-

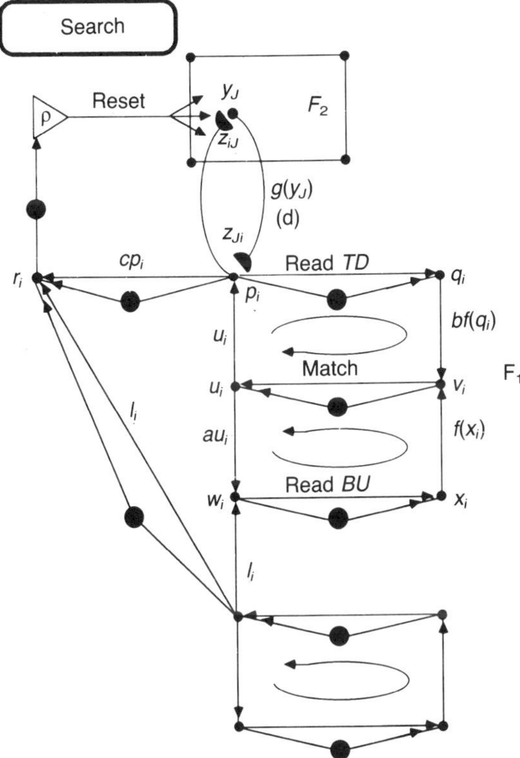

Figure 9.3 Typical ART 2 neural network module, with three-layer F_1 field (Carpenter and Grossberg 1987*b*). Large filled circles are gain control nuclei that nonspecifically inhibit target nodes in proportion to the Euclidean norm of activity in their source fields, as in equation (9.33).

back loop. Both F_1 and F_2 are shunting competitive networks that contrast-enhance and normalize their activation patterns (Grossberg 1982*a*).

9.4 ART BIDIRECTIONAL HIERARCHIES AND HOMOLOGY OF FIELDS

In applications, ART modules are often embedded in larger architectures that are hierarchically organized. Figure 9.4 shows an example of one such hierarchy, a self-organizing model of the perception and production of speech (Cohen, Grossberg and Stork 1988). In Figure 9.4, several copies of an ART module are cascaded upward, with partially compressed codes at each level. Top-down ART filters both within the perception system and from the production system to the perception system serve to stabilize the evolving codes as they are learned. We will now consider how an ART 2 module can be adapted for use in such a hierarchy.

When an ART module is embedded in a network hierarchy, it is no longer possible to make a sharp distinction between the characteristics of the input representation field F_1 and the category representation field F_2. For example, within the auditory perception system of Figure 9.4, the partially compressed auditory code acts both as the category representation field for the invariant feature field, and as the input field for the compressed item code field. In order for them to serve both functions, the basic structures of all the network fields in a hierarchical ART system should be homologous, in so far as possible (Figure 9.5). This constraint is satisfied if all fields of the hierarchy are endowed with the F_1 structure of an ART 2 module (Figure 9.3). Such a design is sufficient for the F_2 field as well as the F_1 field because the principal property required of a category representation field, namely that input patterns be contrast-enhanced and normalized, is a property of the three-layer F_1 structure. The system shown in Figure 9.5 is called an ART bidirectional hierarchy, with each field homologous to all other fields and linked to contiguous fields by both bottom-up and top-down adaptive filters.

9.5 ART CASCADE

For the ART hierarchy shown in Figure 9.5, activity changes at any level can ramify throughout all lower and higher levels. It is sometimes desirable to buffer activity patterns

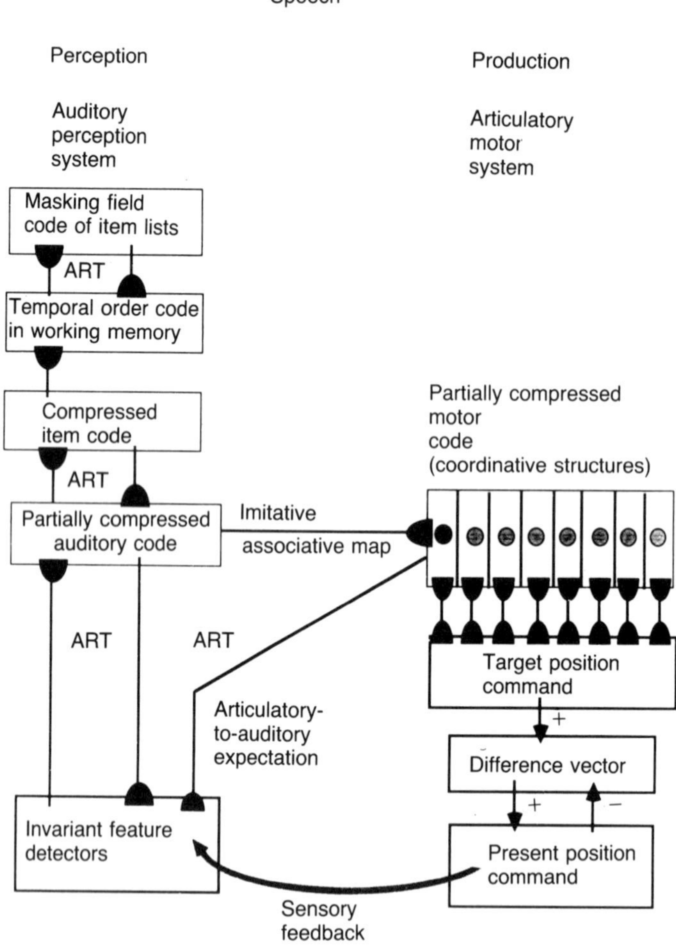

Figure 9.4 Neural network model of speech production and perception (Cohen, Grossberg and Stork 1988).

at lower levels against changes at higher levels. This can be accomplished by inserting a bottom-up pathway between each two-field ART module. Figure 9.6 illustrates a sequence of modules *A*, *B*, *C*, . . . forming an ART cascade. The 'category representation' field F_{2A} acts as the input field for the next field F_{1B}. As in an ART 2 module (Figure 9.3), connections from the input field F_{2A} to the first field F_{1B} of the next module are nonadaptive and unidirectional. Connections between F_{1B} and F_{2B} are adaptive and bidirectional. This scheme repeats itself throughout the hierarchy. Activity changes due to a reset event at a lower level can be felt at higher levels via an ascending cascade of reset events. In particular, reset at the lowest input level can lead to a cascade of input reset events up the entire hierarchy.

9.6 SEARCH IN AN ART HIERARCHY

We now consider the problem of implementing parallel search among the distributed

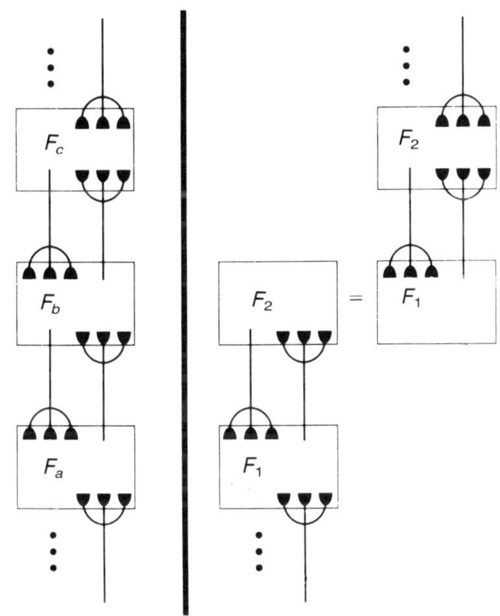

Figure 9.5 Homology of fields F_a, F_b, F_c, . . . in an ART bidirectional hierarchy.

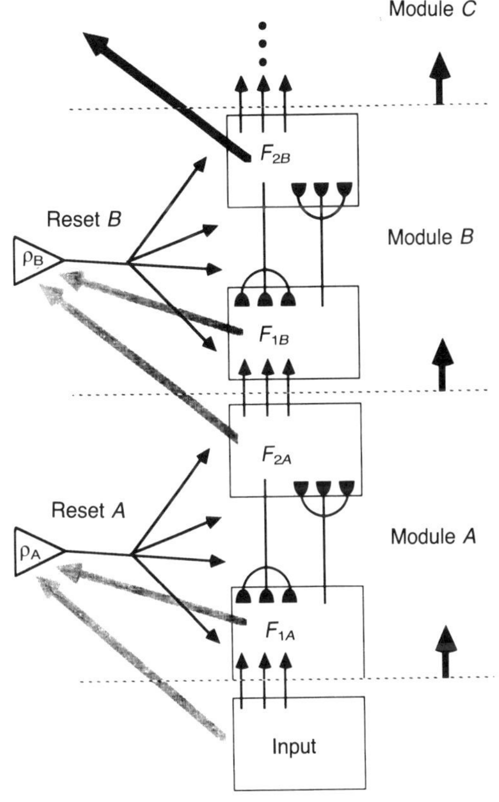

Figure 9.6 An ART cascade. Nonadaptive connections terminate in arrowheads. Adaptive connections terminate in semicircles.

codes of a hierarchical ART system. Assume that a top-down/bottom-up mismatch has occurred somewhere in the system. How can a reset signal search the hierarchy in such a way that an appropriate new category is selected? The search scheme for ART 1 and ART 2 modules incorporates an asymmetry in the design of levels F_1 and F_2 that is inappropriate for ART hierarchies whose fields are homologous. The ART 3 search mechanism described below eliminates that asymmetry.

A key observation is that a reset signal can act upon an ART hierarchy between its fields F_a, F_b, F_c, . . . (Figure 9.7). Locating the site of action of the reset signal between the fields allows each individual field to carry out its pattern processing function without introducing processing biases directly into a field's internal feedback loops.

The new ART search mechanism has a number of useful properties. It: (a) works well for mismatch, reinforcement or input reset; (b) is simple; (c) is homologous to physiological processes; (d) fits naturally into network hierarchies with distributed codes and slow or fast learning; (e) is robust in that it does not require precise parameter choices, timing or analysis of classes of inputs; (f) requires no new anatomy, such as new wiring or nodes, beyond what is already present in the ART 2 architecture; (g) brings new computational power to the ART systems; and (h) although derived for the ART system, can be used to search other neural network architectures as well.

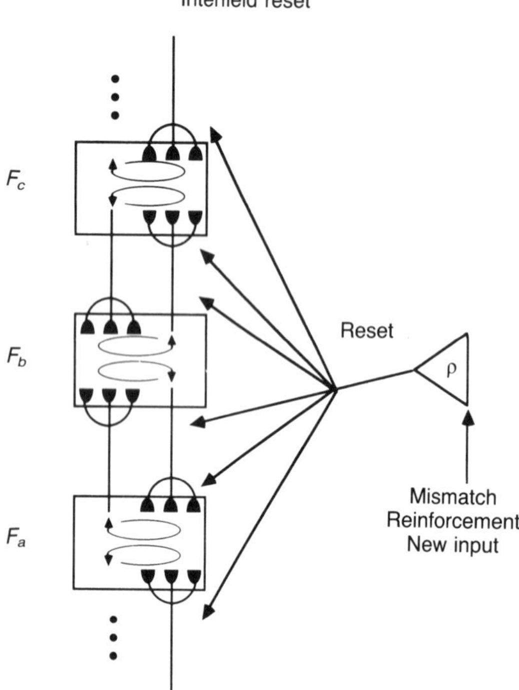

Figure 9.7 Interfield reset in an ART bidirectional hierarchy.

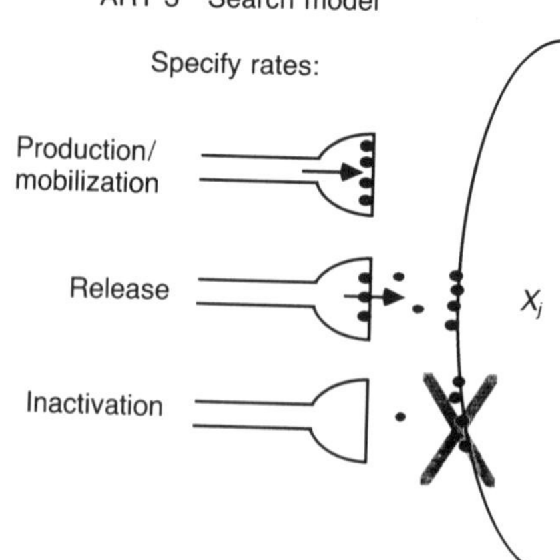

Figure 9.8 The ART search model specifies rate equations for transmitter production, release and inactivation.

9.7 A NEW ROLE FOR CHEMICAL TRANSMITTERS IN ART SEARCH

The computational requirements of the ART search process can be fulfilled by formal properties of neurotransmitters (Figure 9.8), if these properties are appropriately embedded in the total architecture model. The main properties used are illustrated in Figure 9.9, which is taken from Ito (1984). In particular, the ART 3 search equations incorporate the dynamics of production and release of a chemical transmitter substance; the inactivation of transmitter at postsynaptic binding sites; and the modulation of these processes via a nonspecific control signal. The net effect of these transmitter processes is to alter the ionic permeability at the postsynaptic membrane site, thus effecting excitation or inhibition of the postsynaptic cell.

The notation to describe these transmitter properties is summarized in Figure 9.10, for a synapse between the ith presynaptic node and the jth postsynaptic node. The presynaptic signal, or action potential, S_i, arrives at a synapse whose adaptive weight, or long-term memory trace, is denoted z_{ij}. The variable z_{ij} is identified with the maximum amount of available transmitter. When the transmitter at this synapse is fully accumulated, the amount of transmitter u_{ij} available for release is equal to z_{ij}. When a signal S_i arrives, transmitter is typically released. The variable v_{ij} denotes the amount of transmitter released into the extracellular space, a fraction of which is assumed to be bound at the postsynaptic cell surface and the remainder rendered ineffective in the extracellular space. Finally, x_j denotes the activity, or membrane potential, of the postsynaptic cell.

9.8 EQUATIONS FOR TRANSMITTER PRODUCTION, RELEASE AND INACTIVATION

The search mechanism works well if it possesses a few basic properties. These properties can be realized using one of several

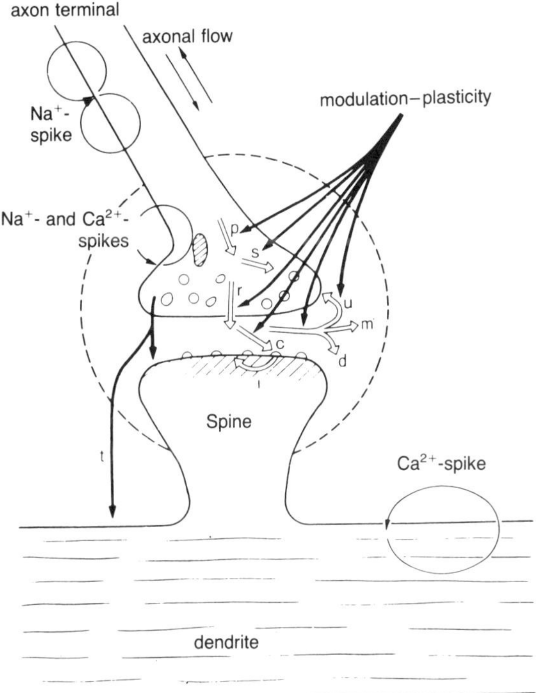

Figure 9.9 Schematic diagram showing electrical, ionic and chemical events in a dendritic spine synapse. Open arrows indicate steps from production of neurotransmitter substance (*p*) to storage (*s*) or release (*r*) to reaction with subsynaptic receptors (*c*), leading to change of ionic permeability of subsynaptic membrane (*i*) or to removal to extracellular space (*m*), enzymatic destruction (*d*), or uptake by presynaptic terminal (*u*)·*t*, action of trophic substance (Ito (1984), p. 52, reprinted with permission).

closely related sets of equations, with corresponding differences in biophysical interpretation. An illustrative system of equations is described below.

Equations (9.1)–(9.3) govern the dynamics of the variables z_{ij}, u_{ij}, v_{ij}, and x_j at the ijth pathway and jth node of an ART 3 system.

Presynaptic transmitter

$$\frac{du_{ij}}{dt} = (z_{ij} - u_{ij}) - u_{ij}[\text{release rate}] \quad (9.1)$$

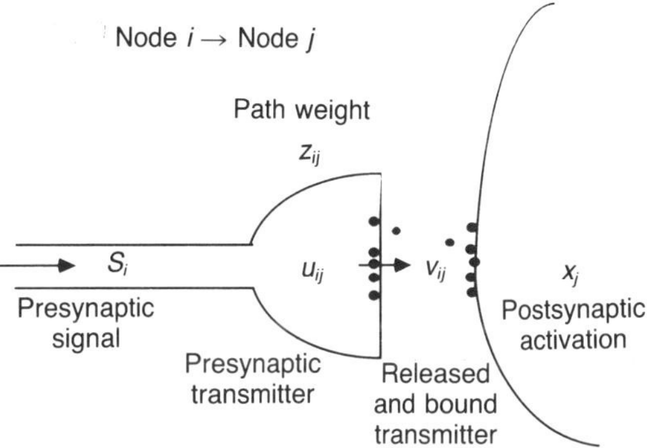

Figure 9.10 Notation for the ART chemical synapse.

Bound transmitter

$$\begin{aligned}\frac{dv_{ij}}{dt} &= -v_{ij} + u_{ij}[\text{release rate}] \\ &\quad -v_{ij}[\text{inactivation rate}] \\ &= -v_{ij} + u_{ij}[\text{release rate}] \\ &\quad -v_{ij}[\text{reset signal}] \end{aligned} \quad (9.2)$$

Postsynaptic activation

$$\begin{aligned}\varepsilon\frac{dx_j}{dt} &= -x_j + (A - x_j)[\text{excitatory inputs}] \\ &\quad - (B + x_j)[\text{inhibitory inputs}] \\ &= -x_j + (A - x_j) \\ &\quad \times \left[\sum_i v_{ij} + \{\text{intrafield feedback}\}\right] \\ &\quad - (B + x_j)[\text{reset signal}] \end{aligned} \quad (9.3)$$

Equation (9.1) says that presynaptic transmitter is produced and/or mobilized until the amount u_{ij} of transmitter available for release reaches the maximum level z_{ij}. The adaptive weight z_{ij} itself changes on the slower timescale of learning, but remains essentially constant on the timescale of a single reset event. Available presynaptic transmitter u_{ij} is released at a rate that is specified below.

A fraction of presynaptic transmitter becomes postsynaptic bound transmitter after being released. For simplicity, we ignore the fraction of released transmitter that is inactivated in the extracellular space. Equation (9.2) says that the bound transmitter is inactivated by the reset signal.

Equation (9.3) for the postsynaptic activity x_j is a shunting membrane equation such that excitatory inputs drive x_j up toward a maximum depolarized level equal to A; inhibitory inputs drive x_j down toward a minimum hyperpolarized level equal to $-B$; and activity passively decays to a resting level equal to 0 in the absence of inputs. The net effect of bound transmitter at all synapses converging on the jth node is assumed to be excitatory, via the term

$$\sum_i v_{ij}. \tag{9.4}$$

Internal feedback from within the target field (Figure 9.3) is excitatory, while the nonspecific reset signal is inhibitory. Parameter ε is small, corresponding to the assumption that activation dynamics are fast relative to the transmitter accumulation rate, equal to 1 in equation (9.1).

The ART 3 system can be simplified for purposes of simulation. Suppose that $\varepsilon \ll 1$ in (9.3); the reset signals in (9.2) and (9.3) are either 0 or $\gg 1$; and net intrafield feedback is excitatory. Then equations (9.1), (9.5), and (9.6) below approximate the main properties of ART 3 system dynamics.

Simplified ART 3 equations

$$\frac{du_{ij}}{dt} = (z_{ij} - u_{ij}) - u_{ij}[\text{release rate}] \tag{9.1}$$

$$\begin{cases} \dfrac{dv_{ij}}{dt} = -v_{ij} + u_{ij}[\text{release rate}] & \text{if reset} = 0 \\ v_{ij}(t) = 0 & \text{if reset} \gg 1 \end{cases} \tag{9.5}$$

$$x_j(t) = \begin{cases} \Sigma_i v_{ij} + [\text{intrafield feeback}] & \text{if reset} = 0 \\ 0 & \text{if reset} \gg 1. \end{cases} \tag{9.6}$$

9.9 ALTERNATIVE ART 3 SYSTEMS

In equations (9.2) and (9.3), the reset signal acts in two ways, by inactivating bound transmitter and directly inhibiting the postsynaptic membrane. Alternatively, the reset signal may accomplish both these goals in a single process if all excitatory inputs in (9.3) are realized using chemical transmitters. Letting w_j denote the net excitatory transmitter reaching the jth target cell via intrafield feedback, an illustrative system of this type is given by equations (9.1), (9.2), (9.7), and (9.8) below.

Presynaptic transmitter

$$\frac{du_{ij}}{dt} = (z_{ij} - u_{ij}) - u_{ij}[\text{release rate}] \tag{9.1}$$

Bound transmitter

$$\frac{dv_{ij}}{dt} = -v_{ij} + u_{ij}[\text{release rate}] - v_{ij}[\text{reset signal}] \tag{9.2}$$

$$\frac{dw_j}{dt} = -w_j + [\text{intrafield feedback}] - w_j[\text{reset signal}] \tag{9.7}$$

Postsynaptic activation

$$\varepsilon\frac{dx_j}{dt} = -x_j + (A - x_j)\Big(\sum_i v_{ij} + w_j\Big). \tag{9.8}$$

The reset signal now acts as a chemical modulator that inactivates the membrane channels at which transmitter is bound. It thus appears in equations (9.2) and (9.7), but not in equation (9.8) for postsynaptic activation.

When the reset signal can be only 0 or $\gg 1$, the simplified system in Section 9.8 approximates both versions of the ART 3 system. However, if the reset signal can vary continuously in size, equations (9.2), (9.7), and (9.8) can preserve relative transmitter quantities from all input sources. Thus this system is a better model for the intermediate cases than equations (9.2) and (9.3).

An additional inhibitory term in the postsynaptic activation equation (9.8) helps to suppress transmitter release, as illustrated in Section 9.25.

9.10 TRANSMITTER RELEASE RATE

To further specify the ART search model, we now characterize the transmitter release and inactivation rates in equations (9.1) and (9.2). Then we trace the dynamics of the system at key time intervals during the presentation of a fixed input pattern (Figure 9.11). We first observe system dynamics during a brief time interval after the input turns on ($t = 0^+$), when the signal S_i first arrives at the synapse. We next consider the effect of subsequent internal feedback signals from within the target field, following contrast-enhancement of the inputs. We observe how the ART 3 model responds to a reset signal by implementing a rapid and enduring inhibition of erroneously selected pattern features. Then we analyse how the ART 3 model responds if the input pattern changes. We will begin with the ART search hypothesis 1, as outlined below.

ART search hypothesis 1

Presynaptic transmitter u_{ij} is released at a rate

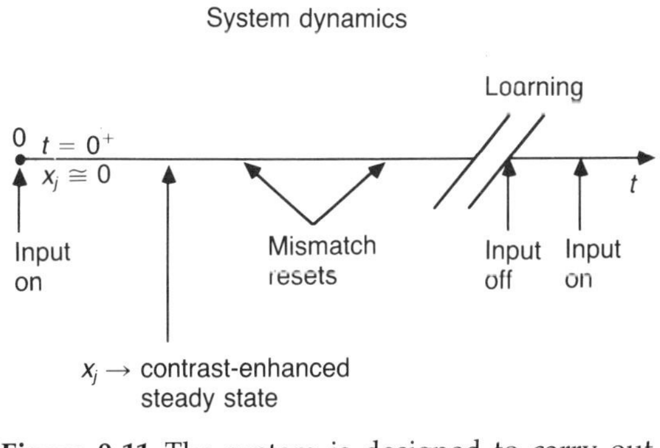

Figure 9.11 The system is designed to carry out necessary computations at critical junctures of the search process.

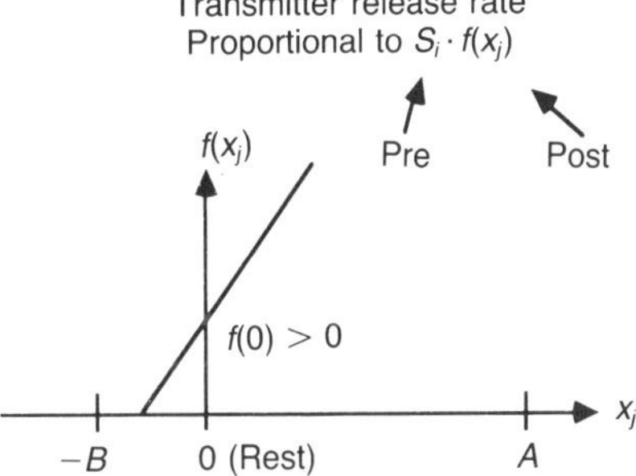

Figure 9.12 The ART search hypothesis 1 specifies the transmitter release rate.

jointly proportional to the presynaptic signal S_i and a function $f(x_j)$ of the postsynaptic activity. That is, in equations (9.1), (9.2), and (9.5),

$$\text{release rate} = S_i f(x_j). \tag{9.9}$$

The function $f(x_j)$ in equation (9.9) has the qualitative properties illustrated in Figure 9.12. In particular $f(x_j)$ is assumed to have a positive value when x_j is at its 0 resting level, so that transmitter u_{ij} can be released when the signal S_i arrives at the synapse. If $f(0)$ were equal to 0, no excitatory signal could reach a postsynaptic node at rest, even if a large presynaptic signal S_i were sent to that node. The function $f(x_j)$ is also assumed to equal 0 when x_j is significantly hyperpolarized, but to rise steeply when x_j is near 0. In the simulations, $f(x_j)$ is linear above a small negative threshold.

The form factor $S_i f(x_j)$ is a familiar one in the neuroscience and neural network literatures. In particular, such a product is often used to model associative learning, where it links the rate of learning in the ijth pathway to the presynaptic signal S_i and the postsynaptic activity x_j. Associative learning occurs, however, on a timescale that is much slower than the timescale of transmitter release. On the fast timescale of transmitter release, the form factor $S_i f(x_j)$ may be compared to interactions between voltages and ions. In Figure 9.9, for example, note the dependence of the presynaptic signal on the Na^+ ion; the postsynaptic signal on the Ca^{2+} ion; and transmitter release on the joint fluxes of these two ions. The ART search hypothesis 1 thus formalizes a known type of synergetic relationship between presynaptic and postsynaptic processes in effecting transmitter release. Moreover, the rate of transmitter release is typically a function of the concentration of Ca^{2+} in the extracellular space, and this function has qualitative properties similar to the function $f(x_j)$ shown in Figure 9.12 (Kandel and Schwartz 1981, p. 84; Kuffler, Nicholls and Martin 1984, p. 244).

9.11 SYSTEM DYNAMICS AT INPUT ONSET: AN APPROXIMATELY LINEAR FILTER

Some implications of the ART search hypothesis 1 will now be summarized. Assume that at time $t = 0$ transmitter u_{ij} has accumulated to its maximal level z_{ij} and that activity x_j and bound transmitter v_{ij} equal 0. Consider a time interval $t = 0^+$ immediately after a signal S_i arrives at the synapse. During this brief initial interval, the ART equations approximate the linear filter dynamics typical of many neural network models. In particular, equations (9.2) and (9.9) imply that the amount of bound transmitter is determined by equation

$$\frac{dv_{ij}}{dt} = -v_{ij} + u_{ij}S_i f(x_j) - v_{ij}[\text{inactivation rate}]. \quad (9.10)$$

Thus at times $t = 0^+$,

$$\frac{dv_{ij}}{dt} \approx z_{ij}S_i f(0) \quad (9.11)$$

and so

$$v_{ij}(t) \approx K(t)S_i z_{ij} \quad \text{for times } t = 0^+. \quad (9.12)$$

Because equation (9.12) holds at all the synapses adjacent to cell j, equation (9.6) implies that

$$x_j(t) \approx \sum_i K(t)S_i z_{ij} = K(t)\mathbf{S} \cdot \mathbf{z}_j \quad \text{for times } t = 0^+. \quad (9.13)$$

Here $\mathbf{S}$ denotes the vector $(S_1, \ldots, S_n)$, $\mathbf{z}_j$ denotes the vector $(z_{1j}, \ldots, z_{nj})$, and $i = 1, \ldots, n$. Thus in the initial moments after a signal arrives at the synapse, the small amplitude activity x_j at the postsynatpic cell grows in proportion to the dot product of the incoming signal vector $\mathbf{S}$ times the adaptive weight vector $\mathbf{z}_j$.

9.12 SYSTEM DYNAMICS AFTER INTRAFIELD FEEDBACK: AMPLIFICATION OF TRANSMITTER RELEASE BY POSTSYNAPTIC POTENTIAL

In the next time interval, the intrafield feedback signal contrast-enhances the initial signal pattern (9.13) via equation (9.6) and amplifies the total activity across field F_c in Figure 9.13(a). Figure 9.13(b) shows typical contrast-enhanced activity profiles: partial compression of the initial signal pattern; or maximal compression, or choice, where only one postsynaptic node remains active due to the strong competition within the field F_c.

In all, the model behaves initially like a linear filter. The resulting pattern of activity across postsynaptic cells is contrast-enhanced, as required in the ART 2 model as well as in the many other neural network models that incorporate competitive learning (Grossberg 1988). For many neural network systems, this combination of computational properties is all that is needed. These models implicitly assume that intracellular transmitter u_{ij} is always accumulated up to its target level z_{ij} and that postsynaptic activity x_j does not alter the rate of transmitter release:

$$u_{ij} \approx z_{ij} \quad \text{and} \quad v_{ij} \approx z_{ij}S_i. \quad (9.14)$$

If the linear filtering properties implied by (9.14) work well for many purposes, why complicate the system by adding additional hypotheses? Even a new hypothesis that

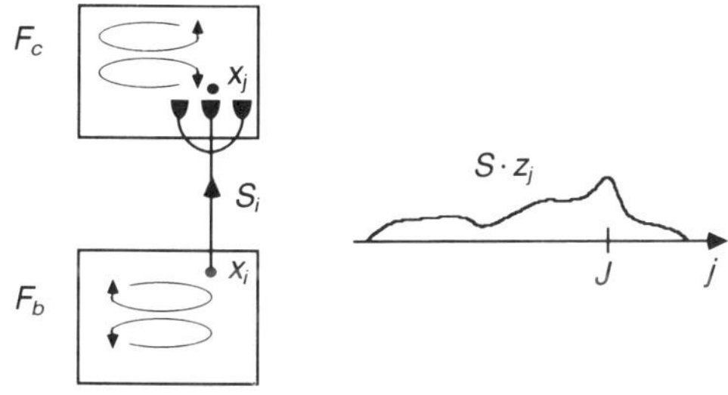

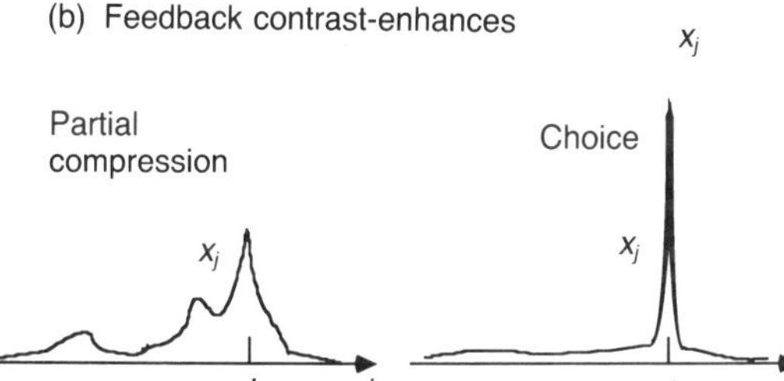

Figure 9.13 (a) If transmitter is fully accumulated at $t = 0$, low-amplitude postsynaptic STM activity x_j is initially proportional to the dot product of the signal vector S and the weight vector z_j. Fields are labeled F_b and F_c for consistency with the ART 3 system in Figure 9.21. (b) Intrafield feedback rapidly contrast-enhances the initial STM activity pattern. Large-amplitude activity is then concentrated at one or more nodes.

makes a neural network more realistic physiologically needs to be justified functionally, or it will obscure essential system dynamics. Why, then, add two additional nonlinearities to the portion of a neural network system responsible for transmitting signals from one location to another? The following discussion suggests how nonlinearities of synaptic transmission and neuromodulation can, when embedded in an ART circuit, help to correct coding errors by triggering a parallel search, allow the system to respond adaptively to reinforcement, and rapidly reset itself to changing input patterns.

In equation (9.10), term

$$u_{ij}S_if(x_j) \tag{9.15}$$

for the amount of transmitter released per unit time implies that the original incoming weighted signal $z_{ij}S_i$ is distorted both by depletion of the presynaptic transmitter u_{ij} and by the activity level x_j of the postsynaptic cell. If these two nonlinearities are significant, the net signal in the ijth pathway depends jointly on the maximal weighted signal $z_{ij}S_i$; the prior activity in the pathway, as reflected in the amount of depletion of the transmitter u_{ij}; and the immediate context in which the signal is sent, as reflected in the target cell activity x_j. In particular, once activity in a postsynaptic cell becomes large, this activity dominates the transmitter release rate, via the term $f(x_j)$ in (9.15). In other words, although linear filtering properties initially determine the small-amplitude activity pattern of the target field F_c, once intrafield feedback amplifies and contrast-enhances the postsynaptic activity x_j (Figure 9.13(b)), it plays a major role in determining the amount of released transmitter v_{ij} (Figure 9.14). In particular, the postsynaptic activity pattern across the field F_c that represents the recognition code (Figure 9.13(b)) is imparted to

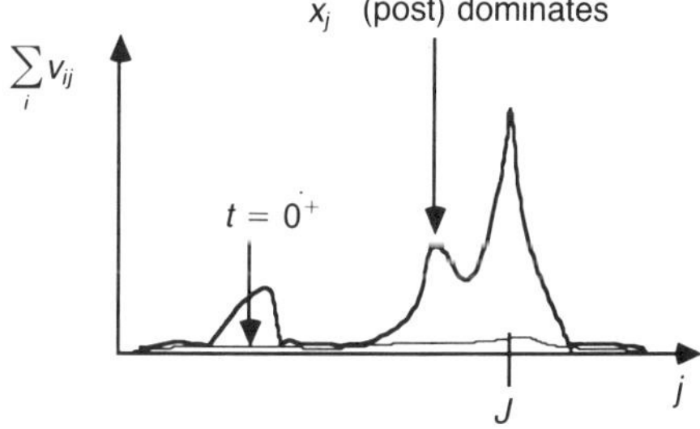

Figure 9.14 The ART search hypothesis 1 implies that large amounts of transmitter (v_{ij}) are released only adjacent to postsynaptic nodes with large-amplitude activity (x_j). Competition within the postsynaptic field therefore transforms the initial low-amplitude distributed pattern of released and bound transmitter into a large-amplitude contrast-enhanced pattern.

the pattern of released transmitter (Figure 9.14), which then also represents the recognition code, rather than the initial filtered pattern $S \cdot z_j$.

9.13 SYSTEM DYNAMICS DURING RESET: INACTIVATION OF BOUND TRANSMITTER CHANNELS

The dynamics of transmitter release implied by the ART search hypothesis 1 can be used to implement the reset process, by postulating the ART search hypothesis 2.

ART search hypothesis 2

The nonspecific reset signal quickly inactivates postsynaptic membrane channels at which transmitter is bound (Figure 9.15).

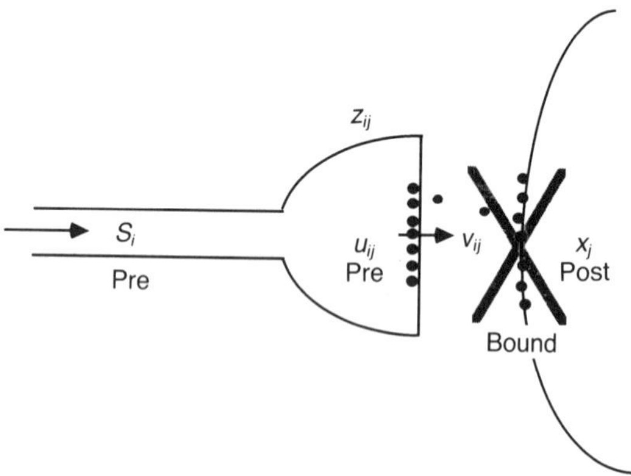

$$\frac{dv_{ij}}{dt} = -v_{ij} + u_{ij}\,[\text{release rate}] - v_{ij}\,[\text{inactivation rate}]$$

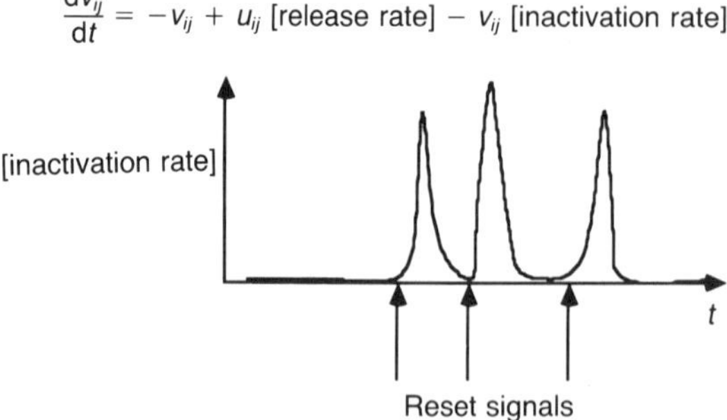

Figure 9.15 The ART search hypothesis 2 specifies a high rate of inactivation of bound transmitter following a reset signal. Postsynaptic action of the nonspecific reset signal is similar to that of a neuromodulator.

The reset signal in equations (9.5) and (9.6) may be interpreted as assignment of a large value to the inactivation rate in a manner analogous to the action of a neuromodulator (Figure 9.9). Inhibition of postsynaptic nodes breaks the strong intrafield feedback loops that implement ART 2 and ART 3 matching and contrast-enhancement (equation (9.3) or (9.6)).

Let us now examine system dynamics following transmitter inactivation. The pattern of released transmitter can be viewed as a representation of the postsynaptic recognition code. The arrival of a reset signal implies that some part of the system has judged this code to be erroneous, according to some criterion. The ART search hypothesis 1 implies that the largest concentrations of bound extracellular transmitter are adjacent to the nodes which most actively represent this erroneous code. The ART search hypothesis 2 therefore implies that the reset process selectively removes transmitter from pathways leading to the erroneous representation.

After the reset wave has acted, the system is biased against activation of the same nodes, or features, in the next time interval: whereas the transmitter signals pattern $S \cdot u_j$ originally sent to target nodes at times $t = 0^+$ was proportional to $S \cdot z_j$, as in equation (9.12), the transmitter signal pattern $S \cdot u_j$ after the reset

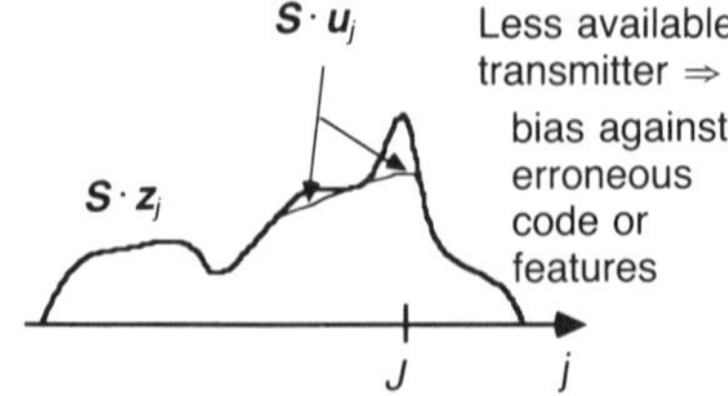

Figure 9.16 Following reset, the system is selectively biased against pathways that had previously released large quantities of transmitter. After a mismatch reset, therefore, the adaptive filter delivers a smaller signal to the previous category representation, the one that generated the reset signal.

event is no longer proportional to $S \cdot z_j$. Instead, it is selectively biased against those features that were previously active (Figure 9.16). The new signal pattern $S \cdot u_j$ will lead to selection of another contrast-enhanced representation, which may or may not then be reset. This search process continues until an acceptable match is found, possibly through the selection of a previously inactive representation.

9.14 PARAMETRIC ROBUSTNESS OF THE SEARCH PROCESS

This search process is relatively easy to implement, requiring no new nodes or pathways beyond those already present in ART 2 modules. It is also robust, since it does not require tricky timing or calibration. How the process copes with a typical slow learning situation is illustrated in Figure 9.17. With slow learning, an input can select and begin to train a new category, so that the adaptive weights correspond to a perfect pattern match during learning. However, the input may not be on long enough for the adaptive weights to become very large. That input may later activate a different category node whose weights are large but whose vector of adaptive weights forms a poorer match than the original, smaller weights.

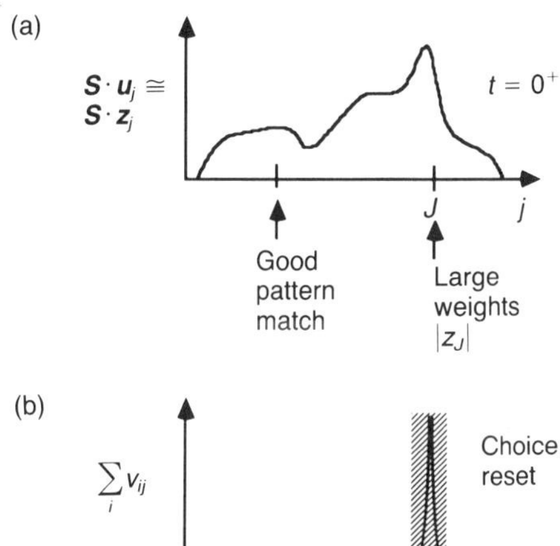

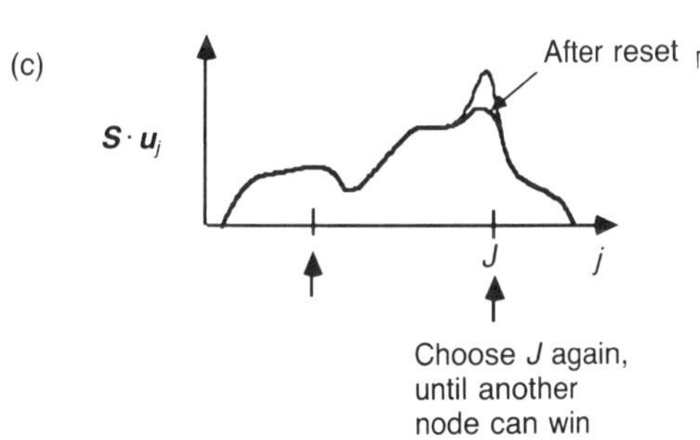

Figure 9.17 An erroneous category representation with large weights (z_{iJ}) may become active before another representation that makes a good pattern match with the input but which has small weights. One or more mismatch reset events can decrease the functional value (u_{iJ}) of the larger weights, allowing the 'correct' category to become active.

Figure 9.17 (a) shows such a typical filtered signal pattern $S \cdot z_j$. During the initial processing interval ($t = 0^+$) the transmitted signal $S \cdot u_j$ and the postsynaptic activity x_j are proportional to $S \cdot z_j$. Suppose that the weights z_{iJ} in pathways leading to the Jth node are large, but that the vector pattern z_J is not an adequate match for the signal pattern S according to the vigilance criterion. Also suppose that dynamics in the target field F_c lead to a 'choice' following competitive contrast-enhancement (Figure 9.17(b)), and that the chosen node J represents a category. Large amounts of transmitter will thus be released from synapses adjacent to node J, but not from synapses adjacent to other nodes. The reset signal will then selectively inactivate transmitter at postsynaptic sites adjacent to the Jth node. Following such a reset wave, the new signal pattern $S \cdot u_j$ will be biased against the Jth node relative to the original pattern. However, it could happen that the time interval prior to the reset signal is so brief that only a small fraction of available transmitter is released. Then $S \cdot u_J$ could still be large relative to a 'correct' $S \cdot u_j$ after reset occurs (Figure 9.17(c)). If this were to occur, the Jth node would simply be chosen again, then reset again, leading to an accumulating bias against that choice in the next time interval. This process could continue until enough transmitter v_{ij} is inactivated to allow another node, with smaller weights z_{ij} but a better pattern match, to win the competition. Simulations of such a reset sequence are illustrated in Figures 9.23–9.26.

9.15 SUMMARY OF SYSTEM DYNAMICS DURING A MISMATCH-RESET CYCLE

Figure 9.18 summarizes system dynamics of the ART search model during a single input presentation. Initially the transmitted signal pattern $S \cdot \boldsymbol{u}_j$, as well as the postsynaptic activity x_j, are proportional to the weighted signal pattern $S \cdot z_j$ of the linear filter. The postsynaptic activity pattern is then contrast-enhanced, due to the internal competitive dynamics of the target field. The ART search hypothesis 1 implies that the transmitter release rate is greatly amplified in proportion to the level of postsynaptic activity. A subsequent reset signal selectively inactivates transmitter in those pathways that caused an error. Following the reset wave, the new signal $S \cdot \boldsymbol{u}_j$ is no longer proportional to $S \cdot \boldsymbol{z}_j$ but is, rather, biased against the previously active representation. A series of reset events ensue, until an adequate match or a new category is found. Learning occurs on a time-scale that is long relative to that of the search process.

9.16 AUTOMATIC STM RESET BY REAL-TIME INPUT SEQUENCES

The ART 3 architecture serves other functions as well as implementing the mismatch-reset-search cycle. In particular it allows an ART system to dispense with additional processes to reset STM at onset or offset of an input pattern. The representation of input patterns as a sequence, $\boldsymbol{I}_1, \boldsymbol{I}_2, \boldsymbol{I}_3, \ldots$, corresponds to the assumption that each input is constant for a fixed time interval. In practice, an input vector $\boldsymbol{I}(t)$ may vary continuously through time. The input need never be constant over an interval, and there may be no temporal marker to signal offset or onset of an 'input pattern' *per se*. Furthermore, feedback loops within a field or between two fields can maintain large amplitude activity even when $\boldsymbol{I}(t) = 0$. Adaptive resonance develops only when activity patterns across fields are amplified by such feedback loops and remain stable for a sufficiently long time to enable adaptive weight changes to occur (Grossberg 1976*b*, 1982*a*). In particular, no reset waves are triggered during a resonant event.

The ART reset system functionally defines the onset of a 'new' input as a time when the orienting subsystem emits a reset wave. This occurs, for example, in the ART 2 module (Figure 9.3) when the angle between the vectors $\boldsymbol{I}(t)$ and $\boldsymbol{p}(t)$ becomes so large that the norm of $\boldsymbol{r}(t)$ falls below the vigilance level $\rho(t)$, thereby triggering a search for a new category representation. This is called an input reset event, to distinguish it from a mismatch reset event, which occurs while the bottom-up input remains nearly constant over a time interval but mismatches the top-down expectation that it has elicited from level F_2 (Figure 9.2).

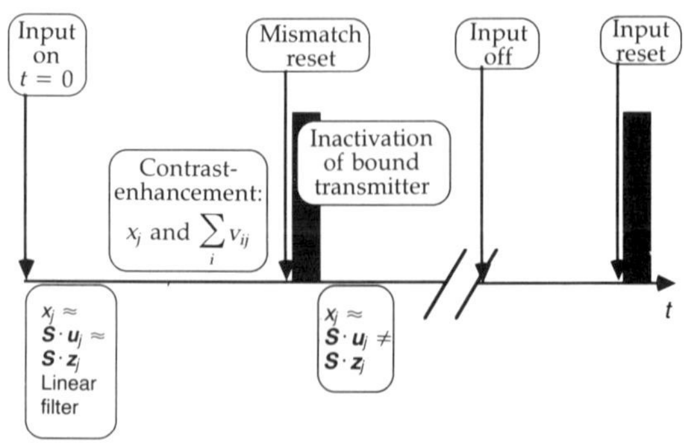

Figure 9.18 ART search hypotheses 1 and 2 implement computations to carry out search in an ART system. Input reset employs the same mechanisms as mismatch reset, initiating search when the input pattern changes significantly.

This property obviates the need to mechanistically define the processing of input onset or offset. The ART search hypothesis 3, which postulates restoration of a resting state between successive inputs (Carpenter and Grossberg 1989), is thus not needed. Presynaptic transmitter may not be fully accumulated following an input reset event, just as it is not fully accumulated following a mismatch reset event. For both types of reset, the orienting subsystem judges the active code to be incorrect, at the present level of vigilance, and the system continues to search until it finds an acceptable representation.

9.17 REINFORCEMENT FEEDBACK

The mechanisms described thus far for STM reset are part of the recognition learning circuit of ART 3. Recognition learning is, however, only one of several processes whereby an intelligent system can learn a correct solution to a problem. We have called recognition, reinforcement and recall the '3 Rs' of neural network learning (Carpenter and Grossberg 1988).

Reinforcement, notably reward and punishment, provides additional information in the form of environmental feedback based on the success or failure of actions triggered by a recognition event. Reward and punishment calibrate whether the action has or has not satisfied internal needs, which in the biological case include hunger, thirst, sex and pain reduction, but may in machine applications include a wide variety of internal cost functions.

Reinforcement can shift attention to focus upon those recognition codes whose activation promises to satisfy internal needs based on past experience. A model to describe this aspect of reinforcement learning was described in Grossberg (1982*a*, 1982*b*, 1984; reprinted in Grossberg 1987*a*) and was supported by computer simulations in Grossberg and Levine (1987; reprinted in Grossberg 1988). An attention shift due to reinforcement

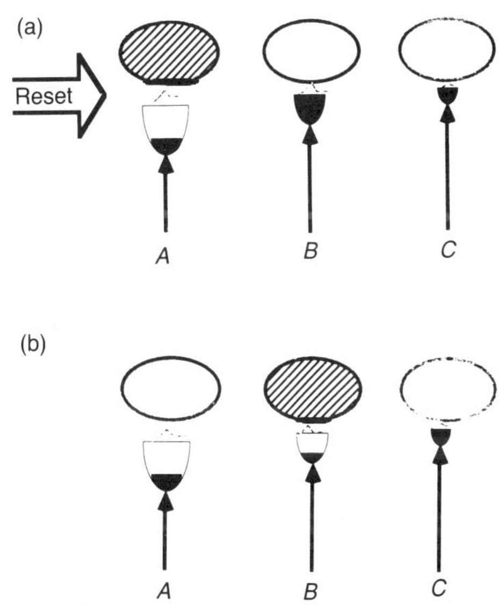

Figure 9.19 A system whose weights are biased towards feature *A* over feature *B* over feature *C*. (a) Competition amplifies the weight bias in STM, leading to enhanced transmitter release of the selected feature *A*. (b) Transmitter inactivation following reinforcement reset allows feature *B* to become active in STM.

can also alter the structure and learning of recognition codes by amplifying (or suppressing) the STM activations, and hence the adjacent adaptive weights, of feature detectors that are active during positive (or negative) reinforcement.

A reset wave may also be used to modify the pattern of STM activation in response to reinforcement. For example, both green and yellow bananas may be recognized as part of a single recognition category until reinforcement signals, contingent upon eating the bananas, differentiate them into separate categories. Within ART 3, such a reinforcement signal can alter the course of recognition learning by causing a reset event. The reset event may override a bias in either the learned path weights (Figure 9.19) or in the input strengths (Figure 9.20) that could otherwise

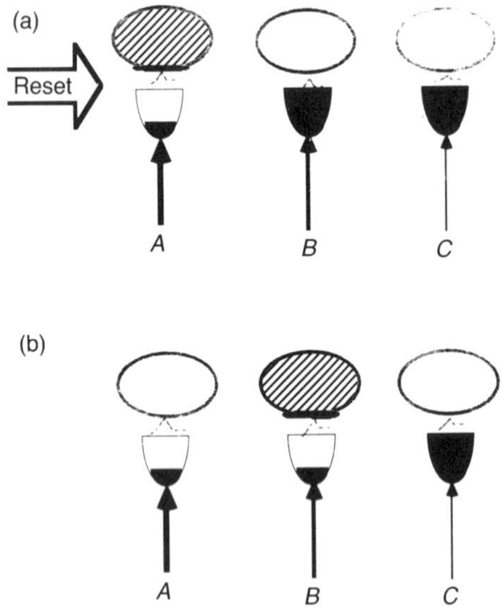

Figure 9.20 A system whose input signals are biased towards *A* over *B* over *C*. (a) Competition amplifies the input bias in STM, leading to enhanced transmitter release of the selected feature *A*. (b) Transmitter inactivation following reinforcement reset allows feature *B* to become active in STM.

prevent a correct classification from being learned. For example, both green and yellow bananas may initially be coded in the same recognition category because features that code object shape (e.g., pathway A in Figures 9.19 and 9.20) prevent features that code object color (e.g., pathway B in Figures 9.19 and 9.20) from being processed in STM. Reset waves triggered by reinforcement feedback can progressively weaken the STM activities of these shape features until both shape and color features can simultaneously be processed, and thereby incorporated into different recognition codes for green bananas and yellow bananas.

In technological applications, such a reset wave can be implemented as a direct signal from an internal representation of a punishing event. The effect of the reset wave is to modify the spatial pattern of STM activation whose read-out into an overt action led to the punishing event. The adaptive weights, or LTM traces, that input to these STM activations are then indirectly altered by an amount that reflects the new STM activation pattern. Such a reinforcement scheme differs from the competitive learning scheme described by Kohonen (1984, p. 200), in which reinforcement acts directly, and by an equal amount, on all adaptive weights that lead to an incorrect classification.

Reinforcement may also act by changing the level of vigilance (Carpenter and Grossberg 1987*a*, 1987*b*). For example, if a punishing event increases the vigilance parameter, then mismatches that were tolerated before will lead to a search for another recognition code. Such a code can help to distinguish pattern differences that were previously considered too small to be significant. Such a role for reinforcement is illustrated by computer simulations in Figures 9.25–9.28.

All three types of reaction to reinforcement feedback may be useful in applications. The change in vigilance alters the overall sensitivity of the system to pattern differences. The shift in attention and the reset of active features can help to overcome prior coding biases that may be maladaptive in novel contexts.

9.18 NOTATION FOR HIERARCHIES

Table 9.1 and Figure 9.21 illustrate notation suitable for an ART hierarchy with any number of fields F_a, F_b, F_c, . . . This notation can also be adapted for related neural networks and algorithmic computer simulation.

Each STM variable is indexed by its field, layer, and node number. Within a layer, x denotes the activity of a node receiving inputs from other layers, while y denotes the (normalized) activity of a node that sends signals to other layers. For example, x_i^{a2} denotes activity at the ith input node in layer 2 of field $F_a (i = 1, \ldots, n_a)$; and y_i^{a2} denotes activity of

Table 9.1 Notation for ART 3 hierarchy

$F_{\text{field}} = F_a$	STM field a
$i = i_a = 1, \ldots, n_a$	node index, field a
$L = 1, 2, 3$	index, 3 layers of an STM field
x_i^{aL}	STM activity, input node i, layer L, field a
y_i^{aL}	STM activity, output node i, layer L, field a
$g^a(y_i^{aL}) \equiv S_i^{aL}$	signal function, field a
p_k^a	parameter, field a, $k = 1, 2, \ldots$
r_i^b	STM activity, reset node i, field b
ρ^b	vigilance parameter, field b
z_{ij}^{bc}	LTM trace, pathway from node i (field b) to node j (field c)
u_{ij}^{bc}	intracellular transmitter, pathway from node i (field b) to node j (field c)
v_{ij}^{bc}	released transmitter, pathway from node i (field b) to node j (field c)

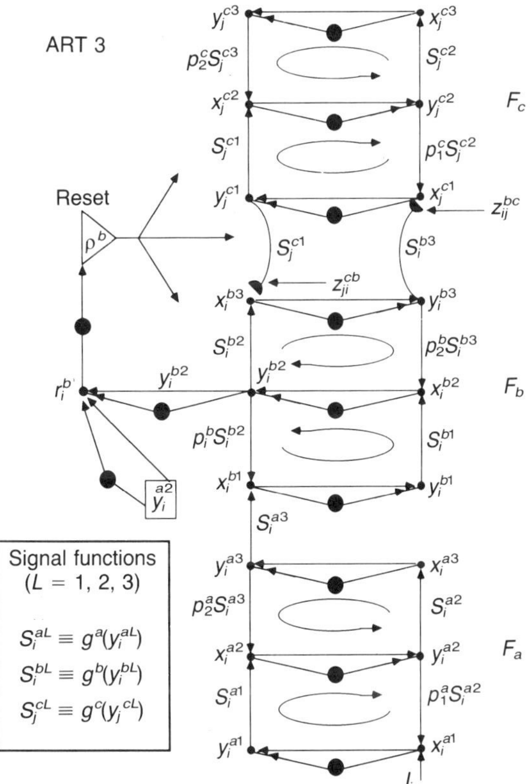

Figure 9.21 ART 3 simulation neural network. Indices $i = 1, \ldots, n_a = n_b$ and $j = 1, \ldots, n_c$. The reset signal acts at all layers 1 and 3 (Section 9.8).

the corresponding output node. Parameters are also indexed by field (p_1^a, p_2^a, ...), as are signal functions (g^a). Variable r_i^b denotes activity of the ith reset node of field F_b, and ρ^b is the corresponding vigilance parameter.

Variable z denotes an adaptive weight or LTM trace. For example, z_{ij}^{bc} is the weight in the bottom-up pathway from the ith node of field F_b to the jth node of field F_c. Variables u_{ij}^{bc} and v_{ij}^{bc} denote the corresponding presynaptic and bound transmitter quantities, respectively. Variables for the top-down pathways are z_{ji}^{cb}, u_{ji}^{cb}, and v_{ji}^{cb}.

Complete simulation equations are specified in Section 9.26.

9.19 TRADE-OFF BETWEEN WEIGHT SIZE AND PATTERN MATCH

The simulations in Sections 9.20–9.24 illustrate the dynamics of search in the ART 3 system shown in Figure 9.21. The simulation timescale is assumed to be short relative to the timescale of learning, so all adaptive weights z_{ij}^{bc} and z_{ji}^{cb} are held constant. The weights are chosen, however, to illustrate a problem that can arise with slow learning or in any other situation in which weight vectors are not normalized at all times, namely, a category whose weight vector only partially matches the input vector may become active because its weights are large. This can prevent initial selection of another category whose weight vector matches the input vector but whose weight magnitudes are small due, say, to a brief prior learning interval.

The search process allows the ART 3 system to reject an initial selection with large weights and partial pattern match, and then to activate a category with smaller weights and a better pattern match. As in ART 2, when weights are very small (nodes $j = 6, 7, \ldots$, Figure

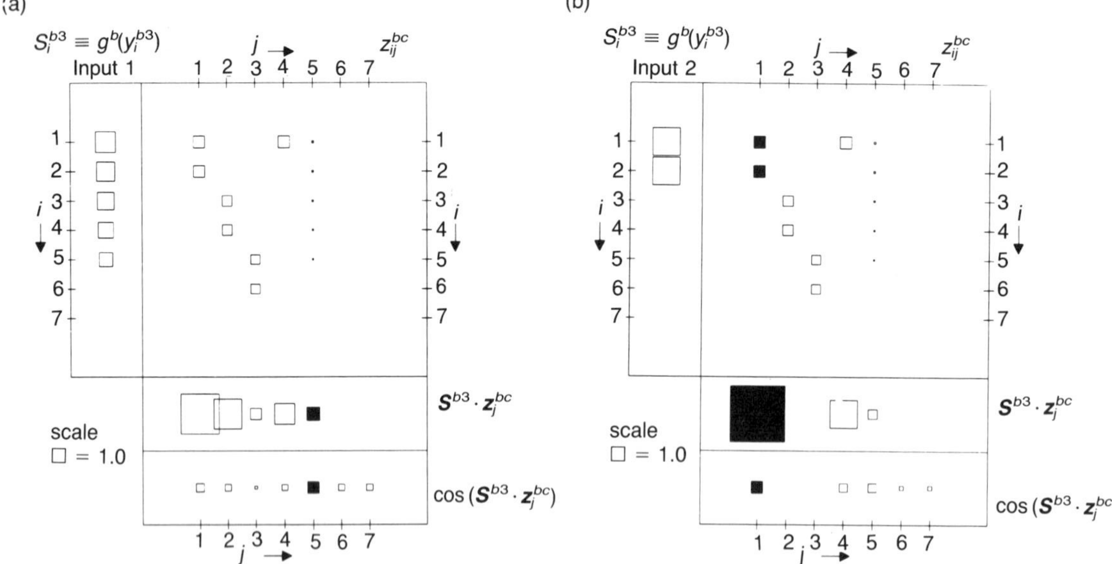

Figure 9.22 The length of a side of the square centered at position i or j or (i,j) gives the value of a variable with the corresponding index. Shown are quantities S_i^{b3}, z_{ij}^{bc}, $S^{b3}\cdot z_j^{bc}$, and $\cos(S^{b3}, z_j^{bc})$. (a) Vector S^{b3} is the signal response to Input 1 in the simulations. Vector z_5^{bc} (filled squares) is parallel to S^{b3}, but $\|z_5^{bc}\|$ is small. Thus $S^{b3}\cdot z_5^{bc}$ is smaller than $S^{b3}\cdot z_j^{bc}$ for $j = 1, 2$, and 4, despite the fact that $\cos(S^{b3}, z_5^{bc})$ is maximal. (b) Vector S^{b3} is the signal response to Input 2 in the simulations. Vector z_1^{bc} (filled squares) is parallel to S^{b3}.

9.22) the ART system tolerates poor pattern matches in order to allow new categories to become established. During learning, the weights can become larger. The larger the weights, the more sensitive the ART system is to pattern mismatch (Carpenter and Grossberg 1987*b*).

Figure 9.22 illustrates the trade-off between weight size and pattern match in the system used in the simulations. In Figures 9.22(a) and 9.22(b), vector $\boldsymbol{S}$ illustrates the STM pattern stored in F_a and sent from F_a to F_b when an input vector $\boldsymbol{I}$ is held constant. The S_i values were obtained by presenting to F_a an input function $\boldsymbol{I}$ with I_i a linearly decreasing function of i. Vector $\boldsymbol{S}$ is also stored in F_b, as long as F_c remains inactive. Initially $\boldsymbol{S}$ is the signal vector in the bottom-up pathways from F_b to F_c. In Figure 9.22(a), $S_1 > S_2 > \ldots > S_5$; for $i = 6, 7, \ldots, 15$ $(= n_a = n_b)$, S_i is small. Each vector z_1, z_2, z_3, and z_4, plotted in columns within the square region of Figure 9.22(a), partially matches the signal vector $\boldsymbol{S}$. These weights are significantly larger than the weights of vector z_5. However z_5 is a perfect match to $\boldsymbol{S}$ in the sense that the angle between the two vectors is 0:

$$\cos(S, z_5) = 1. \tag{9.16}$$

The relationship

$$S \cdot z_j = \|S\|\|z_j\|\cos(S, z_j) \tag{9.17}$$

implies a trade-off between weight size, as measured by the length $\|z_j\|$ of z_j, and pattern match, as measured by the angle between $\boldsymbol{S}$ and z_j. If the initial signal from F_b to F_c is proportional to $S\cdot z_j$, as in (9.13), then the matched node ($j = 5$) may receive a net signal that is smaller than signals to other nodes. In fact, in Figure 9.22(a),

$$S \cdot z_1 > S \cdot z_2 > S \cdot z_4 > S \cdot z_5 > \ldots \tag{9.18}$$

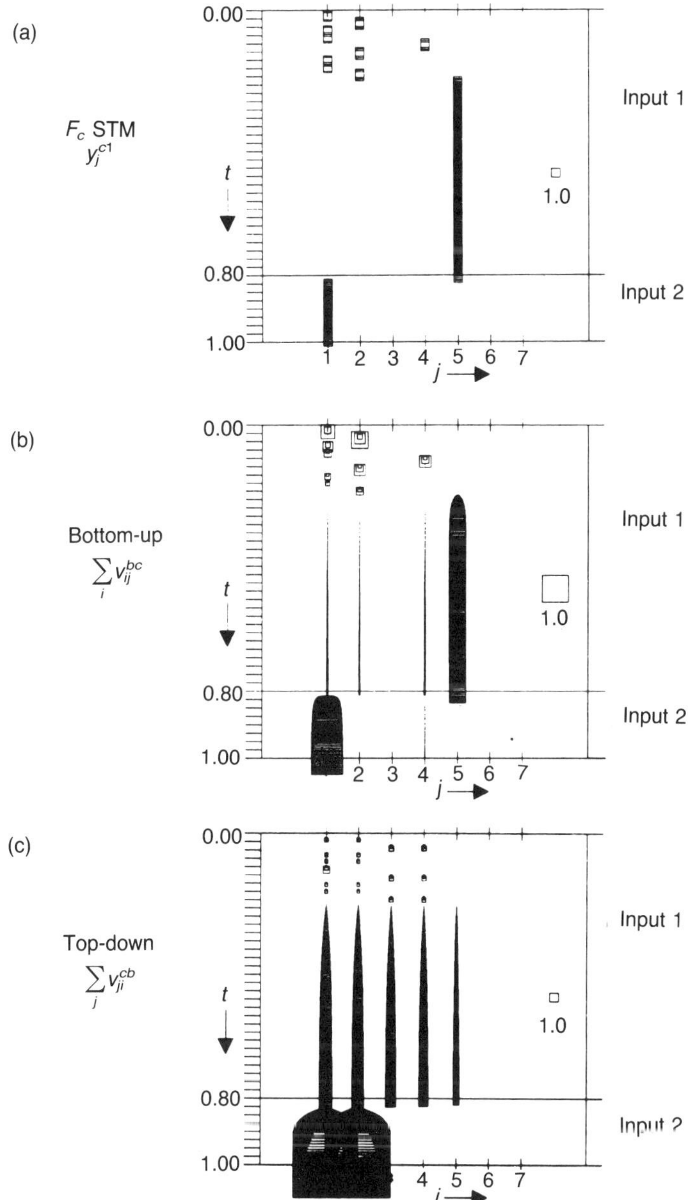

Figure 9.23 ART 3 simulation with $\rho \equiv 0.98$. A series of nine mismatch resets lead to activation of the matched category ($j = 5$) at $t = 0.215$. Input 1 switches to Input 2 at $t = 0.8$ causing an input reset and activation of a new category representation ($j = 1$).

Figure 9.22(b) shows a signal vector S that is parallel to the weight vector z_1.

9.20 ART 3 SIMULATIONS: MISMATCH RESET AND INPUT RESET OF STM CHOICES

The computer simulations summarized in Figures 9.23–9.26 use the inputs described in Figure 9.22 to illustrate the search process in an ART 3 system. In these simulations, the F_c competition parameters were chosen to make a choice; hence, only the node receiving the largest filtered input from F_b is stored in STM. The signal function of F_c caused the STM field to make a choice. In Figure 9.27 a different signal function at F_c, similar to the one used in F_a and F_b, illustrates how the search process reorganizes a distributed recognition code. The simulations show how, with high vigilance, the ART search process rapidly causes a series of mismatch resets that alter the transmitter vectors $\boldsymbol{u}_1$, $\boldsymbol{u}_2$, . . . until $\boldsymbol{S} \cdot \boldsymbol{u}_5$ becomes maximal. Once node $j = 5$ becomes active in STM it amplifies transmitter release. Since the pattern match is perfect, no further reset occurs while Input 1 (Figure 9.22(a)) remains on. Input reset is illustrated following an abrupt or gradual switch to Input 2 (Figure 9.22(b)).

Each simulation figure illustrates three system variables as they evolve through time. The time axis (t) runs from the top to the bottom of the square. A vector pattern, indexed by i or j, is plotted horizontally at each fixed time. Within each square, the value of a variable at each time is represented by the length of a side of a square centered at that point. In each figure, part (a) plots y_j^{c1}, the normalized STM variables at layer 1 of field F_c. Part (b) plots $\Sigma_i v_{ij}^{bc}$, the total amount of transmitter released, bottom-up, in paths from all F_b nodes to the jth F_c node. Part (c) plots $\Sigma_j v_{ji}^{cb}$, the total amount of transmitter released, top-down, in paths from all F_c nodes to the ith F_b node. The ART search hypothesis 1 implies that the net bottom-up transmitter pattern in part (b) reflects the STM pattern of F_c in part (a); and that the net top-down transmitter pattern in part (c) reflects the STM pattern of F_b.

In Figure 9.23, the vigilance parameter is high and fixed at the value

$$\rho \equiv 0.98. \tag{9.19}$$

For $0 \leqslant t < 0.8$, the input (Figure 9.22(a)) is constant. The high vigilance level induces a sequence of mismatch resets, alternating among the category nodes $j = 1$, 2, and 4 (Figure 9.23(a)), each of which receives an initial input larger than the input to node $j = 5$ (Figure 9.22(a)). At $t = 0.215$, the F_c node $j = 5$ is selected by the search process (Figure 9.23(a)). It remains active until $t = 0.8$. Then, the input from F_a is changed to a new pattern (Figure 9.22(b)). The mismatch between the new STM pattern at F_a and the old reverberating STM pattern at F_b leads to an input reset (Figures 9.18 and 9.23). The ART search hypothesis 2 implies that bound transmitter is inactivated and the STM feedback loops in F_b and F_c are thereby inhibited. The new input pattern immediately activates its category node $j = 1$, despite some previous depletion at that node (Figure 9.23(a)).

Large quantities of transmitter are released and bound only after STM resonance is established. In Figure 9.23(b)), large quantities of bottom-up transmitter are released at the F_c node $j = 5$ in the time interval $0.215 < t < 0.8$, and at node $j = 1$ in the time interval $0.8 < t < 1$. In Figure 9.23(c), the pattern of top-down bound transmitter reflects the resonating matched STM pattern at F_b due to Input 1 at times $0.215 < t < 0.8$ and due to Input 2 at times $0.8 < t < 1$.

9.21 SEARCH TIME INVARIANCE AT DIFFERENT VIGILANCE VALUES

Figure 9.24 shows the dynamics of the same system as in Figure 9.23 but at the lower vigilance value

$$\rho \equiv 0.94. \tag{9.20}$$

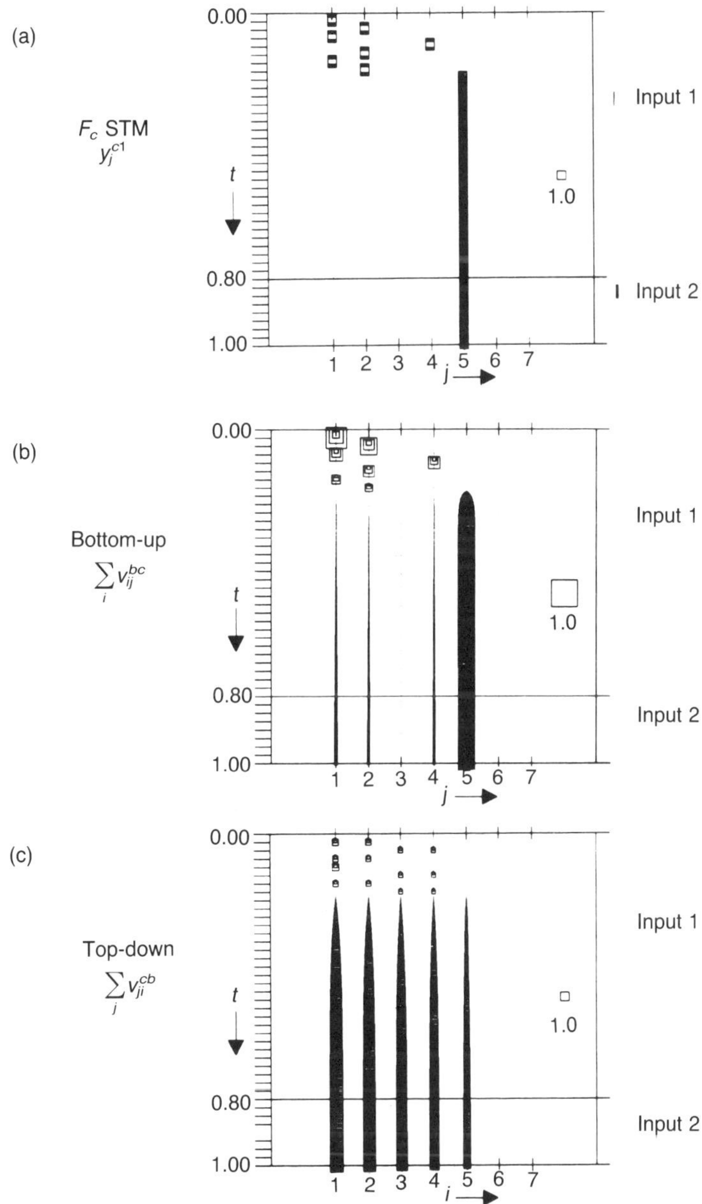

Figure 9.24 ART 3 simulation with ρ = 0.94. A series of seven mismatch resets leads to activation of the matched category ($j = 5$) at $t = 0.19$. Input 1 switches to Input 2 at $t = 0.8$, but no input reset occurs, and node $j = 5$ remains active, due to the lower vigilance level than in Figure 9.23.

The F_c node $j = 5$ becomes active slightly sooner ($t = 0.19$, Figure 9.24(a)) than it does in Figure 9.23(a), where ρ = 0.98. At a lower vigilance, more transmitter needs to be released before the system reacts to a mismatch so that each 'erroneous' category node is

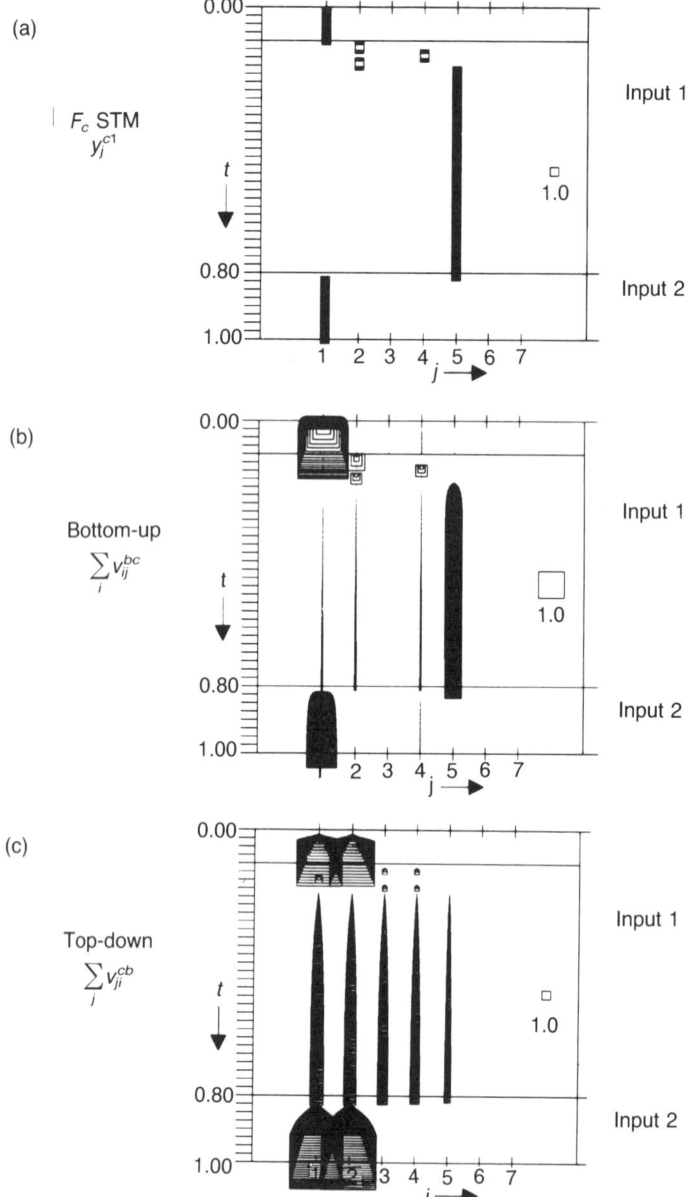

Figure 9.25 ART 3 simulation with $\rho = 0.9$ $(0 \leqslant t < 0.1)$ and $\rho = 0.98$ $(0.1 < t \leqslant 1)$. At low vigilance, activation of node $j = 1$ leads to resonance. When vigilance is suddenly increased due, say, to reinforcement feedback, a series of four mismatch resets leads to activation of the matched category $(j = 5)$ at $t = 0.19$. As in Figure 9.23, switching to Input 2 at $t = 0.8$ causes an input reset and activation of a new category representation $(j = 1)$.

active for a longer time interval than at higher vigilance. When $\rho = 0.98$ (Figure 9.23(b)), node $j = 1$ is searched 5 times. When $\rho = 0.94$ (Figure 9.24(b)), node $j = 1$ is searched only three times, but more transmitter is released during each activation/reset cycle than at comparable points in Figure 9.23(b). Inactivation of this extra released transmitter approximately balances the longer times to reset. Hence the total search time remains approximately constant over a wide range of vigilance parameters. In the present instance, the nonlinearities of transmitter release terminate the search slightly sooner at lower vigilance.

Figure 9.24(a) illustrates another effect of lower vigilance: the system's ability to tolerate larger mismatches without causing a reset. When the input changes at $t = 0.8$, the mismatch between the input pattern at F_a and the resonating pattern at F_b is not great enough to cause an input reset. Despite bottom-up input only to nodes $i = 1, 2$, the strong resonating pattern at nodes $i = 1, \ldots, 5$ maintains itself in STM at F_b (Figure 9.24(c)).

9.22 REINFORCEMENT RESET

In Figure 9.25 vigilance is initially set at value

$$\rho = 0.9, \tag{9.21}$$

in the time interval $0 < t < 0.1$. At this low vigilance level, the STM pattern of F_b does not experience a mismatch reset series. Node $j = 1$ is chosen and resonance immediately ensues (Figure 9.25(a)), as is also reflected in the amplification of transmitter release (Figure 9.25(b)). The simulation illustrates a case where this choice of category leads to external consequences, including reinforcement (Section 9.17), that feed back to the ART 3 module. This reinforcement teaching signal is assumed to cause vigilance to increase to the value

$$\rho = 0.98 \tag{9.22}$$

for times $t \geqslant 0.1$. This change triggers a search that ends at node $j = 5$, at time $t = 0.19$. Note that, as in Figure 9.24, enhanced depletion of transmitter at $j = 1$ shortens the total search time. In Figure 9.23, where ρ also equals 0.98, the search interval has length 0.215; in Figure 9.25, the search interval has length 0.09, and the system never again activates node $j = 1$ during search.

9.23 INPUT HYSTERESIS SIMULATION

The simulation illustrated in Figure 9.26 is nearly the same as in Figure 9.25, with $\rho = 0.9$ for $0 \leqslant t < 0.1$ and $\rho = 0.98$ for $t > 0.1$. However, at $t = 0.8$, Input 1 starts to be slowly deformed into Input 2, rather than being suddenly switched, as in Figure 9.25. The $F_a \rightarrow F_b$ input vector becomes a convex combination of Input 1 and Input 2 that starts as Input 1 ($t \leqslant 0.8$) and is linearly shifted to Input 2 ($t \geqslant 1.7$). Despite the gradually shifting input, node $j = 5$ remains active until $t = 1.28$. Then an input reset immediately leads to activation of node $j = 1$, whose weight vector matches Input 2. Competition in the category representation field F_c causes a history-dependent choice of one category or the other, not a convex combination of the two.

9.24 DISTRIBUTED CODE SIMULATION

Issues of learning and code interpretation are subtle and complex when a code is distributed. However, the ART 3 search mechanism translates immediately into this context. The simulation in Figure 9.27 illustrates how search operates on a distributed code. The only difference between the ART 3 system used for these simulations and the one used for Figures 9.23–9.26 is in the signal function at F_c. In Figures 9.23–9.26, a choice is always made at field F_c. The signal function for Figure 9.26 is, like that at F_a and F_b, piecewise linear: 0 below a threshold, linear above. With its fairly high threshold, this signal function compresses the input pattern; but the compression is not

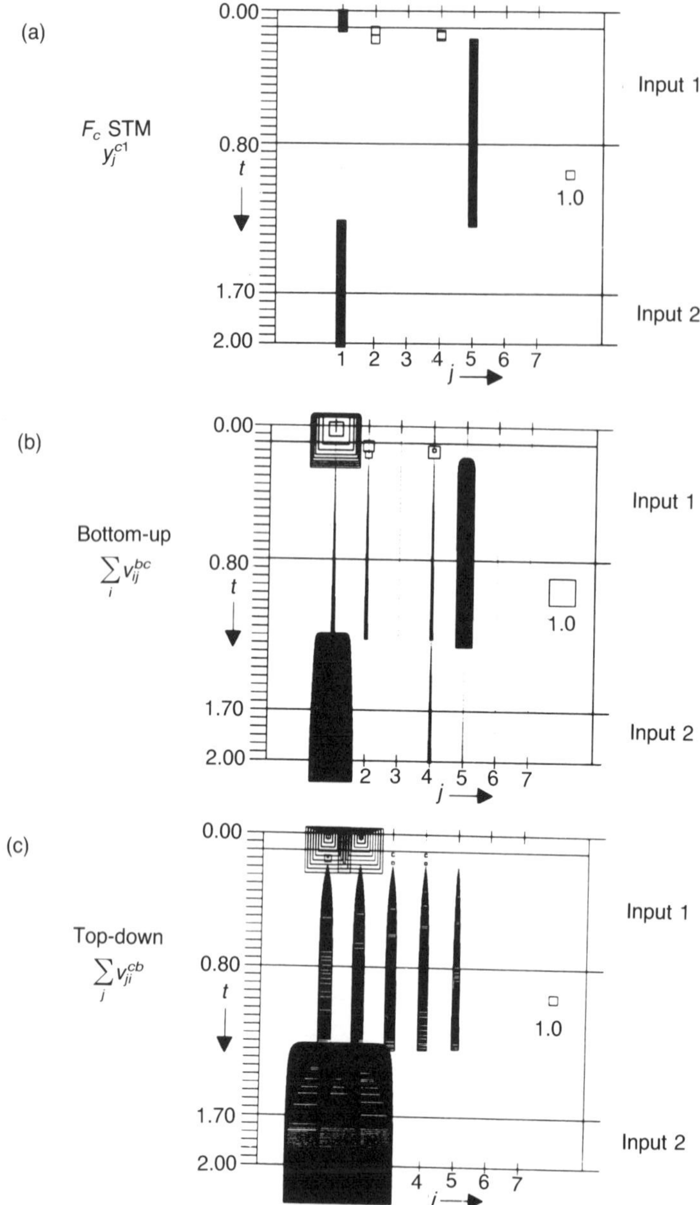

Figure 9.26 ART 3 simulation with $\rho = 0.9$ ($0 \leq t < 0.1$) and $\rho = 0.98$ ($1 < t \leq 2$). Input 1 is presented for $0 \leq t < 0.8$. For $0.8 < t < 1.7$ the input to F_b is a convex combination of Input 1 and Input 2. Then Input 2 is presented for $1.7 < t \leq 2$. At $t = 1.28$ an input reset causes the STM choice to switch from node $j = 5$, which matches Input 1, to node $j = 1$, which matches Input 2.

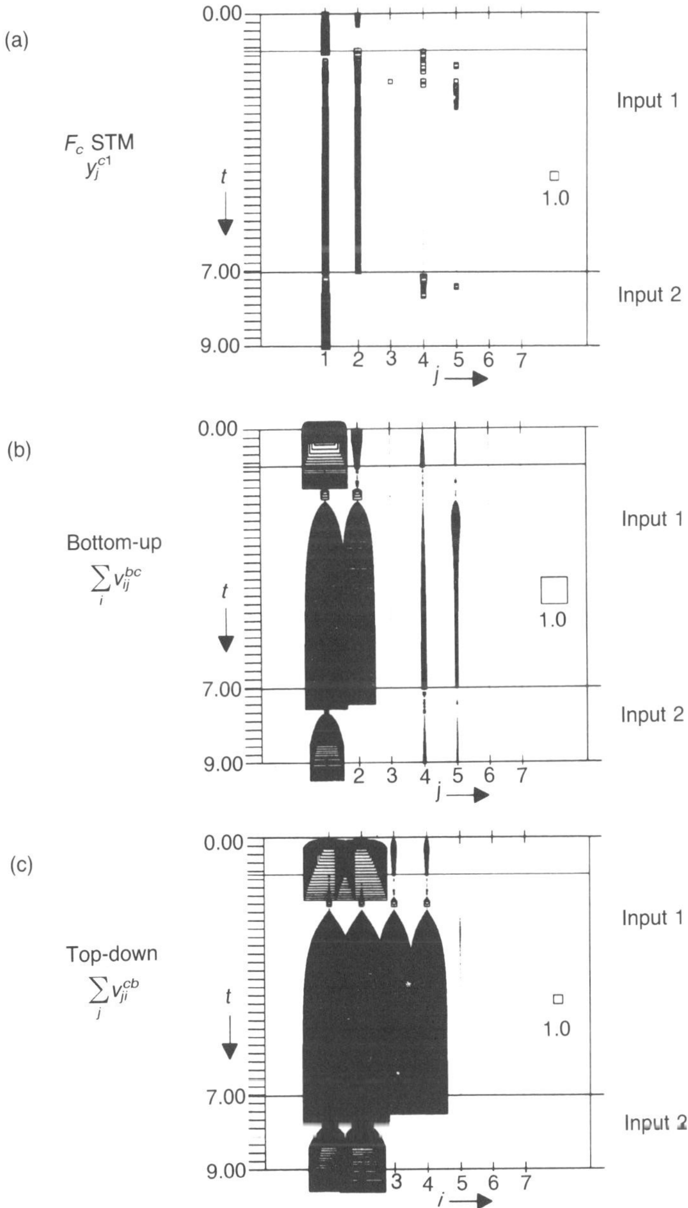

Figure 9.27 ART 3 simulation of a distributed code. Parameter $\rho = 0.9$ for $0 \leqslant t < 1$, and $\rho = 0.98$ for $1 < t \leqslant 9$. Input 1 is presented for $0 \leqslant t < 7$ and Input 2 is presented for $7 < t \leqslant 9$. At resonance, the single node $j = 1$ is active for $t < 1$ and $t > 7.7$. For $2.6 < t < 7$, simultaneous activity at two nodes ($j = 1$ and $j = 2$) represents the category. Top-down weights z_{ji}^{cb} are large for $j = 1$ and $i = 1$ and 2; and for $j = 2$ and $i = 3$ and 4. Together the top-down signals (c) match enough of the bottom-up input pattern to satisfy the vigilance criterion.

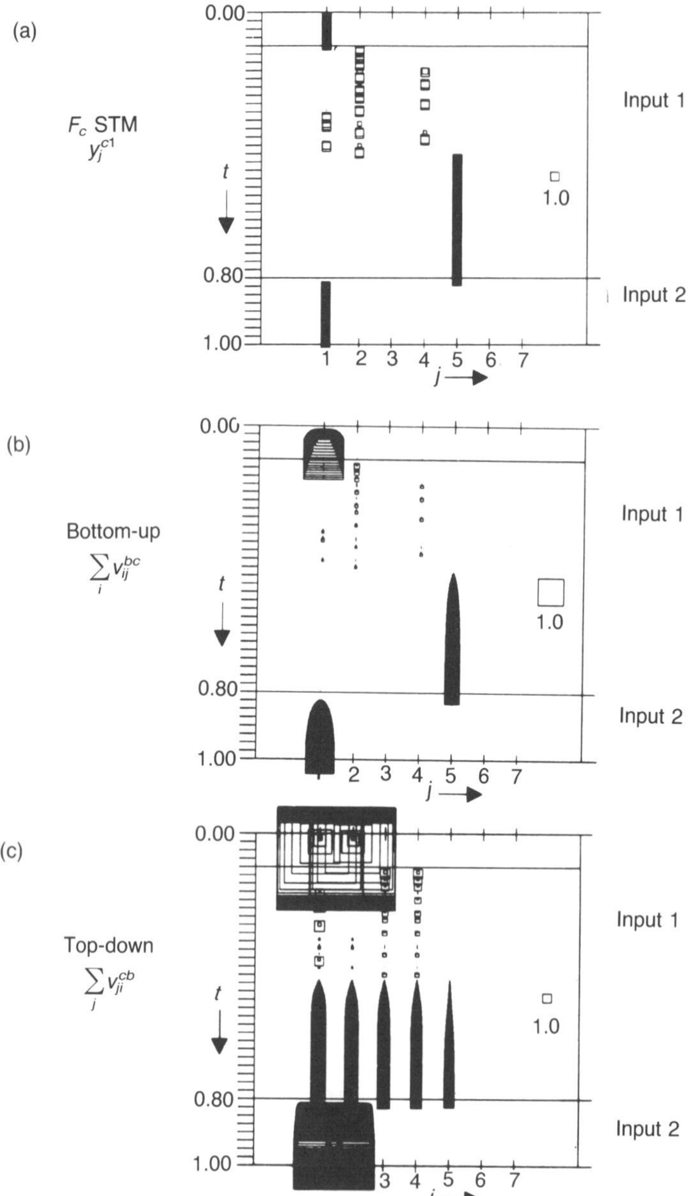

Figure 9.28 An alternative ART 3 model, which allows hyperpolarization of x_j^{c1}, gives results similar to those illustrated in Figure 9.25. As in Figure 9.25, $\rho = 0.9$ for $0 \leqslant t < 0.1$, $\rho = 0.98$ for $t > 0.1$, and Input 1 switches to Input 2 at $t = 0.8$. Category node $j = 5$ becomes active at $t = 0.44$, but immediately switches to node $j = 5$ at $t = 0.8$, when Input 2 is presented.

so extreme as to lead inevitably to choice in STM.

Distributed code STM activity is shown in Figure 9.27(a). At a given time more than one active node may represent a category ($2.6 < t < 7$), or one node may be chosen ($7.7 < t \leqslant 9$).

9.25 ALTERNATIVE ART 3 MODEL SIMULATION

ART 3 systems satisfy the small number of design constraints described above. In addition ART 3 satisfies the ART 2 stability constraints (Carpenter and Grossberg 1987*b*). For example, top-down signals need to be an order of magnitude larger than bottom-up signals, all other things being equal, as illustrated below by equation (9.24) and parameters p_1 and p_2 in Table 9.4 and equations (9.31) and (9.34). At least some of the STM fields need to be competitive networks. However, many versions of the ART systems exist within these boundaries. A simulation of one such system is illustrated in Figure 9.28, which duplicates the conditions on ρ and input patterns of Figure 9.25. However, the system that generated Figure 9.28 uses a different version of the ART 3 STM field F_c than the one described in Section 9.26. In particular, in the STM equation (3), $B > 0$. STM nodes can thus be hyperpolarized, so that $x_j < 0$, by intrafield inhibitory inputs. The transmitter release function $f(x_j)$ (equation (9.9)) equals 0 when x_j is sufficiently hyperpolarized. The system of Figure 9.28 thus has the property that transmitter release can be terminated at nodes that become inactive during the STM competition. Since $f(0)$ needs to be positive in order to allow transmitter release to begin (Figure 9.12), low-level transmitter release by nodes without significant STM activity is unavoidable if nodes cannot be hyperpolarized. Figure 9.28 shows that a competitive STM field with hyperpolarization gives search and resonance results similar to those of the other simulations.

Similarly, considerable variations in parameters also give similar results.

9.26 SIMULATION EQUATIONS

Simulation equations are described in an algorithmic form to indicate the steps followed in the computer program that generated Figures 9.23–9.27.

Timescale

The simulation timescale is fixed by setting the rate of transmitter accumulation equal to 1. The intrafield STM rate is assumed to be significantly faster, and the LTM rate significantly slower. Accordingly, STM equations are iterated several times each time step; and LTM weights are held constant. The simulation time step is

$$\Delta t = 0.005. \tag{9.23}$$

Integration method

Transmitter variables u and v are integrated by first order approximation (Euler's method). The IMSL Gear package gives essentially identical solutions but requires more computer time.

Table 9.2 LTM weights z_{ij}^{bc}

j → 1	2	3	4	5	6		i ↓
1.0	0.0	0.0	1.0	0.176	0.0001	1	
1.0	0.0	0.0	0.0	0.162	0.0001	2	
0.0	0.9	0.0	0.0	0.148	0.0001	3	
0.0	0.9	0.0	0.0	0.134	0.0001	4	
0.0	0.0	0.8	0.0	0.120	0.0001	5	
0.0	0.0	0.8	0.0	0.0	0.0001	6	
0.0	0.0	0.0	0.0	0.0	0.0001	7	
⋮	⋮	⋮	⋮	⋮	⋮	⋮	

$i = 1 \ldots n_a = n_b = 15$
$j = 1 \ldots n_c = 20$

LTM weights

The bottom-up LTM weights z_{ij}^{bc} illustrated in Figure 9.22 are specified in Table 9.2. At 'uncommitted' nodes ($j \geqslant 6$) $z_{ij}^{bc} \equiv 0.001$. Top-down LTM weights z_{ji}^{cb} are constant multiples of corresponding z_{ij}^{bc} weights:

$$z_{ji}^{cb} = 10 \cdot z_{ij}^{bc}. \tag{9.24}$$

This choice of LTM weights approximates a typical state of an ART system undergoing slow learning. Weights do not necessarily reach equilibrium on each presentation, but while the Jth F_c node is active,

$$z_{Ji}^{cb} \rightarrow x_i^{b3} \tag{9.25}$$

and

$$z_{iJ}^{bc} \rightarrow S_i^{b3}. \tag{9.26}$$

Given the parameters specified below, as STM and LTM variables approach equilibrium,

$$x_i^{b3} \cong 10 \cdot S_i^{b3}. \tag{9.27}$$

Equations (9.25)–(9.27) imply that equation (9.24) is a good approximation of a typical weight distribution.

Initial values

Initially,

$$u_{ij}^{bc}(0) = z_{ij}^{bc} \tag{9.28}$$

and

$$u_{ji}^{cb}(0) = z_{ji}^{cb}. \tag{9.29}$$

All other initial values are 0.

Input values

The F_b input values (S_i^{a3}) are specified in Table 9.3. All simulations start with Input 1. Several of the simulations switch to Input 2 either with a jump or gradually. Input 1 values are obtained by presenting a linear, decreasing function I_i to F_a. Input 2 values are obtained by setting $I_1 = I_2 = 1$ and $I_i = 0$ ($i \geqslant 3$).

Table 9.3 $F_a \rightarrow F_b$ input values (S_i^{a3})

i	*Input 1*	*Input 2*
1	1.76	2.36
2	1.62	2.36
3	1.48	0.0
4	1.34	0.0
5	1.20	0.0
6	0.0	0.0
7	0.0	0.0
.	.	.
.	.	.
.	.	.

Implicit in this formulation is the assumption that a changing input vector $\boldsymbol{I}$ can register itself at F_a. This requires that STM at F_a be frequently 'reset'. Otherwise new values of I_i may go unnoticed, due to strong feedback within F_a. Feedback within F_b allows the STM to maintain resonance even with fluctuating amplitudes at F_a.

STM equations

Except during reset, equations used to generate the STM values for Figures 9.23–9.27 are similar to the ART 2 equations (Carpenter and Grossberg 1987*b*). Dynamics of the fields F_a, F_b, and F_c are homologous, as shown in Figure 9.21. Steady-state variables for the field F_b, when the reset signal equals 0, are given by equations (9.31)–(9.36). Similar equations hold for fields F_a and F_c.

Layer 1, input variable

$$\varepsilon \frac{dx_i^{b1}}{dt} = -x_i^{b1} + S_i^{a3} + p_1^b S_i^{b2}. \tag{9.30}$$

In steady state,

$$x_i^{b1} \cong S_i^{a3} + p_1^b S_i^{b2}. \tag{9.31}$$

Table 9.4 specifies parameter $p_1^b, p_2^b, \ldots$ values and the signal function

$$g^b(y_i^{bL}) \equiv S_i^{bL} \tag{9.32}$$

Table 9.4

Parameters		
$p_1^a = p_1^b = p_1^c = 10.0$		
$p_2^a = p_2^b = p_2^c = 10.0$		
$p_3^a = p_3^b = p_3^c = 0.0001$		
$p_4^c = 0.9$		
$p_5^b = p_5^c = 0.1$		
$p_6^b = p_6^c = 1.0$		
Signal functions g^a, g^b, g^c		
F_a, F_b distributed	F_c choice	F_c distributed
$p_7^a = p_7^b = 0.0$	$p_7^c = 1/\sqrt{n_c}$	$p_7^c = 0.0$
$p_8^a = p_8^b = 0.3$	$p_8^c = 0.2$	$p_8^c = 0.4$

Distributed

$$g(w) = \begin{cases} 0 & \text{if } w \leqslant p_7 + p_8 \\ \left(\dfrac{w - p_7}{p_8}\right) & \text{if } w > p_7 + p_8 \end{cases}$$

Choice

$$g(w) = \begin{cases} 0 & \text{if } w \leqslant p_7 \\ \left(\dfrac{w - p_7}{p_8}\right)^2 & \text{if } w > p_7 \end{cases}$$

for layers $L = 1, 2, 3$. Equation (9.31) is similar to the simplified STM equation (9.6), with x_i^{b1} equal to the sum of an interfield input (S_i^{a3}) and an intrafield input ($p_1^b S_i^{b2}$).

Layer 1, output variable

$$y_i^{b1} \cong \frac{x_i^{b1}}{p_3^b + \|\boldsymbol{x}^{b1}\|}. \tag{9.33}$$

Layer 2, input variable

$$x_i^{b2} \cong S_i^{b1} + p_2^b S_i^{b3} \tag{9.34}$$

Layer 2, output variable

$$y_i^{b2} \cong \frac{x_i^{b2}}{p_3^b + \|\boldsymbol{x}^{b2}\|} \tag{9.35}$$

Layer 3, input variable

$$x_i^{b3} \cong S_i^{b2} + p_4^c \sum_j v_{ji}^{cb} \tag{9.36}$$

Layer 3, output variable

$$y_i^{b3} \cong \frac{x_i^{b3}}{p_3^b + \|\boldsymbol{x}^{b3}\|} \tag{9.37}$$

Normalization of the output variables in equations (9.33), (9.35) and (9.37) accomplishes two goals. First, since the nonlinear signal function g^b in equation (9.32) has a fixed threshold, normalization is needed to achieve orderly pattern transformations under variable processing loads. This goal could have been reached with other norms, such as the L^1 norm ($|\boldsymbol{x}| \equiv \Sigma_i x_i$). The second goal of normalization is to allow the patterns to have direct access to category representations, without search, after the code has stabilized (Carpenter and Grossberg 1987*a*, 1987*b*). Equations (9.13) and (9.17) together tie the Euclidean norm to direct access in the present model. If direct access is not needed, or if another measure of similarity of vectors is used, the Euclidean norm may be replaced by L^1 or another norm.

Transmitter equations

When the reset signal equals 0, levels of presynaptic and bound transmitter are governed by equations of the form (9.1) and (9.5), as follows.

Presynaptic transmitter, $F_b \rightarrow F_c$

$$\frac{du_{ij}^{bc}}{dt} = (z_{ij}^{bc} - u_{ij}^{bc}) - u_{ij}^{bc} p_5^c (x_j^{c1} + p_6^c) S_i^{b3} \quad (9.38)$$

Bound transmitter, $F_b \rightarrow F_c$

$$\frac{dv_{ij}^{bc}}{dt} = -v_{ij}^{bc} + u_{ij}^{bc} p_5^c (x_j^{c1} + p_6^c) S_i^{b3} \quad (9.39)$$

Presynaptic transmitter, $F_c \rightarrow F_b$

$$\frac{du_{ji}^{cb}}{dt} = (z_{ji}^{cb} - u_{ji}^{cb}) - u_{ji}^{cb} p_5^b (x_i^{b3} + p_6^b) S_j^{c1} \quad (9.40)$$

Bound transmitter, $F_c \rightarrow F_b$

$$\frac{dv_{ji}^{cb}}{dt} = -v_{ji}^{cb} + u_{ji}^{cb} p_5^b (x_i^{b3} + p_6^b) S_j^{c1} \quad (9.41)$$

Note that equations (9.38) and (9.39) imply that

$$u_{ij}^{bc} + v_{ij}^{bc} \rightarrow z_{ij}^{bc} \quad (9.42)$$

and equations (9.40) and (9.41) imply that

$$u_{ji}^{cb} + v_{ji}^{cb} \rightarrow z_{ji}^{cb}. \quad (9.43)$$

Reset equations

Reset occurs when patterns active at F_a and F_b fail to match according to the criterion set by the vigilance parameter. In Figure 9.21,

$$r_i^b \cong \frac{y_i^{a2} + y_i^{b2}}{p_3^a + \|\boldsymbol{y}^{a2}\| + \|\boldsymbol{y}^{b2}\|}. \quad (9.44)$$

Reset occurs if

$$\|\boldsymbol{r}^b\| < \rho^b, \quad (9.45)$$

where

$$0 < \rho^b < 1. \quad (9.46)$$

As in equations (9.5) and (9.6), the effect of a large reset signal is approximated by setting input variables x_i^{b1}, x_i^{b3}, x_j^{c1}, x_j^{c3}, and bound transmitter variables v_{ij}^{bc}, v_{ji}^{cb} equal to 0.

Iteration steps

Steps 1–7 outline the iteration scheme in the computer program used to generate the simulations.

Step 1. $t \rightarrow t + \Delta t$.
Step 2. Set ρ and S_i^{a3} values.
Step 3. Compute r_i^b and check for reset.
Step 4. Iterate STM equations F_b, F_c five times, setting variables to 0 at reset.
Step 5. Iterate transmitter equations (9.38)–(9.41).
Step 6. Compute sums $\Sigma_i v_{ij}^{bc}$ and $\Sigma_j v_{ji}^{cb}$.
Step 7. Return to Step 1.

9.27 CONCLUSION

In conclusion, we have seen that a functional analysis of parallel search within a hierarchical ART architecture can exploit processes taking place at the chemical synapse as a rich source of robust designs with natural realizations. Conversely, such a neural network analysis embeds model synapses into a processing context that can help to give functional and behavioral meaning to mechanisms defined at the intracellular, biophysical and biochemical levels.

ACKNOWLEDGEMENTS

This research was supported in part by the Air Force Office of Scientific Research (AFOSR F49620-86-C-0037 and AFOSR F49620-87-C-0018), the Army Research Office (ARO DAAL03-88-K-0088), and the National Science Foundation (NSF DMS-86-11959 and IRI-87-16960). We thank Diana Meyers, Cynthia Suchta, and Carol Yanakakis for their valuable assistance in the preparation of the manuscript.

REFERENCES

Carpenter, G. A. and Grossberg, S. (1987*a*). A massively parallel architecture for a self-organizing neural pattern recognition machine. *Computer Vision, Graphics, and Image Processing* **37**, 54–115.

Carpenter, G. A. and Grossberg, S. (1987*b*). ART 2: Self-organization of stable category recognition codes for analog input patterns. *Applied Optics* **26**, 4919–30.

Carpenter, G. A. and Grossberg, S. (1988). The ART of adaptive pattern recognition by a self-organizing neural network. *IEEE Computer: Special issue on Artificial Neural Systems* **21**, 77–88.

Carpenter, G. A. and Grossberg, S. (1989). Search mechanisms for Adaptive Resonance Theory (ART) architectures. *Proceedings of the International Joint Conference on Neural Networks*, June 18–22, Washington, DC, pp. I 201–5.

Carpenter, G. A. and Grossberg, S. (1990). ART 3: Hierarchical search using chemical transmitters in self-organizing pattern recognition architecture. *Neural Networks* **3**, 129–52.

Cohen, M. A., Grossberg, S. and Stork, D. (1988). Speech perception and production by a self-organizing neural network. In *Evolution, Learning, Cognition, and Advanced Architectures* (ed. Y. C. Lee). World Scientific Publishers, Hong Kong, pp. 217–31.

Grossberg, S. (1976*a*). Adaptive pattern classification and universal recoding, I: Parallel development and coding of neural feature detectors. *Biological Cybernetics* **23**, 121–34.

Grossberg, S. (1976*b*). Adaptive pattern classification and universal recoding, II: Feedback, expectation, olfaction, and illusions. *Biological Cybernetics* **23**, 187–202.

Grossberg, S. (1982*a*). *Studies of Mind and Brain: Neural Principles of Learning, Perception, Development, Cognition, and Motor Control*. Reidel Press, Boston.

Grossberg, S. (1982*b*). Processing of expected and unexpected events during conditioning and attention: a psychophysiological theory. *Psychological Review* **89**, 529–72.

Grossberg, S. (1984). Some psychophysiological and pharmacological correlates of a developmental, cognitive, and motivational theory. In *Brain and Information: Event Related Potentials* (ed. R. Karrer, J. Cohen and P. Tueting). New York Academy of Sciences, New York, pp. 58–151.

Grossberg, S. (ed.) (1987*a*). *The Adaptive Brain, I: Cognition, Learning, Reinforcement, and Rhythm*. North-Holland, Amsterdam.

Grossberg, S. (ed.) (1987*b*). *The Adaptive Brain, II: Vision, Speech, Language, and Motor Control*. North-Holland, Amsterdam.

Grossberg, S. (ed.) (1988). *Neural networks and Natural Intelligence*. MIT Press, Cambridge, MA.

Grossberg, S. and Levine, D. S. (1987). Neural dynamics of attentionally modulated Pavlovian conditioning: blocking, inter-stimulus interval, and secondary reinforcement. *Applied Optics* **26**, 5015–30.

Ito, M. (1984). *The Cerebellum and Neural Control*. Raven Press, New York.

Kandel, E. R. and Schwartz, J. H. (1981). *Principles of Neural Science*. Elsevier/North-Holland, New York.

Kohonen, T. (1984). *Self-organization and Associative Memory*. Springer-Verlag, New York.

Kuffler, S. W., Nicholls, J. G. and Martin, A. R. (1984). *From Neuron to Brain*, 2nd edn. Sinauer Associates, Sunderland, MA.

FURTHER READING

Carpenter, G. A., Grossberg, S. and Rosen, D. B. (1991a). Fuzzy ART: Fast stable learning and categorization of analog patterns by an adaptive resonance system. *Neural Networks* **4**, 759–71.

Carpenter, G. A., Grossberg, S., Markuzon, N., Reynolds, J. H. and Rosen, D. B. (1992) Fuzzy ARTMAP: A neural network architecture for incremental supervised learning of analog multidimensional maps. *IEEE Transactions on Neural Networks* **3**, 698–713.

Carpenter, G. A., Grossberg, S. and Reynolds, J. H. (1991c). ARTMAP: Supervised real-time learning and classification of nonstationary data by a self-organizing neural network. *Neural Networks* **4**, 565–88.

Carpenter, G. A., Grossberg, S. and Rosen, D. B. (1991b). ART 2-A: An adaptive resonance algorithm for rapid category learning and recognition. *Neural Networks* **4**, 493–504.

FLUID NEURAL NETWORKS AS A MODEL OF INTELLIGENT BIOLOGICAL SYSTEMS

10

Frank T. Vertosick, Jr.

10.1 INTRODUCTION

Tremendous progress has been made over the last decade in the theory and application of neural networks. Consisting of smaller processors linked by flexible connections, neural networks are excellent pattern recognition devices. By modifying their connectivity according to a defined learning algorithm, neural networks learn to associate patterns of input data with an appropriate output response (Khanna 1990; Jansson 1991; Rummelhart, McClelland *et al.* 1986).

The ability of neural networks to recognize patterns, even if those patterns are incomplete or damaged, has generated tremendous interest from both basic and applied scientists. Basic scientists seek to explain the behavior of living neural networks, for example, the brain, in the context of formal neural network theory, while applied scientists seek to make practical devices, for example, computers that can read different styles of handwriting, based upon neural network technology. The fundamental architecture of parallel networks is so adaptable and appealing, however, that the tenets of parallel distributed processing may underlie many biological processes (Bray 1990; Marijuán 1991; Goodwin *et al.* 1992).

As is apparent from the name 'neural network', the nervous system is considered the premier biological network and there is evidence that aspects of brain function can indeed be explained by neural network theory (Dan and Poo 1992; Lisberger and Sejnowski 1992; Sergent *et al.* 1992). The close relationship of the neurosciences to the broad field of artificial intelligence derives from the belief that 'intelligence' is the product of complex nervous systems. In fact, intelligent networks might be operative at many levels of biological organization, from the cytoplasm to large ecosystems (Paton 1992). I will discuss a specific example in detail here, namely the immune system, and then briefly discuss the wider applicability of network architecture to other living informational constructs.

The basics of neural network theory will be discussed in the next section. Those familiar with neural networks may wish to skip over this review.

10.2 THE NEURAL NETWORK APPROACH

Like all intelligent devices, living or electronic, a neural network is an input/output (I/O) device. The typical neural network consists of a set of processing 'units' linked by weighted connections. Some units are connected only to other units ('hidden' units), while others are also connected to the external environment (input and output units). Each unit receives input and sends output to other units (and the environment in the case of I/O units). The inputs to any given unit are summated to yield the 'activation' of that unit; the unit's output is some function of its state of activation. In most practical networks this function is some simple nonlinear relationship, for example, the sigmoidal I/O relationship:

$$\text{Output} = 1/(1 + e^{-\text{activation}/T}) \quad (10.1)$$

where T represents the 'temperature' of the network.

The connections among units are scaled, or weighted, that is, the output from unit J, O_J, is converted to input to unit K, I_K, according to the formula:

$$I_K = w_{JK} \times O_J \quad (10.2)$$

where w_{JK} is the connection weight between J and K. If w_{JK} is zero, then the units are disconnected, since output from J will not be transmitted to K. The set of all w_{JK} form the connection weight matrix which defines the connectivity of the network at any given time.

A neural network can learn by modifying its I/O behavior according to past experience. Learning is accomplished by changing the connection weights using an algorithm known as the learning rule. There are two broad classes of learning rules: supervised and unsupervised.

Unsupervised learning rules define changes in the connection weights based upon internal, or local, information. For example, the Hebbian learning rule defines the change in the connection weight w_{JK} as a function of the activation state of units J and K, that is,

$$dW_{JK} = f(a_J, a_K) \quad (10.3)$$

such that the connection between very active units is strengthened and the connection between inactive units is weakened (Hebb 1949).

Supervised learning requires a 'trainer' external to the network. The trainer compares real network output to the ideal, or desired, network output and makes adjustments in the connection weights according to the magnitude of the deviation between real and ideal network behavior. The most commonly used supervised learning scheme is the back propagation of errors. Supervised learning requires a training set of data containing correct I/O relationships which the trainer can use to adjust the connection weight matrix.

To summarize, the key features of the neural network approach are as follows:

1. a linked set of independent processing units exhibiting similar, nonlinear I/O behavior;
2. flexible, weighted connections between component units;
3. a capacity to modify connectivity among units, either locally or under external supervision, which allows network adaptation and learning.

10.3 THE NEURAL NETWORK AS A MODEL OF INTELLIGENCE

The phrase 'pattern recognition' does not convey the true power of the neural network approach. In reality, neural networks also perform pattern analysis and pattern extrapolation, in addition to simple pattern recognition. The sequence of pattern recognition, analysis and extrapolation (PRAE) is fundamental to 'intelligent' reasoning.

To illustrate the difference between simple pattern recognition and PRAE, consider a caveman hunter. On one expedition, he and a companion come upon a chimpanzee, which they have never seen before. The caveman's companion attacks the chimpanzee. Although smaller than a man, the chimpanzee is strong and the caveman's companion is killed by the encounter. On a second expedition, the caveman encounters another chimpanzee. Although it looks somewhat different than the first chimpanzee, the caveman recognizes that it is very similar to the animal that killed his companion earlier and he avoids it. This is simple pattern recognition. Although there are some differences between the two chimpanzees, for example, slighter size, different hair patterns (there can be wide variation among individual chimpanzees), the caveman easily concludes that he does not wish to confront the animal.

On a third expedition, the caveman stumbles upon a gorilla. Although a gorilla is noticeably different from a chimpanzee, the caveman literally runs in fright from this creature. Why? Because the caveman realizes (1) that this animal is comparable to a chimpanzee, and (2) if a small chimpanzee can kill a man, this behemoth will kill with even greater certainty. This reasoning illustrates PRAE. The caveman recognizes a pattern (the gorilla looks like a chimpanzee), analyzes the patterns (the main difference between the two animals is their size), and extrapolates information about the new gorilla pattern (larger animals are stronger and more difficult to fight than smaller ones, so this large 'chimpanzee-like' creature is likely to be more dangerous than the chimpanzee).

Finally, the caveman encounters an infant chimpanzee. Although he recognizes that it is very similar to a chimpanzee, he also recognizes that it is very young and not dangerous. Being hungry, he kills the animal. The same sequence of recognition (this creature is similar to a chimpanzee), analysis (this is a tiny, immature-looking chimpanzee), and extrapolation from known experience (very small forms of animals are usually much younger, weaker and more vulnerable than the larger forms) applies here. If pattern recognition alone was operative, however, the caveman would simply recognize that this is a chimpanzee and, despite its small size, avoid it based upon a prior bad experience with chimpanzees.

During pure pattern recognition, similar objects and concepts are classified according to shared features. PRAE goes further. After a pattern is recognized, unique outcomes can be customized based upon the differences between newly encountered and previously encountered patterns. The customization itself draws upon previous experience regarding how these differences should affect outcome; for example, the two animals are similar, but one is larger and larger animals are stronger.

Neural networks use a similar PRAE sequence. For example, consider a simple back propagation network designed to predict the lifetime of an automobile engine when given a pattern of input information, for example, the miles per year the engine was driven, the engine's manufacturer, the frequency of oil changes, the weight of the automobile in which the engine was installed, the color the engine was painted and so on. The network is trained on a set of I/O data, that is, a set of inputs where the output is known. During training, the network performs a kind of multivariate analysis, wherein quantitative relationships between each input parameter and the output are derived from the training set. The network learns that mileage strongly affects the lifetime of the engine, while the manufacturer has a mild influence, and the color of the engine's paint does not matter at all. The network then uses these learned relationships to predict engine lifetimes where the outcome is not known. When presented with two similar engines, one that has never had its oil changed and one that has had its oil changed every three months, the network (if trained on a large enough training set) should predict that the poorly maintained engine will fail sooner. Just like the caveman, the neural network recognizes a pattern (these engines are similar), analyzes the differences between patterns (one engine had no maintenance) and extrapolates an answer (previous experience with the training set shows that poorly maintained engines fail sooner).

PRAE represents the essence of intelligence. What we consider 'problem solving' is in reality a manifestation of pattern extrapolation. PSAE flows naturally from the neural network formalism, and neural network architecture appears to be the ideal format for PSAE within biological systems. Like neural networks, living systems are comprised of many interconnected components. Moreover, as will be discussed below, the connections between components in living systems, whether between proteins and nucleic acids within cells, between cells within organisms or between organisms within an ecosystem, are highly adaptable. It seems unlikely that nature would not have exploited that adaptability to create networks capable of PSAE at many levels of organization. Finding neural networks in many 'nonneural' living systems, however, will require some modification in the conventional view of neural networks as 'hardwired' devices. Although some software approaches to network design do use a fluid adaptability of the connections, the connections are still deterministic, not probabilistic as they are likely to be in cytoplasm, for example.

10.4 FLUID PHASE NEURAL NETWORKS?

As I have argued elsewhere, the immune system might be the ideal organ system in which to find neural network architecture outside of the central nervous system (Vertosick and Kelly 1989, 1991; Vertosick 1992). The immune system consists of many billions of lymphocyte 'units'; although the I/O behavior of lymphocytes *in vivo* is not known with certainty, it is most likely some simple, nonlinear function (Perelson 1989). Lymphocytes can be connected to external antigens (I/O lymphocytes), or to each other via antiidiotypic interactions (hidden lymphocytes).

Unlike neurons and transistors, however, lymphocytes may float freely in blood and lymph. Unlike hardwired electronic networks, the immune system behaves more like a solution filled with different chemical reactants. Individual lymphocytes collide with each other or with immune molecules, for example, lymphokines and immunoglobulins, and interact with one another at a rate which is dependent upon the concentration of the 'reactants' involved (since that determines the collision frequency), the rate of mixing (which is temperature-dependent), and the affinity of the 'reactants' for each other. While it is easy to define the connection weight between two microprocessors, or even two neurons, defining the connection weight between two lymphocytes is more difficult. Consider a clone of lymphocytes J which interacts with an antiidiotypic clone K with affinity A_{JK}. If n_K and n_J are the number of lymphocytes in clones K and J respectively, then the connection weight between the clones might have the form:

$$w_{JK} = f(\text{mixing}) \times n_J \times n_K \times A_{JK} \quad (10.4)$$

where $f(\text{mixing})$ is a function that accounts for lymphocyte mixing and trafficking. If the affinity or clonal sizes are small, then the connection would be weak, while large affinity or clonal sizes would yield a strong connection. The term 'connection' might not be ideal in the context of fluid networks, and could be replaced with the terms 'cross-section' or 'reaction probability', in that communication between fluid phase 'units' is a stochastic process. Connections within a computer or nervous system are deterministic and devoid of any elements of randomness. The connections within a fluid network, on the other hand, represent probability distributions. In other words, the connection weight in a stochastic network would represent the probability that two units will interact, given the number of similar units in the mixture, their mutural affinity, and the 'temperature', or degree of mixing, of the system. Note that this temperature, which affects the connectivity of the network, is different from the temperature in Eqn. (10.1) which determines the shape of the activation I/O curve of individual units. The latter temperature has been employed for optimization of network learning, for example, simulated network annealing. Until formal mathematical models of fluid, stochastic networks are developed, the influence of the mixing temperature on the thermodynamics of learning in fluid phase networks is completely unknown. For a discussion of the role of 'thermodynamics' in network learning, see Hinton and Sejnowski (1986).

Regardless of the precise nature of the connection weights in immune networks, there is little doubt that they are flexible, since the affinities of immune receptors, as well as lymphocytic clonal sizes, undergo major changes during an immune response. The clonal expansion and affinity maturation that accompany an immune response are reminiscent of Hebbian learning (Eqn. (10.3)). Although the connectivity in fluid phase networks radically differs from the connectivity in hardwired networks, the fluid networks should be capable of neural network behavior, for example, PSAE.

In the nervous system, network complexity is achieved topologically, with a rather limited number of cell types linked by a staggering web of fixed connections, magnified by a bewildering array of neurotransmitters. In fluid phase biological networks, however, complexity can only be achieved stereochemically, that is, by having a massive array of stereospecific receptors, for example, immunoglobulins and T cell antigen receptors, since the geometry of the network is less defined.

Perhaps the strongest evidence for some form of neural network architecture in the immune system is the Oudin–Cazenave enigma (Oudin and Cazenave 1971). When an animal is immunized with a large protein expressing multiple different epitopes, multiple immune responses result. Although each antibody formed may recognize a completely different epitope, the responding antibodies often share the same idiotypic markers. This is the Oudin–Cazenave enigma, and it has been observed for a variety of protein antigens. In conventional immunology, the 'enigma' is the apparent coordination of different responding clones at some higher level. In the traditional paradigm, the immune response to multiple epitopes is an uncoordinated array of responding clones, with each clone acting as a 'lone ranger' seeking out its unique target. The sharing of idiotypic markers among different clones implies that the clones do not act alone but are instead informationally linked to some higher authority. But what authority? Even the Jerne model does not fully answer this question.

It is tempting to speculate that idiotypic markers are keywords in a distributed, content-addressable memory which allow the immune network to horizontally integrate individual clonal responses into a broader view of the entire antigen. Immune responses of the network would then represent attractors in 'idiotype space'.

If the immune network is a cognitive neural network based upon idiotypic-antiidiotypic connections, then the formation of antiidiotypes and anti-antiidiotypes should be under rather strict genetic control, so that each animal can reproducibly form a functional network despite exposure to different antigens at different times. Recently, Garcia *et al.* produced an antibody to angiotensin II (MAb 110) and its anti-antiidiotypic antibody (MAb 131). MAb 110, its antiidiotype and the resultant anti-antiidiotype were all developed in different animals, yet the amino acid sequences of MAb 110 and MAb 131 were nearly identical, suggesting that the formation of antiidiotypes and anti-antiidiotypes is indeed tightly controlled (Garcia *et al.* 1992).

The immune network is not wholly fluid phase, since lymphocytes 'condense' in lymph nodes, spleen, thymus, Peyer's patches and areas of chronic inflammation. In these condensations, network dynamics may be closer to hardwired network. Thus, the immune system may be considered a hybrid network, consisting of both fluid and solid phases.

The differences between fluid phase and hardwired neural networks can be summarized as follows:

1. Although both networks are made up of component processing units, hardwired units are immobile with respect to each other while the components of fluid networks are mobile.
2. The connections between units in a hardwired network are deterministic and the network's connectivity is, in large part, topologically defined; the connections between units in a fluid phase network are stochastic and are defined stereochemically.
3. The number of units in hardwired networks is typically fixed; the number of units in a fluid phase network can vary, and this variation may play a major role in the functioning of the network.

Mathematical modelling of fluid phase networks will likely be more difficult than modelling hardwired networks. Factors such as network mixing (temperature), fluctuating numbers of units and regional inhomogeneities must be incorporated. The latter problem, regional inhomogeneity, is not encountered in fixed networks where the units cannot move. In the immune system, for example, lymphocytes and cytokines may converge upon an area of inflammation, altering the local network dynamics. Connectivity may increase as lymphocytes of high mutual affinity congregate in a limited area, increasing collision rates. Furthermore, regional increase in the levels of cytokines, such as interleukin 2 and interferons, may affect the activation behavior of lymphocytes in a small subset of the network. Modelling hydrid networks, where some units are fluid and some are hardwired, will be still more difficult.

Applying neural network formalism to immune network theory is not trivial. Unlike conventional mathematical treatments of immune networks, which are primarily kinetic models (Richter 1975; Perelson 1989), the neural network approach supplies a cognitive model which seeks to expand the immune system's intellect from the cellular level to the organ level. An immune system with neural network architecture could perceive entire antigens, even entire viruses and bacteria, not just single epitopes the width of an immunoglobulin's binding site. Moreover, the temporal patterns of infectious diseases could be stored, in addition to the nature of the antigens involved. Given the number of lymphocytes involved and the diversity of immune receptors, the mammalian immune system may be comparable to the nervous system in cognitive power, albeit in a different realm and operative at a longer timescale. With such capacity, the immune network could participate in somatic tissue morphostasis, in addition to performing its defensive duties, as was suggested a decade ago (Vertosick and Kelly 1983).

An immune system with neural network architecture raises several questions:

How is distributed network learning implemented through the alteration of immune receptor affinity?

The layered network concept of Jerne (idiotype, antiidiotype, anti-antiidiotype and so on) should be abandoned. As the work of Garcia *et al.* has shown, there may be little, if any, difference between the idiotype and the anti-antiidiotype in some cases. I propose that the immune system be considered simply as a complex mixture of immune receptor variable regions (antibodies and T cell antigen receptors), each receptor with an affinity for every other receptor in the mixture ranging from negligible (no interaction) to high (strong interaction). When two receptors interact, it is pure semantics to call one the idiotype and the other an antiidiotype. The important concept here is that the receptors recognize each other via determinantes in or near the variable region. In this receptor mixture, learning would occur through the modification of the number and affinities of the receptors. Although maturation of antibody affinity for exogenous antigen is known, is there evidence that affinities of immune receptors for one another change during an immune response? In fact, work by van der Heijden suggests that such afinity changes do occur (van der Heijden *et al.* 1990).

What is the role of cytokines?

Cytokines can serve as trophic agents, increasing the numbers of responding lymphocyte clones. In stochastic networks, as I have argued above, the size of the clones affects the interclonal collision frequency and thus impacts upon connectivity. I have also suggested (Vertosick and Kelly 1991; Vertosick 1992) the more farfetched notion that cytokines may alter the activation curve of lymphocyte clones and thus may participate in the simulated annealing of the immune network, in the 'Boltzmann machine' sense (see Hinton and Sejnowski 1986). In other network models of the immune response the role of cytokines is excluded, a condition so unrealistic as to cast all such models into serious doubt.

Can the learning and recall of complex spatiotemporal multiepitope cognition be observed?

The most recent immune network models, despite their elegance and sophistication, are untestable. There is no hard evidence that any organ-level informational processing, network-driven or otherwise, is operative in mammalian immune systems. In general, network modelling of the immune system has been kinetic and not cognitive, that is, theorists have been more concerned with why the network can respond to an epitope without exploding into lymphoid neoplasia, and less concerned about the range of the immune system's intellect. For network theory to have a renewed impact upon empirical immunology, we as theorists must direct the experimentalists towards observable phenomena which will convince them that some network integration of information is occurring. For example, carefully designed multiepitope vaccination studies may be able to elicit pattern completion behavior. How would an animal respond if immunized with ten haptens and rechallenged with only eight? Would a secondary antibody response be observed against the two unboosted haptens? To my knowledge, no such studies have been published.

The immune system may be a more fertile ground than the nervous system for studying the biological implementation of neural network architecture at the cellular level. Immune elements are more easily studied *in vitro*, and so dissecting out the network architecture may be readily accomplished in immune networks. The entire paradigm of immunological research would have to shift, however, from the lymphocyte level to the lymphoid organ level, with the goal of understanding the higher level sensory integration of the images seen by the entire set of responding clones, that is, the 'antigen retina'.

10.5 OTHER EXAMPLES OF HYBRID NETWORKS

The immune system is not the only potential example of a hybrid biological network. The cytoplasm, for example, consists of macromolecules floating freely in solution as well as associated with the cytoskeleton, cell membranes and organelles. If these macromolecules are considered units, their stereochemical interactions with other macromolecules may represent connections (see Bray 1990). This connectivity can be influenced by post-translational glycosylation or by phosphorylation and dephosphorylation (Lord *et al.* 1988). The I/O behavior of the macromolecular units would be a function of their biochemical function, for example, enzymatic activity. Farmer has also discussed the possible application of connectionist models to 'autocatalytic' networks of enzymes (Farmer 1990).

Likewise, the nucleus may be a hybrid network of nucleic acid sequences and mobile transcriptional control elements and histones, with the stereochemical connectivity again modified by processes such as phosphorylation/dephosphorylation. The genome, with its restricted geometric organization, would represent the 'hardwired' portion of the network.

Even large ecosystems may possess hybrid network structure. Plants (fixed) and animals (mobile) are connected in an ecological network. Some animals eat (are connected to) certain plants or animals, while ignoring (being disconnected from) others. Some plants exploit animals, for example, honeybees, while ignoring others. Connectivity can be altered rapidly by depletion of species (animals may turn to less desirable plant foods when their first choice becomes scarce), or more slowly through mutational events (evolution). In a preliminary exploration of this concept, Goodwin *et al.* argued that the behavior of ant colonies may be explainable in terms of neural network architecture (Goodwin *et al.* 1992).

Like the immune system, the ecosystem is marked by great regional inhomogeneity. Moreover, the interaction between various units (species) will be a function of the size of that species and their mutural affinity for one another, both of which can change in response to external influences (for example, antigen stimulation in the case of the immune system, drought in the case of the ecosystem).

10.6 CONCLUSION

Neural network architecture derives from a set of independent processing units linked by flexible connections. The parallel nature of a neural network allows it to be robust and fast; the flexibility of the network's connectivity allows it to learn and respond to patterned information, storing these patterns in a distributed fashion which is relatively resistant to damage. Neural networks are practical devices which come up with good solutions quickly, but they are far from ideal for symbolic logic or mathematical operations. Given that living systems consist of multiple component parts linked by flexible connections and need to be fast, robust, practical devices with little use for symbolic logic, I propose that neural network formalism is a good starting point for approaching biological information processing at all levels, not just at the level of the nervous system.

To this end, the concept of a fluid, or stochastic, neural network, as opposed to the conventional deterministic, or hardwired, neural network, was introduced. Most living networks will be hybrids, in that they contain both stochastic and deterministic components. Developing mathematical models of stochastic networks, with particular emphasis on their thermodynamic properties, may be quite useful in understanding complex biological phenomena.

Consider the human oocyte. Contained within this miniscule droplet of proteins and nucleic acids is the blueprint for creating an entire human body. The oocyte is similar to a neuron, a single unit in the nervous system, in size. Yet within the oocyte is the information necessary to create the hardwired brain. Is the oocyte's nucleus simply a disorganized soup of nucleic acids and transcriptional control proteins? Some sophisticated organization of data above the level of simple molecules must be at work. What is it? This is a question which is only recently being addressed systematically, for example, at the Sante Fe Institute (see Waldrop 1992).

The recent advances in molecular biology have enabled researchers to clone ever larger mumbers of hormones, growth factors, growth factor receptors, cell adhesion molecules, transcriptional control elements, kinases, phosphatases and so forth. While this approach initially yielded answers regarding physiology at the cellular and organismal levels, the biochemical picture grows more confusing with time. The number of proteins, enzymes, genes and other players in life's molecular game is staggering, and so the prospect of understanding cancer, mitosis, the immune response or any other process in terms of one (or even a handful) of individual molecular species is dim. We must begin to understand the *system* as well as the components. Stochastic neural networks may be a starting point for this work.

Can an immune system, or an ecosystem, be considered 'cognitive' in the psychological sense? The concepts of cognition and consciousness as they are used by neuroscientists are philosophical issues beyond the scope of this essay. I believe that the architecture underlying networks using parallel distributed processing is the basis of information processing at all levels of biological organization and has been 'highly conserved' throughout evolution, just as the genetic code and the use of phosphorylation to regulate enzyme activity have been highly conserved. At what point the complexity of such networks gives rise to conscious, cognitive entities is a question that remains to be answered.

REFERENCES

Bray, D. (1990). Intracellular signalling as a parallel distributed process. *Journal of Theoretical Biology* **143**, 215–31.

Dan, Y. and Poo, M.-M. (1992). Hebbian depression of isolated neuromuscular synapses *in vitro*. *Science* **256**, 1570–3.

Farmer, J. D. (1990). A rosetta stone for connectionism. *Physica D* **42**, 153–87.

Garcia, K. C., Desiderio, S. V., Ronco, P. M., Verroust, P. J. and Amzel, L. M. (1992). Recognition of angiotensin II: Antibodies at different levels of an idiotypic network are superimposable. *Science* **257**, 528–31.

Goodwin, B. C., Gordon, B. C. and Trainor, L. E. H. (1992). A parallel distributed model of behavior in ant colonies. *Journal of Theoretical Biology* **156**, 293–307.

Hebb, D. O. (1949). *The Organization of Behavior*. Wiley, New York.

Jansson, P. A. (1991) Neural networks: an overview. *Analytical Chemistry* **63**, 357A–362A.

Hinton, G. E. and Sejnowski, T. J. (1986). Learning and relearning in Boltzmann machines. In *Parallel Distributed Processing* (ed. D. E. Rumelhart, J. L. McClelland and the PDP Research Group). MIT Press, Cambridge, MA, pp. 282–317.

Khanna, T. (1990). *Foundations of Neural Networks*. Addison-Wesley, New York.

Lisberger, S. G. and Sejnowski, T. J. (1992). Motor learning in a recurrent network model based on the vestibulo-ocular reflex. *Nature* **360**, 159–61.

Lord, J. M., Bunce, C. M. and Brown, G. (1988). The role of protein phosphorylation in the control of cell growth and differentiation. *British Journal of Cancer* **58**, 549–55.

Marijuán, P. C. (1991). Enzymes and theoretical biology: sketch of an informational perspective of the cell. *BioSystems* **25**, 259–73.

Oudin, J. and Cazenave, P. (1971). Similar idiotypic specificities in immunoglobulin fractions with different antibody function or even without detectable antibody function. *Proceedings of the National Academy of Sciences USA* **68**, 2616–20.

Paton, R. C. (1992). Towards a metaphorical biology. *Biology and Philosophy* **7**, 279–94.

Perelson, A. S. (1989). Immune network theory. *Immunological Reviews* **110**, 5–36.

Richter, P. H. (1975). A network theory of the immune system. *European Journal of Immunology* **5**, 350–4.

Rumelhart, D. E., McClelland, J. L. and the PDP Research Group (eds.) (1986). *Parallel Distributed Processing*. MIT Press, Cambridge, MA.

Sergent, J., Zuck, E., Terriah, S. and MacDonald, B. (1992). Distributed neural network underlying musical sight-reading and keyboard performance. *Science* **257**, 106–9.

van der Heijden, R. W. J., Bunschoten, H., Pascual, V., Uytdehaag, F. C., Osterhaus, A. D. and Capra, J. D. (1990). Nucleotide sequence of a human monoclonal anti-idiotypic antibody specific for a rabies-virus neutralizing monoclonal idiotypic antibody reveals extensive somatic variability suggestive of an antigen-driven immune response. *Journal of Immunology* **144**, 2835–9.

Vertosick, F. T. (1992). The immune system as a neural network: an alternative approach to immune cognition. *SigBio Newsletter* **12**, 4–6.

Vertosick, F. T. and Kelly, R. H. (1983). Autoantigens in an immunological network. *Medical Hypotheses* **10**, 59–67.

Vertosick, F. T. and Kelly, R. H. (1989). Immune network theory: a role for parallel distributed processing? *Immunology* **66**, 1–7.

Vertosick, F. T. and Kelly, R. H. (1991). The immune system as a neural network: a multi-epitope approach. *Journal of Theoretical Biology* **150**, 225–37.

Waldrop, M. M. (1992). *Complexity*. Simon & Schuster, New York.

THE IMMUNE LEARNING MECHANISMS: REINFORCEMENT, RECRUITMENT AND THEIR APPLICATIONS

11

Hugues Bersini and Francisco Varela

11.1 INTRODUCTION

This paper presents an attempt to transpose important points inherent in our understanding and simulations of immune networks into a methodology for adaptive and distributed problem solving. Following a sketchy recapitulation of immune network simulations done in recent years, with greater emphasis on the inspirational sources, an adaptive distributed process control methodology will be the focus of attention including its application to, and results of, well-known canonical problems. Immune networks are members of a large family of biological systems which are structured as networks of interacting units, where environmental events percolate into the network, and which present several forms of adaptive change on different timescales. In the kind of immune networks under study here, there are two types of change which are of relevance: the varying concentration of the immune actors called the 'dynamics' of the system, and the slower recruitment and disappearance of these actors, a plastic change called the 'metadynamics' (Figure 11.1).

Other well-known members of this same family are biological or artificial ecosystems, such as Holland's classifiers systems (Booker *et al.* 1989; Holland *et al.* 1986), and autocatalytic networks (for a more complete description see Farmer (1991)). In an ecosystem, the species population densities vary according to the interactions with other members of the network as well as through environmental impacts. In addition the whole network is subject to structural perturbations through appearance and disappearance of some species. The introduction of new species is caused by crossover and mutations of genetic material already present in the network. A crucial issue is the fact that the network as such, and not the environment, exerts the greatest pressure in the selection of the new actors to be integrated into the network. This ecosystem structural metadynamics characterized by a network-based selectivity is quite reminiscent of what happens in an immune network. However, in the immune case, this introduction of new actors is based on combinatorial and selective mechanisms at the level of the individual organism, and thus distinct from the population-based, longer timescale genetic combinatorials proper to ecosystem metadynamics. In our work, we have designated this metadynamical process the immune recruitment mechanism (IRM).

In general, in this family of biological systems the dynamics of the network can be defined by a set of differential equations

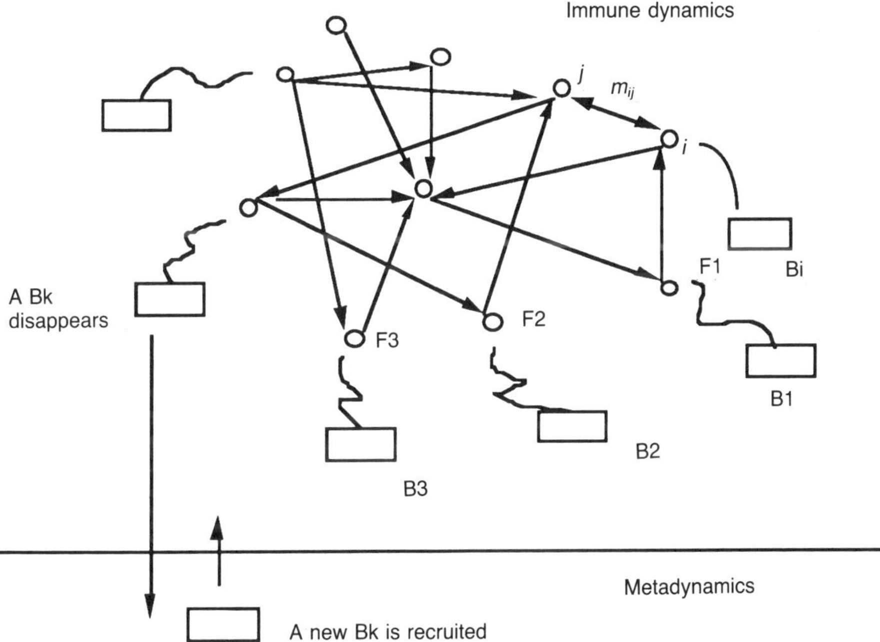

Figure 11.1 Dynamics and metadynamics in the immune network. Rectangle = B lymphocyte; circle = antibody produced by the associated B lymphocyte.

reflecting the network structure through a connectivity matrix m_{ij}, which in the immune case corresponds to the stereochemical affinities of the respective molecular profiles. The specific dynamics observed depends quite heavily on the structure of the matrix. This is true for all kinds of other biological networks. For instance in the case of a fully connected Hopfield neural net, the various dynamical patterns a neuron can be subject to as a function of the matrix are well known (Sompolinsky *et al.* 1988). Roughly, a symmetric synaptic matrix drives the network to a fixed point whereas an asymmetric matrix can generate a chaotic dynamics. Other interesting examples are simulations of individuals playing among themselves a kind of iterated prisoner's dilemma (Lindgren 1992). The connectivity between two individuals can be seen as the score of the first one, characterized by a specific strategy, and playing against the second one, characterized by another strategy. The density of the individuals exhibits interesting dynamics depending on their strategies and, more precisely, on the interactions between these strategies taken in pairs.

The presence of the metadynamics as a continuous generator of novelty has been traditionally understood in the light of the clonal selection theory of immune responses, where the immune system defends the organism against external invaders. A continuous and random production of novel immunoglobins appears necessary to discover and to reinforce the beneficial ones able to bind the unpredictable invaders. However, an important shift in the conceptual framework is emerging in immunology, where the classical clonal selection views are being replaced by a view of the immune system primarily as a means to define an organism's identity, in what we have called its self-assertion role (Varela *et al.* 1988; Varela and Coutinho 1991). In this new light, the metadynamics is a tool for this identity to be preserved while, and somewhat paradoxically, constantly shifting according to the organism's ontogenic changes and in response to the environmental coupling with food and antigens. Thus, the metadynamics could appear as some sort of control mechanism which aims at maintaining the viability of the network by modifying the current population of immunoglobins, in an on-going repertoire shift. The autonomous network regulatory effect will maintain these shifts which consequently appear as a possible support for the learning and memory functions of the immune system. Such support will be referred to in the following as 'population-based memory' in order to differentiate it from the 'parametric-based memory' of neural networks (encoded classically in the value of the synaptic strength).

Section 11.2 will recapitulate on studies and computer simulations which have been developed for several years by our Paris group (in collaboration with J. Stewart and A. Coutinho). The core of these models is a system of differential equations governing the time evolution of the concentration of each immunoglobulin species and the population size of their corresponding B-lymphocytes. This dynamical system is complemented with a mechanism of recruitment of new cells (B lymphocytes produced from the bone marrow) which consequently alters the structure of the differential equations since it changes the list of actors in them. An abundant literature is available about:

1. the biological reasons for this vision of the immune system structured as an idiotypic network (Varela *et al.* 1988, 1991; Lundkvist *et al.* 1989);
2. crucial experimental results supporting this vision (Lundkvist *et al.* 1989; Varela *et al.* 1991);
3. the theoretical justifications of the simple computer model still under study and the first counting results obtained so far (Bersini 1992*b*; Stewart and Varela 1991; Varela and Coutinho 1991);
4. the conceptual message appearing in the background, that the Paris group has spread for a long time now, and which advocates the acceptance of the immune system as a self-assertional system with main finality to maintain a communication and a harmony in the somatic identity of the organism.

Other incisive simulations of the immune network dynamics and metadynamics, from a more classical clonal selection point of view, have been performed by the Santa Fe group (Perelson 1988, 1990; De Boer and Perelson 1991; De Boer *et al.* 1992, 1993).

Section 11.3 leaves the biological world to adopt a more pragmatic perspective and describes how we achieve the transfer of various aspects of the immune network like distributed control, viability, network-based dynamics and metadynamics adaptation, and population-based memory into an adaptive control context. Although these different points might be transposed into various kinds of applications, we will adopt a restrictive attitude and present a complete methodology for adaptive control where:

1. the control is distributed among several simple operators located on a discretized state space (a grid);
2. the objective of the control is to maintain the process viable;
3. the dynamics is very close to the Q_learning method (a very efficient reinforcement learning algorithm) revised within a network context;
4. the recruitment phase (equally sensitive to the network structure) amounts to the generation of new actors on the grid;
5. this recruitment mechanism is based on interesting optimization heuristics;
6. a 'population-based' memory capacity is manifest.

The conception of a system equipped with a distributed controller, interacting with an uncertain and varying environment, and basing its learning on its own experiences entails quite naturally the integration of a reinforcement learning mechanism. The addition of a second slower adaptive mechanism, based on the recruitment of new actors, will be justified as a complement to the previous reinforcement learning which allows us to extend the search space in a parsimonious manner. Memory composed of population samples will be argued as an advantageous substitute for adaptation in an environment subject to unstabilities which nevertheless exhibit some regularity. At different points of the methodology, important similarities to Holland's classifier system principles will be pointed out (Booker *et al.* 1989; Holland 1986); indeed the similarities between the immune recruitment of new species and the genetic algorithms (GA) as well as between the whole immune network and classifiers systems have already been remarked on and discussed (Booker *et al.* 1989; Varela *et al.* 1989; Bersini and Varela 1990, 1991).

As we stated previously, IRM shows interesting optimization heuristics and, taken alone, has been exploited (somewhat like GA when isolated from the classifiers systems) for the optimization of hard multimodal problems. Why should this be used in an optimization perspective? The immune recruitment selects a new species on the basis of the current global state of the system, that is, according to the sensitivity of the network for this candidate species. Concretely a possible candidate will be recruited if its affinity with the different species already present in the network's repertoire exceeds a recruitment threshold fixed *a priori*. This generation respects the simple and common-sense heuristic: 'make the new actors to be similar to the best ones already acting in the population'. Either by GA, which preferentially recombines and mutates the best individuals of the population, or by IRM, which explicitly generates new actors in an area surrounding the best individuals, this simple heuristic is always at work. In the context of these multiactor systems continuously regenerating themselves while engaged in an ongoing interaction with themselves and with the environment, the acceptance of the two generation mechanisms as optimization strategies is a strong engineering drift. In fact, improving on a multiactor system through the disappearance and replacement of some of these actors does not amount to looking for a global maximum. A continuous adaptation of a population of actors favoured by the occasional integration of new actors is quite different from a directed search for the one and only one best actor.

However, the popularity of GA today, when isolated from classifier systems and restricted to another method of optimization, is due to an overstating of three aspects of neo-Darwinist views:

1. the population-selection-based mechanism which favours an implicit parallel search preferable when optimizing a multimodal function;
2. the stochasticity inherent in the generation of new individuals which helps to escape local solutions and to perform some sort of 'lucky hill-climbing';
3. the way new individuals are generated in the population to be tested. Indeed a mixing operation like crossover might be interesting in combinatorial problems in which joining pieces of intermediary satisfactory solutions is a convenient way to improve the quality of this solution. In more formal terms (Forrest and Mitchell 1991), crossover is adequate for problems showing a hierarchical schema-fitness structure.

Working with IRM and getting the results described later in the chapter led us, on the one hand, to strongly support the population-randomness-selection-based mechanism (even if the IRM version is slightly different from the standard version) instead of two or three point local strategies recurrent in classical optimization schemes but, on the other hand, to remain sceptical of the ubiquitous efficiency of recombination operators such as simple crossover (efficiency being discussed by an increasing number of researchers (Forrest and Mitchell 1991; Manderick *et al.* 1991)).

The first objective of the section describing IRM (11.3.6) will be to show that since the way IRM generates a new candidate to be tested is more reminiscent of classical optimization methods than GA, it can improve the speed of optimization. IRM aims for a greater exploitation of the information carried in the comparative nature of the fitness values. It applies a basic heuristic of search which although operational in the majority of classical methods has never been presented in such an abstract and straightforward way:

> each new point to be tested as a potential solution of the problem should be preferentially closer to the best solutions tested so far in the neighborhood of this point.

Very roughly, if in the current population the fitness of point 1 is better than the fitness of point 2, the next point to be tested should be closer to point 1. IRM tries to accelerate and to reinforce the local hill-climbing inherent to any search method without hampering the beneficial exploration resulting from the population-selection-based mechanism. In the exploitation/exploration axis of the global optimization scheme, IRM like evolution strategies (Hoffmeister and Bäck 1990) gets closer to the exploitation pole than GA.

The second objective of the section is a logical consequence of the first. While maintaining the advantageous explorative tendency and the recombination mechanisms inspired from evolution-based algorithms, developments within the GA community should no longer ignore the significant pool of good old-fashioned methods of optimization which remain the most efficient to date for the exploitation stage of the search. Hybridizing evolutionary mechanisms for exploration with classical optimization methods for exploitation will be advocated and illustrated. In order to fulfil these two objectives, Section 11.3.6 will expose two of the main optimization algorithms derived from the inspiration of the immune metadynamics: GIRM (genetic immune recruitment mechanism) and Simplex-GA. Other members of this family of optimization methods are presented in Bersini and Varela (1991) and Bersini and Seront (1992). The two algorithms will be described in detail and the results of their testing on the two functions F1 and F5 of the De Jong test suite (De Jong 1975), as well as on the travelling salesman problem will be presented and commented on. The fourth and final section will describe two simple well-known benchmark cases: the cart-pole and the robot in a maze, with the aim of illustrating the various aspects of the methodology: the network-based reinforcement and recruitment mechanisms for the cart-pole control and the adaptivity and population-based memory for the robot in a maze.

11.2 THE BIOLOGICAL INSPIRATION: A GLANCE AT THE IMMUNE NETWORK

In this section, a rapid description of the immune network under study and some simulation results (with main focus on the metadynamics) are presented. The immune network computer simulation comprises two types of immune actors, the antibodies and their associated producers, the B cell lymphocytes. The model takes into account B cell proliferation, maturation, antibody production, the formation and subsequent elimination of antibody-antibody complexes, the natural death of these two immune actors and an additional source of B lymphocytes

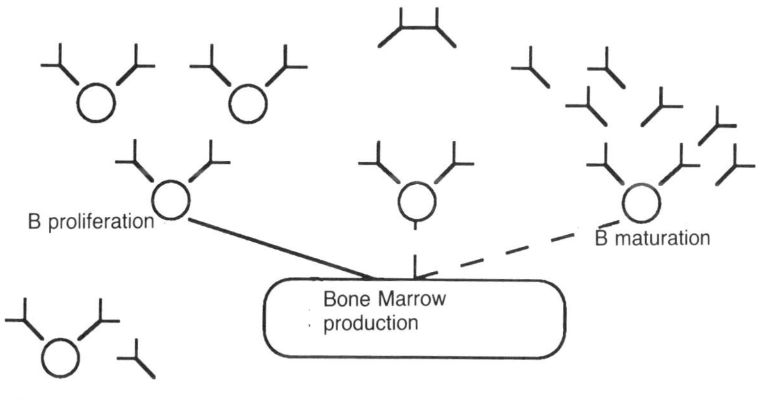

Figure 11.2 The biological inspiration: actors in the immune network.

coming from the bone marrow (Figure 11.2). The model reflects the network structure through a symmetric affinity matrix which affects the dynamics in three different ways:

1. The disappearance rate of an antibody-antibody complex is proportional to the strength of the mutual affinity.
2. A B lymphocyte produces as many associated antibodies, as this later presents a good sensitivity to the whole network.
3. A B lymphocyte replicates itself with a rate, here again, dependent on the sensitivity of its associated antibody to the whole network.

An abundant literature describes the model and its behaviour in greater detail (Bersini 1992*b*; Stewart and Varela 1991; Varela and Coutinho 1991). The full classical spectrum of dynamic patterns – single point attractor, simple and complicated oscillations and chaos – has been obtained while varying the various parameters within biological range, including the structure of connectivity (De Boer *et al.* 1992; Bersini 1992*b*).

As mentioned already, a fundamental property of the immune system that goes beyond the dynamics is the continuous introduction of novel species into the network which provokes continuous structural changes which essentially affect the structure of the connectivity matrix and consequently the dynamics just described. The enormous diversity of immunoglobin species carried on the surface of any inactivated B lymphocyte results from multiple combinations and random mutations of genetic materials extracted from a large pool of gene segments. In order to be integrated in the current network, a novel immunoglobin must show a sensitivity with the whole network which lies in between two threshold values, due to the log-normal shape of the two activation mechanisms: maturation and proliferation. Indeed, if the sensitivity is either below the inferior threshold or above the superior activation threshold, the new immunoglobin will not be able to maturate or to proliferate. The selection process deciding on the permanency of a species in the network needs to be considered on a longer timescale than the dynamics of the network. In effect, a species showing a good sensitivity to the network at the time of its integration can be eliminated later since the continuous development of the network (the uninterrupted introduction and disappearance of species) will have either increased its sensitivity beyond the superior limit or decreased it below the inferior limit.

These last considerations and the resemblance existing between immune networks and ecosystems lead to a slight restatement of the selectionist dogma so well rooted within immunology and evolutionary theory. Selectionism mainly imputes to pressure from the external world interactions the driving force for the system. A common misleading view of adaptation is 'adaptation to the same external world' shared by all species, whereas in reality the impact of the external world is hidden, implicitly accounted by and diffused into the whole network which is then responsible for a continuous reshaping and self-modification of the adaptation criteria. The distinction stressed in Packard and Bedau (1992) between extrinsic and intrinsic adaptation in ecosystems is a corollary to that point. In the view we favour here, selective environmental pressures need to be substituted for the autonomous self-induced pressure of the network itself. In immunology, more than for ecosystems, such a view can be considered as radically new since external antigen input loses the main part in the development of the immune scenario, and it becomes an influence among others with no privileged status to direct the unfolding of immune events. It is the network as a whole which decides which are the candidates sufficiently adapted to the current network for the purpose of harmonizing the cellular and environmental components of the organism. Indeed, the immune recruitment mechanism selects for the generation and the maintenance of a new species on the basis of the current global state of the system, that is, according to the sensitivity of the network for this candidate species.

The project of simply illustrating the effect of the metadynamics on the resulting immune population (i.e., what species are selected by the network to constitute it) by computer graphics is made possible thanks to the concept of 'shape space' (exploited by several authors (Bersini 1992*b*; De Boer *et al.* 1993; Perelson 1990; Stewart and Varela 1991)). The main idea is this: each immunoglobin species is characterized by n physicochemical parameters and is accordingly represented as a point in the real space $\mathbf{R}^n$ which has been called the shape space (and is nothing other than a kind of attribute space); for the sake of graphics we assume $n = 2$ and a clonal species to be a pixel. In one of the simulations (other choices are possible), the mutual affinity between two species m_{ij} is defined by means of a symmetric Gaussian which represents the domain of affinity of the first species. The affinity is obtained by the value of this Gaussian computed at the position of a second species (Figure 11.3): $m_{ij} = \exp(-d_{ij}^2)$ and $d_{ij} = a_{ij}/c$ where a_{ij} is the distance between the symmetrical of i and j, and c is a scaling constant.

The way the metadynamics determines the occupation of the shape space (i.e., the immune repertoire) and the emergence of a certain connectivity structure can be simply obtained by a routine which resembles the

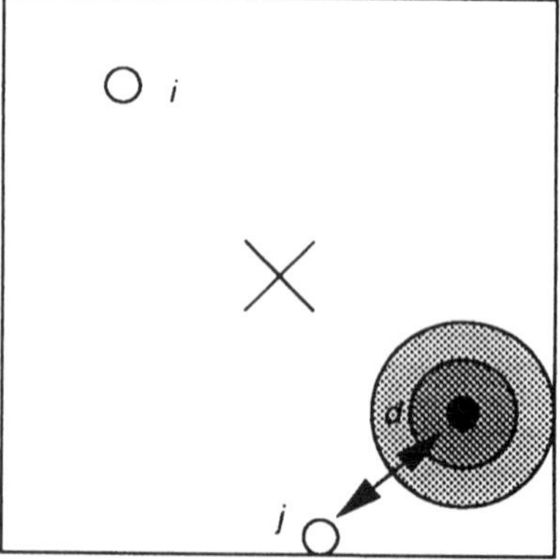

Figure 11.3 How to compute the affinity value between clonal species i and j.

game of life. First, two threshold values of sensitivity must be defined between which the sensitivity of an immunoglobin candidate (a random point in the shape space) must lie in order to be integrated and to stay in the network. The simulation proceeds by random generation of new species in the network if their sensitivity lies between the two thresholds, and by suppression of species out of the network when their sensitivity oversteps either threshold value (in simulations, i can stay in the network as long as $\sigma_{\min} < \Sigma_j m_{ij} < \sigma_{\max}$). Gradually the shape space becomes covered with interesting and very characteristic patterns which slowly stabilize. As represented in Figure 11.4, these stripe-shaped patterns contain cells whose sensitivity with the whole network is comprised between the two thresholds. The real species are marked by a square while their symmetrical location is marked by a cross. An attentive observation of the pattern of occupation shows that each stripe in the space is sustained by another stripe which is slightly shifted with respect to the first one. Indeed the sensitivity exerted by the exact symmetrical stripe is too high to allow the corresponding region to be occupied.

Following a transient of variable length (Figure 11.5), this metadynamic simulation seems to fluctuate around a steady shape space occupation. Why this kind of steady state occupation constitutes an equilibrium point is easily justified by accepting that at a certain moment the sensitivity at each point of the shape space not yet occupied can be either superior to its maximum value or inferior to its minimum value. If this is the case, no further species can be integrated into the network and no species already present needs to disappear.

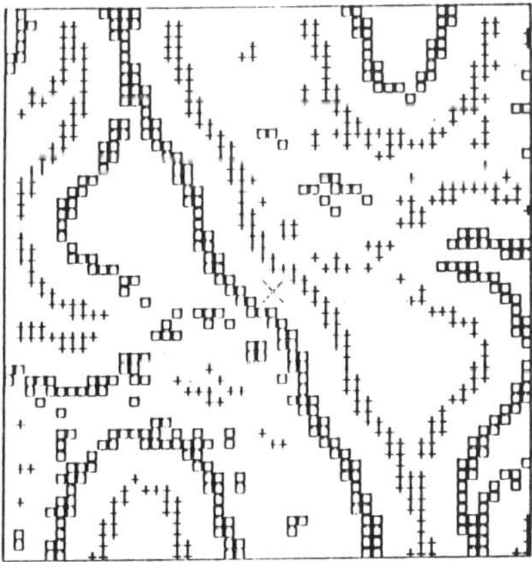

Figure 11.4 A metadynamic simulation in a two-dimensional shape space.

Clearly, what is hard to analyse is the basin of attraction of this steady state, and a very long time might be necessary before reaching this frozen situation. After some time, the total number of species (squares in the space) reaches a maximum value around which it keeps fluctuating. Since the number of species in the space and the average connectivity among them increase together, the existence of a minimum and a maximum value for the repertoire size is not surprising. The whole shape space is not completely filled and the final occupation can be regarded as a self-generated (in the absence of an external antigen) repertoire selection of the cells belonging to the immune network (De Boer and Perelson 1991). Only those cells prone to grow thanks to the network effect are allowed to express themselves in the immune system.

This shape space occupation can play the role of memory, a type of learning and memory support which has been designated by 'population-based memory'. Figure 11.6 shows a modified occupation of the space due to an encounter with a new molecular profile (either an newly expressed self-antigen or an encounter with an external one), here indicated by a black square. Following the suppression of the antigen, the modified occupation shows inertial effect and does not return to its previous nonperturbed state. Perelson (1990) observed a similar phenomenon and described it as a 'repertoire shift'. In some cases, the immune network can fail to integrate the antigenic perturbation. It reacts but this reaction is not compensated

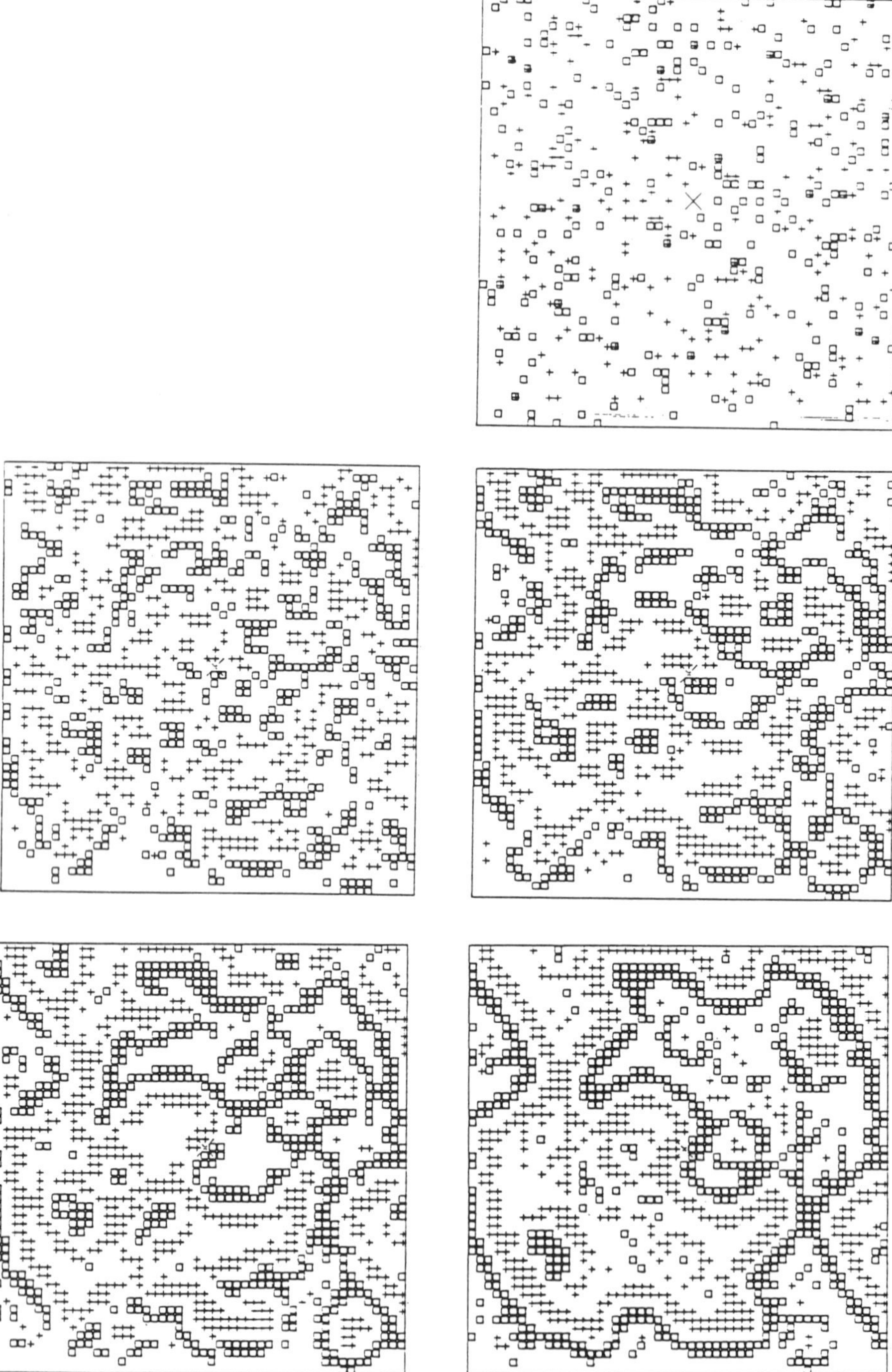

Figure 11.5 The transient occupation of the shape space by the metadynamics at iterations 0, 20, 40, 60, 120. The last occupation is very close to complete stability.

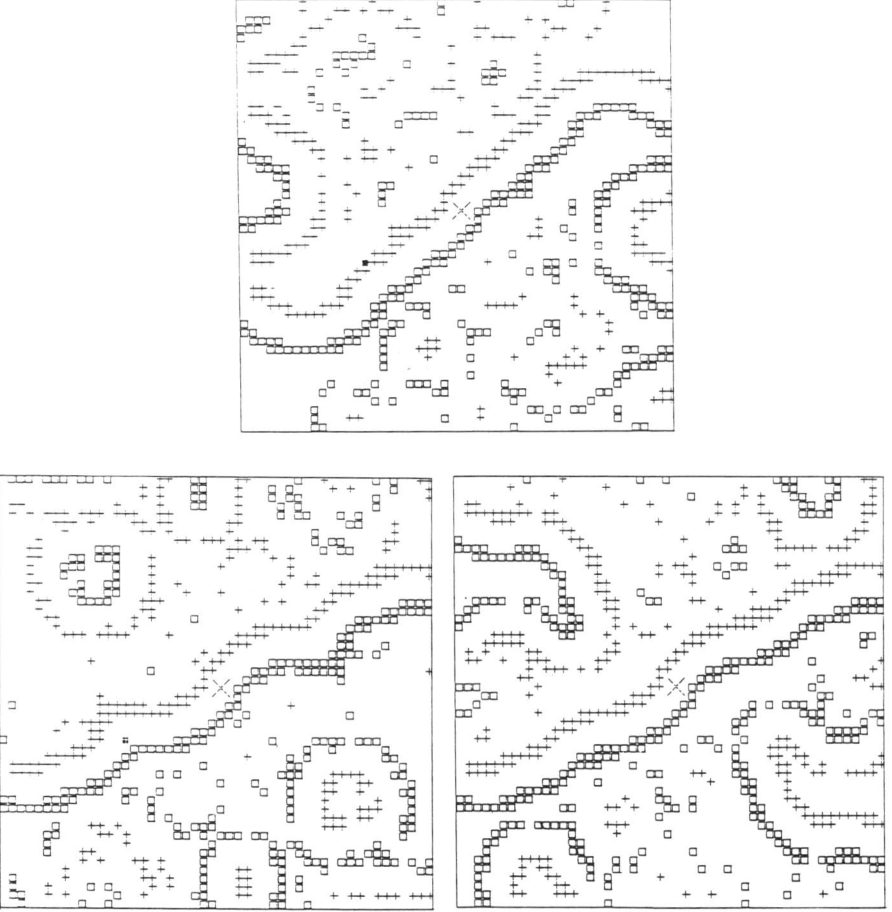

Figure 11.6 An example of the 'population-based' memory following an encounter with an antigen. First the antigen (the little black square) is presented and one can see the slight transformation of the shape space occupation induced by it. Then if the antigen is removed, this alternation is maintained and the occupation pattern does not return to the previous one.

for by an autonomous regulatory effect which is the only way to 'memorize' the encounter.

We believe that given that the whole network is, in the actual organism, embedded in an self-antigenic environment, the *raison d'être* of the metadynamics could be a particular regulatory mechanism maintaining the viability of the network. Simulations have shown that when the immune idiotypic network is isolated from any external interaction, the concentrations do not collapse or explode due to a counterbalancing mechanism inherent to the network itself. However, the situation might be different when coupling the network either to a self- or an external antigen. Then, the supplementary sensitivity caused by this new influence might have a dramatic effect on the network stability and require the intervention of the metadynamics in order to reconfigurate the network so as to compensate for this external perturbation. This understanding of the viability role played by the metadynamics is under active investigation.

11.3 FROM IMMUNE NETWORKS TO ADAPTIVE DISTRIBUTED CONTROL

11.3.1 GENERAL PRINCIPLES

This section will describe an attempt to gather all ideas gained from the knowledge of biological immune systems presented in the previous section for engineering applications. In the wake of connectionist models and the evolutionary based mechanisms (genetic algorithms, classifiers systems and evolution strategies), new lessons addressed to the control of complex processes might derive from understanding aspects of a further biological system the immune network. Distribution, viability, networks parametric and structural adaptation and population-based memory will be transposed, re-interpreted and analysed in an adaptive control perspective. Some of these points have already been integrated into various fruitful methodologies and have been the object of a large but scattered literature. Salient links with the Barto and Sutton reinforcement learning work (Barto and Singh 1990; Sutton 1990) as well as with Holland's classifier systems need to be pointed out. The methodology we propose is based on seven general principles.

Principle 1

The control of the process is executed by a set of small and simple operators distributed in time and/or space. This set of operators is organized into a network structure represented by an affinity matrix.

Principle 2

The aim of the controller is to maintain the viability of the process to control.

Principle 3

The controller learns to maintain this viability and self-adapts to keep the process viable despite perturbations affecting this process. This learning and adaptation relies on two different types of plasticity: a parametric plasticity which, while keeping constant the population of operators, modifies some parameters associated to them with the effect of tuning their actions; and a structural plasticity based on a recruitment mechanism which modifies the current population of operators.

Principle 4

The learning and adaptation are realized by mechanisms of the reinforcement type.

Principle 5

Both temporal changes, dynamics and the metadynamics, are sensitive to the network structure through the affinity matrix.

Principle 6

Isolated from the whole methodology, the immune recruitment mechanism can stand alone as an interesting optimization algorithm.

Principle 7

The controller presents a 'population-based memory' capacity which appears to be precious in a changing environment but subject to recurrent changes.

These principles are indeed very general and some of them are already in play in the connectionist community for the automatic discovery of optimal neural network architectures (Deffuant 1992) but also in the classifiers systems framework. We will now present one possible implementation of all these principles into an adaptive control methodology.

11.3.2 PRINCIPLE 1: DISTRIBUTED CONTROL ON A GRID

Suppose the process to be controlled is characterized by a state vector $X(t) = \{x_i(t)\}$

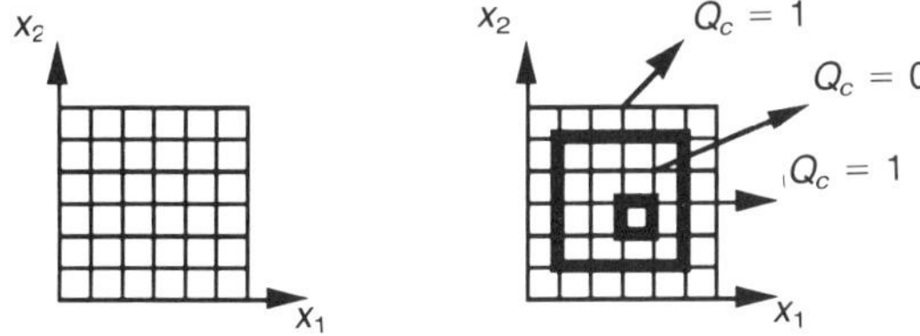

Figure 11.7 The partition of the state space and the viability and preferential zones.

represented in the state space $\mathbf{R}^n$. This state space is partitioned into several subzones or cells. Each variable x_i has its domain of variation divided into K_i intervals: $[x_i^m(k),\ x_i^M(k)]$ with $x_i^M(k) = x_i^m(k+1)$. Then, the total number of cells is $\Pi_i K_i$. Figure 11.7(a) illustrates such a partition for a two-variable process.

In the present methodology each cell lodges a certain number of operators responsible for control as soon as the process accesses the specific cell and as long as the process stays in it. The control is then in the hands of a large number of small operators, their efficacy being restricted to their cells. Since the Selfridge's Pandemonium, the idea of distributed control is not very new. It shares in the recent enthusiasm for a self-organizing system displaying emergent functionalities (a complete account of our understanding of 'emergence' is given in Forrest (1991)) like a Hopfield net or a colony of insects. It is also related to Minsky's views of cognition as a society of minds (1986), and to Brooks' subsumption architecture (1990). Roughly each actor has a very simple and localized responsibility and a very localized access to the information, both when acting and when improving itself from the results of its action. The larger the distribution, the easier a future hardware parallel implementation will be. The local communication existing among the actors often amounts to trivial excitatory or inhibitory signals circulating through a network structure. The improvement of each operator depends on the behaviour of its most affine neighbours, that is, the system exhibits local plasticity. The resulting behaviour is a whole system's phenomenon or property. No operator has any privileged view on the final objective it contributes to satisfy. Distributed control can achieve two different kinds of objective:

1. cooperative control where different operators can act simultaneously on different parameters of the process. The responsibility is distributed in space.
2. sequential control where at any time one and only one operator is acting on the whole process. The responsibility is distributed in time.

Even if the general principles presented above are implemented in this contribution just for the second category of problems, namely a single controller distributed in time, Bersini (1993) shows how this methodology can be extended for controllers distributed in space and acting simultaneously while sharing a same reinforcement.

11.3.3 PRINCIPLE 2: VIABILITY

A domain of viability V is a specific zone of the state space in which the process must indefinably remain: $X(t)\ C\ V$. Such constraint of viability might relax the objective to follow a referential trajectory for processes either too complex or interacting with an open, hard-to-formalize and unpredictable environment. While smoothing classical engineering objectives which are more prone to optimize than to stay viable, it convenes more adequately to biological reality. On the other hand such constraints fit the objectives of reliability and safety which are generally stated by the establishment of a bounded zone for some variables of the process. This idea is to a large extent inherited from Ashby (Kauffman 1989*a*). J. P. Aubin (1991) has constructed a complete mathematical framework in order to formalize the notion of viability in control theory (the design of control laws for nonlinear processes with state constraints) and its possible exploitation for macrosystems arising typically in economics, biology and complex engineering plant.

Actually, viability is more meaningful for biological systems than for applications which require the introduction of an *a priori* finality and of an associated fitness function. Here the aim of the control is no longer to drive the system to a precise target by means of feedback gradient ascent mechanisms but rather to keep it within certain limits. No error-based feedback is required but just a negative reinforcement at the frontiers of the viable zone instead. In the previous more biological section, we have discussed the connections between the immune network's autonomous dynamics and a related notion of viability. A viable immune network is not something easy to define except by roughly speaking of a noncollapsing (all the species concentration going to 0) and nondiverging (at least one species growing infinitely) network. The viewpoint advocated demands a move beyond the view of the immune system as a basic defence system, to one where the internal self-consistency is the central issue.

In contrast with the immune case, in the present approach the zone of viability is predefined and taken to be one of the cells or a certain union of these cells (Figure 11.7(b)). For each cell c, a quality measure Q_c is defined. For instance $Q_c = 0$ for each cell in the zone of viability, and $Q_c = f(d(c, V))$ (where $d(c, V)$ is a certain distance between the cell c and the zone of viability, and $f < 0$ is inversely related to d) for each cell outside the zone. For instance, revised with this notion of viability, Barto *et al.*'s control of the cart-pole (1983) fixes a zone of viability for both the angle and the position of the pole. Outside the zone the quality is -1. A third zone can be defined as a preferential zone, contained in the zone of viability, with the quality of its cells defined in $[0, 1]$. In another example, Sutton's Dyna method (1990) teaches a robot to reach a target while avoiding bumping into obstacles. Not viable are all those zones containing obstacles (their quality is negative) while the quality of the preferential zone, that is, the target cell, is 1. We will come back to these two specific applications in Section 11.4.

11.3.4 PRINCIPLES 3, 4, 5: NETWORK-BASED PARAMETRIC AND STRUCTURAL PLASTICITY

Reinforcement learning

Initially, in the methodology being presented, a random set of n operators is given in each cell. At the end of the learning stage, one and only one operator (as long as the situation is stationary) should be preferentially selected in each cell. The way this selection is done relies on a reinforcement learning strategy which adjusts the selection probability of each operator on-line according to a mechanism called Q_learning and originally developed by Watkins (Barto and Singh 1990). As stated by Barto (1990), when the performance measure for a 'weakly supervised' system is not defined in terms of a set of targets by means of a known error criterion, reinforcement learning addresses the problem of improving performance as evaluated by any measure informative enough and whose value can be supplied to the learning system. Roughly, systems exhibiting reinforcement learning adapt in response to only moderately informative feedback such as: 'it's good', 'it's bad'. While it is difficult to imagine any different type of learning for biological or behaviouristic systems (indeed in the immune system two species do or do not show mutual affinity, their only source of reinforcement), the increasing attention raised by this learning approach in the control community is a direct consequence of the interest for more and more complex and autonomous systems.

The complexity can be inherent to the control mechanism itself when the precise knowledge of how to tune the control parameters in order to obtain a satisfactory control behaviour is lacking, that this, when the controller does not possess sufficient self-knowledge. This is the case (and it is indeed our case) for distributed control when the interesting behaviour is an 'emergent' phenomenon. The effect of one operator modification on global behaviour is not easy to predict and the learning must be more of a 'trial and error' than of a 'supervised' type. However the complexity can also be inherent to the environment the controller is interacting with. This environment can be uncertain, varying and hard to specify. In such cases, the precise impact a change in the control parameters provokes in the environment is unknown and there again, the controller needs to adjust its parameters, not according to a gradient completed by a sensitivity analysis of the environment, but through trial and error. Thus reinforcement learning is partly due to an uncertain knowledge of both the controller and what is controlled.

On the other hand, an autonomous system exhibits autonomy with respect to two different things: the human operator or programmer and the system environment (Bersini 1991*b*). Firstly, the controller should not rely on continuous human supervision, providing instructions on the exact way it must behave. Secondly, it must interact with an environment whose feedback will never be as precise as a quadratic error to minimize by any gradient descent, but purely qualitative and poorly informative instead. For instance, a robot bumps into an obstacles or it does not, it reaches its targets or it does not. The environment must not be regarded any more as an instructor but rather as a means to trigger or to select potential ways of behaving for the controller. In short, the conception of a system equipped with a complex controller, interacting with an uncertain and varying environment, and basing its learning on its own experiences, without external guidance, entails quite naturally the integration of a reinforcement learning mechanism.

In addition, it is very rare to see the qualitative feedback following the control actions being supplied without delay. In our specific case, a negative reinforcement is received only when crossing the viability zone and a positive one when reaching the preferential zone. In general a sequence of control actions is fully executed before receiving any information on the quality of the whole sequence, and once received, the impact of this information on each step of the sequence (a kind of credit assignment problem) can be a serious difficulty. Recently Barto, Sutton and Watkins (Barto and Singh 1990; Sutton 1990) have derived a new type of reinforcement learning algorithm in order to account for this feedback delay. The idea to be shared with Holland's bucket brigade algorithm in the context of rule-based system (Booker *et al*. 1989; Holland *et al*. 1986) is ingenious although simple. When the effective impact of the environment is delayed, it is still possible to provide the controller with an intermediary reinforcement measure: the strength or quality of the subsequent actors. This information is obtained while the controller is acting and it allows the successive choice of actions to improve on-line.

A decision policy at a certain stage of the control sequence is adjusted only on the basis of what will happen at a subsequent stage. The nature of this intermediary reinforcement measure (the whole methodology is called temporal difference (TD) and is presented in Sutton (1990)) has led to a family of algorithms whose simplest member is the Q_learning algorithm which we will now present. Since in dynamic programming too, a decision at a certain stage is only a function of data retrieved at a subsequent stage, the connections of the delayed reinforcement learning family with dynamic programming makes such learning highly promising for control in general (Barto and Singh 1990).

Q_learning

An initial random set of n operators is generated in each cell, and each operator can act on the process as soon as the process reaches the concerned cell. An operator acting in a cell c: O_c is characterized by k values and can then be described by a vector W in a k-dimensional space $W = \{w_1, w_2, \ldots, w_k\}$. For instance such an operator might be a single value ($k = 1$) or a linear combination of the process state variables (k = the dimension of the state space). Each operator can then be represented by $O_c(W_i)$ with c indexing the cell and i indexing the operator (from 1 to n). Associated with each operator is a quantity called the Q_value $QO_c(W_i)$ which measures its quality and weights its selection probability. The selection of one of the n operators to act when the process accesses its specific cell is performed through a probabilistic Boltzmann distribution given by:

$$P_i = \frac{\exp(QO_c(W_i)/T)}{\sum_{j=1}^{n}\exp(QO_c(W_j)/T)}.$$

T is a temperature parameter which tunes the degree of randomness during the choice (an 'annealing' procedure). Obviously, exploration in the space of solution should be initially favoured and then be gradually decreased to promote deterministic choice.

Once an operator i has been selected in cell c, it will control the process as long as this one stays in the same cell. When the process leaves cell c to get into a neighbouring one $c + 1$, the quality of the acting operator is updated by:

$$\Delta QO_c(W_i) = \alpha(R - QO_c(W_i)) \quad (|\alpha| < 1).$$

Figure 11.8 shows the various elements present in this formula.

R is called the critic in the reinforcement learning literature and is the only feedback the operator receives:

$$R = (Q_{c+1} - Q_c) + \beta \operatorname{Max}_j(QO_{c+1}(W_j)).$$

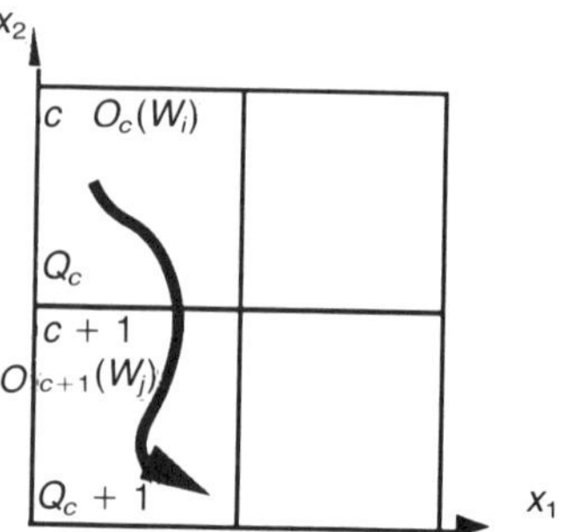

Figure 11.8 What is involved in Q_learning.

Q_c and Q_{c+1} are the qualities of the respective cells, $\beta < 1$ is the discount rate and $\operatorname{Max}_j(QO_{c+1}(W_j))$ is the quality of the best operator, the one with the highest probability to be selected in the next cell.

The formula for R clearly indicates the connection with DP (dynamic programming). The selection policy of DP is conserved, based on the assumption that the best strategy is already discovered in the next cell. However a fundamental difference resides in the addition of a learning phase which approximates, while acting, the values upon which the decision has to be taken. $Q_{c+1} - Q_c = 0$ as long as the process stays in the viability zone (no external reinforcement is received); $Q_{c+1} - Q_c = -1$ when the process leaves the viability zone (the control is punished); $Q_{c+1} - Q_c = 1$ when the process accesses the target cell (the control is rewarded).

It has been shown (Barto and Singh 1990; Sutton 1990) that the value of the operator quality $QO_c(W_i)$ tends to $\Sigma_{j=1}^{m}\beta^j(Q_{c+j} - Q_{c+(j-1)})$ (where m is the number of cells to cross first to receive an external feedback positive or negative) which is called the cumulative reward. According to this, the quality of the operator will be proportional to the degree to which its action contributes to keep the process viable as long as possible (i.e., $Q_{c+m} - Q_{c+(m-1)} = -1$; the process leaves the viability zone for large m) and to access the target zone as fast as possible (i.e.,

$Q_{c+m} - Q_{c+(m-1)} = 1$; the process gets inside the target zone for small m).

When applying Q_learning in the field of process control, the kind of problem to be faced is highly stochastic, that is to say the exit cell $c + 1$ is not only dependent on the action taken in the current cell but also on the way the process accesses that specific cell. Applying exact dynamic programming to this kind of Markovian problem would require the knowledge of the state-transition probabilistic matrix i.e., $p(c + 1/c, i)$: the probability of accessing cell $c + 1$ when operator i acts in cell c. This would require either an important amount of prior simulation or a preliminary analyis of the problem. In Barto and Singh (1990), it is explained why Q_learning avoids these very computationally expensive preliminary data and compares favourably with dynamic programming despite this absence of the explicit probabilistic transition matrix. In Bersini (1992*a*), several additions to Q_learning are proposed, in particular: (1) for accelerating the search; and (2) matching classical control requirement such as (a) relying only on observable and not on all state variables and (b) favouring a smooth control over an abrupt one.

Parallel Q_learning

The fundamental limitation of Q_learning is the size of a search space explored in a very unsupervisory way. Imagine a grid of n cells each containing m operators. The size of the associated search space turns out to be m^n. This rapidly becomes an enormous number for realistic problems which cannot profit from efficient heuristics. In such situations, every addition or change which aims at accelerating the search is welcome. One way is to allow the search to be achieved simultaneously in different zones of the search space (Figure 11.9). In process control, this amounts to initiating the learning with different initial conditions. Now this can appear as a 'fictitious' parallelism, since only one simulation of the process can be done at any time and the different departures must be achieved sequentially. But it can also express a true parallelism. For instance, imagine a robot searching for its path in a maze. Several robots might circulate simultaneously in the environment and play with the same set of $QO_c(W_i)$ values. This would be highly reminiscent of the way a colony of ants gradually traces a very short path to reach a target. The way Q_learning poses the problem and executes its search allows for such decomposition (see Singh (1992) for an interesting example of the exploitation of this decomposability).

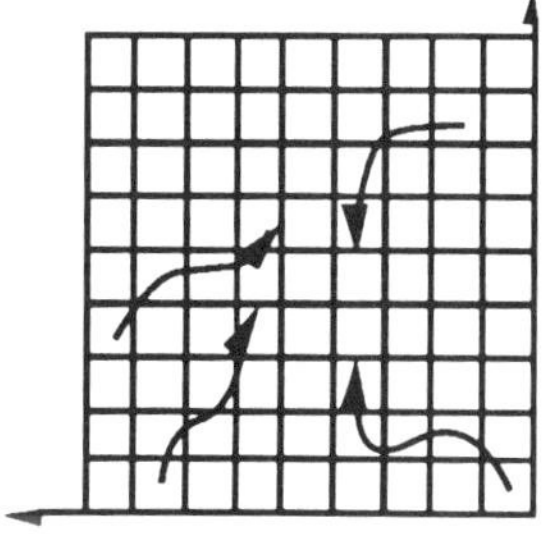

Figure 11.9 Parallel Q_learning.

Network-based type of Q_learning

In Section 11.3.1 where we discussed the general principles of the methodology, the fifth principle stated that the parametric plasticity must integrate a network structure represented by an affinity matrix. So far, such an affinity matrix has not appeared in the reinforcement learning strategy based on Q_learning since the different actors in whatever cells they act do not present any connection. Recently McCallum (1992) filled this gap and improved the basic algorithm by the introduction of such a matrix. He proposed a new version of the algorithm call transitional proximity Q_learning where the Q_values are updated in parallel so as to speed up the learning phase. When an operator i acts in cell c with the effect of driving the process into cell $c + 1$, its Q_value is updated through the classical mechanism. However, all the inactive operators j contained in another cell c' are equally updated by the new algorithm:

$$\Delta QO_{c'}(W_j) = \alpha m_{ij}(R_i - QO_{c'}(W_j)) \quad (|\alpha| < 1)$$

with

$$R_i = (Q_{c+1} - Q_c) + \beta\,\mathrm{Max}_k(QO_{c+1}(W_k))$$

and m_{ij} the affinity value between the operators i and j given by $\psi^{d(j,c',c)}$ where $d(j, c, c')$ is the minimum number of cells to cross in order to reach c from c' provided the operator being selected in cell c' is j, and $\psi < 1$ (clearly the greater this distance the less impact one operator modification has on the other one). These affinity values are learned on-line while acting, otherwise a deeper knowledge of the problem would be needed and would thus challenge the *raison d'être* of reinforcement type of learning.

This reformulation of Q_learning in a network-based form, in which when an operator modifies itself in response to a given feedback, other connected operators (not directly informed by the external reinforcement) modify themselves as well, results in an acceleration of the learning phase. The notion of affinity proposed by McCallum is not unique and another way of defining affinity has been proposed in Mahadevan and Connel (1992) where two operators are similar and can mutually influence each other while learning (by sharing their respective reinforcement) when their local environment is similar. Indeed, one can easily understand, for instance, for the robot in a maze, that what is being learned facing an obstacle in a certain location could be transferred to another location presenting the same obstacle configuration. In the first case, affinity is synonymous with complementarity, in the second case with similarity.

Combining the network effect and the parallelism described in the previous section might lead to a further formulation of Q_learning as follows:

$$\Delta QO(W_j) = \alpha \sum_{i=1}^{n} \frac{m_{ij}}{n}(R_i - QO(W_j))$$

with

$$R_i = (Q_{c+1} - Q_c) + \beta\,\mathrm{Max}_k(QO_{c+1}(W_k))$$

and n the number of parallel searches ($i = 1, \ldots, n$) (n robots in the maze).

Adaptation in nonstationary environments

In its original description Q_learning is not very adaptive. Given an environment which is not completely stationary (for instance a process with possible failure states or a robot wandering in an environment cluttered with moving obstacles), the Q_learning (even with Boltzmann selection) slowly freezes in a certain solution state and cannot escape to face a modified situation. Sutton (1990) has improved the adaptive capacity of Q_learning by increasing its explorative tendencies. A new parameter n_i^c called 'exploration bonus' has been added to the initial method which is the number of time steps the action i in cell c has not been selected computed from the last time it was selected. The new decision policy chooses the action i which maximizes:

$$QO_c(W_i) + e\sqrt{n_i^c}$$

where e is a small positive parameter and nothing is modified in the way Q_values are updated. n_i^c is re-initialized to 0 each time the action i is taken in cell c. It is obvious how this addition increases the exploration to the detriment of exploitation. Obviously in the case of a stationary environment, the learning process turns out to be much slower for discovering a solution and such a solution never really stabilizes.

Depending on the stability of the environment, from rare perturbations to frequent ones, alternative algorithms are equally possible. We implemented an algorithm which still relies on an annealing selection of action, but with the temperature increasing again (thus promoting the exploration again) in case of breakdowns (Bersini 1990). A breakdown here is an expectation-failure if, after a certain lapse of time, the critic received by an operator is all at once completely different from its current Q_value. In such a case, something occurs in the environment and a new way of facing the problem needs to be discovered by releasing the Boltzmann exploration (thus increasing the temperature). This new adaptive algorithm will be preferable in environments subject to rare and lasting modifications like general industrial processes appear to be. Section 11.3.7 will discuss why, in applications showing recurrent perturbations, a memory capacity should be added in addition to the adaptive one. In Section 11.4, the adaptive and 'population-based' memory capacities will be illustrated by the robot in a maze.

11.3.5 STRUCTURAL PLASTICITY BASED ON THE RECRUITMENT MECHANISM

In process control, the operators to find in each cell are frequently real-valued vectors. Except when a drastic solution such as 'bang bang control' (only two values in each cell) is allowed, it is preferable to search over a large spectrum of possible integer or real values. Even in case of the robot finding its path in a grid maze, more than four possible moves could be contemplated. A possible addition could be, for instance, diagonal moves which would multiply by 2 the number of possible operators in each cell. Since the search space increases exponentially with the number of possible operators by cells, Q_learning performs very poorly when this number exceeds reasonable limits. In order to tackle such a potentially enormous search space, we suggest a new type of reinforcement learning algorithm combining two adaptation mechanisms: Q_learning + intermittent recruitment of new operators (Figure 11.10). This second slower mechanism is akin to the immune metadynamics and makes the resulting methodology reminiscent of Holland's strategy for classifier systems which merges the 'bucket brigade algorithm' (a reinforcement learning weighting the existing rules in a way equivalent to the Q_learning algorithm) with GA responsible for the generation of new rules in the system.

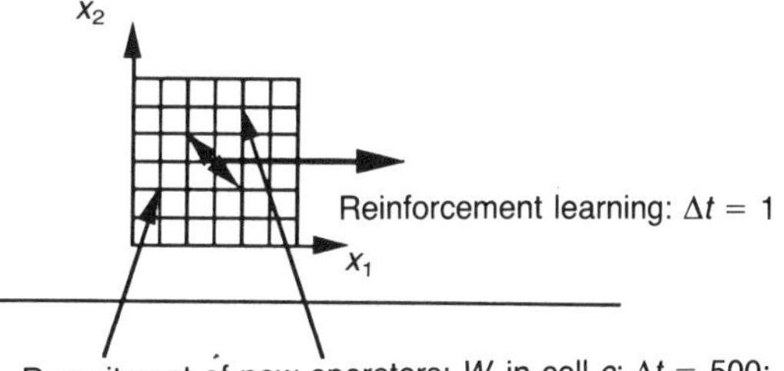

Figure 11.10 Q_learning plus the recruitment mechanism.

The new reinforcement learning is as follows:

1. Initially n operators (n is a small number, in our experiments, between 2 and 5) with random values are generated in each cell. Q_learning is released for a certain number of steps supposed to be approximately the time needed for convergence. In general what happens with this initial set is that (a) either one does not find a satisfactory solution (for instance, the current operators cannot keep the pole balancing) or (b) one does find a solution but it is not very satisfactory (for instance, the robot avoids the obstacles and reaches the target but the path is unacceptably long).
2. At that stage (following a period required for Q_learning convergence, for instance, $\Delta t = 500$ time steps in the cart-pole application), a certain number of cells are selected to replace the worst operators (with the lowest Q_values) contained in them by fresh ones. These new operators are associated with the Q_values of the best ones in the cell so as to make them operational as soon as they get in. The new operators are not generated totally randomly and some of them need to satisfy

a network-based recruitment test. In order to succeed the test, these new operators have to be closer (in the *W* space characterizing each operator) to the best operators with which they present high affinity. The metric on which the affinity is based is the same as the one shaping the parametric plasticity. Roughly, the search proceeds by investigating new solutions close to the best ones obtained so far. However some randomness is still accepted during this refreshing phase: a small fraction of the new operators is generated randomly to induce exploration in new zones of the solution space.

3. Q_learning is released again and again, that is, the cycle goes back to (1) until the discovery of a satisfactory solution.

We believe this two-level learning method to be the only viable alternative for infinite search space when an important number of potential operators is present in each cell. Q_learning is still being run for applications of a reasonable size even if large extensions of the search space become possible. In addition, this revised version handles adaptivity better than the algorithm described in Section 11.3.4 which requires the presence in each cell of all the operators needed to face all the possible situations. The new adaptive mechanism will recruit fresh operators as soon as a breakdown occurs with the possibility of focusing this recruitment on specific cells closer to the breakdown location. As with classifier systems, these two adaptive mechanisms allow the controller to escape the brittleness problem characteristic of a large set of problem solvers constrained by an *a priori* limited representation of the situation to face. However, this recruitment-based adaptive mechanism still deserves deeper analysis and further experiments.

11.3.6 PRINCIPLE 6: THE IMMUNE RECRUITMENT MECHANISM ALONE AS AN OPTIMIZER

This section will present two optimization algorithms: GIRM and Simplex-GA which are directly inspired by immune metadynamics. In previous papers (Bersini and Varela 1990, 1991), IRM was presented as a possible optimization method for functions defined in $\mathbf{R}^n$ where each possible solution is a point in $\mathbf{R}^n$. Here, adopting the classical way to introduce GA, we will restrict ourselves to two methods which aim at optimizing combinatorial problems and functions defined in a Hamming space.

GIRM (genetic immune recruitment mechanism)

Like GA, the algorithm starts with a random population of bit strings and modifies this population in an iterative way expecting to include finally a very satisfactory solution in the last generation of the population. Two preliminary aspects need to be defined:

1. an affinity measure between two bit strings of the population which is inversely related to the distance between the two bit strings. In general this affinity is defined in the solution space (if the genotypic space is different from the phenotypic one, that is the solution space): $m(i, k) = 1 - \text{dist}(i, k)/L$ with $\text{dist}(i, k)$ being the Euclidian distance of the two points i and k encoded by the bit strings and L a scaling parameter.
2. a domain of affinity defined by its radius d_a (if $\text{dist}(i, k) > d_a$ then $m(i, k) = 0$). All further operations to be described will be limited to the individuals having some affinity with the candidate.

The algorithm runs as follows:

1. A candidate k is proposed to be recruited into the current population. Once recruited the fitness value of the candidate will be computed.

2. The individuals presenting some affinity with the candidate k are retrieved: $\{x_i\}$ ($\text{dist}(i, k) < d_a$). Let us call this subpopulation the affine group.
3. The recruitment threshold T is calculated as the mean of the n best points of the affine group.
4. In order to be recruited into the population the candidate k needs to succeed in the recruitment test. This will happen if its average affinity with the m best points (with $m \geqslant n$) of the affine group is superior to the recruitment threshold:

$$\frac{\sum_{i=1}^{m} m(i, k) f_i}{\sum_{i=1}^{m} m(i, k)} \geqslant \frac{\sum_{j=1}^{n} f_j}{n}$$

with f_i being the fitness of point i of the affine group.

Two important parameters m (the number of points in the affine group) and n (the number of points to calculate the recruitment threshold) characterize the recruitment test and in consequence the whole strategy represented by GIRM-m-n. The strategies we tested most frequently were GIRM-2-2 and GIRM-3-2.

1. If the test succeeds, the candidate is locally integrated into the population. It replaces the worst point within the affine group. This type of integration strategy is very similar to the one used in GENITOR (Whitley *et al.* 1991) and called 'one at a time'. Indeed the local best solutions found so far will be held undisturbed in the population until a better solution is obtained.
2. If the search progresses too slowly or if no more candidates succeed in the recruitment test, in order to accelerate and to refine the optimization, the domain of affinity d_a can be reduced through successive generations.
3. In the same way and again expecting to accelerate the search, the algorithm can concentrate gradually on the best regions of the search space, rejecting through the successive generations the worst points of the whole population.

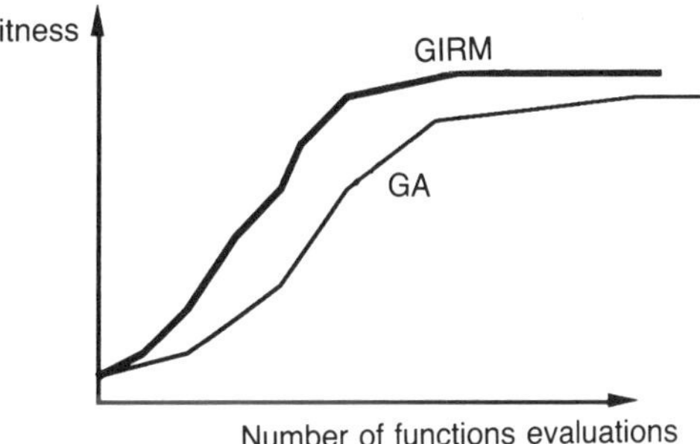

Figure 11.11 Comparing GA and GIRM for F1 and F5.

Figure 11.11 is representative of the way GIRM-2-2 compares favourably with GA both for the unimodal F1 and the multimodal F5 of the De Jong test suite (1975) in terms of the number of times the objective function is calculated. The GA algorithm we used is elitist with a ranking selection strategy. On the Y-axis, the fitness value of the best candidate is indicated. On the X-axis, the number of objective function calculations is indicated.

GIRM can be applied for all kinds of combinatorial optimization problems, provided a metric has been defined in the search space. For instance, for the travelling salesman problem such a metric can be defined as the number of common edges between two tours. Relying on this metric, GIRM has been shown to challenge the best GA version when the same genetic operators were exploited to propose new solutions (Bersini 1991*a*).

Simplex-GA

Since we hypothesized that one explanation for the GIRM improvements in comparison with GA was its respect of basic principles of classical optimization methods (such as a more efficient exploitation of the local fitness values already available), we decided to illustrate this claim by an attempt to combine GA with one of the most famous optimization methods: the simplex. A clear description of the simplex method is provided in Walsh (1980). The use of the crossover operator implicitly assumes what could be called a combinatorial linearity, that is to say the fitness of the combination of two schema is expected to be the sum of the respective fitnesses of each schema (such an assumption is also discussed in Palmer (1991)). As a consequence, it might be a good idea to combine the simplex optimization strategy with GA in order to profit from the best of each: the population + randomness + crossover ideas of evolutionary strategies and good clues coming from optimization methods for the exploitation of the fitness landscape in the neighbourhood of available points.

We have implemented an algorithm called Simplex-GA whose basic difference from GA is the application of a modified crossover to three instead of two bit strings. The new crossover we consider is of an uniform type on account of its good performances and because its extrapolation to the simplex case was easier to derive. A lot of different possibilities for this simplex crossover based on other forms of genetic operators like the family of multipoints crossover might easily be found.

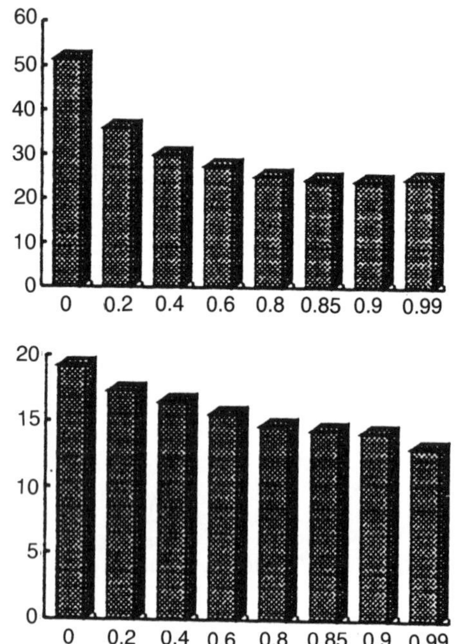

Figure 11.12 Increasing the use of the simplex crossover for F1 and F5.

The simplex crossover

Let I1, I2 and I3 be three bit strings in the current population and their respective fitness values such that I1 > I2 > I3. A new candidate I4 will be generated as follows:

1. If the *n*th bit of I1 is equal to the *n*th bit of I2 then it will be the *n*th bit of I4.
2. If the *n*th bit of I1 is different from the *n*th bit of I2 then the *n*th bit of I4 will be the inverse of the *n*th bit of I3.

Relying on the same kind of metaphor we used originally for presenting GIRM (Bersini and Varela 1990, 1991), I1 and I2 exert an attractive effect on the next point to test while I3 exerts a repulsive effect. In Figure 11.12, the probability of applying in the same GA run simplex-crossover instead of the normal two-individuals crossover is increased from 0 to 0.99 (both for F1 and F5). The graphics show the average number of generations needed to generate an individual above a given high fitness value. Here again the Simplex-GA performs better than the standard GA and facilitates the justification of the GIRM recruitment test, while being easier to implement and faster to run (the recruitment test is now absent). This very simple attempt to marry evolutionary ideas with clues coming from standard optimization methods finds thus a convincing validation. We claim, encouraged by the results just shown, that the emphasis laid by GA users on the exploration side needs to be lessened and that more exploitation can still be compatible with the exploration demanded by multimodal problems. Such a fruitful combination should be pursued further. For instance, in a higher-dimensional space, the simplex involves more than three points ($n + 1$ for an n-dimensional space). Applying a modified crossover on more than three points but still respecting the simplex scheme is worth a try. We also expect that similar attempts can help to fill the gap between the optimization and GA communities which today demonstrate a mutual indifference, although interested in the same kind of problems.

11.3.7 PRINCIPLE 7: POPULATION-BASED MEMORY

Interacting with an unstable but regulatory environment (i.e., the same perturbations occur regularly), an interesting question concerns the profits to be gained by adding a memory capacity to complement the adaptive one. Memory requires space and time, and in very unstable and unpredictable environments the time and the cost to search in memory for an existing solution could exceed the time required to find a new adapted solution (even if this situation occurs repeatedly). Since memory is a fundamental component of immune and cognitive systems, we decided to study the addition of a 'population-based' memory capacity in the whole methodology and to illustrate it on the robot in a maze problem (as the next section will show).

Adding a memory capacity into the original Q_learning implies:

1. A solution for a specific situation is memorized as soon as a satisfactory one is discovered.
2. A solution is the list of operators with a highest Q_value acting in each cell.
3. Each solution must be indexed by the situation it applies to. The way situations are organized in memory is highly dependent on the problem being treated.
4. This indexing must allow for a similarity measure. In case of a situation switch, the solution found for the most similar situation can be a very good solution for this new situation too or might constitute an excellent departure point for the new searching phase.
5. The different situations can be ordered in a list (to be scrolled during its access) based on frequency and/or recency criteria, the most frequent or the most recent situation being the first to be tested.
6. A further criterion for ordering the list can be the quality of the solution associated with each situation (for instance the length of the path for the robot problem). In such a case the first solution retrieved in the list for the current problem will be the best solution obtained so far for this problem.
7. The list is scrolled entirely until a possible solution for a new problem is found.
8. When the memorized solution is not a satisfactory one for the new situation, the adaptive Q_learning can still be released and searched for a new solution. However, initiating the search from a solution found for a similar situation (rather than from the current solution) can accelerate the discovery of the new solution.

The benefit of this memory addition with respect to the basic adaptive mechanism will only depend on the type of unstability and regularity characterizing the problem to be treated.

11.4 SOME RESULTS FOR TWO ILLUSTRATIVE CASES

In this section we describe the application of the methodology to two classical toy problems: the cart-pole and the robot in a maze. Not all the aspects of the method have been applied to both problems. Rather each of the problems will enable us to focus on one original aspect of the method: the recruitment mechanism for the cart-pole and the population-based memory for the robot in the maze.

11.4.1 THE CART-POLE

This problem is a very well-known control application frequently described in the literature (for instance in Barto *et al.* (1983)). It is characterized by four variables: x, θ, dx/dt, $d\theta/dt$ (Figure 11.13). The distribution and the partition are achieved in a four-dimensional space. The aim of the method is to find in each cell a force F between -10 N and 10 N (it is no longer a bang bang control) to exert on the cart so as to keep it viable as long as possible (the viability zone is a specific zone defined on the x and θ coordinates).

The original Q_learning was run and not the network-based version due to the difficulty in that case of retrieving the affinity values between two operators and because this problem was tackled mainly in order to illustrate the contribution of the recruitment mechanism.

1. Initially five operators with random values in [-10 N, 10 N] are generated in each cell. Q_learning runs for 500 steps (one step takes place as long as the pole stays viable).
2. Each 500 steps, five new operators are recruited, one per cell (the five cells are selected randomly), according to the recruitment mechanism which forces the new value in [-10, 10] to be closer to the two best operators than to the three other ones currently acting in the selected cell (it is one possible very simple version of IRM-5-2). This new operator replaces the worst one in that cell and takes the Q_value of the best one.
3. Q_learning is released once again and cycles back to the first stage.

The partition of the state space was adequately assumed to be a critical feature of the problem. Indeed in the case of a partition identical with the one in Barto *et al.* (1983) (i.e., 162 cells: three for x, six for θ, three for dx/dt and three for $d\theta/dt$), no recruitment was needed to find a solution. On the whole, one solution was always in the preliminary random set and ±400 steps sufficed to find it and keep the cart-pole viable. Nevertheless, by decreasing the partition of the state space, the difficulty of the problem increased and the experimental results testified to the need for the complementary recruitment mechanism. For instance, for a 36-cell partition (three for x, four for θ and three for $d\theta/dt$), a satisfactory solution was found only after several recruitment generations. In order to justify the usefulness of the recruitment instead of leaving Q_learning running for longer, a statistical analysis was made which indicated that the final solution always contained several of the recruited operators, thus confirming the need for this complementary mechanism. In addition further tests are still being done to validate the quality of the recruitment based on the immune mechanism with respect to just a random recruitment. It appears that the greater selectivity of the renewal through the recruitment test might be responsible for faster results than simple random renewal; further testing and analysis are needed.

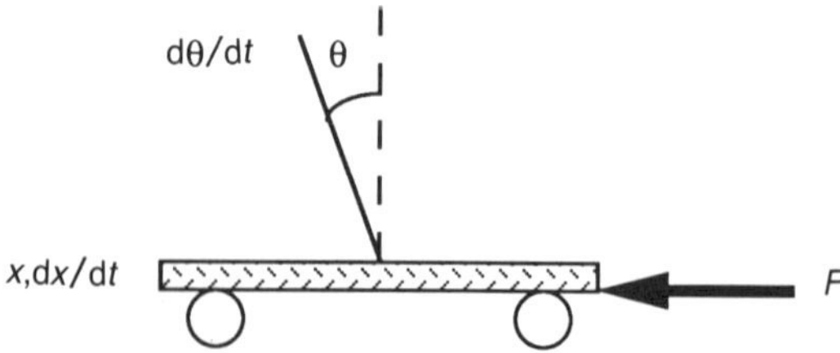

Figure 11.13 The cart-pole problem.

11.4.2 THE ROBOT IN A MAZE

This problem shown in Figure 11.14 was initially presented by Sutton (1990) as an interesting and very illustrative test case for Q_learning. A robot has to find a very short path to reach a target while avoiding bumping into obstacles. In each cell, the robot can move according to the four cardinal directions. Once in the target, it receives a positive

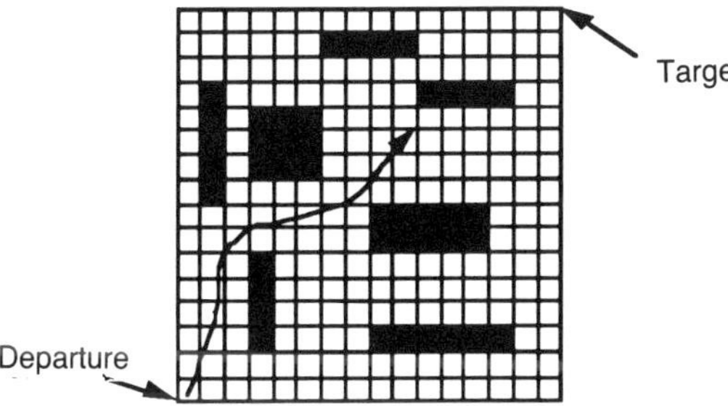

Figure 11.14 The problem of the robot in a maze.

reinforcement and, when bumping an obstacle, receives a negative reinforcement. The network-based Q_learning has been shown by McCallum (1992) to thus accelerate the search in comparison with the basic Q_learning. We are working on a parallel version of the network Q_learning (with many robots departing from different locations and searching together) and we can confirm such results. However this case will enable us to focus more on the adaptive and memory capacities which were introduced in the previous section.

Two adaptive algorithms were implemented: the one described in Sutton (1990), and the one described in Section 11.3.4 which increases exploration (and thus the temperature in the annealing mechanism) just in case of a sudden breakdown. The obstacles were being moved each time a solution was found for a current configuration, and indeed each time a solution was found and stabilized for this situation, it was stored. The various situations being encountered by the robot, in our case the obstacle configuration, were indexed on the basis of the obstacle positions. Thus, two situations were similar if they had a lot of obstacles situated in the same locations.

Now the algorithm with additional memory runs as follows. When a breakdown occurs (i.e., following a reconfiguration of the obstacles, the robot bumps into one of them), the first step is to search in memory for situations already resolved which have the same obstacle (the one causing the breakdown) in the same location. The encoded situations are ranked according to the quality of the solution (the length of the path). If a similar situation is found, the solution associated to this situation is applied. If it is not a satisfactory solution for the problem, the Q_learning is released again, but now based on this new initial retrieved solution. Then once a new satisfactory solution is obtained by improving the retrieved one, the memory is updated with this new solution. Important analogies of this method with case-based reasoning are apparent.

11.5 CONCLUSIONS

This paper presents ongoing developments of an adaptive control methodology whose basic notions are inspired from a model of the immune network. These basic notions are: the distribution of the control, the existence of a viability zone, reinforcement learning and adaptation based on a parametric and structural plasticity sensitive to the network structure, a population-based memory and the isolation and transposing of the immune metadynamics in an optimization scheme. Numerous connections have been underlined with existing methods like reinforcement, namely Q_learning and the classifier system.

This type of work, along with that of others whose common inspirations come from biological systems, appears to us to be fruitful, as two dimensions of the research are well attended to. The first is an emphasis on enlarging the scope of formalism inspired from biological systems. Good examples are the connection of Q_learning with dynamic programming, the combination of GA with well-proven optimization techniques like the simplex method or simulated annealing (Mahfoud and Goldberg 1992), and the exploitation of Aubin's theoretical analysis of the notion of viability (Aubin 1991). Obviously all these methods will be easier to advocate and to utilize if accompanied with solid analytical results, such as proof of convergence.

The second emphasis is likely to be more controversial. To some extent it goes counter to the first one. It consists of an important revision of the classical conception and motivations of the performances of control and engineering systems. For instance viewing control as the optimization of a certain criterion is largely relaxed by the introduction of a viability domain in which the process needs to be maintained rather than driven to a specific target. In the same way, reinforcement learning characterized by a minimum guidance is not the kind of parametric adjustment frequently met in the control literature. The conception of controllers showing important autonomy, intended for interacting with realistic environments impossible to model and to predict with precision, will naturally require a substitution of optimization criteria by 'satisfying' ones, of precise targets by fuzzy ones, of teaching by weakly supervised guidance. This in fact demands highly flexible systems, both at a parametric and structural level, capable of important internal and undirected reorganization. Like the environment they will be designed to interact with, these controllers will keep a large degree of autonomy, an important emancipation with respect to the designer, a potentiality slowly revealed through their interaction with the world to control, and an identity not predetermined but constantly in the making.

REFERENCES

Aubin, J. P. (1991). *Viability Theory*. Birkhauser.

Barto, A. (1990). Connectionist learning for control. In *Neural Networks for Control* (ed. Thomas Miller III, Richard Sutton and Paul Werbos). MIT Press, Cambridge, MA, pp. 5–58.

Barto, A. and Singh, S. P. (1990). Reinforcement learning and dynamic programming. In *Proceedings of the Sixth Yale Workshop on Adaptive and Learning Systems* (ed. K. S. Narendra). New Haven, CT.

Barto, A., Sutton, R. and Anderson, C. (1983). Neuronlike adaptive elements that can solve difficult learning control problems. *IEEE Transactions on Systems, Man and Cybernetics* **13**(5), 834–47.

Bersini, H. (1990). A cognitive model of goal-oriented automatisms and breakdowns. In *Proceedings of the Eighth SSAISB Conference on Artificial Intelligence* (ed. Steels and Smith), pp. 51–61.

Bersini, H. (1991*a*). *Immune network and adaptive control*. IRIDIA Internal Report.

Bersini, H. (1991*b*). Animat's I. In *Proceedings of the First European Conference on Artificial Life* (ed. Varela and Bourgine). MIT Press, Cambridge, MA, pp. 456–65.

Bersini, H. (1992*a*). Reinforcement and recruitment learning for adaptive process control. In *Proceedings of the 1992 IFAC/IFIP/IMACS on Artificial Intelligence in Real Time Control*, pp. 331–7.

Bersini, H. (1992*b*). The interplay between the dynamics and the metadynamics of the immune network. In *The Third Conference on Artificial Life*, Santa Fe 15–19 June.

Bersini, H. (1993). *DORA: A distributed optimizer based on reinforcement algorithms*. IRIDIA Internal Report.

Bersini, H. and Seront, G. (1992). In search of a good evolution-optimization crossover. In *Parallel Problem Solving 2* (ed. B. Manderick and Männer). Elsevier, Amsterdam, pp. 479–87.

Bersini, H. and Varela, F. (1990). Hints for adaptive problem solving gleaned from immune networks. In *Proceedings of the First Conference on Parallel Problem Solving from Nature* (ed. Schwefel and Männer). Springer-Verlag, Berlin, pp. 343–54.

Bersini, H. and Varela, F. (1991). The immune recruitment mechanism: a selective evolutionary mechanism. In *Proceedings of the Fourth Conference on Genetic Algorithms* (ed. Belew and Booker). Morgan Kaufmann, Los Altos, CA, pp. 520–6.

Booker, L. B., Goldberg, D. E. and Holland, J. H. (1989). Classifier systems and genetic algorithms. *Artificial Intelligence 40*, Elsevier, Amsterdam, pp. 235–82.

Brooks, R. (1990). Challenges for complete creature architectures. In *Proceedings of the First SAB Conference* (ed. Meyer and Wilson). MIT Press, Cambridge, MA, pp. 434–43.

De Boer, R. J. and Perelson, A. (1991). Size and connectivity as emergent properties of a developing immune network. *J. Theoretical Biology* **149**, 381–424.

De Boer, R. J., Hogeweg, P. and Perelson, A. S. (1992). Growth and recruitment in the immune network. In *Theoretical and Experimental Insights into Immunology* (ed. A. S. Perelson, G. Weisbuch and A. Coutinho). Springer, New York.

De Boer, R. J., Kevrekidis, I. G. and Perelson, A. S. (1993). Immune network behavior I: from stationary states to limit cycle oscillation; Immune network behavior II: from oscillations to chaos and stationary states, *Bull. Math. Biol.* **55**(4), 745–816.

Deffuant, G. (1992). Reseaux connectionistes autoconstruits. Thèse d'Etat.

De Jong, K. (1975). An analysis of the behaviour of a class of genetic adaptive systms. Ph.D. thesis, University of Michigan. *Diss. Abstr. Int.* **36**(10), 5140B, University Microfilms No. 76-9381.

Farmer, D. (1991). A Rosetta stone to connectionism. In *Emergent Computation* (ed. S. Forrest). MIT Press, Cambridge, MA.

Forrest, S. (ed.) (1991). *Emergent Computation*. A Bradford Book, MIT Press, Cambridge, MA.

Forrest, S. and Mitchell, M. (1991). The performance of genetic algorithms on Walsh polynomials: some anomalous results and their explanation. In *Proceedings of the Fourth Conference on Genetic Algorithms*. Morgan Kaufmann, Los Altos, CA, pp. 182–9.

Goldberg, D. E. (1989). *Genetic Algorithms in Search, Optimization and Machine Learning*. Addison-Wesley, Reading, MA.

Hoffmeister, F. and Bäck, T. (1990). *Genetic Algorithms and Evolution Strategies: Similarities and Differences*. Internal Report of Dortmund University, Bericht Nr. 365.

Holland, J. H., Holyoak, K. J., Nisbett, R. E. and Thagard, P. R. (1986). *Induction: Processes of Inference, Learning and Discovery*. MIT Press, Cambridge, MA.

Kauffman, S. A. (1989*a*). Principles of adaptation in complex systems. In *Lectures in the Sciences of Complexity* (ed. D. Stein). SFI Series on the Science of Complexity, Addison-Wesley, Reading, MA, pp. 618–60.

Kauffman, S. A. (1989*b*). Adaptation on rugged fitness landscapes. In *Lectures in the Sciences of Complexity* (ed. D. Stein). SFI Series on the Science of Complexity, Addison-Wesley, Reading, MA, pp. 527–618.

Lindgren, K. (1992). Evolutionary phenomena in simple dynamics. In *Artificial Life II* (ed. Langton, Taylor, Farmer and Rasmussen). Addison-Wesley, Santa Fe, NM.

Lundkvist, I., Coutinho, A., Varela, F. and Holmberg, D. (1989). Evidence for a functional idiotypic network among natural antibodies in normal mice. *Proc. Nat. Acad. Sci. USA* **86**, 5074–8.

Mahadevan, S. and Connel, J. (1992). Automatic programming of behavior-based robots using reinforcement learning. In *Artificial Intelligence 55*. Elsevier, Amsterdam, pp. 311–65.

Mahfoud S. W. and Goldberg, D. E. (1992). *Parallel recombinative simulated annealing: a genetic algorithm*. IlliGAL Report No. 92002.

Manderick, B., de Weger, M. and Spiessens, P. (1991). The genetic algorithm and the structure of the fitness landscape. In *Proceedings of the Fourth Conference on Genetic Algorithms* (ed. Belew and Booker). Morgan Kaufmann, Los Altos, CA, pp. 143–50.

McCallum, R. A. (1992). Using transitional proximity for faster reinforcement learning. In *Proceedings of the Ninth International Workshop of Machine Learning* (ed. Sleeman and Edwards). Morgan Kaufmann, Los Altos, CA.

Minsky, M. (1986). *The Society of the Mind*. Simon and Schuster, New York.

Packard, N. H. and Bedau, M. A. (1992). Measurement of evolutionary activity, teleology and life. In *Artificial Life II* (ed. Langton, Taylor, Farmer and Rasmussen). Addison-Wesley, Reading MA.

Palmer, R. (1991). Optimization on rugged landscapes. In *Molecular Evolution on Rugged Landscapes: Proteins, RNA and the Immune System* (ed. Perelson and Kauffman). Addison-Wesley, Reading, MA.

Perelson, A. S. (1988). Towards a realistic model of the immune system. In *Theoretical Immunology, Part Two* (ed. A. S. Perelson). SFI Studies in the Sciences of Complexity, Vol. 3. Addison Wesley, Reading, MA, pp. 377–401.

Perelson, A. S. (1990). Theoretical immunology. In *Lectures in Complex Systems* (ed. Erica Jen). SFI Studies in the Sciences of Complexity, Vol. 2. Addison-Wesley, Reading, MA, pp. 465–500.

Singh, S. P. (1992). Transfer of Learning by Composing Solutions of Elemental Sequential Tasks. *Machine Learning* **8**, 323–39.

Sompolinsky, H., Crisanti, A. and Sommers, H. J. (1988). Chaos in random neural networks. *Physical Review Letters* **61**(3), 259–62.

Stewart, J. and Varela, F. (1991). Morphogenesis in shape space: elementary metadynamics in a model of the immune network. *J. Theoret. Biol.* **153**, 477–98.

Sutton, R. S. (1990). Reinforcement learning architectures for animats. In *Proceedings of the First SAB Conference* (ed. Meyer and Wilson). MIT Press, Cambridge, MA, pp. 288–96.

Varela, F. J. and Coutinho, A. (1991). Second generation immune network. *Immunology Today* **12**(5), 159–66.

Varela, F., Coutinho, A., Dupire, B. and Vaz, N. (1988). Cognitive networks: immune, neural and otherwise. In *Theoretical Immunology*, Vol. 2 (ed. A. Perelson). SFI Series on the Science of Complexity, Addison-Wesley, Reading, MA, pp. 359–75.

Varela, F., Sanchez, V. and Coutinho, A. (1989). Adaptive strategies gleaned from immune networks. In *Evolutionary and epigenetic order from complex systems: a Waddington memorial volume* (ed. B. Goodwin and P. Saunders). Edinburgh University Press, UK.

Varela, F., Anderson, A., Dietrich, G., Sundblad, A., Holmberg, D., Kazatchkine, M. and Coutinho, A. (1991). Population dynamics of natural antibodies in normal and autoimmune individuals. *Proc. Nat. Acad. Sci. USA* **88**, 5917–21.

Walsh, G. R. (1980). *Methods of Optimization.* Wiley, London.

Whitley, D., Mathias, K. and Fitzhorn, P. (1991). Delta coding: an iterative search strategy for genetic algorithm. In *Proceedings of the Fourth Conference on Genetic Algorithms* (ed. R. K. Belew and L. B. Booker). Morgan Kaufmann, Los Altos, CA, pp. 77–84.

ARTIFICIAL TISSUE MODELS 12

W. Richard Stark

Computational properties of use to biological organisms or to the construction of computers can emerge as collective properties of systems having a large number of [communicating] simple equivalent components . . . The collective properties are only weakly sensitive to details of the modeling or the failure of individual [components].

Hopfield (1982)

Hopfield's remarks hint at the importance of the study of biological information processing, particularly artificial tissue, to computation theory.

Artificial tissues are abstract models of distributed information processing which formalize some of the distributed information processing that biologists see in living tissues composed of large numbers of communicating cells. For example, Hopfield's famous neural network is an 'artificial tissue' with communication and local processing as seen in neural tissue. The purpose of this chapter is to examine some of these models from computational and mathematical points of view. Special emphasis will be placed on non-neural tissues.

The study of tissue models may be viewed as a part of artificial life as defined by Chris Langton (1989); but it is older and mathematically more tractable than most of the models currently studied in this domain. In *The Chemical Basis of Morphogenesis* (1952), Alan Turing described a model of the information processing involved in the morphogenesis of non-neural tissue. Such non-neural cells communicate through the intercellular fluid and, more directly, through gap junctions (defined in Section 12.2.2). In *The Theory of Self-reproducing Automata* (1966), John von Neumann, using a suggestion of Stanislaw Ulam's, developed the cellular automata model as a formalism for multicellular systems with self-reproductive behavior. In 'Neural networks and physical systems with emergent collective computational abilities' (1982), John Hopfield modified an earlier model by McCulloch and Pitts to allow feedback, irregular lines of communication, and to de-emphasize global synchronization of neurons. The result is the very successful model of neural networks. Aristid Lindenmayer's L-systems, described in *The Algorithmic Beauty of Plants* (Prusinkiewicz and Lindenmayer 1990), are dynamic approaches to exploring plant morphogenesis. Ray Paton, in 'Computing at the tissue/organ level (with particular reference to the liver)' (1991), has begun work on a model of the heterogeneous information processing mechanisms of the liver (a very complex non-neural tissue). These investigators, and others like them, have drawn heavily from biology, computation theory and mathematics, in order to gain insights into the extreme forms of distributed information processing seen in tissues.

Non-tissue models of biological distributed

information processing include: gene regulation networks (Kauffmann 1971), the immune system (Forrest 1990; Perelson 1990), social systems (Stark and Kotin 1989; Waldrop 1984; Reed and Lesser 1984; Chandrasekaran 1981), and ecological systems (Ulanowicz 1989; Kephart, Hogg and Hubermann 1989). These approaches distinguish themselves from tissue models in their representation of information processing, communication and storage; as well as in the problems addressed.

12.1 TISSUES

> Most of the cells in a multicellular animal are organized into cooperative assemblies called tissues which in turn are associated in[to] . . . larger functional units called organs . . . In the space of a few days or weeks, a single fertilized egg gives rise to a complex multicellular organism consisting of differentiated cells arranged in a precise pattern . . . [In] almost every tissue . . . the organization must be preserved even though, in most adult tissues, cells are constantly dying and being replaced.
>
> Alberts *et al.* (1989)

This excerpt provides us with a notion of what a tissue is, and a description of two global computational processes of tissues. Tissues include epithelial tissue, connective tissue, muscle tissue, neural tissue and so on. In each, the cells are of a specialized and relatively homogeneous type and function. (Actually, most tissues are an intricate mixture of secondary cell types, ancillary to the specialized function of the primary type, which originate outside of the tissue and invade it during the early stages of development. See Chapter 17 of Alberts *et al.* (1989).) The global computational processes mentioned above are seen as morphogenesis and maintenance. (A computational process is one in which information is communicated, processed and stored by an algorithm which is, in some sense, repeatable, i.e., not random.) Other processes include the global pulsing (the cardiac pacemaker), pattern formation as in the formation of leopards' spots – see Turing (1952) and Pool (1991), hormone production, the antibiotic reactions of epithelial cells (Loewenstein and Kanno 1964) and possibly the amplification of received hormonal signals.

12.2 BIOLOGICAL INFORMATION PROCESSING

> Cells in a multicellular organism need to communicate with one another in order to regulate their development and organization into tissues, to control their growth and division, and to coordinate their functions. Animal cells communicate in three ways: (1) they secrete chemicals that signal to cells some distance away; (2) they display plasma-membrane-bound signaling molecules that influence other cells [which come into] direct physical contact; and (3) they form gap junctions that directly join the cytoplasms of the interacting cells, thereby allowing the exchange of small molecules.
>
> Alberts *et al.* (1989)

These biologists base tissue processes on intercellular communication and the information processing abilities of cells. To a computer scientist, the system being described is an enormous (about 10^{12} cells/kg) distributed computation; to a mathematician it is a complex dynamical system.

12.2.1 STATES AND MEMORY

Like finite state automata, cells have memory in the form of variable internal states. They respond to external signals by a state change of some sort, for example, the activation/deactivation of genes and progress through the cell cycle. Even if a cell's internal memory were determined only by patterns of gene activation, then the number of discrete states could be astronomical (Kauffmann 1971).

Thus, I feel safe in assuming that the internal computational complexity of individual cells is non-trivial.

12.2.2 COMMUNICATION

> The individual cells within a multicellular organism do not operate autonomously but instead interact together in a controlled and co-ordinated manner. In order that an organism can function . . . there must be some form of communication between the component cells . . . [It] may take the form of direct communication between adjacent cells . . . through permeable membrane channels known as gap junctions.
>
> MacDonald (1985)

In most animal tissues (including neural tissue, but excepting blood) cell-to-cell communication also occurs through gap junctions – tiny channels directly connecting the cytoplasms of neighboring cells and capable of transmitting ions and small (molecular weight <2000) molecules – (Schwartzmann *et al.* 1981). Gap junctions are simpler, faster and metabolically less expensive than synaptic communication.

Figure 12.1(a) is taken from the point of view of one cell's nucleus looking along the axis of the junctions' channels toward the neighboring cell. Figure 12.1(b) shows individual gap junctions crossing the plasma membranes to connect the cytoplasms of neighboring cells.

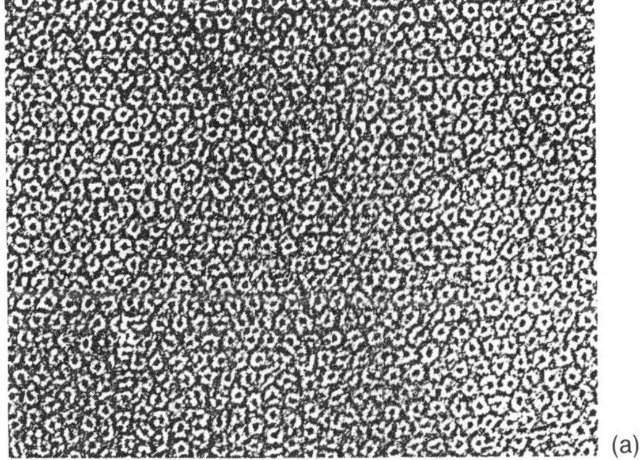

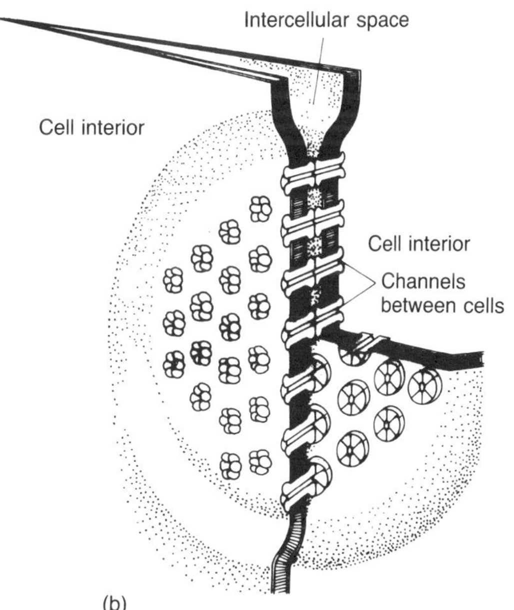

Figure 12.1 Gap junctions.

In term of abstract function, there are several different types of cell-to-cell communication. Communication between neurons via synapses is specific (i.e., a given synapse supports communication between two fixed neurons), directed (the presynaptic neuron may signal the postsynaptic neuron, but not vice versa), and not anonymous. Communication is anonymous if the sender is not identified, or in other words, a receiver's response does not depend on which neighbor sent the signal. Communication via gap junctions is specific, but not usually directed (it communicates as well in one direction as in the other) and seems to be anonymous (Yamashita and Kameda 1989). Hormonal communication is directed and anonymous (at the cellular level), but not specific (at the cellular level).

All forms of cell-to-cell communication are alike in that they are asynchronous at the cellular level. Some tissues exhibit an interesting rough synchronization (Gray *et al.*

1990), but I am using the term in a stronger sense. A distributed system is synchronous if the individual component processes act in strict lock-step. In other words, the internal information processing cycles of all individuals begin and end at exactly the same moment, and communication occurs strictly between these cycles. Something equivalent to broadcasting the ticks of a global clock is necessary to achieve synchronization in this strong sense. This strict notion is a major bifurcation point in the mathematical theory of distributed processes (in strictly synchronous processes, classical algebraic techniques are applicable to the analysis of the dynamics of global states).

12.2.3 VOCABULARY AND GLOBAL SIGNAL SPEED

The size of the cell-to-cell vocabulary is also an indication of the potential power of tissue processes. A typical cell has the following small molecule profile: H_2O, 20 different inorganic ions, 200 carbohydrates and precursors, 100 amino acids and precursors, 200 nucleotides and precursors, 50 lipids and precursors, and 200 others (all with molecular weight <2000). Thus, cells in animal tissues have a potential gj vocabulary of 800 words.

The gap junction coordinated contraction of the heart muscle (DeMello 1982) shows that the global signals may travel as fast as $10^{1.5}$ cm/sec in some animal tissues.

12.2.4 MEASURING BIOLOGICAL INFORMATION

Given a vocabulary of 800 molecules and ions (mi) and assuming that 1% of them carry information, a signal molecule is counted as $3 = \log_2(8)$ bits of information. A single junction can transmit up to 10^6 mi/sec – so 10^2 mi/sec with a redundancy of 10^{-2} signals/mi gives an estimated 1 signal/gj/sec. Using these numbers, the information rate per junction is

$$3 \text{ bits/signal} \times 10^0 \text{ signals/gj/sec} = 10^{0.5} \text{ bits/gj/sec.}$$

Neighboring cells are joined by arrays of hundreds (say $10^{2.8}$) junctions. Assuming that a cell has an average of eight neighbors, and that each pair of neighboring cells is, on the average, joined by one array of junctions; a typical cell in a typical tissue has about $2500 = 10^{3.4}$ junctions. Since each gap junction is counted twice (once for each cell that uses it), the information rate per cell is approximately

$$10^{0.5} \text{ bits/gj/sec} \times 10^{3.4} \text{ gj/cell} \div 2 = 10^{3.6} \text{ bits/cell/sec.}$$

Finally, there are about 10^{12} cells per kilogram of tissue, giving a estimated gj information rate of

$$10^{3.6} \text{ bits/cell/sec} \times 10^{12} \text{ cells/kg} \approxeq 10^{15.6} \text{ bits/kg/sec}$$

in non-neural tissue similar to skin or liver. (These calculations are based on what seem to be reasonable estimates of the magnitudes involved – estimates which I will consider successful of they are within $10^{\pm 1.5}$ of the actual value. In spite of their crudeness, the estimates are convincing evidence that non-neural tissues are computationally non-trivial.)

> The human cerebral cortex . . . is comprised of 10^{11} neurons with each having an average of 1000 dendrites that form 10^{14} synapses: given that this system operates at 100 Hz, it functions at about 10^{16} interconnections per second.
>
> DARPA (1988, p. 77)

At 10 bio-bits per interconnection, neural tissue processes about 10^{17} bits/sec/kg. At this point one may ask why are certain non-neural tissues apparently capable of moving (and processing?) information at a rate which is one or two orders of magnitude less than that of the brain?

12.2.5 BIOLOGICAL INFORMATION PROCESSING

So far, our concepts apply to tissue-based information processing in general, with neural tissue as a special case. However, from now on, we will focus on non-neural tissue with gap junction (gj) communication only. We will assume that cell states and cell-to-cell signals are discrete rather than continuous. This assumption is supported by the views of theoretical biologists, for example, Kauffmann (1971). The local context is represented in cell state or the multiset of neighboring cell states. Time is considered to be discrete. The effect of this assumption is not entirely clear – maybe it is similar to the effect to replacing analog computers by digital computers.

In this section, I have viewed cells as being finite state automata, and tissues as large communication networks of cells. In the next section, this bio/computational view will be given a mathematical formalism.

12.3 ABSTRACT MODELS – ARTIFICIAL TISSUES

> Will some . . . great discoveries or theories in biology be found and formulated on the basis of abstruse mathematical theories . . . [and] mathematical deductions from general principles? . . . [P]erhaps . . . forms of life might exist . . . that embody general properties of what we call life, and may yet not be based on all the special processes of chemistry that seem so universal or exclusive on our earth.
>
> S. M. Ulam (1972)

A cell X is represented as a finite state automaton, a tissue as a large but finite collection C of cells joined, by communication edges, into a graph (C, G). The graph is defined by a neighborhood function $G: C \rightarrow \mathscr{P}(C)$ specifying the set $G(X)$ of neighbors of X. Naturally, a degree of randomness or irregularity in (C, G) is to be expected. Communication is in the style of gap junctions – local, anonymous, non-directed and asynchronous. (C, G) for a gj tissue is shown in Figure 12.2.

12.3.1 OUR STATIC MODEL

> [A] mathematical model of a growing embryo [is] described . . . [it] will be a simplification and an idealization . . . cells are geometrical points . . . One proceeds as with a physical theory and defines 'the [global] state of the system'.
>
> Turing (1952)

Turing's models were to be used in a study of the formation of global patterns expressed by cell states (rather than changes in (C, G)) and so were static. Static models contrast with the dynamic models studied by Lindenmayer in which cell division, death and even migration were primary variables.

The tissue structure, as expressed by (C, G), is fixed and somewhat random for each instance of a tissue. Thus, even though (C, G) can be chosen from infinitely many possible connected graphs, it will not change during processing, and the theory will not be sensitive to the details of (C, G) structure. This leads to a novel computation theory in which (C, G) is, for the most part, a dummy variable. In other words, computational properties will not depend on (C, G), as long as it has certain basic assumed properties such as finiteness, connectedness and non-directedness.

Each cell's response to gj signals is programmed as a finite-state automaton with states and inputs from a finite set L of cell states, and state transition function $\alpha: L^2 \rightarrow L$. The initial state q_0 and accepting state set F are not used. Assuming only that (C, G) is connected and non directed; the model is specified by

$$\alpha: L^2 \rightarrow L, \quad \text{and} \quad \nabla: \mathscr{M}(L) \rightarrow L,$$

where $\mathscr{M}(L)$ is the set of all multisets of values from L, and ∇ is a function which converts a multiset of neighboring cell states

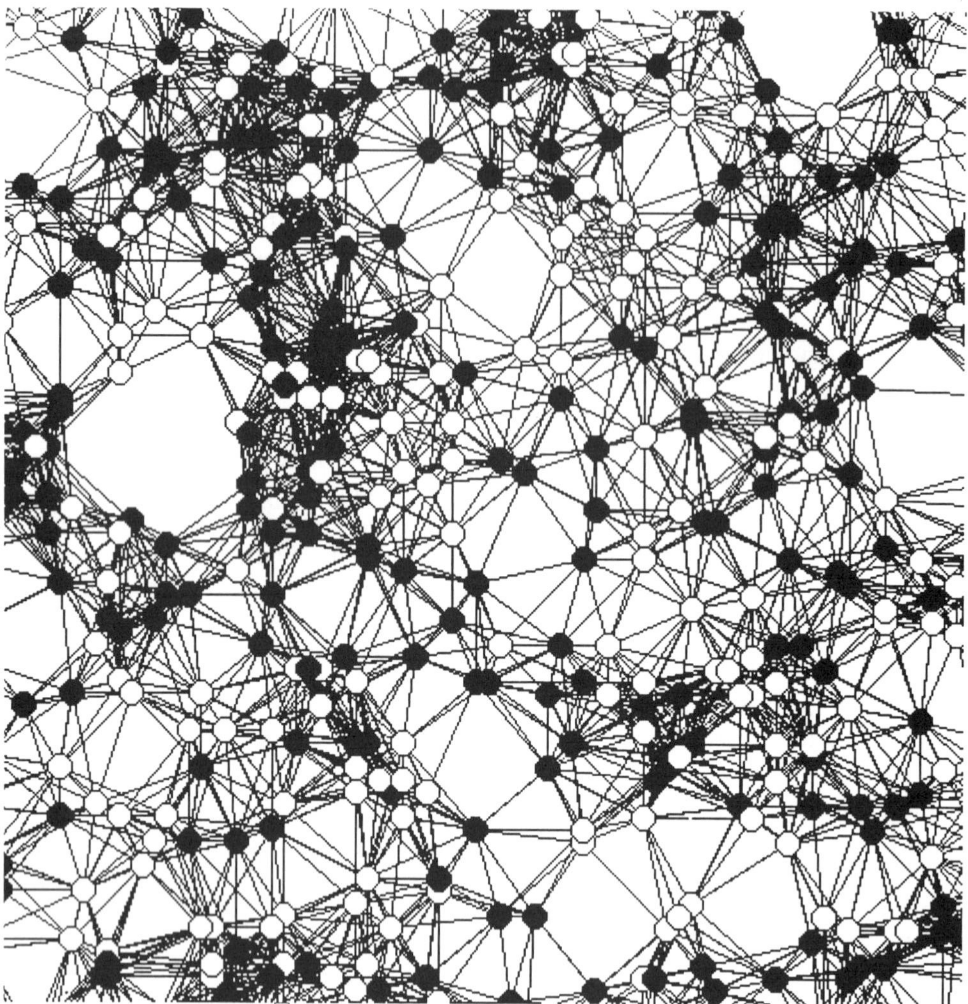

Figure 12.2 (*C*, *G*) for a gj tissue (cell shade indicates state).

into the input signal $m \in L$. Specifically, each active cell *X*, has its activity defined by

$$s'_X = \alpha(s_X, m_X)$$

$$\text{where} \quad m_X = \nabla(\eta_x)$$

$$\text{and} \quad \eta_x = \{s_Y \mid Y \in G(X)\}.$$

s_X is *X*'s current state, m_X is *X*'s input, and s'_X is *X*'s next state. The point of ∇ is to compute an input from the states of a variable number of anonymous neighbors (see Figure 12.3).

A schedule is a random function $\sigma: \mathbb{N} \to \mathscr{P}(C)$ specifying cell activity as a function of time. Activity is asynchronous in that at any moment *i* in $\mathbb{N}$ (the non-negative integers) a random set $\sigma(i) \subseteq C$ of cells is actively reading gj signals from neighbors and recomputing their cell states *s*. (Since *C* is finite, a random $\sigma(i) \subseteq C$ can be determined from the uniform distribution on *C*.) Synchronous activity is not important here, but if it were then it would be produced by the constant schedule – $\sigma(i) = C$ for all $i \in \mathbb{N}$.

The tissue's global states are a function

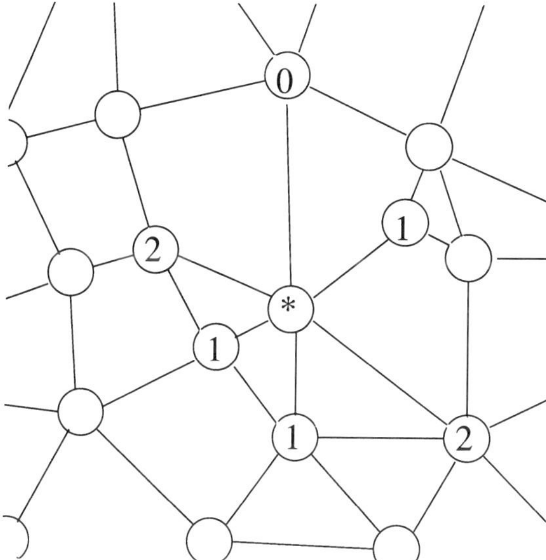

Figure 12.3 If X is the * cell, then $\eta = \{0, 1, 1, 1, 2\}$ is the value used in $m = \nabla(\eta)$.

$$\bar{s}: C \to L \quad \text{defined by } \bar{s}(X) = s_X \quad \text{for each} \quad X \in C.$$

In state $\bar{s}_i$, at time i, the effect of activity by the cells of $\sigma(i)$ is to produce the next tissue state

$$\bar{s}_{i+1}(X) = \begin{cases} \alpha(\bar{s}_i(X), m(X)) & \text{if } X \in \sigma(i), \\ \bar{s}_i(X) & \text{otherwise.} \end{cases}$$

Computational notation for this relation is $\bar{s}_i \rightarrow_{\sigma(i)} \bar{s}_{i+1}$. Every initial tissue state $\bar{s}_0$ and every schedule $\sigma \in \mathscr{P}(C)^N$ generates a global computation

$$\bar{\bar{s}}^{\sigma}: \bar{s}_0 \Rightarrow_{\sigma(0)} \ldots \bar{s}_i \Rightarrow_{\sigma(i)} \bar{s}_{i+1} \Rightarrow_{\sigma(i+1)} \ldots$$

Notice that, since the activity is non-deterministic, computation is non-deterministic. However, in Section 12.5, important global properties of these processes will be shown to be deterministic. The mathematical theory for this model will also be introduced in Section 12.5.

12.4 PROBLEMS, RESULTS AND SIMULATIONS

With so much chaos at the local level (i.e., communication that is asynchronous, anonymous and irregular) it is difficult to imagine how well-behaved global processes could result. How does a heart beat without a signal from the brain? How are a leopard's spots computed during its development? How are the cells of a developing fetus coordinated to produce a well-formed adult? These are all instances of the famous coordination problem of biological information processing.

In this section I consider instances of the coordination problems as well as a naive automorphism argument that highlights some of the reasons for believing that the coordination problem is non-trivial. Specifically, the automorphism-based argument seems to indicate that significant global organization in a model such as I have described is impossible.

12.4.1 LEOPARD'S SPOTS AND GLOBAL PATTERNS

> It is suggested that a system of chemical substances, called morphogens, reacting together and diffusing through a tissue, is adequate to account for the main phenomena of morphogenesis. Such a system, although it may originally be quite homogeneous, may later develop a pattern or structure due to an instability of the homogeneous equilibrium, which is triggered off by random disturbances. Such reaction-diffusion systems are considered in some detail . . . The most interesting [systems] form stationary waves . . . [which], in two dimensions, give rise to patterns reminiscent of dappling.
>
> Turing (1952)

Imagine the kilogram or so of skin in a developing leopard fetus. This two-dimensional tissue is composed of roughly 10^{12} cells, all programmed alike, all communicating with a few neighbors, all doing their job in relative isolation and without the aid of an unnatural form of synchronization. Turing suggests that at this late point in its development, the tissue begins to compute a (previously undeveloped) parameter describing this leopard-to-be's spots. Until then, the epidermal tissue lacked a data structure corresponding (in scale and distribution) to the spots. But now, the 10^{12} cells cooperatively develop a pattern of about 10^2 similar spots – each containing billions of cells with a few hundred billion cells between the spots.

How can the cells be programmed to generate the global computations of a spot-generating process? Turing developed an analytic program based on diffusion-reaction instabilities. His model was a graph (C, G), with C a set of reacting (i.e., computing) cells and G the set of diffusion (i.e., communicating) edges. Reaction-diffusion mechanisms were introduced as the basis for morphogenesis (and spot formation) and an analysis verified their behavior for simple graphs. Turing went on to point out the mathematical similarities between the methods used on this spatial morphogenesis problem and the temporal oscillator problems. Turing's ideas are nicely redeveloped by Murray (1989), and, recently, chemically confirmed by Lengyel and Epstein (1991).

12.4.2 AN AUTOMORPHISM ARGUMENT FOR SPECIAL SCHEDULES

At first, the following argument seems to suggests that such a spot computation is impossible, but the reasoning is valid only for synchronous and other special schedules. The argument does, however, show why such coordination problems are considered difficult.

Let (C, G) be a non-directed connected graph of 10^{12} cells which is homogeneous enough to have an automorphism* f such that, for all $X \in C$ and some $k < 10^{12}$, f's orbit

$$\{X, f(X), f^2(X), \ldots, f^{k-1}(X)\}$$

is non-repeating until $f^k(X) = X$. Let $\bar{s}_0$ be a state which is constant, or at least homogeneous enough to satisfy

$$\bar{s}_0(X) = \bar{s}_0(f(X)) \quad \text{for all } X \in C;$$

and σ be a schedule† satisfying

$$\sigma(i) = f(\sigma(i)) \quad \text{for all } i.$$

An argument by induction on time shows that the resulting computation

$$\bar{s}_0 \Rightarrow_{\sigma(0)} \ldots \bar{s}_i \Rightarrow_{\sigma(i)} \bar{s}_{i+1} \Rightarrow_{\sigma(i+1)} \ldots \Rightarrow_{\sigma(j)} \ldots$$

consists of global states satisfying

$$\bar{s}_i(X) = \bar{s}_i(f(X)) = \bar{s}_i(f^2(X)) = \ldots = \bar{s}_i(f^{k-1}(X)) = \ldots$$

Proof. The result for $\bar{s}_0$ is given in the previous paragraph, so let $\bar{s}_i(X) = \bar{s}_i(f(X))$ be our induction hypothesis. Since f is an automorphism, $G(f(X)) = f(G(X))$.

$$\bar{s}_{i+1}(f(X)) = \alpha(\bar{s}_i(f(X)), \nabla(\bar{s}_i(G(f(X))))) = \alpha(\bar{s}_i(X), \nabla(\bar{s}_i(G(X)))) = \bar{s}_{i+1}(X).$$

Suppose, without loss of generality, that the spot computation begins at i, that the spots are fully developed at j, and that at time j the spot cells are indicated by $\bar{s}(X) = 0$ and the intermediate cells by $\bar{s}(Y) = 1$. In other words, X is in a spot and Y is between spots, if and only if $\bar{s}_j(X) \neq \bar{s}_j(Y)$. The result of the previous paragraph and the edge-preserving property of automorphisms indicate that,

*f: $(C, G) \to (C, G)$ is an automorphism on (C, G) if it is one-one, onto, and preserves edges – i.e., $(\forall X, Y \in C) \times [X \in G(Y) \text{ iff } f(X) \in G(f(Y))]$.

†For example, given a graph (D, H) of order $10^{12}/k$ with a non-void subset $E \in D$, define (C, G) to be the result of joining k copies $(D_1, H_1), \ldots, (D_k, H_k)$ of the given graph at E (by adding edges (X_i, X_{i+1}) between corresponding cells X in D_i and D_{i+1}); define f by $f(X_i) = X_{(i+1)\text{mod}(k)}$ and define σ to be the synchronous schedule $\sigma(i) = C$.

in $\bar{s}_j$, f takes spots to spots. So $\bar{s}_j$ has at least $(k - 1)$ spots.

What does this prove? Intuitively, k should be quite variable, resulting in large differences in the number of spots possible. For very small k, the leopard should be patchy like a dog. For values of a million or greater, the leopard will go from a finely textured pattern to a uniform blend of spot and background colors. Obviously, neither nature nor Turing had such variability in mind.

Are our assumptions theoretically acceptable? The algorithm is assumed to be robust – that is to say it works properly in all connected graphs and from all $\bar{s}_0$ – so the assumptions of homogeneity, while very special, will not disrupt the algorithm. The assumption that $\bar{s}_0$ is homogeneous implies the tissue, at some time, was of a single color. Everything in the argument was acceptable until I described the schedule. In the space of all schedules, there are infinitely many schedules satisfying the orbit-preserving condition on σ, but they are countably infinite and so occur with probability 0. These schedules eliminate the true randomness that is a source of homogeneity-erasing instabilities – an important part of Turing's process.

Thus, this automorphism argument is valid for synchronous and other highly controlled schedules of cell activity, but they are schedules that occur in nature (and in our model) with probability 0.

12.4.3 PULSING TISSUES AND COUPLED OSCILLATORS

Living organisms are full of biological oscillators – tissues whose job it is to orchestrate cellular activity into global periods. Such oscillators generate a rough global synchronization different from the previously mentioned precise synchronization of external origin. Among the best paradigms is the heart muscle. Under the direction of internally generated signals, it contracts at a rate of approximately 10^6 beats/week. Its activity is orchestrated, without the help of neural tissue, by its own computationally active muscle cells – cells which use gj communication to transmit an information wave back and forth, between two small reflecting/generating subtissues – see 'cardiac pacemaker' in Epstein (1991); DeMello (1982).

Can a large, homogeneous (without structures such as the pacemaker's reflecting subtissues), and loosely coupled set of oscillators become globally synchronized and then pulse? Alan Hobson (1975) asserts: '[We have] the hypothesis that reciprocal interactions between functionally interconnected cell populations may determine the cyclic alternation of [global] behavior states.' He is describing a tissue in which the cells (acting as local oscillators) respond to a feedback mechanism to generate global oscillations. It is a tissue in which each cell reads the state of the majority of its neighbors, then enters a different state – say, the most different.

A model of this type, **H**[k], is described (on a static graph) with states $L = \{0, \ldots, k - 1\}$ and the cell program

$$\nabla(\eta) = \begin{cases} a & \text{if } a \text{ occurs in } \eta \text{ more than any other value,} \\ 0 & \text{otherwise.} \end{cases}$$

$$\alpha(s, m) = \left(m + \left\lfloor \frac{k}{2} \right\rfloor\right) \bmod k,$$

where $\eta \subseteq L$ is a multiset and $m = \nabla(\eta) \in L$.

For $k = 2$, information processing in this tissue is proved (next section) to reach a global halting state ($\bar{s}$ is a halting state if $\bar{s}(X) = \alpha(\bar{s}(X), (\nabla \cdot \bar{s}G)(X))$ for all $X \in C$) with probability 1. In other words, it does not pulse. This is shown in Figure 12.4 (of simulation data) and rigorously proved in the next section. However, for $k > 2$ processing in this gj model has been proved (next section) to never halt; and so it could pulse. For $k = 2$ with directed (i.e., neural) communication, the author has proved it to never halt (Figure 12.5).

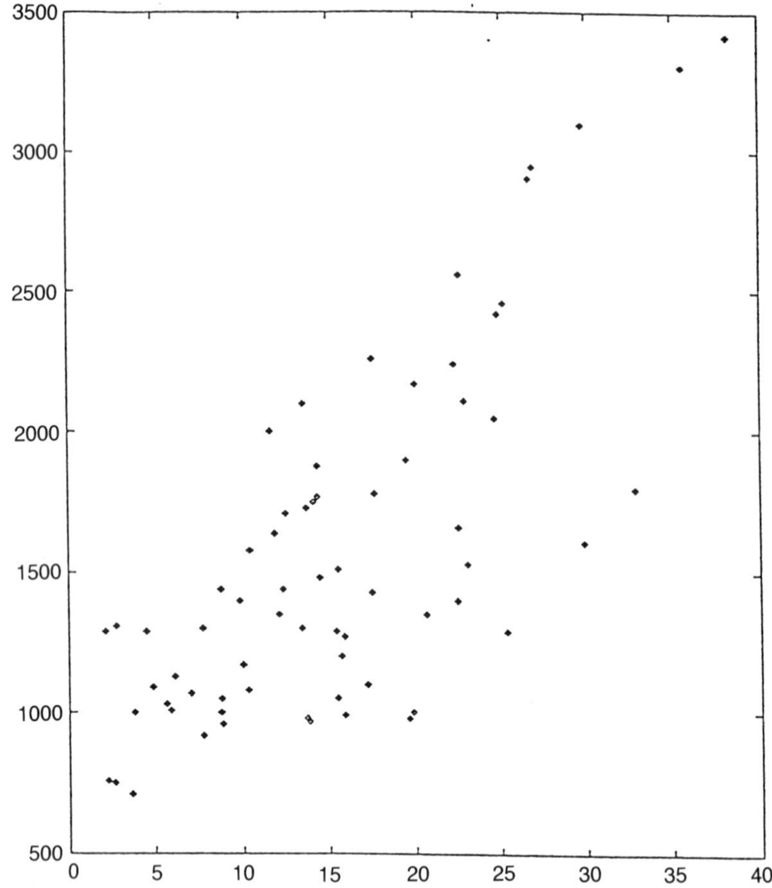

Figure 12.4 **H**[2]: the number of steps to halting as a function of connectivity.

A global computation is said to pulse if the value of

$$z(n) = \frac{1}{\| C \|} \cdot \| \{X \in C \,|\, \bar{s}_n(X) = 0\} \|$$

varies with large amplitude and approximately regular periods. **H**[2] rarely exhibits pulsing behavior. Pulsing behavior seems to be linked to a decrease in value for the global variance $\mathrm{var}_{(C,G)}(\bar{s}_n)$ (defined next section).

This suggests that programs which decrease local variance might have tendencies toward self-organization which would result in global pulsing. Constant states $\bar{s}_n$ have the lowest possible variance, but without local change there can be no global pulsing. States with a form of continuity should have low variance without halting. Let $(L, H_m \,|\, m \in L)$ be the state-transition graph for the cells' automaton, that is, $H_m = \{(r, s) \,|\, s = \alpha(r, m)\}$. The global state $\bar{s}$: $(C, G) \rightarrow (L, H)$ is continuous if the image of every edge $(X, Y) \in G$ is either a single value $\bar{s}(X) = \bar{s}(Y)$ in L or an edge $(\bar{s}(X), \bar{s}(Y)) \in H \subseteq L^2$.

The cell program **I**[k], on $L = \{0, \ldots, k-1\}$ promotes and preserves continuity

$$\nabla(\eta) = \begin{cases} a + 1 & \text{if } \eta \subseteq [a, a], \\ b & \text{if } \eta \subseteq [a, b] \text{ where} \\ & a < b \leqslant (a + 2)\text{mod}(k), \\ \text{nil} & \text{otherwise;} \end{cases}$$

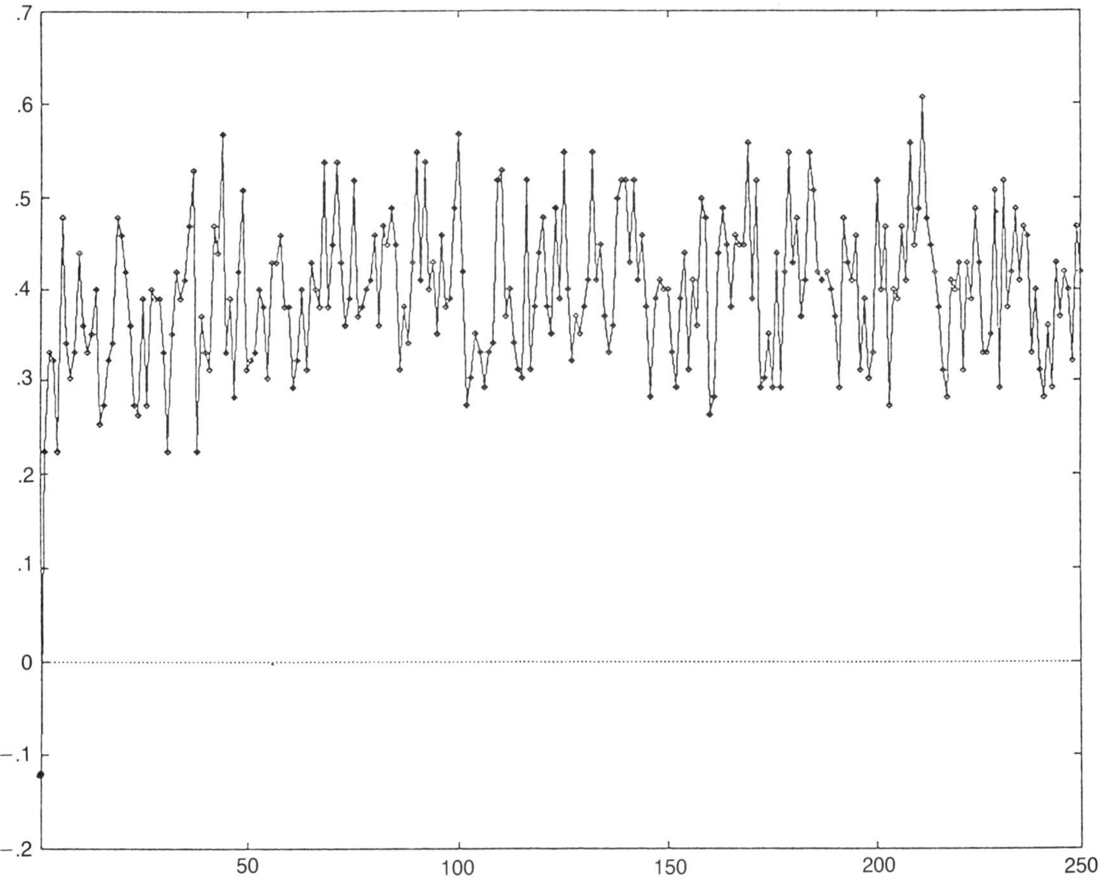

Figure 12.5 **H**[3] never halts, but it is not clear that it pulses.

where $[a, b]$ represents the smallest interval, in the circle of values of L, which contains the neighboring states η; and

$$u(s, m) = \begin{cases} m & \text{if } m \neq \text{nil} \\ s + 1 & \text{otherwise.} \end{cases}$$

For theoretical reasons, I expected the variance of this process to decrease dramatically, and, as it did, to see global pulsing begin. Simulations confirm these expectations (Figure 12.6).

The discontinuity measure is

$$d(n) = \frac{1}{ed} \cdot \| \{(X, Y) \in G \mid 2 < | \bar{s}_n(X) - \bar{s}_n(Y) | \} \|$$

where *ed* is the expected number of discontinuities in a random state. At (approximately) steps 50, 550 and 700 discontinuity decreases dramatically. These steps correspond to the process falling into wells and wells-within-wells of permanent continuity for large parts of (C, G). Pulsing behavior (as in the second graph) begins when the global states enter one of the wells. The graphs for $\mathbf{I}[k]$ are from a tissue with $\| L \| = 6$, $\| C \| = 100$, each cell having an average of 5.14 neighbors, and running for 2000 steps with a random set of about nine cells active at each step. The pulses show $z(n)$ $(500 < n)$ varying from 0.1 to 3.0 of its expected value with a period of about 200 steps or 1800 cell activations. Cells

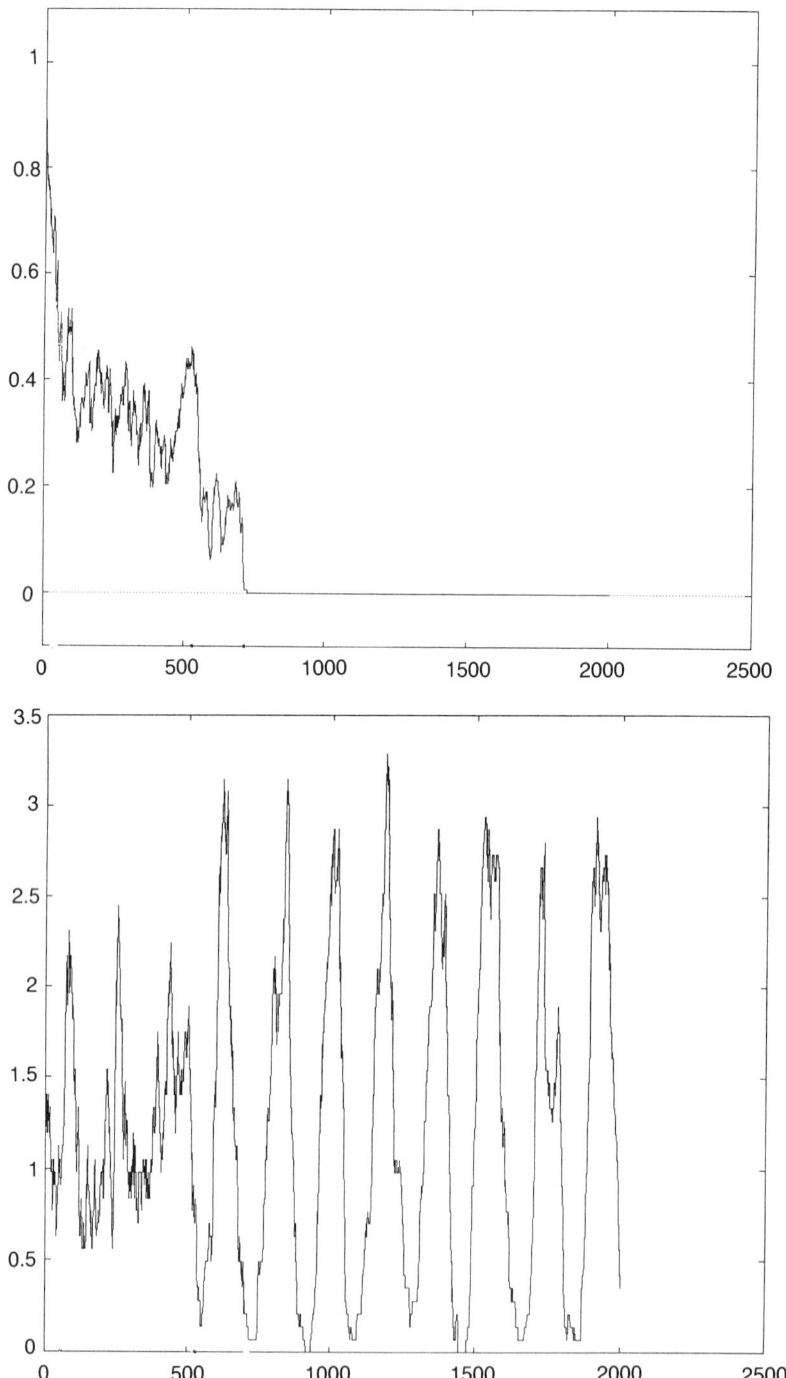

Figure 12.6 **I**[k] ($2 < k$): graphs of $d(n)$ and $z(n)$.

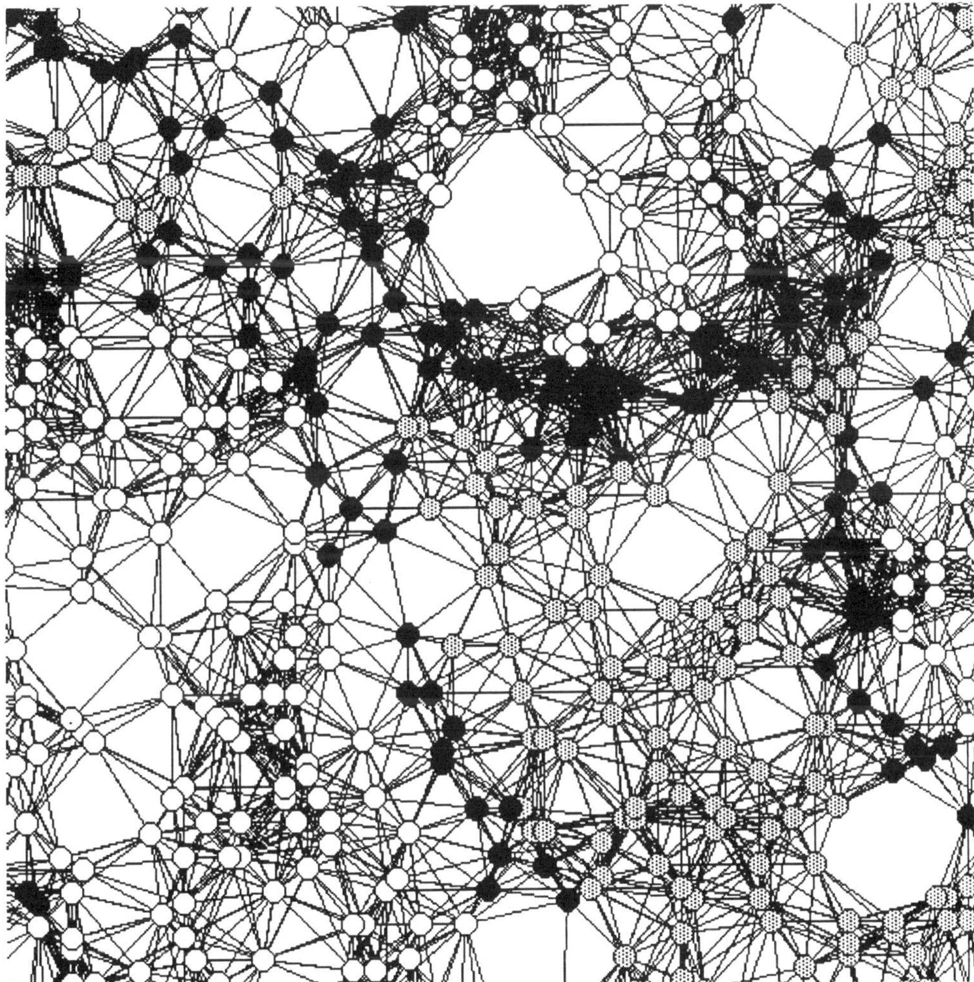

Figure 12.7 Organ with three concatenated tissues (A = grey, B = black C = white).

are, in effect, attempting to reduce, and never increase, the entropy around them. I[k] eventually produces a global state which looks like a smooth wave. Since α is forcing the wave to move around L, a pulsing in $z(n)$ results.

12.4.4 TISSUE COMPOSITION AND THE EVOLUTION OF MEMBRANES

An organ is composed of three or more concatenated tissues as shown in Figure 12.7 – say A (for the interior cells), B (the membrane cells surrounding A), and C (exterior cells touching B cells but not A cells). Suppose that they have a common vocabulary $L = L_A \cup L_B \cup L_C$ (with L_A, L_B, L_C pairwise disjoint). Cells in A have states in, and read states from, L_A only; B cells have states in L_B and reads states from $L_A \cup L_B$; and C cells have states in L_C but reads states from $L_B \cup L_C$ only.

Organ mechanism

Global states in A can be read by B, and then read and altered by C, but not vice versa. This is an information diode which seems to provides a means of composing A, B and C to compute C·A.

Problem

Does this mechanism work? What other types of gross structure could play a role in large-scale information processing? Could such mechanisms have played a role in biological evolution (Stark 1992)?

12.5 MATHEMATICS

The following approach to understanding properties of our non-deterministic static model is related to the famous Baire category theorem of real analysis. Let $\zeta_n(x)$ $n \in \mathbb{N}$ be infinitely many properties of real numbers $x \in [0, 1]$. If each $\mathcal{O}_n = \{x \in [0, 1] | \zeta_n(x)\}$ is a uniformly dense, open set of reals; then, with probability 1.0, a random real $r \in [0, 1]$ will satisfy every property. In this section, we will treat global computations and schedules as if they were real numbers. A topology and measure will be developed that will support this type of reasoning for the random computations of our model. Then, results form the previous section will be proved.

12.5.1 THEORY

Let $\mathcal{S}$ be the set of all schedules. The length of a finite schedule ρ is $|\rho| = \|\text{domain}(\rho)\|$. Concatenation between finite $\rho \in \mathcal{S}$ and $\sigma \in \mathcal{S}$ is defined as usual – $(\rho\sigma)(i) = \rho(i)$ if $i < |\rho|$ and $(\rho\sigma)(i + |\rho|) = \sigma(i)$ otherwise. For finite ρ, $\bar{s}^{\rho}$ denotes the terminal state of the finite computation

$$\bar{\bar{s}}^{\sigma}: \bar{s}_0 \Rightarrow_{\rho(0)} \ldots \Rightarrow_{\rho(|\rho|-1)} \bar{s}_k^{\bar{\rho}}.$$

τ extends ρ, in symbols $(\rho \sqsubseteq \tau)$, if $(\exists\sigma)(\rho\sigma) = \tau]$.

The schedules are partially ordered in $(\mathcal{S}, \sqsubseteq)$. The set $\mathcal{C} = \mathcal{C}^{\bar{s}_0}$ of all computations from a given $\bar{s}_0$ is partially ordered in $(\mathcal{C}, \sqsubseteq)$. The empty schedule $\varphi = ()$ is $\sqsubseteq$-minimal, the infinite schedules are $\sqsubseteq$-maximal, and every finite ρ has $\aleph_0$ finite extensions and $2^{\aleph_0}$ infinite extensions.

$(\mathcal{S}, \sqsubseteq)$ is given the product topology as follows. For finite $\sigma \in \mathcal{S}$

$$\mathcal{O}^{\sigma} = \{\tau \in \mathcal{S} | \sigma \sqsubseteq \tau\}$$

is a basic open set. An open set is either the union $\mathcal{O} = \bigcup_{\sigma \in \mathcal{R}} \mathcal{O}^{\sigma}$ of basic open sets for $\mathcal{R} \subseteq \mathcal{S}^{\text{finite}}$, or the empty set. $\mathcal{D} \subseteq \mathcal{S}$ is dense in $\mathcal{S}$ if every finite σ has an extension $\sigma+ \in \mathcal{D}$. It is uniformly dense if $|\sigma+| \leqslant |\sigma| + k$, for some k depending on $\mathcal{D}$.

$(\mathcal{C}, \sqsubseteq)$ is also given the product topology, and so the function execute: $(\mathcal{S}, \sqsubseteq) \to (\mathcal{C}, \sqsubseteq)$ defined by execute$(\sigma) = \bar{\bar{s}}^{\sigma}$ is continuous. Further execute is onto but not usually one-one.

An immediate consequence of (C, G) and L being finite is that the set L^C of all global states is finite, and so every infinite computation must repeat a global state. Since the values of $\sigma(i) \subseteq C$ are random, this does not

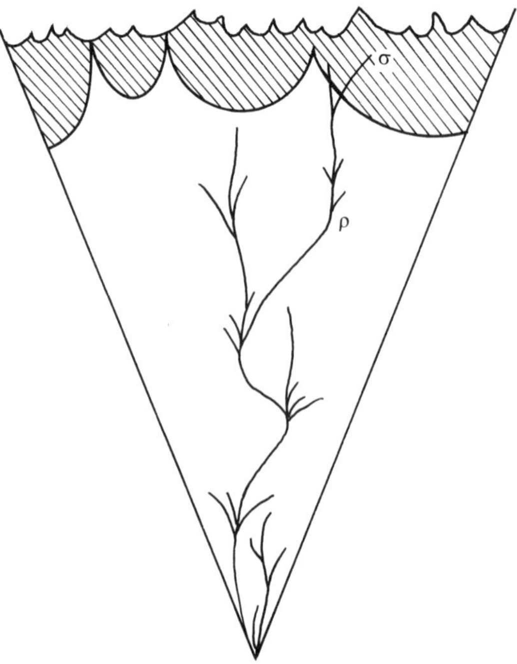

Figure 12.8 The schedules as a po-set. A shaded dense and open subset of $(\mathcal{S}, \sqsubseteq)$.

imply that they are looping. However, it does imply that if $\bar{s} \Rightarrow \ldots \Rightarrow \bar{s}'$, then there is a computation from $\bar{s}$ to $\bar{s}'$ of length at most $\| L^C \|$, which, in turn, implies the following.

UNIFORMITY LEMMA. EVERY DENSE $\mathscr{D}$ IN $(\mathscr{S}, \sqsubseteq)$ IS UNIFORMLY DENSE

Since it is the schedule, not the computation, that is the product of a random process, the probability measure is most naturally developed in $(\mathscr{S}, \sqsubseteq)$. First, let prob: $\mathscr{P}(C) \rightarrow [0, 1]$ be the probability prob $(\mathscr{X})$ of $\mathscr{X} \subseteq C$ being the set of active individuals at any time. For example, if the probability is uniform, then $\text{prob}(\mathscr{X}) = 2^{-\|C\|}$ for every $\mathscr{X} \subseteq C$. Now extend prob to a measure μ: $\mathscr{P}(\mathscr{S}) \rightarrow [0, 1]$ by

$$\mu(\mathscr{O}^{\varphi}) = \mu(\mathscr{S}) = 1;$$

and given ρ with $|\rho| = 1$ and $\sigma \in \mathscr{S}$

$$\begin{aligned} \mu(\mathscr{O}^{\rho}) &= \text{prob}(\rho(0)) \\ \text{and} \quad \mu(\mathscr{O}^{\rho\sigma}) &= \mu(\mathscr{O}^{\rho}) \cdot \mu(\mathscr{O}^{\sigma}); \end{aligned}$$

and for countable $\mathscr{R} \subseteq \mathscr{S}$

$$\mu(\mathscr{R}) = 0.0;$$

and, for more complex sets, by induction using

$$\mu(\mathscr{S} - \mathscr{R}) = 1.0 - \mu(\mathscr{R})$$

and $$\mu(\cup_{n \in \mathbb{N}} \mathscr{R}_n) = \sum_{n \in \mathbb{N}} \mu(\mathscr{R}_n)$$

if the $\mathscr{R}_n$ $(n \in \mathbb{N})$ are pairwise disjoint.

The probability of a random computation $\bar{\bar{s}}$ being in a set $\mathscr{B} \subseteq \mathscr{C}$ is

$$\text{prob}(\mathscr{B}) = \mu(\{\sigma \in \mathscr{S} \mid \bar{\bar{s}}^{\sigma} \in \mathscr{B}\}).$$

Global properties (of global computations) correspond to sets of schedules which are measurable (i.e., in the domain of μ). For example, the property

$$\text{halt}(\bar{\bar{s}}): (\exists i)(\forall X \in C)[\bar{\bar{s}}_i(X) = \alpha(\bar{\bar{s}}_i(X), \bar{\bar{m}}_i(X))]$$

corresponds to the measurable set

$$\mathscr{H} = \{\sigma \mid \text{halt}(\bar{\bar{s}}^{\sigma})\} \subseteq \mathscr{S}.$$

Properties of computations which occur with probability 1.0 are said to be generic. Using the uniformity lemma the following theorem follows easily.

GENERIC PROPERTY THEOREM

Let $\zeta_n(\bar{\bar{s}}) n \in \mathbb{N}$ be properties of computations $\bar{\bar{s}}^{\sigma} \in \mathscr{C}$. If each $\mathscr{R}_n = \{\sigma \mid \zeta_n(\bar{\bar{s}}^{\sigma})\}$ is dense and open; then, with probability 1.0, a randomly scheduled computation $\bar{\bar{s}}$ will satisfy the conjunction

$$\bigwedge_{n \in \mathbb{N}} \zeta n(\bar{\bar{s}}).$$

In other words, if each of the properties is generic then the conjunction is generic.

12.5.2 APPLICATIONS OF THE THEORY TO H[2] AND H[k]

We now apply this theory to prove properties of the **H**[k] models of the previous section. Define the variance of a global state by

$$\text{var}_{(C,G)}(\bar{s}) = \frac{1}{2} \sum_i \sum_j \delta(\bar{s}(i), \bar{s}(j))$$

where $\delta(x, y) = 0$ if $x = y$, $\delta(x, y) = 1$ otherwise. Let edges be $\frac{1}{2} \Sigma_{X \in I} \| G(X) \|$ – the number of edges in (C, G).

H[2]-Variance lemma

In the **H**[2] model $\text{var}_{(C,G)}$ behaves as follows.

(a) If $\sigma(i) \subset I$ is edge-free and $\bar{s}_i \Rightarrow_{\sigma(i)} \bar{s}_{i+1}$, then

$$\text{var}_{(C,G)}(\bar{s}_i) \leqslant \text{var}_{(C,G)}(\bar{s}_{i+1}) \leqslant \text{edges}.$$

(b) The inequality is strict if $\bar{s}_i(X) = 1$ and $\alpha(\bar{s}_i, \bar{m}_i)(X) = 0$ for some $X \in \sigma(i)$.

(c) If $\sigma(i)$ is not edge-free, then $\text{var}_{(C,G)}(\bar{s}_i) > \text{var}_{(C,G)}(\bar{s}_{i+1})$.

(d) If $\text{var}_{(C,G)}(\bar{s}_i)$ is maximal, then $\text{halt}(\bar{s}_{i+1})$.

The proof if this lemma is trivial, and so is left to the reader.

Halting lemma

In **H**[2], the set $\mathscr{H}$ of halting schedules is dense and open.

Proof. Let $\mathscr{L}$ be a linear ordering of $\mathscr{P}(C)$ and

$$\text{unstable}(\bar{s}) = \{X \mid \bar{s}_{i+1}(X) \neq \alpha(\bar{s}_i(X), \overline{m}_i(X))\}$$

be the set of individuals which will change state if activated in $\bar{s}$. For every $\bar{s}^{\sigma}$ with unstable($\bar{s}^{\sigma}$) non-void, define $J(\bar{s}^{\sigma})$ to be the $\mathscr{L}$-largest, edge-free subset of unstable ($\bar{s}^{\sigma}$). Let σ^+ be the one-step extension of σ defined by $\sigma^+ (|\sigma|) = J(\bar{s}^{\sigma})$. By (a) of the variance lemma and $J(\bar{s}^{\sigma})$ being edge-free and non-void, we have

$$\text{var}_{(C,G)}(\bar{s}^{\sigma}) \leqslant \text{var}_{(C,G)}(\bar{s}^{\rho^+})$$

with equality just in case $\bar{s}^{\sigma}(X) \neq 0$ for all $X \in J(\bar{s}^{\sigma})$. Thus, the variance must increase by at least 1 if σ is extended $\| I \|$ times. Since edges is an upper bound on $\text{var}_{(C,G)}$, at most edges $\| C \|$ extensions, $\sigma \sqsubseteq \sigma^+ \sqsubseteq \ldots \sqsubseteq \sigma^{+\ldots+} = \sigma^*$, are required for unstable ($\bar{s}^{\sigma}$) to be empty, and $\bar{s}^{\sigma*} \in \mathscr{H}$. Every finite $\sigma \in \mathscr{S}$ has an extension σ^* in $\mathscr{H}$ with $\mathscr{O}^{\sigma*} \subseteq \mathscr{H}$. $\mathscr{H}$ is dense and open in $(\mathscr{S}, \sqsubseteq)$.

H[2]-Halting theorem

For every finite (C, G), **H**[2] halts generically. Proof. This is immediate from the generic property theorem and the halting lemma (applied to one property ζ_0 = halt).

H[k]-Liveliness theorem

If $2 < k$, then the **H**[k] processes do not halt. Proof. Each global state $\bar{s}$ is proved not to be a halting state as follows. Partition C by $X \equiv Y$ iff $\bar{s}(X) = \bar{s}(Y)$ with equivalence classes $[i] = \{X \in C \mid \bar{s}(X) = i\}$. If $\bar{s}$ were halting, then most of the edges from cells of [0] would connect to $[k-1]$, most of the edges from $[i+1]$ would connect to $[i]$, for some i. This is clearly impossible, so $\bar{s}$ cannot be halting.

12.6 CONCLUSION

> People investigating neural nets, parallel processes, the immune response, the economic system, pattern recognition, the evolutionary system, autocatalytic reactions – disparate but related problem areas – are discovering a common ground for understanding these problems . . . The force behind this change is the computer . . . A new world view is laboring to be born as scientists move to examine the vast realm of complex systems.
>
> Heinz Pagels (1988)

Biologically inspired models of distributed information processing are a major approach to research in this important frontier of computer science. In the past mathematical computation theory, recursion theory and complexity theory have given us an almost complete understanding of serial computation; but nothing as perfect as this seems likely for asynchronous distributed computation. The relation of distributed computation to serial computation could be as revolutionary as that of quantum mechanics to Newtonian mechanics. Will insights into biological information processing lead us, again as with neural networks, to a new insights into computation theory? I believe they will

> Enormous polymers . . . grow from a tiny core that might be used in drug delivery, filtering, catalysis – even artificial cells . . . the porosity of the outer membrane can be carefully controlled . . . [a]nd since control of the porosity of the membrane is a central function of living cells, Tomalia has charted his course to go after nothing less than artificial cells. '. . . this is the first step toward making artificial cells,' he [D. Tomalia] says.
>
> Alper (1991)

Artificial cells (actually, computationally-inert, cell-like, polymer-grown structures) have been mass produced recently by Donald Tomalia (1990). It is reasonable to expect to see these cell-like objects given an internal chemistry capable of primitive state changes. When this is accomplished, an understanding of tissue processes (such as I am discussing here) could provide the understanding for coupling artificial cells, by primitive communication, into a computationally active material.

REFERENCES

Alberts, B., Bray, D., Lewis, J., Raff, M., Roberts, K. and Watson, J. (1989). *Molecular Biology of the Cell*. 2nd edn. Garland, New York.

Alper, J. (1991). Rising chemical stars could play many roles. *Science* **251**, 1562–5.

Barr, L., Dewey, M. M. and Berger, W. (1965). Propagation of action potentials and the structure of the nexus in cardiac muscle. *J. Gen. Physiol.* **48**, 13–25.

Chandrasekaran, B. (1981). Natural and social system metaphors for distributed problem solving: introduction to the issue. *IEEE Trans. Syst. Man Cyber.* **11**, January.

DARPA (1988). *DARPA Neural Network Study*. AFCEA International Press, Fairfax, VA.

DeMello, W. C. (1982). Cell-to-cell communication in the heart and other tissues. *Prog. Biophys. Mol. Biol.* **39**, 147–82.

Epstein, I. (1991). Nonlinear oscillations in chemical and biological systems. *Physica D* **51**, 152–60.

Forrest, S. (1990). Emergent computation: self-organizing, collective, and cooperative phenomena in natural and artificial computing networks. *Physica D* **42**, 1–11.

Gray, C., Engel, A. K., Konig, P. and Singer, W. (1990). Stimulus-dependent neuronal oscillations in cat visual cortex. *Euro. J. Neurosci.* **2**, 607.

Hobson, A. (1975). Sleep cycle oscillation: reciprocal discharge two brainstem neuronal groups. *Science* **189**, 5–58.

Hopfield, J. J. (1982). Neural networks and physical systems with emergent collective computational abilities. *Proc. Nat. Acad. Sci. USA* **79**, 2554–8.

Hopfield, J. J. and Tank, D. (1986). Computing with neural networks: a model. *Science* **233**, 625–33.

Kauffmann, S. (1971). Gene regulation networks: a theory for their global structure and behaviors. In *Current Topics in Developmental Biology* (ed. A. A. Moscona and A. Monroy). Academic Press.

Kephart, J. O., Hogg, T. and Huberman, B. (1989). Dynamics of computational ecosystems. *Physical Review A* **40**, 404–21.

Langton, C. G. (ed.) (1989). *Artificial Life I.* Santa Fe Institute Studies in the Sciences of Complexity, VI. Addison-Wesley, Redwood City, CA.

Lengyel, I. and Epstein, I. (1991). Modeling Turing structures. *Science* **251**, 650.

Loewenstein, W. R. and Kanno, Y. (1964). Studies on an epithelial (gland) cell junction. I. Modification of surface membrane permeability. *J. Cell Biol.* **22**, 565–86.

Loewenstein, W. R. *et al.* (1965). Intercellular communication: renal, urinary bladder, sensory, and salivary gland cells. *Science* **149**, 295–8.

MacDonald, C. (1985). Gap junctions and cell-cell communication. *Essays in Biochemistry*, 21. The Biochemical Society.

Mehta, P. P., Bertram, J. S. and Loewenstein, W. R. (1989). The actions of retinoids on cellular growth correlate with their actions on gap junctional communication. *J. Cell Biol.* **108**, 1053–65.

Mirollo, R. E. and Strogatz, S. H. (1990). Synchronization of pulse-coupled biological oscillators. *SIAM J. Appl. Math.* **50**, 1645–62.

Murray, J. D. (1989). *Mathematical Biology*. Springer-Verlag, Berlin.

Pagels, H. R. (1988). Editor's comments. *The Sciences*, New York Academy of Sciences, NY.

Paton, R. *et al.* (1991). Computing at the tissue/organ level (with particular reference to the liver). TM, Department of Computer Science, University of Liverpool, UK.

Perelson, A. (1990). Theoretical immunology. *1989 Lectures on Complex Systems* (ed. E. Jen). Santa Fe Institute Studies in the Sciences of Complexity Lectures, II, Addison-Wesley, Redwood City, CA.

Pool, R. (1991). Did Turing discover how the leopard got its spots? *Science* **251**, 627.

Prusinkiewicz, P. and Lindenmayer, A. (1990). *The Algorithmic Beauty of Plants*. Springer, New York.

Reed, S. and Lesser, V. (1980). Honey bees and the distributed focus of attention. CIS memo, Sept., University of Massachusetts, Amherst, MA.

Schwartzmann, G., Rose, B., Zimmerman, A., Ben-Haim, D. and Loewenstein, W. R. (1981). Diameter of the cell-to-cell junctional membrane channels as probed with neutral molecules. *Science* **213**, 551–3.

Simpson, I., Rose, B. and Loewenstein, W. R. (1977). Size limit of molecules permeating the junctional membrane channels. *Science* **195**, 294–6.

Stark, W. R. (1992). Tissue models with programming problems from God's notebook. *ACM SIGBio*. Newsletter **12**(2), 7–13.

Stark, W. R. and Kotin, L. (1989) The social metaphor for distributed processing. *J. Par. Dist. Comp.* **7**, 125–47.

Tomalia, D., Naylor, M. and Goddard, W. III (1990). Starburst dendrimers: molecular-level control of size, shape, surface chemistry, topology, and flexibility from atoms to macroscopic matter. *Angewandte Chemie* (English edition) **29**, 138–75.

Turing, A. M. (1952) The chemical basis of morphogenesis. *Phil. Trans. Roy. Soc.* **237**, 5–72.

Ulam, S. (1972). Some ideas and prospects in biomathematics. *Annual Review of Biophysics and Bioengineering*. AMS, Providence RI.

Ulanowicz, R. (1989). A phenomenology of evolving networks. *Systems Research* **6**, 209–17.

von Neumann, J. (1966) *The Theory of Self-reproducing Automata* (ed. A. W. Burks). University of Illinois Press, Urbana, IL.

Waldrop, M. (1984). The intelligence of organizations. *Science* **225**, 1136–8.

Yamashita, M. and Kameda, T. (1988). Computing on anonymous networks. *Proceedings of the Seventh Annual Symposium on the Principles of Distributed Computing*, ACM Press, New York.

STREAMING ORGANISM: THE TISSUE AUTOMAT 13

Gershom Zajicek

13.1 INTRODUCTION

From the day of its creation, the digital computer was applied to simulating complex medical systems. Medical modelling was applied mainly to simulation of physiological systems, for example, heart-lung, biochemical kinetics and in medical decision making. Models were generally compartmental and compartments interacted by first-order kinetics. Both approaches were essentially linear and applicable only to relatively simple systems. Most medically interesting problems could not be simulated by these means.

Recent innovations in programming, for example, neural nets and artificial life, open new opportunities for simulating complex non-linear medical systems. These programming tools have the following properties:

1. Elements in the system operate in parallel.
2. They interact in a non-linear fashion.
3. The system evolves in an unpredictable way and its structure emerges.

Although a major effort is invested in neural computing today, a small number of researchers favour a more general approach, known as biocomputing. Paton *et al.* (1991) developed the idea of 'biologically motivated computing' and its merit for describing organs like the liver. Ebeling (1992) illustrated how competing aging elements in an evolving parallel system are applicable for solving optimization problems. Stark (1992) reviewed some interesting mathematical tools for modelling tissues in the organism.

The present study introduces yet another approach, medical computing, that may be regarded as biocomputing of streaming systems that maintain an invariable structure, known as homeorrhesis (maintenance of equilibrium in a streaming system) (Dorland 1992). The elements of medical systems are streams that operate in parallel, yet unlike cellular automata or genetic algorithms, they evolve in a predictable fashion (Zajicek 1986, 1992*c*). Recent developments in non-linear dynamics provide new tools for describing medical systems. Diseases are represented by attractors, and the system proceeds reversibly from one attractor to another, maintaining homeorrhesis.

The ideas form the essence of the present chapter. The first section describes the changing representation of the organism, introducing the notion of the 'streaming organism'. The following section illustrates a prototype of a streaming tissue unit, the crypt villus. Section 13.4 describes the elementary computational unit of the medical system, the tissue automat (TA), whose medical implications are discussed in Section 13.5.

13.2 THE CHANGING REPRESENTATION OF THE ORGANISM

Until the beginning of the twentieth century the human organism was conceived as an engine immersed in its own liquid environment. In the past, this watery envelope assisted primordial marine organisms that left the sea to inhabit the continents, and is therefore necessary for the proper function of the human engine. It was assumed that all engine constituents exist as long as the organism does. Only the watery envelope turns over, maintaining a steady state that is called homeostasis (maintenance of equilibrium in a dynamic system) (Dorland 1988).

The first experiments with radioactive isotopes that were introduced in the 1930s revealed that all molecules of an organism continuously turn over and are randomly replaced with new ones. In spite of this, microscopic constituents, for example, cells and tissues, were assumed to be permanent and replaced only when damaged. The 1950s revealed that molecular turnover is not at all random. Molecules are replaced according to the first-in-first-out rule, the first that are incorporated are also the first to leave, which was first demonstrated in protein synthesis (Ross *et al.* 1989). The protein starts its existence when its DNA code is translated into RNA that serves as a template for the assembly of amino acids into short protein chains that grow with time and stream away from the nucleus. They stream through submicroscopic cavities of the endoplasmic reticulum toward the cell membrane where they are secreted. The protein is born near the nucleus and ages while streaming. Its age may be read off from its position in the cytoplasm. The more distant it is from the nucleus the older it gets.

Experiments with tritiated thymidine revealed that cells turn over in an orderly fashion. Thymidine is a building block of DNA. Before a cell divides it synthesizes new DNA and so increases its DNA content. If at the same time tritiated thymidine is injected into the body it is incorporated into the DNA. Cells that do not divide do not synthesize DNA and remain unlabelled.

Using this tool, Leblond (1964) divided cells into three groups:

1. renewing populations, for example, skin or the inner lining of the gut;
2. static populations, for example, the liver;
3. non-dividing cells, for example, nerve cells that in an adult do not divide.

Initially he assumed that turnover in dividing cells proceeds in a random fashion, but then discovered that in renewing populations turnover is oriented in space and time. Dividing cells occupy the unit interior; following division one cell replaces the parent while the other advances outward. Together with other newly formed cells, the outwardly advancing cell participates in a stream that is directed from the tissue interior outwards. Although it appears as if the stream is propelled by the pressure of dividing cells it also continues after their destruction. When dividing cells are irradiated with X-rays, that selectively inhibit cell division, non-dividing cells that are more remote in relation to the division locus continue advancing. They are neither pushed nor pulled in a mechanical sense, they simply stream, in the same way as water molecules do in the river. By this analogy the cells are carried on a 'metabolic' stream.

Recently we have shown that even static populations, for example, salivary glands and the kidney and adrenal cortex, also stream, only much more slowly than Leblond's renewing cells (Zajicek *et al.* 1985, 1986, 1987, 1988; Arber and Zajicek 1990; Zajicek and Arber 1991). Thus, all cells except nerve cells continuously stream from the tissue interior outward. In spite of this, the organism maintains homeorrhesis. The idea that everything streams, or *panta rhei*, dates back to the Greek philosopher Heraclitus, who lived in the fifth century BC. He said: 'You never enter the same river twice' since it continuously changes, to which we may now add: 'You never meet the same individual twice' since all its constituents change.

This extremely complex system has been studied by medicine from the dawn of mankind. With time, medicine developed means to manipulate the organism in spite of its complexity. Diagnosis, for instance, is the identification of a minimal set of statements sufficient to recognize a disease, and treatment is the implementation of a minimal set of means, or actions, necessary to restore health. Medicine is thus a science of a complex system.

This important facet of medicine was hitherto misunderstood by the physical sciences. Medical reasoning was regarded as non-scientific and medicine was viewed as an art. Physics and mathematics postulated that any system, simple or complex, should be reconstructed from its elements, which should be applicable also to the human organism. Yet attempts to structure biology and medicine in the same way failed. Apparently the human organism is more than the sum of its parts and cannot be reconstructed from its elements. That was the main reason for regarding medicine as an art. Recent advances in mathematics have revealed that its traditional approach is inadequate for understanding complex systems.

Henri Poincaré (see Ekeland 1988) was the first to show that Newton's laws of motion do not suffice to predict even the motion of three heavenly bodies, much less the entire planetary system. The behaviour of the planetary system is more complex than predicted by the laws of motion. This seemingly heretical view is obvious to any rocket launcher. If Newton were correct the entire rocket trajectory could be derived from its initial condition. In reality, since non-linear phenomena interact with the rocket during its flight, its trajectory cannot be computed in advance and has to be corrected from time to time. Thus the outcome of a relatively simple physical system like rocket launching, much simpler than the organism, cannot be foreseen.

Poincaré is regarded as the father of chaos theory, according to which nearly all natural phenomena are non-linear and complex and their behaviour cannot be reconstructed from their elements. Chaos proclaims first of all a crisis in mathematics that fails to cope even with simple equations like the logistic equation. For years this quadratic function was mathematically tame, continuous and differentiable, and yet when manipulated recursively it generates unpredictable number sequences, known as orbits, that dance chaotically on the computer screen. This baffling complexity is generated by a simple algorithm operating on a simple quadratic function. Even a minute change of its initial state ends in a different chaotic number sequence. This is a phenomenon known as sensitivity to initial conditions (Schuster 1984; Zajicek 1991*a*, 1991*b*).

Medical computing adopts from chaos theory concepts describing streams and turbulences. From this viewpoint the streaming organism may be regarded as a complex dynamic chaotic system and its appearance as an attractor in a multi-dimensional chaotic space. When examining a patient the physician actually interacts with a complex system. He applies simplifying rules that enable him to interact with the organism in a meaningful way. These rules may now be expressed by the language of chaos theory. Chaos is a new and a rapidly developing field. Most of its concepts are still rudimentary and sometimes inconsistent, yet their prospect for medicine is enormous. It appears as if chaos might provide the necessary mathematical framework for medicine. The present study illustrates its great potential in analysing streaming tissues. The cell is regarded here as the elementary unit of the organism. Several cells form a tissue unit, and several tissues make an organ. The kinetic structure of all proliferating tissues in the organism is similar to that of the crypt-villus unit (CU) of the inner gut lining and will therefore be introduced first (Wright and Alison 1984).

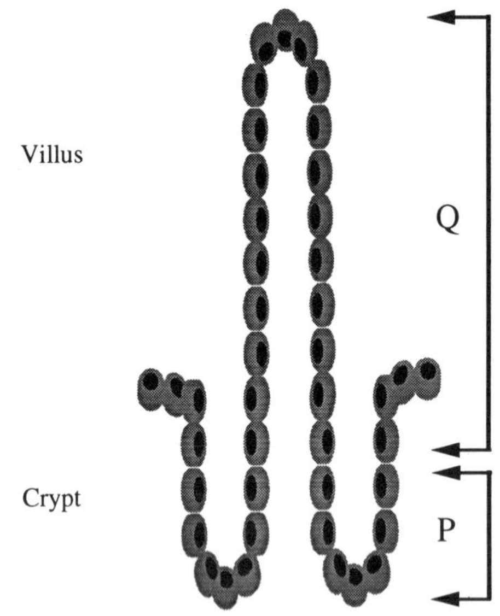

Figure 13.1 A diagram of the crypt-villus unit.

13.3 THE CRYPT-VILLUS UNIT

Crypt-villus units pave the gut cavity. Each unit consists of a finger-like structure, the villus, that protrudes into the cavity, surrounded by cylindrical indentations, or crypts. Figure 13.1 depicts a longitudinal cross-section through this unit. Crypt cells, known as progenitors (P in the figure), continuously divide and the newly formed cells stream into the villus where they lose their capacity to reproduce and turn into Q-cells (Q in the figure).

Figure 13.2 is a scheme of the crypt-villus unit. All cells in the unit are descendants of one cell, called a determined stem cell (DS). When the DS divides, one of its progeny replaces the parent and remains a DS, while the other starts streaming outward. The streaming cell may differentiate along four cell lines; absorptive, goblet, endocrine and Paneth.

When entering a differentiation pathway it is called a committed stem cell (CS). The CS differentiates while streaming, gradually maturing until reaching the unit outer boundary where it dies. If, for example, the DS is scheduled to generate a mucus-producing cell line, one of its progeny becomes a mucus CS. It is equipped with a big nucleus and scanty cytoplasm. At this stage nothing reveals that it will turn into a secretory cell since it still lacks the typical organelles. As the cell advances through the unit the organelles appear and the cell starts secreting its product. This process is called differentiation. Differentiation is time dependent. The older the cell the more differentiated it becomes. In the crypt-villus unit the entire differentiation process lasts about two days.

Cells in the unit differentiate while

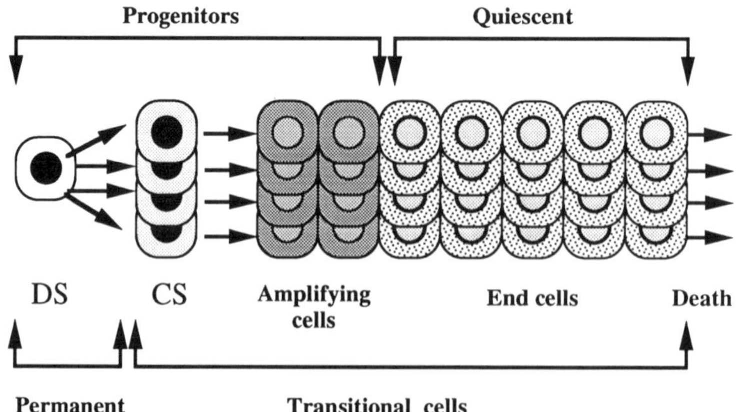

Figure 13.2 The tissue unit.

streaming. Initially they continue multiplying and are called amplifying progenitors. Then they lose the capacity to synthesize DNA and become non-dividing (quiescent) end cells. When reaching the outer unit boundary they die. Cell death in the healthy tissue unit is a differentiation state. The time of cell death is programmed in the genetic code and the cell is eliminated by apoptosis (Arends and Wyllie 1991). This differentiation dependent cell death has to be distinguished from random cell death that occurs only in disease.

All cells except the DS are transitional and short-lived (they live about two days); only the DS is permanent. It is the source of a cell stream from the tissue interior outward. After destroying the tissue unit it is repopulated by new cells that are formed by the DS. Yet when the DS dies transitional cells continue streaming and since they are not being replaced by newcomers, after two days the entire unit disappears.

When the DS divides one of its progeny replaces the parent and remains a DS and the other turns into a streaming transitional cell. Thus a DS division generally results in two different cells, a stem and a transitional. It is therefore called an asymmetric division. A transitional cell may divide several times (if it is an amplifying cell) and both its descendants are always transitional. A cell division that generates two equal cells is called symmetric. Since all transitional cells are descendants of one DS, they belong to one cell clone and the entire unit is mono-clonal.

13.4 THE TISSUE AUTOMAT (TA)

The model will be described in two stages. Firstly the laws governing the tissue unit are specified in a form that is easily implemented in a higher level programming language. The laws describe three states; normal, pathological (non-cancerous) and cancerous. They were implemented in part in a model that runs on the CM2 (Sipcic *et al.* 1991). Next a simplistic computational model of the TA is described together with an implementation in the Mathematica programming language.

13.4.1 THE FRAMEWORK FOR THE TISSUE UNIT

The normal tissue unit (TU)

1.0 **Genealogy**
Describes the relationships of cells to their ancestor, the determined stem cell (DS).

1.1 A determined stem cell (DS) gives rise to one committed stem cell (CS): that generates a maturation pathway.

1.2 The DS divides asymmetrically. One of its progeny replaces the parent and becomes a DS. The other generates a clone of transitional cells which differentiate as they go.

1.3 The TU is monoclonal.

2.0 **Differentiation**
Describes the changes that are observed on the DS progeny during their streaming.

2.1 All changes observed in a transitional cell are defined as differentiation. The set of differentiation markers observed in a cell at a given instant is its differentiation state. The degree of differentiation of a cell is expressed in relation to the DS. Some important differentiation markers:

2.2 Cell location.

2.3 Cell fecundity, which is the cell's probability to enter the cell cycle and divide. A TU cell starts as a progenitor (P) and differentiates into a non-proliferating Q cell (know also as a terminally differentiated TU cell). Fecundity depends on the cell cycle phases: G0, G1, S, G2 and M.

2.4 Cell death is confined only to terminally differentiated cells. It is regarded as the ultimate differentiation state. Cells are eliminated by two mechanisms:

(a) apoptosis that is differentiation dependent;
(b) necrosis which operates randomly, independently of differentiation (it is not a differentiation marker).

2.6 Necrosis does not operate in the healthy TU.

3.0 **Population dynamics**
TU cells exist as two major populations, progenitors and quiescent cells.

3.1 TU cells are divided into P and Q cells. The two populations are mutually exclusive. All P become Q yet Q never become P ($P \rightarrow Q$) ($Q \nrightarrow P$).

3.2 P cells of a given differentiation state may replicate several times and amplify their number.

3.3 TU cells are divided into permanent (DS) and transitional cells (the rest).

4.0 **Topology**
The TU has several typical topological landmarks, e.g., origin and periphery.

4.1 All descendants of a DS form the TU.

4.2 The DS is located at the TU interior. The site where cells are eliminated is defined as the TU periphery. The typical trajectory of a TU cell connecting its birth site with the TU periphery is called tissue radius.

4.3 P cells occupy the TU interior, Q cells its exterior.

5.0 **Cell streaming**
Cell streaming in the TU has to be distinguished from other kinds cell streaming, e.g., that of red blood cells, that are driven by the pumping heart. Here cell streaming is driven by cell proliferation at tissue origin.

5.1 Transitional cells stream from the TU interior to its periphery. Since cells are neither pulled nor pushed in a mechanical sense their motion is regarded as cell streaming.

5.2 Streaming is linked with proliferation. Each mitosis at the interior part of the tissue radius promotes all peripheral cells by one location.

5.3 Streaming velocity depends on mitotic rate. If expressed in cell location units it equals mitotic rate. Along the tissue radius, one cell division is associated with the displacement of all distal cells by one location. Generally the mitotic rate of P cells equals the velocity of Q cells.

6.0 **Time dimension**
Two time measures may be applied for describing phenomena in the TU; chronological and biological (Zajicek 1987, 1992*a*). The two generate, respectively, two time frames or dimensions, linear and non-linear. The non-linear time frame is chaotic, and is the only adequate time frame for describing biological phenomena. Biological time is closely related to differentiation.

6.1 The asymmetrical division of DS marks the birth of a TU cell.

6.2 Cell age is expressed in relationship to the time of the asymmetric DS division.

6.3 Two types of age measures may be applied to describe events in the TU: chronological and biological.

Chronological age is the time which has elapsed since the asymmetrical division of the DS that generated the clone to which a particular cell belongs.

Biological age is the location of a cell on the tissue radius. For example, a cell at location 12 was formed 30 hours ago (when its ancestor split from the DS). Its chronological age is 30 hours while its biological age is 12 locations. A cell at location 15 is three locations older than the cell at location 12.

6.4 Since at each location the cell is more differentiated, cell location may be used as a differentiation measure. We may say that the cell at location 15 is by three locations more differentiated that a cell at location 12. Actually biological age and differentiation are equivalent.

6.5 A cell at location $i - 1$ represents the immediate past of a cell at location i, and the cell at location $i + 1$, its immediate future.

6.5a A cell at location $i - 1$ represents the previous differentiation state of a cell at location i, and the cell at location $i + 1$, its next differentiation state.

6.6 The tissue radius summarizes the entire differentiation history of the TU.

6.7 If biological age is taken to describe events on the TU radius, one cell division at location i equals a:

displacement	of all cells in locations $> i$ by one location
aging	of all cells in locations $> i$ by one location (unit)
differentiation	of all cells in locations $> i$ by one location (unit)

This is regarded as a fundamental relationship in the normal TU.

6.8 In order to estimate the chronological age of a cell it has to be labelled with tritiated thymidine and followed with time, while its biological age may be read from the tissue radius and requires only one observation.

7.0 **Homeorrhesis**

7.1 The adult TU maintains a constant cell population. Since cells are continuously formed and eliminated, the TU maintains homeorrhesis in which the birth rate $n(b)$ (= mitotic rate) equals death rate $n(d)$, $n(b) = n(d)$. In the equation $n(d)$ is independent while $n(b)$ depends on $n(d)$. Stated otherwise, $n(d)$ controls $n(b)$. The information on the death rate $n(d)$ is conveyed to all P cells. In other words, death rate affects the fecundity of each P cell.

8.0 **TU pathology**

Laws 8–9, specify TU behavior in non-cancerous conditions.

8.1 TU pathology is caused by a noxa that is external to the TU.

8.2 The noxa interacts with the TU solely by changing birth $[n(b)]$ and death $[n(d)]$ rates.

8.3 In addition birth rate is always controlled by death rate (see 7.1).

9.0 **Perturbation**

In health and in disease the TU always maintains homeorrhesis. In other words, it occupies a given attractor state (Zajicek 1991*d*). A perturbation drives the TU out of its stable state and endangers its existence. The organism always attempts to minimize perturbations and settle as soon as possible at an attractor state. During disease it proceeds from one attractor to another.

9.1 Perturbations may be classified into acute and chronic.

9.2 Each may further may be classified into $n(b)$ and $n(d)$ perturbations.

9.3 In acute perturbations, as long as the DS is unharmed, original homeorrhesis will be restored. Once the DS is eliminated the TU will disappear.

9.4 Following chronic perturbations the TU proceeds through a transitional phase until a new homeorrhesis state has been established.

9.5 **Birth rate [$n(b)$] perturbations:**

9.5.1 Acute increased birth rate $n(b)$ ⇑ is followed by TU elongation and a return to its initial state.

9.5.2 Chronic increased birth rate $n(b)$ ⇑ is followed by TU elongation and the establishment of a new homeorrhesis state.

9.5.3 Acute decreased birth rate $n(b)$ ⇓ is followed by TU shortening and a return to its initial state.

9.5.4 Chronic decreased birth rate $n(b)$ ⇓ is followed by TU shortening and the establishment of a new homeorrhesis state.

9.6 **Death rate [$n(d)$] perturbations:**

9.6.1 Acute increased death rate $n(d)$ ⇑ is followed by TU shortening and a return to its initial state.

9.6.2 Chronic increased death rate $n(d)$ ⇑ is followed by TU shortening and the establishment of a new homeorrhesis state.

9.6.3 Acute decreased death rate $n(d)$ ⇓ .

9.6.4 Chronic decreased death rate $n(d)$ ⇓ is followed by TU elongation the and establishment of a new homeorrhesis state.

10.0 **Neoplastic progression of the TU**
Cancer has two kinds of manifestation: localized and systemic (Zajicek 1991*c*, *d*). Localized manifestations are the cell mass that appears in the diseased organ, known as neoplasia. Systemic manifestations, for example, weight loss and prostration, are known as paraneoplasia and cachexia. Here we are concerned only with phenomena occurring in the TU, that is, neoplasia.

10.1 Neoplasia evolves through short and discrete phases during which it maintains homeorrhesis. Although the disease continuously evolves during short periods of time it behaves as if maintaining homeorrhesis and may be viewed as if progressing stepwise from one homeorrhesis state to another. Each step is initiated by a chronic birth or death rate perturbation which then settles to a new homeorrhesis state.

10.2 The TU proceeds through the following phases.

10.2.1 **Phase 1.** Increased Q cell death rate [$n(d)$ ⇑] accompanied by increased birth rate $n(b)$ ⇑ leading to a shortening of Q and elongation of P (P > Q). TU length does not change. Q shortening is called maturation arrest. Since birth and death rates increase, cell production, streaming and turnover increase as well.

10.2.2 **Phase 2.** Increased Q cell death rate [$n(d)$ ⇑] accompanied by an increased birth rate $n(b)$ ⇑ leading to a shortening of Q and elongation of P (P > Q). The TU elongates.

10.2.3 **Phase 3.** Incomplete TU fission and formation of buds. It is assumed that each bud contains its own DS. All features which were observed in the two previous phases become more pronounced and are accompanied by DS accumulation.

10.3 Since all TU cells except the DS are transitional cells, neoplastic transformation had to occur first in the DS and is then inherited by its transitional progeny (the TU is monoclonal).

10.4 The terminology which was applied above to the normal TU, for example, genealogy, topology and differentiation, is applicable also to describe phenomena observed during neoplastic progression.

The present framework, however, is inadequate for describing manifestations observed in TU populations, such as metastasis, TU turnover and multiplication. It is adequate solely for describing phenomena observed in the single TU.

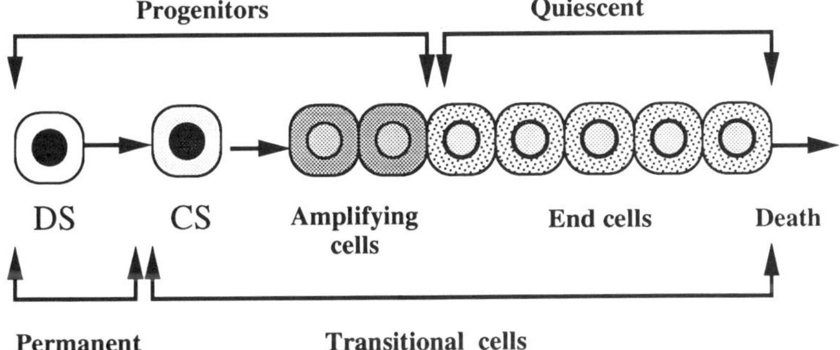

Figure 13.3 The tissue automat.

The crypt-villus unit (Figure 13.2) consists of one DS that generates four differentiation pathways. The DS is the generator of the entire unit and can restore it even if all transitional cells are destroyed. The TA to be represented here consists of one differentiation pathway (Figure 13.3) and was programmed in Mathematica (Wolfram 1991).

13.4.2 THE COMPUTATIONAL TA

This simple cell automat reproduces the salient features of streaming tissues, namely:

1. The stem cell generates the entire column and all transitional cells are its progeny.
2. TA cells are divided into P and Q cells. The two populations are mutually exclusive. All P become Q yet Q never become P ($P \rightarrow Q$) ($Q \nrightarrow P$).
3. Cells stream outward. They age and differentiate as they go. The stem cells are least differentiated.
4. TA size (or its cell content) is controlled solely by cell death which is manifested by two mechanisms:
   (a) programmed cell death or apoptosis (represented in the program by 'death');
   (b) random cell death or necrosis (represented in the program by 'kill').
5. Cell production is proportional to cell elimination. Q cells produce a substance that inhibits P cell proliferation. The more abundant the Q cells the fewer P cells are formed and vice versa.
6. Cells in the TA exhibit two differentiation markers:
   (a) loss of fecundity in Q cells;
   (b) secretion of P cell inhibitor by Q cells.
7. The TA has one attractor state of 35 cells.

The cell column starts its existence as an undifferentiated stem cell. Since it lacks Q cells that secrete the P inhibitor the column elongates exponentially until the first Q cells are formed, start producing the inhibitor and cell production declines. While P cell production rate is controlled by the number of Q cells, TA size is controlled solely by cell death.

Some simulations by this model are now described. Figure 13.4 depicts one run of the program. Initially the TA is a stem cell that proliferates and the unit elongates. Its first progeny are dividing progenitors (P cells). As the column elongates some remote cells become Q cells, they mature and at a certain stage start dying. The rate of cell death is proportional to the size of the Q compartment. The longer it becomes the more cells are eliminated until at time step = 600, when cell death rate equals cell formation rate, the column maintains a constant length of about 35 cells. All this time, cells stream from the bottom, and since the column maintains an invariant length it is in a state of homeorrhesis. This state occupies the centre of the phase space diagram (Figure 13.5). The values

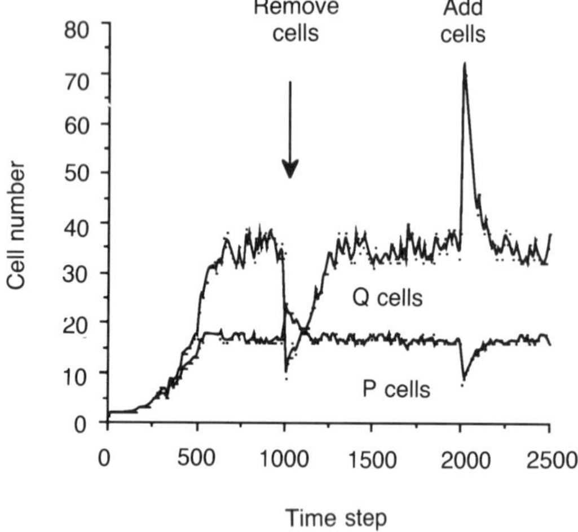

Figure 13.4 Results from one run of tissue automat simulation.

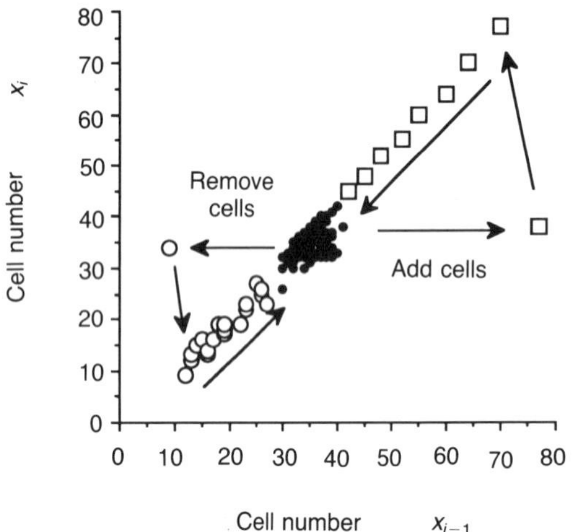

Figure 13.5 Phase space diagram for tissue automat.

oscillate chaotically within a certain region, known as an attractor.

At time step 950 all Q cells and some P cells are eliminated and the column length drops to about 12 cells. The progenitor compartment elongates and produces more cells until the damage has been restored and the column returns to its attractor state. At time step = 2000 the column receives an 'infusion' of about 40 cells. The P compartment shrinks and produces fewer cells. The infused cells are rapidly eliminated and the TA returns to its attractor state.

Figure 13.5 depicts the phase space diagram of the TA. The abscissa depicts TA cell content during the previous time step (x_{i-1}), and the ordinate the present time step (x_i). The black cluster in the centre represents the TA attractor of about 35 cells. The first perturbation that was initiated by cell removal is represented by empty circles that gradually return to the attractor state (homeorrhesis). The TA responds similarly to a perturbation that results from adding cells to the TA (empty squares).

13.5 BIOMEDICAL IMPLICATIONS OF THE TA MODEL

13.5.1 CELL AND UNIT DIFFERENTIATION

There are two differentiation scales, one describing the differentiation of cells and the other of the entire unit. As the cell advances it differentiates. It is born a P cell and differentiates into a Q cell. These changes also reshape the differentiation of the unit. Initially the TA is undifferentiated since it is made of one stem cell. During the first time steps it is made solely of P cells and only after 500 time steps does it become fully developed or differentiated. TA differentiation is given by the Q/P ratio. Initially Q/P = 0. It gradually increases until reaching its attractor state of Q/P = 35/16. Following cell killing (time step = 950), TA differentiation drops to Q/P = 0 (Figure 13.6). This apparent loss of TA differentiation is also known as de-differentiation, after which the TA gradually differentiates again to its differentiation-attractor state of Q/P = 35/16. Cell differentiation is always irreversible, since cells continuously stream outward. Tissue unit differentiation, on the other hand, may appear as reversible.

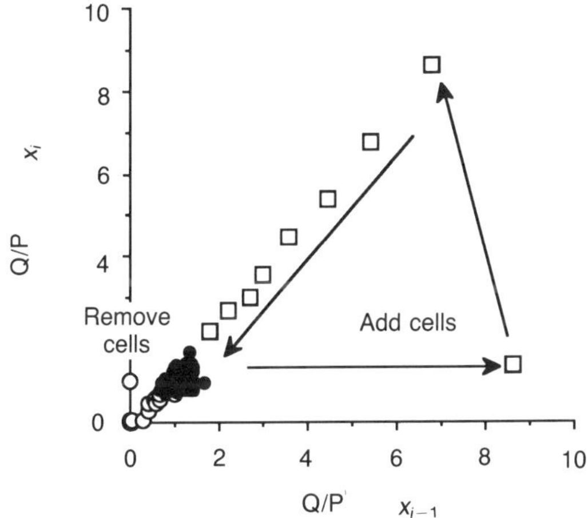

Figure 13.6 The tissue automat's response to damage.

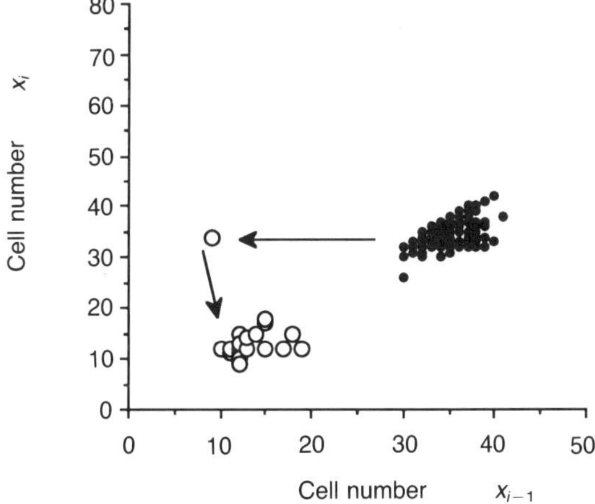

Figure 13.7 The tissue automat is a chronic pathology.

13.5.2 ACUTE AND CHRONIC PATHOLOGY

The response of the unit to cell destruction is a pathology. A pathology is an interaction between an external factor and a strategy mobilized by the unit in response to the factor. Here the external factor ('kill') eliminates cells, and the strategy is the TA's response to the damage. Q/P fluctuation (Figure 13.6) represents the unit strategy mobilized against the external factor. Following an acute pathology the system returns to its 'healthy' attractor state of 35 cells (Figure 13.5). When cells are removed continuously the TA settles at a new attractor state (Figure 13.7), that represents a chronic pathology.

The manifestation of a pathology depends on the differentiation state of the TA. The pathology will differ whether or not the TA is undifferentiated, consisting solely of P cells with Q/P = 0, or fully differentiated with a Q/P = 35/16. Since the TA starts as a stem cell and gradually differentiates, its pathology may thus be regarded as an emergent property. Generally the global behaviour of a TA emerges from local interactions between its constituents, and since it is generated by the TA program, it is an emergent computation (Forrest 1991).

Differentiation and pathology are two independent TA features with typical attractor states. Actually the TA attractor is a multicomponent vector in which each component stands for a feature. The different TA features may be derived from the TA framework.

13.5.3 CARCINOGENESIS

Cancer starts with the transformation of a normal cell into a malignant cell. If this event hits a transitional cell it is soon washed out of the unit. Only if a DS is transformed does the event remain permanently recorded. Thus a single DS can generate a cancer and the disease can be regarded as the emergent behaviour of a 'mutated' DS. After being triggered, the process evolves on its own and is unrelated to the trigger. Cancer is an emerging pathology that traverses well-defined attractor states. In the model, cancer is actually an emergent computation.

The set of all DS cells in the organism is capable of generating the entire organism. It is estimated that 0.2% of crypt-villus cells are DS. Extrapolated to other tissue units this means that 99.8% of all cells in the body are transitional and the entire organism can be reconstructed from a relatively small number of DS. Only this cell population lives as long as the organism does.

13.5.4 EMBRYOGENESIS

The same arguments also apply to embryogenesis. The fertilized egg is the primordial DS. Initially it divides symmetrically and generates only DS cells. Apparently the first transitional cells appear during gastrulation. From then onwards all embryonal tissue units consist of the two cell types, DS and transitional, and can be simulated by a TA. The first divisions of the fertilized egg, known as cleavage, are symmetric and generate only DS cells. After the first division each DS can still generate an entire embryo. As the embryo grows the potential of a DS to form different tissue units narrows. This change is called determination (Wessels 1977). With time the DS becomes more and more determined until in the adult it is capable of generating only one kind of unit.

We may thus distinguish between two differentiation scales, global and local. The first describes the differentiation of the DS and its rising determination from the fertilized egg until adulthood. The local differentiation scale describes the change of the cells in each tissue unit.

13.6 CONCLUDING REMARKS

The TA represents the smallest element of the organism that is medically meaningful. Its significance may be appreciated if compared with other artificial life constructs, for example, cellular automata, the game of life or the Lindenmayer system. They lack a key ingredient of the TA – long-term non-linear memory. TA memory is manifested by the age distribution of its cells. After killing some of the TA cells ('Remove cells' in Figure 13.4), the unit replenishes its cells rapidly. The age distribution on the other hand, returns to its initial state much more slowly. Immediately after cell killing it is made of young and undifferentiated cells that age only gradually. After infusing the TA with old and over-differentiated cells ('Add cells' in Figure 13.4) the TA returns to its normal size quite rapidly. Its age distribution contains predominantly old cells and it takes some time to return to its initial state. This simple experiment illustrates that the age distribution 'remembers' how it was perturbed. If the age distribution is 'young' it was perturbed by 'kill' and if it is relatively old, it was perturbed by 'add cells'.

The TA also illustrates a possible solution to the reductionism-holism problem. Medicine will have to follow the example of the physical sciences and define a proper unit from which the organism may be reconstructed. Yet, unlike in physics and chemistry, the medical unit will not be absolute. Its definition will depend on the context within which it is studied. Such a unit was called, by Koestler, a holon (see Barlow 1991). The holon, like the Roman god Janus, has two faces looking in opposite directions. To the outside world it appears as an indivisible unit, or atom, yet by itself it is an entire universe. A cell, for instance, may be regarded as a unit of the organism, while within the sub-microscopic realm it encompasses an entire universe. The medical holon streams and maintains homeorrhesis.

The holon concept simplifies the way medicine tackles the complexity of the organism. The complexity of a system depends on the definition of its elements. It is nearly impossible to describe the functioning of the organism when using cellular organelles as units. This task becomes more feasible when adopting a higher order holon, for example, tissue or organ. Thus the complexity of the system can be reduced by selecting appropriate holons for describing it.

The TA depicts a new kind of computing element. Its memory is non-linear and long term. Input is known in medicine as noxa (Zajicek 1991*d*), represented here by 'kill', and its output is represented here by TA differentiation. The TA is computationally inefficient; it is, however, the only medically meaningful element.

REFERENCES

Arber, N. and Zajicek, G. (1990). Streaming liver VI. Streaming intra-hepatic bile ducts. *Liver* **10**, 205–8.

Arends, M. I. and Wyllie, A. H. (1991). Apoptosis: mechanisms and roles in pathology. *Internat. Rev. Exper. Path.* **32**, 223–40.

Barlow, C. (1991). *From Gaia to Selfish Genes*. MIT Press, Cambridge, MA.

Dorland's Illustrated Medical Dictionary, 27th edn (ed. W. A. N. Dorland). W.B. Saunders, Philadelphia, PA.

Ebeling, W. (1992). The optimization of a class of functionals based on developmental strategies. In *Parallel Problem Solving from Nature* (ed. R. Manner and B. Manderick). Elsevier, Amsterdam, pp. 463–8.

Ekeland, I. (1988). *Mathematics and the Unexpected*. The University of Chicago Press, Chicago, IL, p. 36.

Forrest, S. (ed.) (1991). *Emergent Computation*. MIT Press, Cambridge, MA.

Leblond, C. P. (1964). Classification of cell populations on the basis of their proliferative behavior. *Natl. Cancer Institute Monogr.* **14**, 119–45.

Paton, R, Nwana, H., Shave, M. and Bench-Capon, T. (1991). Computing at the tissue/organ level (with particular reference to the liver). TM, Department of Computer Science, University of Liverpool, UK.

Ross, M. R., Reith, E. J. and Romrell, L. J. (1989). *Histology*. Williams and Wilkins, Baltimore, MD.

Schuster, H. G. (1984). *Deterministic Chaos*. Physic Verlag, Weinheim, Germany.

Sipcic, S. R., Deutsch, D. and Zajicek, G. (1991). Simulation of cancer progression in the colon on a massively parallel processor (CM-2). *Proceedings of the Sixth SIAM Conference on Parallel Processing* (ed. J. Dongawa).

Stark, W. R. (1992) Tissue models, with programming problems from God's notebook. *SigBio Newsletter* **12**, 7–13.

Wessels, N. K. (1977). *Tissue Interactions and Development*. Benjamin-Cummings, Menlo Park, CA.

Wolfram, S. (1991). *Mathematica*. Addison-Wesley, New York.

Wright, N. and Alison, M. (1984). *The Biology of Epithelial Cell Populations*, Vols. 1, 2. Clarendon Press, Oxford.

Zajicek, G. (1986). The application of kinematic equations for the study of cell turnover. *J. Theoret. Biol.* **120**, 141–9.

Zajicek, G. (1987). The time dimension in histology. *Methods Inform. Med.* **26**, 1–2.

Zajicek, G. (1991*a*). Chaos and biology. *Meth. Inform. Med.* **30**, 1–3.

Zajicek, G. (1991*b*). Meta-analysis and chaos. *Cancer J.* **4**, 152.

Zajicek, G. (1991*c*). What is neoplasia? *Cancer J.* **4**, 228.

Zajicek, G. (1991*d*). What is a disease? *Cancer J.* **4**, 296.

Zajicek, G. (1992*a*). Time dimension in histopathology. *Path. Res. Pract.* **188**, 410–12.

Zajicek, G. (1992*b*). Artificial life. *Meth. Inform. Med.* **31**, 167–8.

Zajicek, G. (1992*c*). The streaming organism. *ACM SigBio Newsletter* **12**, 18–25.

Zajicek, G. and Arber, N. (1991). Streaming kidney. *Cell Prolif.* **24**, 433–4.

Zajicek, G., Yagil, C. and Michaeli, Y. (1985). The streaming submandibular gland. *Anat. Rec.* **213**, 150–6.

Zajicek, G., Ariel, I. and Arber, N. (1986). The streaming adrenal cortex. *J. Endocrin.* **111**, 447–82.

Zajicek, G., Bartfeld, E., Schwartz-Arad, D. and Michaeli, Y. (1987). Computerized extraction of the time dimension in histological sections. *Appl. Optics* **26**, 3408–12.

Zajicek, G., Ariel, I. and Arber, N. (1988). The streaming liver III. Littoral cells accompany the streaming hepatocyte. *Liver* **8**, 213–18.

PART THREE

GENETICS AND EVOLUTION

EVOLUTIONARY ALGORITHMS: COMPARISON OF APPROACHES

14

Thomas Bäck

14.1 BACKGROUND

Over three decades of research and applications have clearly demonstrate that mimicking the search process of natural evolution can yield very robust, direct computer algorithms, although these imitations are crude simplifications of biological reality. The resulting **evolutionary algorithms** are based on the collective learning process within a population of individuals, each of which represents a search point in the space of potential solutions to a given problem. In order for an evolutionary algorithm to work, the population is arbitrarily initialized, and it evolves towards increasingly better regions of the search space by means of randomized processes of selection (which is deterministic in some algorithms), mutation and recombination (which is completely omitted in some algorithmic realizations). The environment (given the aim of the search) delivers a quality information (fitness) value for new search points, and the selection process favours the reproduction of individuals of higher fitness more often than worse individuals. The recombination mechanism allows the mixing of parental information while passing it to their descendants, and mutation introduces innovation into the population.

In concrete applications, the aim of such a search is to identify an approximation of a global optimum of an objective function $f: M \to \mathbb{R}$, that is, to evolve a solution $\vec{x}^* \in M$ such that $\forall \vec{x} \in M: f(\vec{x}) \geqslant f(\vec{x}^*)$ (in case of a minimization problem; maximization is completely analogous). The structure of M depends on the application and is in the following restricted to $M \subseteq \mathbb{R}^n$, that is, parameter optimization problems are considered. The representation of potential solutions by individuals depends on the particular evolutionary algorithm and is discussed in detail for each instance presented in the following. However, though they are in most cases used for solving optimization tasks, evolutionary algorithms are sometimes more generally viewed as algorithms which model a dynamic adaptation process by maintaining a certain level of diversity and fluctuation rather than focusing completely on a special solution (for example, De Jong (1992)).

The informal description presented above leads to the following general outline of an evolutionary algorithm (t denotes the generation counter, $P(t) = \{\vec{a}_1(t), \ldots, \vec{a}_\mu(t)\}$ is the population at generation t, consisting of individuals a_i, and μ denotes the population size):

ALGORITHM 1 (OUTLINE OF AN EVOLUTIONARY ALGORITHM)

```
t := 0;
initialize(P(t));
evaluate(P(t));
while not terminate(P(t)) do
```

$P'(t) := recombine(P(t));$
$P''(t) := mutate(P'(t));$
$evaluate(P''(t));$
$P(t+1) := select(P''(t) \cup Q);$
$t := t+1;$
od

Here, Q is a set of individuals which are additionally taken into account during the selection step (e.g., $Q = P(t)$). The evaluation process yields a multiset (i.e., a set that may contain multiple copies of identical elements) of fitness values, which in general are not required to be identical to objective function values. However, since the selection criterion operates on fitness instead of objective function values, fitness values are used here as a result of the evaluation process. The evaluation of objective function values is always necessary during the calculation of fitness, such that the information is available and can easily be stored in an appropriate data structure.

Three main streams of instances of this general algorithm, developed independently of each other, can be identified:

1. **evolutionary programming** (EP), originally developed by Fogel, Owens and Walsh (1966) in the United States and recently refined by Fogel (1991);
2. **evolution strategies** (ESs), developed in Germany by Rechenberg (1965, 1973) and Schwefel (1965, 1981); and
3. **genetic algorithms** (GAs), developed by Holland (1975), originating also in the United States.

Based on genetic algorithms, some evolutionary algorithms, working on much more complex domains than those discussed here, were also developed. The most important approaches are **classifier systems** as proposed by Holland (1975) and the evolution of computer programs (unsuccessfully) tried for the first time by Friedberg (1958) and nowadays further developed by Koza (1992) using the terminology **genetic programming paradigm** (see also his chapter in this book).

Classifier systems are essentially rule-based systems capable of inductive learning by means of a low-level rule representation syntax that incorporates a 'don't care' symbol to achieve generalization possibilities and default hierarchies. The rule set of classifier systems undergoes evolution by a GA. An overview of classifier systems and their relation to the GA is given by Booker, Goldberg and Holland (1989). For more detailed information, the reader may consult Chapter 6 of Goldberg (1989) or Serra and Zanarini (1990) (Chapter 6).

Genetic programming goes one step further towards automatic programming by applying an evolutionary algorithm to a population of computer programs written in syntactically simple programming languages. Examples include the binary encoded machine languages as proposed by Friedberg (1958), integer-list representations of programs as formulated by Cramer (1985), or LISP programs as used for the first time by Fujiko and Dickinson (1987) and currently by Koza (1992).

These more complex instances of evolutionary algorithms as well as the numerous hybrid approaches of GAs, ESs and EP will not be discussed here. Instead, we focus on the comparison of the basic state-of-the-art variants of genetic algorithms, evolution strategies and evolutionary programming, when they are mainly used for parameter optimization tasks. These algorithms have each clearly demonstrated their capability of yielding good approximations to the extrema of complicated multimodal, discontinuous, non-differentiable and even noisy or moving response surfaces in optimization problems. An example of a typical multimodal surface occurring in the case of nonlinear parameter estimation problems is shown in Figure 14.1.

A variety of applications has been presented in conference proceedings edited by Grefenstette (1985, 1987*a*), Schaffer (1989), Belew and Booker (1991) (GAs); Fogel and

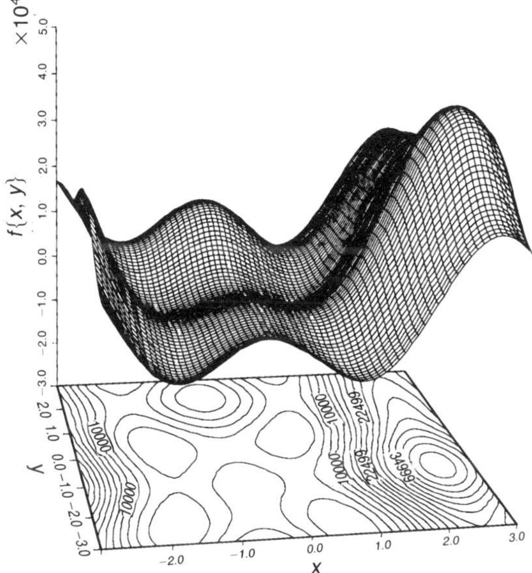

Figure 14.1 Two-dimensional surface of a typical multimodal parameter optimization problem.

Atmar (1992) (EP); Schwefel and Männer (1991), Männer and Manderick (1992) (GAs and ESs as well as other natural metaphors); and an annotated bibliography collected at the University of Dortmund by Bäck, Hoffmeister and Schwefel (1992) actually contains 260 references to applications of EAs.

Until recently, however, no personal contact between these research communities existed. Since 1990, contacts have been established between the GA and the ES communities, confirmed by collaborations and scientific exchange during now regularly alternating conferences in the United States (International Conference on Genetic Algorithms and their Applications, ICGA) since 1985 and in Europe (International Conference on Parallel Problem Solving from Nature, PPSN) since 1990. A first contact of researchers in the GA field to the EP community was also established in 1990, and the ES and EP community became known to each other not earlier than in 1992. For algorithms bearing so many similarities due to their reliance on organic evolution, this is a surprising fact. It is a central hope of the author that this chapter may provide an overview of the strong similarities between these algorithms and stimulate further discussions between these groups.

14.2 COMPARISON

To simplify the comparison of the algorithms, some categories of information guided along the algorithm's outline (algorithm 1) are presented for each of the three algorithms. Each subsection is introduced by a short historical overview of the algorithm and there follows a description of fitness evaluation and the representation of search points, initialization of the population, mutation, recombination and finally selection mechanisms. Concluding each subsection, the algorithms are presented in an abstract, high-level notation emphasizing specific concepts of and differences between the algorithms. Because of their similar representation of search points, we start with discussing ESs and EP, then turn to GAs.

In order to simplify the explanation of the algorithms, the following mathematical notations will be used: $N(0, 1)$ denotes a realization of a normally distributed one-dimensional random variable z having expectation zero and standard deviation 1, that is, with probability density function

$$p(z) = \frac{1}{\sqrt{2\pi}} \cdot \exp\left(-\frac{1}{2}z^2\right). \qquad (14.1)$$

Using an index for such a realization, the notation $N_i(0, 1)$ indicates that the random variable is sampled anew for each value of the counter i. A realization of a normally distributed one-dimensional random variable having expectation zero and standard deviation σ is then given by $\sigma \cdot N(0, 1)$.

Furthermore, $f: \mathbb{R}^n \rightarrow \mathbb{R}$ denotes the objective function to be optimized. Without losing generality, we assume a minimization task in the following. In general, fitness and objective function values of an individual are not required to be identical, such that $\Phi: I \rightarrow \mathbb{R}$ is used to denote the fitness function, where the symbol I denotes the space of individuals. In general, $\vec{a} \in I$ is used to indicate an individual, while $\vec{x} \in \mathbb{R}^n$ denotes an object variable vector.

14.2.1 EVOLUTION STRATEGIES

First efforts towards an ES took place in 1964 at the Technical University of Berlin (TUB) in Germany. Applications were mainly experimental and dealt with hydrodynamical problems like shape optimization of a bent pipe and of a two-phase (steam/liquid) jet nozzle. Different versions of the strategy were also simulated on the first available computer at TUB by Schwefel (1965). Rechenberg (1973) developed a theory of convergence velocity for the so-called $(1 + 1) - \mathrm{ES}$, a simple mutation-selection mechanism working on one individual which creates one offspring per generation by means of normally distributed mutations (see equation (14.1)). He proposed a theoretically confirmed rule for changing the standard deviation of mutations (1/5-success rule). He also proposed the first multimembered ES, a $(\mu + 1) - \mathrm{ES}$ where $\mu \geqslant 1$ individuals recombine to form one offspring, which after mutation eventually replaces the worst parent individual. This strategy, though never widely used, provided the basic idea to facilitate the transition to $(\mu + \lambda) - \mathrm{ES}$ and $(\mu, \lambda) - \mathrm{ES}$ introduced and investigated by Schwefel (1977, 1981), the resulting strategies (especially the latter one) being representative of the current ES research. In their most general form, these strategies are described here.

Fitness evaluation and representation

Search points in ESs are n-dimensional vectors $\vec{x} \in \mathbb{R}^n$, that is, objective functions $f: \mathbb{R}^n \rightarrow \mathbb{R}$ mapping search points to function values are expected. The fitness value of an individual is identical to its objective function value, that is, $\Phi(\vec{a}) = f(\vec{x})$ where $\vec{x}$ is the object variable component of $\vec{a}$. Additionally, each individual may include up to n different variances $c_{ii} = \sigma_i^2$ $(i \in \{1, \ldots, n\})$, as well as up to $n \cdot (n - 1)/2$ covariances c_{ij} $(i \in \{1, \ldots, n - 1\}, j \in \{i + 1, \ldots, n\})$ of the generalized n-dimensional normal distribution with expectation vector $\vec{0}$, having probability density function

$$p(\vec{z}) = \sqrt{\frac{\det \mathbf{A}}{(2\pi)^n}} \exp\left(-\frac{1}{2}\vec{z}^{\mathrm{T}} \mathbf{A} \vec{z}\right), \quad (14.2)$$

where $\mathbf{A}^{-1} = (c_{ij})$ represents the covariance matrix, and $\vec{z}$ denotes a random variable. Altogether, up to $w = n \cdot (n + 1)/2$ strategy parameters can be combined with object variables to form an individual $\vec{a} \in I = \mathbb{R}^{n+w}$. Often, however, only the variances are taken into account, such that $\vec{a} \in I = \mathbb{R}^{2n}$, and it is sometimes also useful to work with only one variance valid for all object variables, that is, $\vec{a} \in I = \mathbb{R}^{n+1}$.

Since the mutation mechanism is required to preserve orthogonality of the coordinate system or, equivalently, positive definiteness of the covariance matrix $\mathbf{A}^{-1}$, the algorithm uses rotation angles α_{ij} $(\tan 2\alpha_{ij} = 2c_{ij}/(\sigma_i^2 - \sigma_j^2))$ rather than the covariances themselves. Correlated mutations are then realized by performing the coordinate system rotations accordingly. Recently, Rudolph (1992) presented a formal proof of the validity of this mechanism, demonstrating that the orthogonal transformations can be generated exactly by means of the mutation mechanism of rotation angles chosen in evolution strategies (see also the section on mutation).

Furthermore, mutation step sizes are represented in the implementation by standard deviations σ_i rather than variances σ_i^2, which is, of course, also equivalent. In the following, we use the notation $\vec{N}(\vec{0}, \vec{\sigma}, \vec{\alpha})$ to denote a realization of a random vector distributed according to the generalized n-dimensional normal distribution having expectation $\vec{0}$ and a covariance matrix determined by the standard deviations $\vec{\sigma}$ and rotation angles $\vec{\alpha}$. Finally, $\vec{a} = (\vec{x}, \vec{\sigma}, \vec{\alpha}) \in \mathbb{R}^{n+w}$ denotes a complete individual.

Initalization

The initial population $P(0) = \{\vec{a}_1(0), \ldots, \vec{a}_\mu(0)\}$ is generated by means of mutation from one initial starting point. This starting point may be given by the user of the algorithm, if he or she wishes to do so, or can be chosen from the feasible region according to a uniform, n-dimensional probability distribution.

Mutation

In its most general form the mutation operator $m: I \rightarrow I$ (where $I = \mathbb{R}^{n+w}$) yields a mutated individual $m(\vec{a}) = (\vec{x}', \vec{\sigma}', \vec{\alpha}')$ by first mutating the standard deviations and rotation angles and then mutating the object variables according to the now modified probability density function of individual $\vec{a}$, i.e., $\forall i \in \{1, \ldots, n\}$, $\forall j \in \{1, \ldots, n \cdot (n-1)/2\}$:

$$\begin{aligned} \sigma_i' &= \sigma_i \cdot \exp(\tau' \cdot N(0, 1) + \tau \cdot N_i(0, 1)) \\ \alpha_j' &= \alpha_j + \beta \cdot N_j(0, 1) \\ \vec{x}' &= \vec{x} + \vec{N}(\vec{0}, \vec{\sigma}', \vec{\alpha}'). \end{aligned} \tag{14.3}$$

Mutations of the object variables may now be linearly correlated according to the values of $\vec{\alpha}$, and $\vec{\sigma}$ provides a scaling of the metrics. The global factor $\tau' \cdot N(0, 1)$ allows for an overall change of the mutability, whereas $\tau \cdot N_i(0, 1)$ allows for individual changes of the 'mean step sizes' σ_i. The factors τ, τ' and β are rather robust parameters, which Schwefel, based on a theoretical derivation for τ and τ' (Schwefel 1977, pp. 165–9) and empirical investigations for β, suggests should be set as follows:

$$\begin{aligned} \tau &\propto \left(\sqrt{2\sqrt{n}}\right)^{-1} \\ \tau' &\propto (\sqrt{2n})^{-1} \\ \beta &\approx 0.0873, \end{aligned} \tag{14.4}$$

where usually the proportionality constants for τ and τ' have the value 1. The value suggested for β (in radians) equals 5°. If only one standard deviation is incorporated into the genotype, that is, $\vec{a} \in \mathbb{R}^{n+1}$, the modification rule for standard deviations reduces to $\sigma' = \sigma \cdot \exp(\tau'' \cdot N(0, 1))$, where $\tau'' = (\sqrt{n})^{-1}$.

Altogether, this special mutation mechanism enables the algorithm to evolve its own strategy parameters (standard deviations and covariances) during the search, exploiting an implicit link between an appropriate internal model and good fitness values. The resulting evolution and adaptation of strategy parameters according to the topological requirements has been termed **self-adaptation** by Schwefel (1987).

Recombination

Different recombination mechanisms are used in ESs either in their usual form, producing one new individual from two randomly selected parent individuals, or in their global form, allowing components to be taken for one new individual from potentially all individuals available in the population. Furthermore, recombination is performed on strategy parameters as well as on the object variables, and the recombination operator may be different for object variables, standard deviations and rotation angles. Recombination rules for an operator $r: I^\mu \rightarrow I$ creating an individual $r(P(t)) = \vec{a}' = (\vec{x}', \vec{\sigma}', \vec{\alpha}') \in I$ are given here representatively only for the object variables ($\forall i \in \{1, \ldots, n\}$):

$$x_i' = \begin{cases} x_{S,i} & \text{without recombination} \\ x_{S,i} \text{ or } x_{T,i} & \text{discrete recombination} \\ x_{S,i} + \frac{1}{2} \cdot (x_{T,i} - x_{S,i}) & \text{intermediate recombination} \\ x_{S_i,i} \text{ or } x_{T_i,i} & \text{global, discrete} \\ x_{S_i,i} + \frac{1}{2} \cdot (x_{T_i,i} - x_{S_i,i}) & \text{global, intermediate} \end{cases} \tag{14.5}$$

Indices S and T denote two parent individuals selected at random from $P(t)$. For the global variants, for each component of $\vec{x}$ the parents S_i and T_i are determined anew, that is, in the case of global intermediate recombination the mechanism proceeds as follows.

For each component x_i' of the offspring vector:

1. Select two parents S_i, T_i at random from $P(t)$.
2. Calculate x_i' as the average of the corresponding components $x_{S_i,i}$, $x_{T_i,i}$ of S_i and T_i.

For global discrete recombination, step (2) is substituted by:

2. Set $x_i' = x_{S_i,i}$ or $x_i' = x_{T_i,i}$, where both events occur with probability one half.

A generalization of the intermediate mechanism to allow for arbitrary weight factors $\chi \in [0, 1]$ instead of $\chi = \frac{1}{2}$ is obvious, but has not been tested so far. The standard (non-global) variants of recombination determine the parent individuals just once per creation of one offspring individual.

Empirically, discrete recombination on object variables and intermediate recombination on strategy parameters have been observed to give the best results, and recombination of strategy parameters has shown to be mandatory for the self-adaptation mechanism of strategy parameters to work properly (Schwefel 1987, 1988).

Selection

Selection in ESs is completely deterministic, selecting the best μ ($1 \leqslant \mu < \lambda$) individuals out of the set of λ offspring individuals ((μ, λ)-selection, denoted by an operator $s_{(\mu,\lambda)}: I^{\lambda} \to I^{\mu}$) or out of the union of parents and offspring (($\mu + \lambda$)-selection, $s_{(\mu+\lambda)}: I^{\mu+\lambda} \to I^{\mu}$). Though being elitist (i.e., survival of the best individual is assured) and therefore guaranteeing a monotonously improving performance, the latter strategy is unable to deal with changing environments and hinders the self-adaptation mechanism with respect to the strategy parameters (internal model). Therefore, the (μ, λ)-selection is currently recommended, investigations by Schwefel (1987) indicating a ratio for $\mu/\lambda \approx 1/7$ being optimal.

Conceptual algorithm

Combining the previous topics, the following conceptual algorithm of a ($\mu + \lambda$) − ES and (μ, λ) − ES, respectively, results in algorithm 2, below.

ALGORITHM 2 (OUTLINE OF AN EVOLUTION STRATEGY)

$t := 0;$
initialize $P(0) := \{\vec{a}_1(0), \ldots, \vec{a}_\mu(0)\} \in I^\mu$
  where $I = \mathbb{R}^{n+w}$
  and $\vec{a}_k = (x_i, \sigma_i\ \forall i \in \{1, \ldots, n\},\ \alpha_{ij} = \alpha_{ji}\ \forall i \in \{1, \ldots, n-1\},\ j > i);$
evaluate $P(0)$: $\{\Phi(\vec{a}_1(0)), \ldots, \Phi(\vec{a}_\mu(0))\}$ **where** $\Phi(\vec{a}_k(0)) = f(\vec{x}_k(0));$
while not *terminate*$(P(t))$ **do**
  recombine $\vec{a}_k'(t) := r(P(t))\ \forall k \in \{1, \ldots, \lambda\};$
  mutate $\vec{a}_k''(t) := m(\vec{a}_k'(t))\ \forall k \in \{1, \ldots, \lambda\};$
  evaluate $P''(t) := \{\vec{a}_1''(t), \ldots, \vec{a}_\lambda''(t)\}$:
    $\{\Phi(\vec{a}_1''(t)), \ldots, \Phi(\vec{a}_\lambda''(t))\}$ **where** $\Phi(\vec{a}_k''(t)) = f(\vec{x}_k''(t));$
  select $P(t+1) :=$ **if** (μ,λ)–ES
    then $s_{(\mu,\lambda)}(P''(t));$
    else $s_{(\mu+\lambda)}(P''(t) \cup P(t));$
  $t := t + 1;$
od

14.2.2 EVOLUTIONARY PROGRAMMING

In the early and mid-1960s, Fogel (1962) and Fogel, Owens and Walsh (1966) described EP for the evolution of finite state machines in order to solve sequence prediction tasks. The state transition tables of these machines were modified by uniform random mutations on the corresponding discrete, finite alphabet and on the size and connection of the machines' states. Evaluation of fitness took place according to the number of symbols predicted correctly or any other selected payoff function operating on the meaning of the symbols. Each machine in the parent population generated one offspring by means of mutation, and the best half number of parents and offspring were selected to survive, which would be called a $(\mu + \mu)$-strategy in ES-terminology. EP has been extended by D. B. Fogel to work on real-valued object variables based on normally distributed mutations, and the following description is based on Fogel (1992*a*) and Fogel (1992*b*).

Fitness evaluation and representation

In real-valued optimization problems, EP assumes a restricted subspace $I = \Pi_{i=1}^{n}[u_i, v_i] \subset \mathbb{R}^n$ $(u_i < v_i)$ of $\mathbb{R}^n$ as the domain of optimization, that is, individuals are object variable vectors $\vec{a} = \vec{x} \in I$. In Fogel (1992*b*), a concept of **meta-evolutionary programming** is presented which is essentially identical to self-adaptation of standard deviations in ESs. To incorporate a vector $\vec{v} \in [0, c]^n$ of variances rather than standard deviations ($v_i = \sigma_i^2$), the space of individuals is extended to $I' = \Pi_{i=1}^{n}[u_i, v_i] \times [0, c]^n \subset \mathbb{R}^{2n}$, where $c > 0$ is a constant. Since EP is probably not known very widely, we will discuss both approaches in the following and therefore use the notation I for the standard EP individual space and I' for the space in meta-EP. In a third approach called **Rmeta-EP**, Fogel (1992*b*, pp. 287–9) additionally incorporates a vector of correlation coefficients into the genotypes for self-adaptation, a mechanism that is essentially identical to correlated mutations in ESs. This approach, however, was implemented and tested by Fogel only for a restricted, two-dimensional case such that it is not discussed here in detail. As in ESs, a general method has to be developed here which guarantees the possibility of generating just the positive definite covariance matrices by means of mutation acting on the representation.

The fitness values $\Phi(\vec{a})$ are obtained from objective function values $f(\vec{x})$ by scaling them to positive values and possibly by imposing some random alteration κ: $\Phi(\vec{a}) = \delta(f(\vec{x}), \kappa)$, where δ denotes the scaling function. To meet the requirement of a useful scaling, it is of the general form $\delta: \mathbb{R} \times S \rightarrow \mathbb{R}^+$ where S is an additional set of parameters involved in the process.

Initialization

Initialization in EP and meta-EP is done by sampling from a uniform distribution ranging over $[u_i, v_i]$ for object variables x_i, and ranging over $[0, c]$ for variances v_i $(i \in \{1, \ldots, n\})$. In contrast to ESs, where only one initial point is given or chosen at random in the feasible region and a population is created from this point by means of mutation with given starting values for the vectors $\vec{\sigma}$ and $\vec{\alpha}$, in EP a complete initial population is generated in this way.

Mutation

In the case of a standard EP, the Gaussian mutation operator $m: I \rightarrow I$, $m(\vec{x}) = \vec{x}'$, uses a standard deviation which is obtained for each component x_i as the square root of a linear transformation of the fitness value $\Phi(\vec{a}) = \Phi(\vec{x})$, that is, $(\forall i \in \{1, \ldots, n\})$:

$$x_i' = x_i + \sigma_i \cdot N_i(0, 1)$$
$$\sigma_i = \sqrt{\beta_i \cdot \Phi(\vec{x}) + \gamma_i}. \qquad (14.6)$$

Here the proportionality constants β_i and the offsets γ_i are $2n$ parameters which must be tuned for a particular task. Often, however, β_i and γ_i are set to one and zero, respectively, such that $x_i' = x_i + \sqrt{\Phi(\vec{x})} \cdot N_i(0, 1)$.

To overcome the tuning difficulties with this approach, meta-EP self-adapts n variances per individual quite similar to ESs. Mutation m': $I' \to I'$, $m'(\vec{a}) = (\vec{x}', \vec{\nu})$ works as follows ($\forall i \in \{1, \ldots, n\}$):

$$\begin{aligned} x_i' &= x_i + \sqrt{\nu_i} \cdot N_i(0, 1) \\ \nu_i' &= \nu_i + \sqrt{\alpha \nu_i} \cdot N_i(0, 1). \end{aligned} \tag{14.7}$$

Here, α denotes a parameter assuring that ν_i tends to remain positive. Whenever a variance would become negative or zero by means of mutation, it is set to a small value $\varepsilon > 0$. Although the idea is the same as in ESs, the underlying stochastic process is fundamentally different. The log-normally distributed alterations in ESs automatically guarantee positivity of σ_i, as well as no drift in case of zero selection pressure, whereas fluctuations in the meta-EP algorithm are expected to be much larger than in the ES and are expected to drift in case of missing selective pressure. This mechanism surely deserves further investigation.

Recombination

In EP, no recombination is used at all.

Selection

After creating μ offspring from μ parent individuals by mutating each parent once, a variant of stochastic q-tournament selection ($q \geq 1$ being a parameter of the algorithm) selects μ individuals from the union of parents and offspring, that is, a randomized $(\mu + \mu)$-selection is used. In principle, for each individual $\vec{a}_k \in P(t) \cup P'(t)$, where $P'(t)$ is the population of mutated individuals, q individuals are chosen at random from $P(t) \cup P'(t)$ and compared to $\vec{a}_k$ with respect to their fitness values. Then, for $\vec{a}_k$ it is counted how many of the q individuals are worse than $\vec{a}_k$, resulting in a score $w_k \in \{0, \ldots, q\}$. After doing so for all 2μ individuals, the individuals are ranked in descending order of the score values w_i ($i \in \{1, \ldots, 2\mu\}$), and the μ individuals having highest score w_i are selected to form the next population. More formally, we have ($\forall i \in \{1, \ldots, 2\mu\}$):

$$w_i = \sum_{j=1}^{q} \begin{cases} 1 & \text{if } \Phi(\vec{a}_i) \leq \Phi(\vec{a}_{\chi_j}) \\ 0 & \text{otherwise.} \end{cases} \tag{14.8}$$

$\chi_j \in \{1, \ldots, 2\mu\}$ is a uniform integer random variable, sampled anew for each comparison. As the tournament size q is increased, the mechanism becomes more and more a deterministic $(\mu + \mu)$-scheme (the scores w_i are binomially distributed with a variance tending towards zero as q is increased). Since the best individual is assigned a guaranteed maximum fitness score q, survival of the best is guaranteed, that is, the selection mechanism is elitist.

Conceptual algorithm

In the framework of our general algorithm outline, the formulation for a meta-EP algorithm is derived as algorithm 3.

ALGORITHM 3 (OUTLINE OF AN EVOLUTIONARY PROGRAMMING ALGORITHM)

$t := 0$;
initialize $P(0) := \{\vec{a}_1(0), \ldots, \vec{a}_\mu(0)\} \in I^\mu$
  where $I = \mathbb{R}^{2n}$
  and $\vec{a}_k = (x_i, \nu_i \; \forall i \in \{1, \ldots, n\})$;
evaluate $P(0)$: $\{\Phi(\vec{a}_1(0)), \ldots, \Phi(\vec{a}_\mu(0))\}$ **where** $\Phi(\vec{a}_k(0)) = \delta(f(\vec{x}_k(0)), \kappa_k)$;
while not *terminate*$(P(t))$ **do**

$mutate\ \vec{a}'_k(t) := m(\vec{a}_k(t)) \quad \forall k \in \{1, \ldots, \mu\};$
$evaluate\ P'(t) := \{\vec{a}'_1(t), \ldots, \vec{a}'_\mu(t)\}:$
$\{\Phi(\vec{a}'_1(t)), \ldots, \Phi(\vec{a}'_\mu(t))\}$) **where** $\Phi(\vec{a}'_k(t)) = \delta(f(\vec{x}'_k(t)), \kappa_k);$
$select\ P(t+1) := s(P(t) \cup P'(t), q);$
$t := t + 1;$
od

14.2.3 GENETIC ALGORITHMS

In the 1960s, the research interests of Holland (1962) were devoted to the study of general adaptive processes, concentrating on the idea of a system receiving sensory inputs from the environment by binary detectors. Structures in the search space were progressively modified in this model by operators selected by an adaptive plan, judging the quality of previous trials by means of an evaluation measure. In Holland (1975), he points out how to interpret the so-called reproductive plans in terms of genetics, economics, game-playing, pattern recognition and parameter optimization (Chapter 3). His genetic plans or genetic algorithms were applied to parameter optimization for the first time by De Jong (1975), who laid the foundations of this application technique. Currently, numerous modifications of the original GA are applied to all (and more) of the fields Holland had indicated. However, many of these applications show enormous differences to the basic GA as explained in the following, such that the boundary to the other algorithms discussed above becomes blurred.

Fitness evaluation and representation

As indicated, GAs work on bit strings of fixed length l, that is, $I = \{0, 1\}^l$. For pseudo-Boolean objective functions, this representation can be used directly. In order to apply GAs to continuous parameter optimization problems of the form $f: \prod_{i=1}^n [u_i, v_i] \rightarrow \mathbb{R}$ ($u_i < v_i$), the bit string is logically divided into n segments of (in most cases) equal length l_x (i.e., $l = n \cdot l_x$), and each segment is interpreted as the binary code of the corresponding object variable $x_i \in [u_i, v_i]$. A segment decoding function $\Gamma^i: \{0, 1\}^{l_x} \rightarrow [u_i, v_i]$ typically looks like

$$\Gamma^i(a_{i1} \ldots a_{il_x}) = u_i + \frac{v_i - u_i}{2^{l_x} - 1}\left(\sum_{j=0}^{l_x - 1} a_{i(l_x - j)} \cdot 2^j\right), \tag{14.9}$$

where $(a_{i1} \ldots a_{il_x})$ denotes the ith segment of an individual $\vec{a} = (a_{11} \ldots a_{nl_x}) \in I^{n \cdot l_x} = I^l$.

Nowadays, however, it is common practice to use a **Gray code** interpretation of the bit string segments rather than the simple base two representation indicated in equation (14.9). This encoding technique represents adjacent integer values by bit strings of Hamming distance one, that is, differing by one bit. Assuming $i_1, i_2 \in \{0, \ldots, 2^{l_x} - 1\}$ to denote the integer values represented by Gray-coded bit-string segments $\vec{a}_2, \vec{a}_2 \in \{0, 1\}^{l_x}$, we have

$$|i_1 - i_2| = 1 \Rightarrow \rho_H(\vec{a}_1, \vec{a}_2) = 1 \tag{14.10}$$

where ρ_H denotes the Hamming distance, that is, $\rho_H(a_{11} \ldots a_{1l_x}, a_{21} \ldots a_{21x}) = \sum_{i=1}^{l_x} |a_{1i} - a_{2i}|$. It is widely believed that using a Gray code is advantageous for a GA due to the avoided Hamming cliffs, that is, situations where in order to achieve a small change in the decoded space a simultaneous change of several bits in the binary space would be necessary. This is illustrated by comparing a three-bit standard binary code to a Gray code in Table 14.1.

Utilization of a Gray code in GAs was proposed by Bethke (1981) and became a standard by the dissemination of the GENESIS genetic algorithm software by Grefenstette (1987*b*).

Table 14.1 Comparison of a three-bit standard binary code and the corresponding Gray code

Integer	*Standard*	*Gray*
0	000	000
1	001	001
2	010	011
3	011	010
4	100	110
5	101	111
6	110	101
7	111	100

Though the Hamming cliffs existing for the standard code between the codes for integer 3 (011) and 4 (100), disappear in the Gray code, it should be noted that (14.10) is not an equivalence: even for a Gray code a change of a single bit may cause arbitrarily large changes to the corresponding integer values. This highly discontinuous link between the genotypic effect of mutations and their phenotypic effect has still not been investigated sufficiently on a level beyond that of experimental comparisons as performed by Caruna and Schaffer (1988).

Conversions between Gray and standard code and vice versa additionally complicate the representation mechanism. Assuming $\vec{a} = (a_1 \ldots a_{l_x})$ to denote a standard-coded bitstring segment and $\vec{b} = (b_1 \ldots b_{l_x})$ its Gray-coded counterpart, the conversions can be defined on the basis of an operator $\oplus$ that denotes addition modulo 2 on $\{0, 1\}$ (i.e., $0 \oplus 0 = 0$, $0 \oplus 1 = 1 \oplus 0 = 1$, $1 \oplus 1 = 0$) following Wright (1991) ($\forall_i \in \{1, \ldots, l_x\}$):

1. standard to Gray:

$$b_i = \begin{cases} a_i, & \text{if } i = 1 \\ a_{i-1} \oplus a_i, & \text{if } i > 1. \end{cases} \tag{14.11}$$

2. Gray to standard:

$$a_i = \bigoplus_{j=1}^{i} b_j. \tag{14.12}$$

Then, for a GA using Gray code, the decoding function given in equation (14.9) must be extended to

$$\Gamma^i(a_{i1} \ldots a_{il_x}) = u_i + \frac{v_i - u_i}{2^{l_x} - 1}\left(\sum_{j=0}^{l_x-1}\left(\bigoplus_{k=1}^{l_x-j} b_{ik}\right) \cdot 2^j\right). \tag{14.13}$$

Combining the segment-wise decoding function Γ^i to an individual-decoding function $\Gamma = \Gamma^1 \times \ldots \times \Gamma^n$, fitness values are obtained by setting $\Phi(\vec{a}) = \delta(f(\Gamma(\vec{a})))$, where again δ denotes a scaling function assuring positive fitness values such that the best individual receives the largest fitness. Most commonly, a linear dynamic scaling is used which takes the worst individual of the population $P(t - \omega)$ ω time steps before ($t - \omega < 0 \Rightarrow t - \omega := 0$) into account:

$$\delta(f(\Gamma(\vec{a})), P(t - \omega)) = \max\{f(\Gamma(\vec{a}_j)) \mid \vec{a}_j \in P(t - \omega)\} - f(\Gamma(\vec{a})), \tag{14.14}$$

where ω is called the scaling window. Some scaling methods are discussed by Goldberg (1989, pp. 122–4).

In biological terms, there is an analogy to the discrete alphabet of nucleotide bases, double strands of which form the deoxyribonucleic acid (DNA). This information chain (the genotype) is decoded in several steps to amino acids (by means of the genetic code), to proteins, and finally to the phenotype, that is, the physical embodiment of the organism, by means of the epigenetic apparatus. Therefore, the GA representation scheme emphasizes the genotype rather than the phenotype from a bottom-up perspecitive. However, the impact of this additional representational level on the suitability for parameter optimization has still to be investigated further. The epigenetic apparatus has been completely ignored until now.

In contrast to this, the representation and mutation mechanisms in ESs and EP are based on the phenotype, since they consider mutation to model phenotypical changes, that is, changes of the object variable vector $\vec{x}$ (see also Section 14.3), directly rather than operating on a genotypical level.

Initialization

Working on bit strings, the initialization of a start population in GAs is performed by sampling a binary uniform random variable $l \cdot \mu$ times (independent coin tosses). In this way each bit of the start population $P(0) = \{\vec{a}_1(0), \ldots, \vec{a}_\mu(0)\}$ is determined at random.

Mutation

Mutation in GAs works on the bit-string level and is traditionally referred to as a 'background operator' according to Holland (1975). It works by occasionally inverting single bits of individuals, with the probability p_m of this event usually being very small ($p_m \approx 1 \cdot 10^{-3}$ per bit). The parameter p_m can be interpreted as the mutation step size of GAs. Often, it neither depends on the number n of object variables, nor on the total length l of the bit string. Recent theoretical investigations, however, give clear evidence that a mutation rate $p_m = 1/l$ is generally recommendable and, in the case of unimodal pseudo-Boolean functions, is an optimal setting (Mühlenbein 1992; Bäck 1992*b*). On a single individual, mutation $m: I \times [0, 1] \rightarrow I$, $m((s_1, \ldots, s_l), p_m) = (s'_1, \ldots, s'_l)$ works as follows ($\forall_i \in \{1, \ldots, l\}$):

$$s'_i = \begin{cases} s_i & \chi_i > p_m \\ 1 - s_i & \chi_i \leq p_m. \end{cases} \tag{14.15}$$

Here $\chi_i \in [0, 1]$ is a uniform random variable, sampled anew for each bit. The reader should be very conscious about mutation rate values given in the literature, because originally mutation was defined by Holland (1975) as a substitution of a bit by a random element from the alphabet $\{0, 1\}$, such that mutation rates according to our definition are twice as large as mutation rates according to that original definition. However, because it is more natural to interpret mutation as a real change (instead of a random toss with probability one-half), it is used as a reversal event by most GA researchers.

Recombination

In GAs, emphasis is mainly concentrated on crossover, the recombination operator of GAs, as the main variation operator which hopefully recombines useful segments from different individuals. Crossover $r: I^\mu \times [0, 1] \rightarrow I$ is again an operator working entirely on the bit representation, completely ignoring the epigenetic apparatus. A parameter p_c (crossover rate) indicates the probability per individual of undergoing recombination. Typical values for p_c are in the range [0.6, 1.0]. When two parent individuals $\vec{s} = (s_1, \ldots, s_l)$, $\vec{v} = (v_1, \ldots, v_l)$ have been selected (at random) from the population, crossover forms two offspring individuals $\vec{s}'$, $\vec{v}'$ according to the following scheme:

$$\begin{aligned} \vec{s}' &= (s_1, \ldots, s_{\chi-1}, s_\chi, v_{\chi+1}, \ldots, v_l) \\ \vec{v}' &= (v_1, \ldots, v_{\chi-1}, v_\chi, s_{\chi+1}, \ldots, s_l). \end{aligned} \tag{14.16}$$

As before, $\chi \in \{1, \ldots, l\}$ denotes a uniform random variable, and one of both offspring individuals is randomly selected to be the overall result of crossover. This one-point crossover can be extended naturally to a generalized m-point crossover by sampling more than one breakpoint and exchanging each second resulting segment, an extension performed for the first time by De Jong (1975). Actually, however, there is neither clear theoretical nor empirical evidence to decide upon the question of which crossover operator is most appropriate, although several investigations have tried to shed light on these questions, for example Caruna, Eshelman and Schaffer (1989), Eshelman, Caruna and Schaffer (1989), and Schaffer, Caruna, Eshelman and Das (1989).

Selection

Just as in EP, selection in GAs emphasizes a probabilistic survival rule mixed with a fitness dependent chance to have (different) partners for producing more or less offspring. By deriving an analogy to the game-theoretic multi-armed bandit problem, Holland (1975) identifies the necessity to use proportional selection in order to optimize the trade-off between further exploiting promising regions of the search space while at the same time also exploring other regions (Chapter 5). For proportional selection $s: I^{\mu} \to I^{\mu}$, the survival probabilities of individuals $\vec{a}_i$ are given by their relative fitness, i.e. ($\forall_i \in \{1, \ldots, \mu\}$):

$$p_s(\vec{a}_i) = \frac{\Phi(\vec{a}_i)}{\sum_{j=1}^{\mu} \Phi(\vec{a}_j)}. \qquad (14.17)$$

Sampling μ individuals according to this probability distribution yields the next generation of parents. Obviously, this mechanism fails in the case of negative fitness or minimization tasks, which explains the necessity to introduce a scaling function δ as mentioned in the section on fitness evaluation and representation above.

Conceptual algorithm

As before, the traditional GA is embedded here in the framework of our conceptual algorithm:

ALGORITHM 4 (OUTLINE OF A GENETIC ALGORITHM)

$t := 0$;
initialize $P(0) := \{\vec{a}_1(0), \ldots, \vec{a}_\mu(0)\} \in I^\mu$
    where $I = \{0,1\}^l$;
evaluate $P(0)$: $\{\Phi(\vec{a}_1(0)), \ldots, \Phi(\vec{a}_\mu(0))\}$ **where** $\Phi(\vec{a}_k(0)) = \delta(f(\Gamma(\vec{a}_k(0))), P(0))$;
while not *terminate*$(P(t))$ **do**
    recombine $\vec{a}'_k(t) := r(P(t), p_c)\ \forall k \in \{1, \ldots, \mu\}$;
    mutate $\vec{a}''_k(t) := m(\vec{a}'_k(t), p_m)\ \forall k \in \{1, \ldots, \mu\}$;
    evaluate $P''(t) := \{\vec{a}''_1(t), \ldots, \vec{a}''_\mu(t)\}$:
        $\{\Phi(\vec{a}''_1(t)), \ldots, \Phi(\vec{a}''_\mu(t))\}$ **where** $\Phi(\vec{a}''_k(t)) = \delta(f(\Gamma(\vec{a}''_k(t))), P(t-\omega))$;
    select $P(t+1) := s(P''(t))$
        where $p_s(\vec{a}''_k(t)) = \Phi(\vec{a}''_k(t)) / \sum \Phi(\vec{a}''_j(t))$;
    $t := t + 1$;
od

14.3 SUMMARY

The main characteristic similarities and differences between the algorithms presented in this article are summarized in Table 14.2.

It is a remarkable fact that each algorithm emphasizes different features as being most important to a successful evolution process. In analogy to repair-enzymes which give evidence of a biological self-control of mutation rates of nucleotide bases in DNA (Gottschalk 1989, pp. 269–71), both ESs and meta-EP use self-adaptation processes for the mutation rates. In GAs, this concept was successfully tested only recently by Bäck (1992*a*), but will need more time to be recognized and applied. Both ESs and EP concentrate on mutation as the main search operator, while the role of mutation in GAs is usually seen to be of secondary (if any) im-

Table 14.2 Main characteristics of evolutionary algorithms

	ES	*EP*	*GA*
Representation	Real-valued	Real-valued	Binary-valued
Self-adaptation	Standard deviations and covariances	Standard deviations (meta-EP) and correlation coefficients (Rmeta-EP)	None
Fitness is	Objective function value	(Scaled) objective function value	Scaled objective function value
Mutation	Main operator	Only operator	Background operator
Recombination	Different variants, important for self-adaptation	None	Main operator
Selection	Deterministic, extinctive	Probabilistic, extinctive	Probabilistic, preservative

portance. On the other hand, recombination plays a major role in GAs, is not emphasized in EP, and is absolutely necessary for use in connection with self-adaptation in ESs. One of the characteristics of EP is the strict denial of recombination being important for general searches, although, as indicated in Fogel and Atmar (1990), recombination may be useful in particular problems. Finally, both GAs and EP emphasize a probabilistic selection mechanism, while from the point of view of ESs selection is completely deterministic without any evidence for the necessity of incorporating probabilistic rules. In contrast, both ESs and EP definitely exclude some individuals from being selected, that is, they use **extinctive** selection mechanisms, while GAs generally assign a non-zero selection probability to each individual, which we term a **preservative** selection mechanism. A more detailed classification of selection mechanisms in evolutionary algorithms and an overview of the impact of selection on the search process was given by Bäck and Hoffmeister (1991).

Very naturally, the reader can deduce several interesting questions for future research. It is interesting to see very different, sometimes contrasting, design principles for evolutionary algorithms being emphasized by the different research communities. A clear goal of future research should be to identify the rational as well as the not-so-rational reasons for this fact and to extract the general rules for designing components of new and maybe even better EAs.

Finally, concluding this summary of the algorithmic components, we return to the biological background by trying to relate the representational mechanisms of evolutionary algorithms to their biological counterparts, similar to the effort of Goldberg (1989, pp. 21–2) for GAs. In biology, the basic building blocks of DNA are four different nucleotide bases that form the alphabet of genetic information. Triples of nucleotide bases, called codons, forms the units of translation that are mapped to the twenty or so amino acids by means of the (highly redundant) genetic code. Chains of amino acids, that is, proteins, are synthesized during the translation process at the ribosomes. Commonly, a gene on the DNA is identified as a unit that encodes just one protein. At the next higher organizational level, chromosomes are large collections of genes. The complete collection of genetic material of an organism is the genotype. Up to the genotype level, a clearly hierarchical structure of information exists, bearing similarities to a text which is structured into sentences, words and letters. Table 14.3 tries

Table 14.3 Relations of artificial and natural information representation

Biology	*ES*	*EP*	*GA*
Nucleotide base	–	–	Binary digit $a_i \in \{0, 1\}$
Codon	–	–	–
Gene	Single x_i, σ_i, α_j	Single x_i, ν_i	Bit-string segment $(a_{i1}, \ldots, a_{il_x}) \in \{0, 1\}^{l_x}$
Chromosome	Complete vectors $\vec{x}, \vec{\sigma}, \vec{\alpha}$	Complete vectors $\vec{x}, \vec{\nu}$	Complete bit string $\vec{a}$
Genotype	Collection of chromosomes, $(\vec{x}, \vec{\sigma}, \vec{\alpha})$	Collection of chromosomes, $(\vec{x}, \vec{\nu})$	Complete bit string $\vec{a}$
Phenotype	Component $\vec{x}$	Component $\vec{x}$	Decoded structure $\vec{x} = \Gamma(\vec{a})$

to relate components of the representations used in evolutionary algorithms to the biological origin.

In GAs, we can very naturally identify a correspondence between the binary alphabet and the quarternary one of nucleotide bases. Then, one can either view a bit-string segment as a codon, thus implying an analogy of a single (decoded) object variable to a single amino acid, or we can view a bit-string segment as a complete gene by leaving out the codon level, which implies an analogy of a single object variable to a protein. Both points of view are equally inexact, the former because it leaves out the gene level, the latter because it does not supply us with the obvious interpretation of segment decoding. However, we tend to prefer the latter because the highly non-linear code in genetic algorithms seems to correspond better to the gene-protein relation than to the codon-amino acid relation and is also a bit more consistent with viewing the decoded vector $\vec{x}$ as a phenotype. Consequently, the genotype is made up of the complete binary vector that in turn corresponds to a single chromosome.

Since the phenotype is surely represented by that unit which undergoes an evaluation of its fitness in the environment, the object variable vector $\vec{x}$ can for all these evolutionary algorithms be interpreted as the phenotype. It is derived from the genotype in the case of evolution strategies and evolutionary programming by a rather trivial mapping, since no explicit code is used. The complete genotype consists of an object variable vector and one or two strategy parameter vectors, and each vector represents a chromosome. It is straightforward to interpret single components of these vectors as genes.

However, the similarities between the algorithms and their natural origin are still crude, and even the seemingly closer correspondence of genetic algorithms gives no evidence for a more successful modelling. But biological terminology is established for use in the field of evolutionary algorithms, such that an identification of and agreement on the meaning of biological terms in this context is urgently needed. Of course, other assignments are possible, but they should in any case use biological terms in a correct context.

14.4 A GUIDE TO THE LITERATURE

Since this article contains references to the basic literature on each type of evolutionary algorithms, it seems appropriate to guide the reader by providing a hint to further reading.

For each algorithm discussed above, we can identify two generations of books, that is:

1. ESs: Rechenberg (1973) discusses the (1 + 1) − ES and its theory (in German only). A look at the book of Schwefel (1981), where the (μ + λ) − ES and (μ, λ) − ES are introduced and compared to traditional optimization methods, may be more useful. Furthermore, a convergence rate theory for these algorithms is developed for the test functions introduced by Rechenberg.
2. EP: For historical reasons it is worth looking at the early work of L. J. Fogel, for example, Fogel, Owens and Walsh (1966), since his ideas were vehemently criticized at that time. For the modern EP algorithm, the reader is referred to Fogel (1991), a book on using EP for system identification, for meta-EP and Rmeta-EP to the thesis of Fogel (1992b).
3. GAs: Holland (1975) gives a detailed discussion of adaptive systems in general and theory as well as basic algorithms of a GA in his first book. The modern state of the art and a detailed historical overview are presented in the more practically oriented book by Goldberg (1989).

For those readers interested in a more detailed, implementation-oriented introduction to the algorithms, we recommend a look at Schwefel (1981) (ESs); Fogel (1992b) (EP); and Goldberg (1989) (GAs). Practical applications and unusual extensions of GAs, sometimes crossing the boundaries to ESs and EP, are presented by Davis (1991) and Michalewicz (1992). Specialized articles can be found in the conference proceedings (mentioned in Section 14.1) and in the journal *Evolutionary Computation* (MIT Press).

ACKNOWLEDGEMENTS

The author gratefully acknowledges financial support by the DFG (Deutsche Forschungsgemeinschaft), grant Schw 361/5-1. Special thanks to David B. Fogel for helpful discussions about evolutionary programming.

REFERENCES

Bäck, T. (1992a). Self-adaptation in genetic algorithms. In *Proceedings of the First European Conference on Artificial Life*, December 11–13, 1991, Paris, France. MIT Press, Cambridge, MA, pp. 263–71.

Bäck, T. (1992b). The interaction of mutation rate, selection, and self-adaptation within a genetic algorithm. In *Parallel Problem Solving from Nature*, Vol. 2 (ed. R. Männer and B. Manderick). Elsevier, Amsterdam, pp. 85–94.

Bäck, T. and Hoffmeister, F. (1991). Extended selection mechanisms in genetic algorithms. In *Proceedings of the Fourth International Conference on Genetic Algorithms and their Applications* (ed. R. K. Belew and L. B. Booker). Morgan Kaufmann, San Mateo, CA.

Bäck, T., Hoffmeister, F. and Schwefel, H.-P. (1992). Applications of evolutionary algorithms. Report of the Systems Analysis Research Group (LS XI) SYS-2/92, University of Dortmund, Department of Computer Science.

Belew, R. K. and Booker, L. B. (eds.) (1991). *Proceedings of the Fourth International Conference on Genetic Algorithms and their Applications*. Morgan Kaufmann, San Mateo, CA.

Bethke, A. D. (1981). Genetic algorithms as function optimizers. Ph.D. thesis, University of Michigan.

Booker, L. B., Goldberg, D. E. and Holland, J. H. (1989). Classifier systems and genetic algorithms. In *Machine Learning: Paradigms and Methods* (ed. J. G. Carbonell). MIT Press, Cambridge, CA, pp. 235–82.

Caruna, R. A., Eshelman, L. J. and Schaffer, J. D. (1989). Representation and hidden bias II: Eliminating defining length bias in genetic search via shuffle crossover. In *Eleventh International Joint Conference on Artificial Intelligence* (ed. N. S. Sridharan). Morgan Kaufmann, San Mateo, CA, pp. 750–5.

Caruna, R. A. and Schaffer, J. D. (1988). Representation and hidden bias: Gray vs. binary coding for genetic algorithms. In *Proceedings of the Fifth International Conference on Machine Learning* (ed. J. Laird). San Mateo, Morgan Kaufmann, San Mateo, CA, pp. 153–61.

Cramer, M. L. (1985). A representation for the adaptive generation of simple sequential programs. In *Proceedings of the First International Conference on Genetic Algorithms and their Applications* (ed. J. J. Grefenstette). Lawrence Erlbaum, Hillsdale, NJ, pp. 183–7.

Davis, L. (ed.) (1991). *Handbook of Genetic Algorithms*. Van Nostrand Reinhold, New York.

De Jong, K. (1975). An analysis of the behaviour of a class of genetic adaptive systems. Ph.D. thesis, University of Michigan.

De Jong, K. (1992). Are genetic algorithms function optimizers? In *Parallel Problem Solving from Nature*, Vol. 2 (ed. R. Männer and B. Manderick). Elsevier, Amsterdam, pp. 3–13.

Eshelman, L. L., Caruna, R. A. and Schaffer, J. D. (1989) Biases in the crossover landscape. In *Proceedings of the Third International Conference on Genetic Algorithms and their Applications* (ed. J. D. Schaffer). Morgan Kaufmann, San Mateo, CA, pp. 10–19.

Fogel, L. J. (1962). Toward inductive inference automata. In *Proceedings of the International Federation for Information Processing Congress*, Munich, pp. 395–9.

Fogel, D. B. (1991). *System Identification through Simulated Evolution: A Machine Learning Approach to Modeling*. Ginn Press, Needham Heights.

Fogel, D. B. (1992*a*). An analysis of evolutionary programming. In *Proceedings of the First Annual Conference on Evolutionary Programming* (ed. D. B. Fogel and W. Atmar). Evolutionary Programming Society.

Fogel, D. B. (1992*b*). Evolving Artificial Intelligence. Ph.D. thesis, University of California, San Diego.

Fogel, D. B. and Atmar, J. W. (1990). Comparing genetic operators with Gaussian mutations in simulated evolutionary processes using linear systems. *Biological Cybernetics* **63**, 111–14.

Fogel, D. B. and Atmar, W. (eds.) (1992). *Proceedings of the First Annual Conference on Evolutionary Programming*, La Jolla, CA, February 21–2. Evolutionary Programming Society, San Diego, CA.

Fogel, L. J., Owens, A. J. and Walsh, M. J. (1966). *Artificial Intelligence through Simulated Evolution*. Wiley, New York.

Friedberg, R. M. (1958) A learning machine: Part I. *IBM Journal* **2**(1), 2–13.

Fujiko, C. and Dickinson, J. (1987). Using the genetic algorithm to generate lisp source code to solve the prisoner's dilemma. In *Proceedings of the Second International Conference on Genetic Algorithms and their Applications* (ed. J. J. Grefenstette). Lawrence Erlbaum, Hillsdale, NJ, pp. 236–40.

Goldberg, D. E. (1989). Genetic algorithms in search, optimization and machine learning. Addison-Wesley, Reading, MA.

Gottschalk, W. (1989). *Allgemeine Genetik*, 3rd edn. Georg Thieme Verlag, Stuttgart.

Grefenstette, J. J. (ed.) (1985). *Proceedings of the First International Conference on Genetic Algorithms and their Applications*. Lawrence Erlbaum, Hillsdale, NJ.

Grefenstette, J. J. (ed.) (1987*a*). *Proceedings of the Second International Conference on Genetic Algorithms and their Applications*. Lawrence Erlbaum, Hillsdale, NJ.

Grefenstette, J. J. (1987*b*). *A User's Guide to GENESIS*. Navy Center for Applied Research in Artificial Intelligence, Washington, DC.

Holland, J. H. (1962). Outline for a logical theory of adaptive systems. *Journal of the Association for Computing Machinery* **3**, 297–314.

Holland, J. H. (1975) *Adaptation in Natural and Artificial Systems*. The University of Michigan Press, Ann Arbor.

Koza, J. R. (1992). The genetic programming paradigm: Genetically breeding populations of computer programs to solve problems. In *Dynamic, Genetic, and Chaotic Programming* (ed. B. Souček and the IRIS Group). Wiley, New York, Chapter 10, pp. 203–321.

Männer, R. and Manderick, B. (eds.) (1992). *Parallel Problem Solving from Nature*, Vol. 2. Elsevier, Amsterdam.

Michalewicz, Z. (1992). *Genetic Algorithms + Data Structures = Evolution Programs*. Springer, Berlin.

Mühlenbein, H. (1992). How genetic algorithms really work: I. mutation and hillclimbing. In *Parallel Problem Solving from Nature*, Vol. 2 (ed. R. Männer and B. Manderick). Elsevier, Amsterdam, pp. 15–25.

Rechenberg, I. (1965). Cybernetic solution path of an experimental problem. Royal Aircraft Establishment libr. transl. 1122. Farnborough, Hants, UK.

Rechenberg, I. (1973). *Evolutionsstrategie: Optimierung technischer Systeme nach Prinzipien der biologischen Evolution*. Frommann-Holzboog Verlag, Stuttgart.

Rudolph, G. (1992). On correlated mutations in evolution strategies. In *Parallel Problem Solving from Nature*, Vol. 2 (ed. R. Männer and B. Manderick). Elsevier, Amsterdam, pp. 105–14.

Schaffer, J. D. (ed.) (1989). *Proceedings of the Third*

International Conference on Genetic Algorithms and their Applications. Morgan Kaufmann, San Mateo, CA.

Schaffer, J. D., Caruna, R. A., Eshelman, L. J. and Das, R. (1989). A study of control parameters affecting online performance of genetic algorithms for function optimization. In *Proceedings of the Third International Conference on Genetic Algorithms and their Applications* (ed. J. D. Schaffer). Morgan Kaufmann, Los Altos, CA, pp. 51–60.

Schwefel, H.-P. (1965). Kybernetische Evolution als Strategie der experimentellen Forschung in der Strömungstechnik. Diploma thesis, Technical University of Berlin.

Schwefel, H.-P. (1977). *Numerische Optimierung von Computer-Modellen mittels der Evolutionsstrategie. Volume 26 of Interdisciplinary Systems Research*. Birkhäuser, Basel.

Schwefel, H.-P. (1981). *Numerical Optimization of Computer Models*. Wiley, Chichester, UK.

Schwefel, H.-P. (1987). Collective intelligence in evolving systems. In *Ecodynamics, Contributions to Theoretical Ecology* (ed. W. Wolff, C.-J. Soeder and F. R. Drepper). Springer, Berlin.

Schwefel, H.-P. (1988). Evolutionary learning optimum-seeking on parallel computer architectures. In *Proceedings of the International Symposium on Systems Analysis and Simulation 1988, I: Theory and Foundations* (ed. A. Sydow, S. G. Tzafestas and R. Vichnevetsky). Akademie der Wissenschaften der DDR. Akademie-Verlag, Berlin, pp. 217–25.

Schwefel, H.-P. and Männer, R. (eds.) (1991). *Parallel Problem Solving from Nature – Proc. 1st Workshop PPSN I*. Lecture Notes in Computer Science, Vol. 496. Springer, Berlin.

Serra, R. and Zanarini, G. (1990). *Complex Systems and Cognitive Processes*. Springer-Verlag, Berlin.

Wright, A. H. (1991). Genetic algorithms for real parameter optimization. In *Foundations of Genetic Algorithms* (ed. G. J. E. Rawlins). Morgan Kaufmann, San Mateo, CA, pp. 205–18.

ARTIFICIAL EVOLUTION AND THE PARADOX OF SEX

15

Robert J. Collins

This chapter introduces **artificial evolution**, and applies this bottom-up simulation technology to a long-standing problem in evolutionary biology: the paradox of sexual reproduction. Artificial evolution is a subset of genetic algorithms, and is distinguished from genetic algorithms primarily by a strong emphasis on biologically accurate simulated evolution. Artificial evolution provides a unique opportunity to study in detail the evolution of realistically sized populations through thousands of generations.

We usually apply artificial evolution to large populations of artificial organisms. The simulations are low-level and quite detailed, separately representing each gene and each chromosome in each of the tens of thousands of individuals that make up our population. The transmission of the genetic material from one generation to the next is modeled after natural genetic systems, and includes sexual reproduction with recombination and point mutations. A typical simulation run lasts for several thousand generations, allowing us to study macroevolutionary phenomena. Artificial evolution is computationally expensive, and we use a Connection Machine-2 supercomputer for all of our simulations.

After an introduction to artificial evolution, we demonstrate its use by exploring the paradox of sexual reproduction via a simulation called **Parasite**. The prevalence of sex is considered a paradox because sexual reproduction in many species is nearly twice as expensive as asexual reproduction would be. In addition, evolution theory states that in the absence of selection on other traits, there should be selection for asexual reproduction. The main genetic consequence of sexual reproduction is that each offspring receives a combination of the genetic information of the two parents. This effect is referred to as **mixis**. The search for a strong selective force favoring sexual reproduction has proceeded for more than 100 years, but has resulted in little concrete evidence. In this chapter, we test the **parasite hypothesis**, which is one of the many hypotheses that have been proposed to explain sex.

The parasite hypothesis suggests that mixis results in faster evolution, and that fast evolution provides a selective advantage to individuals in a host species that are in competition with parasites. In the usual case, parasite species reproduce with a shorter generation time than their hosts. All other things being equal, a shorter generation time allows parasites to evolve more quickly than their hosts. It therefore follows that host–parasite coevolution should be a selective force that favors greater mixis in the host species, and thus sexual reproduction. In Parasite, our simulation model, we place the amount of mixis in the sexually reproducing host species under genetic control via a recombination rate modifier gene. By observing the results of selection on the modifier gene, we have confirmed that host–parasite coevolution results in selection for sexual reproduction (non-zero mixis) in the host's species.

15.1 THE FIELD OF ARTIFICIAL LIFE

The style of artificial evolution introduced in this chapter is often classified as part of the field of artificial life, which is broadly defined as the study of systems that exhibit life-like behavior. Artificial life researchers have varying goals, including increasing our understanding of the nature of life, creating new life forms (or at least near-life forms), and exploiting life-like processes in engineering domains. For hundreds of years, biologists have pursued an understanding of natural life as we find it here on Earth, but the field of artificial life is directed toward the study of life as it could be (Langton 1989). Biologists have focused their attention on DNA-based living systems. Those who study 'life as it could be' seek to understand the properties and processes that will be characteristic of any living systems that might exist.

The scientific methodologies underlying biological and artificial life research are quite different. Biologists have traditionally used a top-down approach, analyzing complex living systems by breaking them into subsystems, and further analyzing those subsystems, deducing the form and function of each component. In sharp contrast, the basic artificial life approach is bottom-up, attempting to synthesize complex, life-like systems. The analytic and synthetic approaches are complementary, and together form a set of very powerful scientific tools.

The dichotomy between the top-down analysis in biology and the bottom-up synthesis in artificial life studies is very important. Biological analysis involves the deduction of the mechanisms underlying known living systems in great detail, leading to inferences about general principles underlying the types of life with which we are familiar. It is difficult to separate the features of natural living systems that are general properties of life from those features that are simply quirks of the way life evolved on Earth. In contrast, the synthetic approach allows us to explore systems composed of the simplest underlying mechanisms that result in the emergence of life-like behavior.

Biology and artificial life are complementary fields. Most artificial life studies are based on knowledge gained from the study of natural life, and the synthetic approach of artificial life can be applied to many questions about natural life (see Langton 1989; Langton *et al.* 1991; Meyer and Wilson 1991 for a number of examples). To date, most of the information flow has been in the direction from biology to artificial life. Hopefully the synthetic approach to the study of life will prove useful to biologists as the field of artificial life matures.

15.2 COMPARISON AMONG EVOLUTION MECHANISMS

In this section, we present an overview of three distinct evolution mechanisms: **natural evolution**, **genetic algorithms** and **artificial evolution**. In general, genetic algorithms are an optimization technique inspired by natural evolution, but in most genetic algorithms there has been little or no effort to model natural evolution in detail. We define artificial evolution as a kind of genetic algorithm designed with the explicit intent of biological realism.

15.2.1 NATURAL EVOLUTION

Natural evolution is the progressive change in genetic composition of a population over many generations. This change need not be adaptive. In fact, many population geneticists believe that most evolution may be adaptively neutral (Kimura 1968; Hartl and Clark 1989). The two principal components of the evolutionary process are natural selection and random genetic drift (Wright 1931; Mayr 1983). Natural selection is the effect by which more 'fit' individuals leave more offspring (on average) in succeeding generations than do the less 'fit' individuals in the population (Darwin 1859). (We discuss fitness in detail below.) Random genetic drift results from random events in the lives of the individuals making up the population. Such events include random mutations in the genetic material, the accidental death of an apparently highly fit individual before it gets an opportunity to reproduce, and so on. Apparently well-adapted individuals can get unlucky and leave few offspring in the next generation, and apparently poorly adapted individuals can get lucky and leave many offspring. The stochastic effects of genetic drift are most noticeable in small populations, where they can swamp the effects of even strong selection. In large populations the stochastic effects tend to 'even out', so drift is less noticeable.

The idea of an adaptive landscape or fitness surface is a useful way of visualizing how selection acts on an evolving population (Wright 1932; Kauffman and Levin 1987). The basic idea is to plot fitness as a function over the space of possible genetic combinations for one species in a particular environment. The resulting graph forms the fitness 'surface'. The space of possible organisms may have thousands of dimensions, and the fitness surfaces are thought typically to have a huge number of peaks and valleys, varying in height, because some gene combinations are more adaptive than others.

A population is represented as a cloud of points on the fitness surface, one for each organism in the population; the more variation in the population, the more scattered it is on the adaptive landscape. The offspring of the more fit individuals in the population will tend to be both more numerous and more fit than the offspring of the less fit individuals in the population. In this way, over many generations natural selection tends to move populations uphill (higher fitness) in the adaptive landscape. Drift due to chance events, the other component of the evolutionary process, moves the population randomly upon the fitness surface. If selection dominates the evolution of a population, it will tend to move up gradients in the adaptive landscape; if chance dominates, evolution tends to proceed in random directions.

In biology, fitness is defined as the relative ability to survive and reproduce in the context of a particular environment and gene pool. In nature, the fitness of an organism is often very difficult to determine, and the fitness contribution of a particular trait is even more difficult to calculate.

Fitness should be regarded as an attribute of an entire genotype (or resulting phenotype), rather than of any particular gene, trait or phenotypic phase of an organism's life cycle. Although it is not unusual to refer to the fitness of a particular gene without regard to the rest of the genotype or the environment, this is technically an ill-defined concept. As Dobzhansky put it (1956, p. 340):

> It cannot be stressed too often that natural selection does not operate with separate 'traits'. Selection favors genotypes. The reproductive success of a genotype is determined by the totality of the traits and qualities which it produces in a given environment.

We might attempt to make sense of the notion of fitness of a single gene by determining the mean fitness of the genotypes in the population that contain a particular gene, and this might help us understand how the frequency of the gene might change. For example, if the genotypes containing the gene are of above average fitness, the gene might become more abundant due to natural selection. Unfortunately, this 'mean fitness of a gene' calculation may actually tell us very little about the actual effect of the gene on fitness in general. In essence, this is simply a calculation of the correlation between fitness and the presence of the gene, not the causal contribution of that gene toward the organism's overall fitness. The gene may have no influence on the fitness of the genotypes in which it is found, and simply be there by chance; it may have effects that are strongly dependent on the presence or absence of certain other genes in the genotype; or it may be in linkage disequilibrium (Hartl and Clark 1989) with other genes that have a much stronger influence on fitness. The term 'linkage' usually refers to the association of genes on the same chromosome, but 'linkage disequilibrium' refers to *any* non-random association of genes, even if they reside on different chromosomes. Such associations can result from the interacting effects of the genes (or interacting effects of genes that are located nearby on the chromosome). These interactions can take a number of forms such as mating preferences, or lethal gene combinations.

The fitness of a particular genotype is determined by events at all stages of the organism's life cycle. The most visible part of fitness is viability, which is the ability of an organism to develop and survive from a zygote (fertilized egg) to an adult. The culling of less viable organisms by natural selection is called viability selection. Another component of fitness is the ability to form mating pairs, known as sexual selection. Sexual selection can take the form of male–male competitions (e.g., head butting in bighorn sheep), male choice, female choice or mutual choice via male–female interactions. Another component of fitness is called meiotic drive, which is due to non-random differences in the production of the various genetic combinations of gametes (unfertilized sperm and eggs) during meiosis (gamete formation). Another form of selection is gametic selection, which results from differential fertilization success among the gametes that have been produced. The fecundity of an individual, the number of zygotes (fertilized gametes) produced, is still another component of the overall fitness of a genotype. The combined effects of all of these components determine fitness. For instance, an individual may be very viable (well adapted to its environment) and healthy as an adult, but fail to find a compatible mate due to sexual selection. Such an individual has zero fitness, despite its apparent health and vitality.

15.2.2 GENETIC ALGORITHMS

Genetic algorithms (Holland 1975; Goldberg 1989) are a class of optimization algorithms loosely based on natural evolution. They are typically used to search for good or optimal solutions to complex optimization problems (Goldberg 1989). A solution is represented as a string of arguments that (more or less) maximizes the particular function to be optimized. This function is called the objective or fitness function. The string of arguments is analogous to the genome of a natural organism. The location of each parameter in the string is analogous to a locus on a chromosome, and the value of the parameter is analogous to the gene at that locus. The function is analogous to the environment faced by a natural organism, which determines its relative fitness.

A genetic algorithm evolves a population by assigning a numeric score to each string (genome) based on how well the genetically encoded arguments maximize the objective function. This is roughly analogous to viability fitness (the ability to survive to adulthood) in a natural organism. New strings are then created through a process of sexual reproduction between relatively high scoring strings that are currently in the population. The likelihood that a particular string will be chosen for mating is a function of its own score and those of the rest of the population. Sexual reproduction provides the offspring with a mixture of the genetic material from its two parents. In addition, small random changes in the genetically encoded arguments (analogous to mutations) are made to the offspring's string. The new string usually displaces a low-scoring individual from the population, or the whole population is simultaneously replaced by newly created strings.

The way biologists think of 'fitness' is very

different from the way the term is used in the context of genetic algorithms for optimization. The notion of fitness used in genetic algorithms is a numeric score used to determine the number of offspring that should be produced. To biologists, however, fitness is an emergent property that can only be measured *after* reproduction has already occurred, and is based on the number of viable, fertile offspring that actually were produced.

15.2.3 ARTIFICIAL EVOLUTION

We define artificial evolution to be a particular class of genetic algorithms. While the design of traditional genetic algorithms is driven by the goal of optimization, the intent of artificial evolution is biological realism. Artificial evolution genetic algorithms require a clear separation between genotype and the information encoded in the genotype. The genotype is represented as a linear string, and the genetic operators of recombination and mutation operate randomly at the lowest level of organization of the string, without reference to any syntactic or semantic structure that may be encoded there. We view the genotype as encoding a program; the fitness of the genotype is determined by decoding and executing the program, perhaps in an environment that is shared by the other members of the population. Often, the selection and mating process of the artificial evolution genetic algorithm will include spatial structure (the formation of mating pairs is influenced by geographic distance).

Methodology

Simulations based on the observation that the execution of a computer program is very similar to the life of an organism have emerged in recent years (Taylor *et al.* 1989*a*; Taylor *et al.* 1989*b*; Fry *et al.* 1989; Coulson *et al.* 1987; Werner and Dyer 1991; Jefferson *et al.* 1991). In these 'life-as-process' simulations, each organism is represented by a program, as are the various environmental factors. Only the local interactions between the individual organisms and environmental factors are modeled directly, but based on them, the complex global behavior of the population emerges.

These 'life-as-process' simulations separately represent each individual organism and environmental effect, and the biologically significant events in an organism's life are all separately simulated in detail. Each organism in the population is represented as a program, and its life is represented by a process (i.e. the execution of a program), a detailed sequence of events, including its birth, its interactions with a dynamic environment (potentially including many other organisms and environmental factors), its mating and reproduction (if any) and its death.

An artificial evolution simulation is a genetic algorithm that applies the low-level, detailed 'life-as-process' model of each organism to a large, evolving population. The simulation consists of three main parts:

1. the genetic algorithm, which drives the evolution;
2. the organisms, each represented as a computer program of some kind; and
3. the environment in which the organisms live, represented by zero or more additional programs.

All of the thousands of programs that make up the simulation execute in parallel. The organism programs are heritable because they are encoded in the organism's chromosome. A genetic algorithm is typically used to drive the evolutionary process, but the behavioral repertoire of the organisms may include death, differential reproduction, mate choice and so on, in which case many of the components of the genetic algorithm are omitted. The environment of a particular organism may consist not only of one or more environmental processes, but also all of the organism programs from its own (and possibly other) species. In other cases the environment might be extraneous to the hypothesis that is being tested, and thus might not be explicitly simulated at all.

By executing an organism's program, the fitness of that individual is determined either implicitly (fitness is emergent) or explicitly (fitness is assigned by an external formula). Sometimes the processes execute in a shared environment, allowing for interactions with the other organism processes. Fitness evaluation may be either implicit or explicit. In Parasite, fitness is evaluated explicitly based on the interactions between an individual in the host species and its infecting parasite.

Advantages and limitations

The major tools that are available for the study of natural evolution are the fossil record, molecular analysis, observational and experimental studies and mathematical analysis. These approaches are inherently limited in some ways that computer simulations are not (Taylor *et al.* 1989*a*). The fossil record is notoriously biased, incomplete, and difficult to interpret. Also, while molecular studies can determine the underlying genetic similarities and difference of various species, the results are often as difficult to interpret as the fossil record. Evolutionary experiments in the laboratory or field are usually limited to small populations and at most a few dozen generations because natural organisms (other than microbes) grow and reproduce slowly compared to the time scale of human experiments. In addition, such experiments are difficult to control and repeat, because of the complexity of the interactions between organisms and their environments. Taylor (1983) points out that many of the experiments you might want to perform have the potential to be very damaging to the ecology. And finally, mathematical analysis can completely describe only the simplest genetic systems.

In contrast, artificial evolution makes it possible to study non-trivial models of evolving systems over thousands of generations (macroevolution). Although these models are inevitably idealized in some ways, they are much more complex and realistic than those that can be attacked mathematically. Computer simulations are also easily repeated and varied, with all relevant parameters under the full control of the experimenter. Some biologists saw the potential for computer simulations even 30 years ago (Crosby 1963, p. 415):

> For some biologists, experiments with living organisms are hardly practicable. For example, many problems of evolution would obviously need too much time. As an alternative, experiments with realistic models of evolutionary systems would go far towards overcoming this difficulty, if a sufficient speed of operation could be achieved. This is where the electronic computer can become a valuable tool for population genetics.

Of course, a weakness of computer simulation is the inability to attain the full complexity of natural life and environments, although the degree of complexity that is feasible scales with improvements in available computer technology.

Most computer simulations in biology (including most previous attempts at simulated evolution) are based on solving differential equations from mathematical models (Swartzman and Kaluzny 1987; Taylor 1983), where the equations specify the dynamics of the system. Due to the difficulty of mathematical analysis, models of evolving systems are usually simple and unrealistic. Complex models that incorporate a large number of both intrinsic factors (e.g., the life history of the organisms) and extrinsic factors (e.g., weather, competitors) are more accurate and useful, but unfortunately, such complex evolutionary models are difficult or impossible to solve analytically. Differential equation-based models that are tractable are usually linear and of low order, and thus do not properly capture the discrete nature of the dynamics of real populations.

The distinctions between equation-based and the artificial evolutionary simulation paradigm are not new. They were described well by Crosby (1963, p. 415):

> The computer can simulate, in mathematical terms, complex genetical and evolutionary systems [equation-based simulations], or mathematics can largely be eliminated by forming, within a computer, electronic 'organisms' possessing electronic 'genes' [artificial evolution simulations]. These organisms can reproduce themselves in any way we choose, obeying the ordinary laws of genetics or any other desired pattern of heredity; while the natural variability of real biological systems can be imitated with fair realism by the use of computer-produced numbers (pseudo-random numbers) in sequences which effectively imitate true randomness.

Parasite, the macroevolution simulation that we describe in this chapter contains tens of thousands of organisms and environmental processes, and lasts for thousands of generations. Before the introduction of massively parallel computers such as the Connection Machine-2 (Hillis 1985; Hillis and Steele 1986; Hillis and Barnes 1987), such a detailed simulation was computationally infeasible. Where previous studies that resemble the artificial evolution style of simulation presented in this chapter have been applied to evolving systems, the simulations have been simple and operate on small populations and for a relatively small number of generations (Crosby 1963; Schull and Levin 1964; Ohta 1987; Taylor *et al.* 1989*a*; Ohta 1989; Keightley and Hill 1989; Ohta and Tachida 1990). While these simulations produced interesting results, the computational costs have been generally the limiting factor on the complexity of the simulation models that have been attempted. The costs apparently were too great to attempt artificial macroevolution.

15.3 ARTIFICIAL EVOLUTION APPLICATIONS

15.3.1 BIOLOGICAL APPLICATIONS

While we cannot expect artificial evolution to reconstruct an actual scenario in the history of natural life, we can explore particular hypotheses about evolution, eliminating some and giving credence to others. Simulations provide an artificial world in which to perform evolutionary experiments that can be fully controlled and repeated, and can span thousands of generations.

There are a large number of important macroevolutionary problems that traditional biological techniques cannot readily address. Artificial evolution experiments might be used to shed light on a number of open evolutionary problems, many of which are touched on by the Parasite simulation study described in this chapter. The kinds of open evolutionary problems to which artificial evolution can readily be applied include the following.

1. Modes of speciation. Dozens of hypotheses have been proposed over the years, but which ones are most likely to occur in particular sexual systems and ecological situations? In simulation, we can place a population in an environment that is relevant to one or more speciation hypotheses and observe the evolutionary results.
2. Evolution of mutation and recombination rates: Evolution theory states that mutation and recombination rates in a population that lives in a stable environment should tend to decrease to zero under the influence of natural selection. Despite this result, natural populations maintain non-zero mutation and recombination rates. Under what conditions is there selection for non-zero rates?
3. Evolution of information processing behavior. Under what conditions do sensory-motor integration, communication, and so

on evolve? What are the dynamics of the evolution of language and culture?

4. Evolution of sexual reproduction. Sexual reproduction is generally much more expensive than asexual reproduction. What selective advantages does sexual reproduction provide? Under what conditions is it maintained in competition with asexual reproduction? Why are there usually only two genders?
5. Punctuated equilibria. The fossil record seems to indicate that most evolution occurs at speciation events, and not within species. Is this really the case? At the genotypic level? At the level of gross morphology (which is typically all that is preserved as fossils)? What about the rate of evolution of behavior and culture?
6. Predator–prey relationships. What are the dynamics of competitive 'arms races'? Are they generally stable? Under what conditions does one species defeat the other? Does the defeated species generally become extinct, or move to another niche?
7. Host–parasite interaction. Similar to predator–prey relationships. How important are parasites as a driving force in evolution?
8. Stability of ecosystems. Are ecosystems generally stable? How much damage can an ecosystem stand before mass extinctions and/or dramatic reorganizations occur?
9. Sexual selection and the evolution of mal adaptive characteristics. In many species (such as the peacock), the adult male exhibits grossly exaggerated secondary sexual characteristics to the point of reducing the male's ability to survive to adulthood. Presumably, the females of the species prefer these features and provide those males with more prominent displays with more mating opportunities. Is the intensity of the male's display correlated with the viability of his offspring (both male and female), or do the females just have a random preference that drives the male's evolution?

15.3.2 EVOLVING ARTIFICIAL ORGANISMS

We have also used artificial evolution simulation to evolve artificial organisms that live and reproduce in relatively complex environments and that possess many sensors (both internal and external) and effectors. The organisms may possess some amount of internal memory, allowing their behavior to be history sensitive. In the course of its life, each organism is born, makes thousands of decisions (eat, move, mate, etc.), and eventually dies. As in natural life, the reproductive success of an artificial organism is affected by its behavior throughout its lifetime. Although at UCLA we have focused our attention towards the evolution of the behavior of the organisms (Taylor *et al.* 1989*a*; Collins and Jefferson 1991*c*; Collins and Jefferson 1992; Collins and Jefferson 1991*b*; Collins and Jefferson 1991*a*; Werner and Dyer 1991; Jefferson *et al.* 1991; Collins 1992), artificial evolution also allows for heritable morphological features, which might be sensed by nearby organisms or affect the behavior of the organism.

15.3.3 ENGINEERING APPLICATIONS

Engineers have used genetic algorithms for many years to search for optimal solutions to a variety of functions. Artificial evolution suggests the use of large, spatially structured (local mating) rather than small, unstructured (panmictic) populations that are typical of conventional genetic algorithms. Local mating leads to a more robust genetic algorithm, even on function optimization problems (Collins and Jefferson 1991*d*; Collins 1992). In addition, multiple solutions can be simultaneously supported by the spatially structured population and optimal solutions are discovered faster, both in terms of number of generations to an optimal solution and computation time in a particular massively parallel implementation on the Connection Machine-2.

15.4 THE PROBLEM OF SEXUAL REPRODUCTION

One of the most significant outstanding problems in the study of evolution is the evolution and maintenance of sex (Bell 1982; Michod and Levin 1987). In particular, why did sexual reproduction evolve, and why has it remained so prevalent? This is such a perplexing problem, because sexual reproduction is usually much more energetically and genetically expensive than asexual reproduction. Sex must provide significant adaptive advantages to overcome this cost. We can, to some extent, measure the costs. However, measurable benefits have remained somewhat elusive.

15.4.1 THE DEFINITION OF SEX

In this chapter, the term 'sex' refers neither to gender nor the act of mating. Instead, 'sex' refers only to the production of offspring that possess a combination of the genetic material of two or more parent organisms.

The main effect of sexual reproduction is mixis, the mixing of the genes of the two parents. In the most common sexual systems (Bell 1982), mixis occurs at two levels: the independent assortment and segregation of the different chromosomes, and the process of recombination via crossover within individual chromosomes. Independent assortment provides the offspring with one haploid copy of each chromosome from each parent. On average, one fourth of the genetic material is derived from each of the four grandparents, one eighth from each great-grandparent, and so on. But a chromosome is not necessarily derived as a whole from a distant ancestor. Crossover events swap portions of a chromosome, so within the haploid copy that is passed to an offspring there may be portions of each of the grandparent copies of the chromosome. Assortment provides a mixing at the level of whole DNA molecules, while recombination mixes within DNA molecules.

15.4.2 THE COSTS OF SEX

One of the largest costs of sex is the so-called two-fold genetic cost (Shields 1987). The fitness of a genotype is a function of the mean number of viable offspring produced by those individuals carrying that genotype (Section 15.2.1). To be somewhat more precise, the fitness of a genotype is defined in terms of the representation of portions of that genotype in offspring in the next generation. Sexually reproducing individuals contribute only half of their genetic information to each of their offspring. Consider what would happen if a parthenogenic (asexually reproducing) mutant female that produced only parthenogenic female offspring arose in an otherwise sexually reproducing population. Rather than producing unfertilized eggs, she would self-fertilize her eggs and produce offspring without the aid of a male. Assuming that she could produce as many offspring as the wild type (unmutated, sexually reproducing) females, she would produce twice as many copies of her genome as the wild type. By mutating to asexuality, the parthenogenic female would double the frequency of her genotype each generation, and her lineage would take over the population in only a few generations. This genetic cost of sex can be mitigated somewhat by inbreeding (mating with genetically related individuals) (Williams 1980; Shields 1987); the closer the relation between the mates, the closer the relation between parent and offspring.

Another source of inefficiency in sexual reproduction is the cost of males (Ghiselin 1987; Seger and Hamilton 1987), which is sometimes referred to as the two-fold ecological cost of sex. In most sexual species that have two genders, the males contribute little or nothing to their offspring other than gametes. The half of the offspring in each generation that are males are essentially a drain on the ecosystem: half of the ecological resources of the species' niche are spent on males that are non-productive (in the sense that they do not invest their resources in their offspring). The advantage of the parthenogenic mutant female here is that she does not waste any of her offspring on males, producing twice as many females as her sexual counterparts. If the males provide some resources to their offspring, then the ecological cost of males will be somewhat lower (Seger and Hamilton 1987; Shields 1987). An unequal sex ratio (in favor of more females) will also mitigate part of this cost.

Other costs of sexual reproduction include the energy needed to find a mate, more complex reproductive systems, mechanisms to avoid mating with the wrong species, and so on.

15.4.3 THE BENEFITS OF SEX

What is it about mixis that is so beneficial that it can overcome the two-fold genetic and the two-fold ecological cost of sex? Many hypotheses have been proposed to explain why, and under what conditions, sexual reproduction might be advantageous (Bell 1982; Muller 1964; Maynard Smith 1987; Felsenstein and Yokoyama 1976; Haigh 1978; Seger and Hamilton 1987; Hamilton 1990; Rennie 1992). Most of these hypotheses are probably right to some degree, with their relative importance varying from species to species and through time for any given taxon.

The hypothesis that we investigate in this chapter is that mixis is beneficial in the context of rapidly changing environments, and in particular in the host species in host–parasite coevolution systems. Mixis can be beneficial in changing environments when the phenotypic traits that are subject to selection are coded for by many interacting genes (i.e., are polygenic), and when at equilibrium, the genes remain unfixed (variation is maintained). Many traits are polygenic, with non-extreme equilibria. The effect of mixis on such a trait is to produce individuals of many different gene combinations, and thus a range of phenotypes. If the environment is constantly changing so that the 'optimal' phenotype keeps shifting, the population will be able to evolve relatively quickly to maintain a mix of combinations centered on the 'optimal' value. In contrast, an asexual population can change the trait only in small increments due to mutations, and may not be able to track environmental changes quickly enough to avoid extinction.

15.5 THE PARASITE HYPOTHESIS

The hypothesis that mixis allows rapid adaptation is plausible, but for this effect to have a significant influence on selection in favor of sexual reproduction, environments characterized by fairly rapid and sustained change must be common. The physical environment may or may not change quickly relative to the generation time of a species, but the biotic environment often undergoes rapid change. The parasite hypothesis suggests that parasites present their host species with a rapidly changing and challenging environment over long periods of time (Seger and Hamilton 1987; Hamilton 1990; Rennie 1992). Furthermore, resistance to parasites in the host species is usually highly polygenic (a precondition to beneficial effects of mixis; see Section 15.4.3).

For many years, biologists have realized the prevalence and importance of parasitic species (May 1983), but only recently have they begun to appreciate the dominant role they may play in the adaptation of their hosts. From an evolutionary point of view, the interactions of parasite and host species are ecologically similar to those in predator–prey relationships. However, the term 'parasite' is used when the species has a shorter generation time than its host, whereas the term 'predator' is used when generation time is as long or longer than its prey's. Its comparatively long generation time puts the host species at an evolutionary disadvantage, because the parasites may be able to evolve new methods of attack much faster than the host can evolve new defenses. With a constantly changing set of defenses and attacks, the hosts and parasites each provide the other with a rapidly changing environment, although the change seen by the host may be more dramatic, due to the shorter generation time of the parasites. It is likely that a good host strategy is to maintain diversity in the population and recombine each generation, challenging the parasites with a different combination of defenses each generation. This might offset the speed advantage of the parasites (Levin 1975; Glesener and Tilman 1978; Jaenike 1978; Bremermann 1980; Bremermann and Pickering 1983; Hamilton 1980; Hamilton 1982; Hamilton 1986; Hamilton *et al.* 1981; Anderson and May 1982; Bell 1982; Price and Waser 1982; Tooby 1982; Rice 1983).

15.6 TESTING THE PARASITE HYPOTHESIS

One of our goals in this chapter is to demonstrate that artificial evolution can be used to study significant biological problems such as the evolution and maintenance of sex. We are not trying to solve the problem of sex once and for all, as there may be no single explanation. Instead, we have designed and implemented Parasite, a simple model of host–parasite coevolution, in order to test the plausibility of the parasite hypothesis.

One approach to testing this hypothesis would be to model the host species as having the ability to reproduce either sexually or asexually, under the control of a heritable gene. We could then observe the evolution of this 'sex' gene under the influence of fitness based on interactions with a coevolving parasite species. While this model is not too complex to simulate, we have chosen a somewhat simpler model that will get at many of the same issues.

Instead of observing the evolution of a 'sex' gene, we use a recombination rate modifier gene in our artificial evolution simulation. Evolutionary control of crossover rates has been used in evolutionary algorithms on a number of occasions, such as Rosenberg, (1967), Bäck *et al.* (1991) and Schaffer and Morishima (1987). While a recombination modifier gene does not directly attack the problem of the evolution of sex, it measures one of the components of mixis. In our simulation, we place all of the loci involved in the host–parasite interactions on one chromosome, so crossover is the *only* component of mixis in our system. The recombination rate modifier gene specifies the rate of mixis, and thus we can directly measure its evolution on a continuum from no mixis (effectively asexual reproduction) to significant mixis (strong sexual effects).

Placing the recombination rate under genetic control is a biologically realistic thing to do; there are heritable variations in the rate of recombination within many populations (Brooks 1987). It is a well-known analytic result that in a stable environment (i.e., at equilibrium), there is selection for decreased recombination rates (Fisher 1930; Feldman *et al.* 1986; Felsenstein 1987), and thus reduced mixis. The parasite hypothesis suggests a counterbalancing selective force that favors higher recombination rates and higher rates of mixis.

Despite the variety of analytic models for the evolution of recombination rates, none of them include realistic population dynamics (Brooks 1987), and in most cases are limited to two loci, each with two genes. Our model simultaneously includes mutation, a heritable recombination rate (with the probability of a crossover varying from 0.0 to 0.1 between each pair of loci), selection via competition with a coevolving parasitic population, spatial structure (isolation by distance), a large but finite and drifting population, and 100 loci of host–parasite interaction, each with two genes.

The evolution in our model is driven by a genetic algorithm. The host genome consists of two chromosomes, one for the host–parasite interaction loci, and one for the recombination modifier gene. These two chromosomes segregate independently. The parasite has only one chromosome, which contains the loci involved with host–parasite interactions. All chromosomes in both species are subject to recombination and mutations (bit flips).

Some simulation studies suggest that selection for an increased recombination rate is likely to be stronger in models that incorporate more loci that are subject to selection (Martin and Cockerham 1960; Lewontin 1964; Franklin and Lewontin 1970). For this reason, we model the host–parasite interactions with 100 loci in each species. (100 loci is an enormous number compared to the one or two loci that are typical of population genetics models.) There are two possible genes for each of these 100 loci. As we noted above, all 100 loci are placed on the same chromosome, so the only possibility for mixis among them is by recombination.

The host's recombination modifier gene is a 10-bit gray-coded unsigned integer, with 1024 genes (possible values). Gray-coding is a way of encoding integers in bit-strings so that all consecutive integers differ by at most one bit position. This coding scheme allows a smooth progression from one value to another via mutation, although some mutations can still cause large changes in the value of the encoded number. This integer modifier gene is scaled linearly to the range [0.0, 0.1] to specify the probability per bit of a crossover (allowing the expected number of crossovers on the 100-bit chromosome to range from zero to ten per individual).

Both populations (the hosts and parasites) are placed in a two-dimensional, toroidal grid, with exactly one host and one parasite in each grid location. The grid structure both matches a parasite to each host, and defines a neighborhood for local mating. Both populations are the same size, each consisting of 65 536 individuals in each generation.

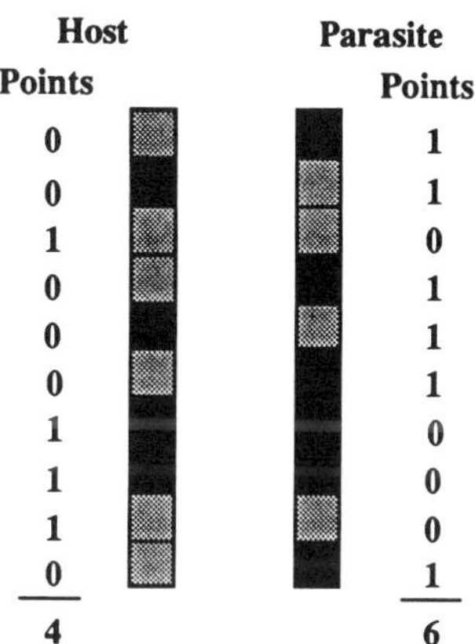

Figure 15.1 Host–parasite fitness calculation. The two chromosomes are aligned; where they match the host gets a point, and where they differ the parasite gets a point. Here, the host has four points and the parasite has six points. Equation (15.1) is then applied. Here $l = 10$, so the fitness of the host is -1 and the fitness of the parasite is 1.

The host–parasite competition is very simple, consisting of a complementary gene-for-gene system for interaction. Each species has a 100-bit haploid chromosome (the 100 loci). When a parasite infects its host, their chromosomes are aligned (Figure 15.1). Each gene in the host is a 'defense' and the parasite's gene at the corresponding loci is its corresponding 'attack'. If the defense matches the attack, the host scores a point, while if the defense does not match, the parasite scores a point. This calculation is performed for all 100 loci, and the points for each individual are added and normalized, producing a fitness score s_i for individual i:

$$s_i = \sum_{j=0}^{l-1} \text{point}(j) - \frac{l}{2} \qquad (15.1)$$

where l is the number of loci (in this case 100) and point(j) is 1 if the individual scores a point at locus j, otherwise it is 0. If the host and its infecting parasite are equally balanced (match on half the loci), both will have a fitness of 0, while if the parasite has the upper hand, it will have a fitness score greater than 0 and its host will have a score less than 0.

Both the hosts and parasites reproduce sexually; the parasites have a constant recombination rate, while the recombination rate in the hosts is heritable. Once the fitness value for each individual is determined, two parents are chosen for each location to produce an offspring organism in that square. The first parent is chosen by randomly sampling two individuals within the 5 × 5 region centered on the offspring location with an approximately normal distribution of parent–offspring distances. The higher scoring of the two individuals that are examined is selected as the first parent. The second parent is determined in the same way (with replacement, so the same individual could be chosen as both parents of a new individual). The genetic material of the two parents is copied and the recombination rate of one of the parents (chosen randomly) is used to recombine the two chromosomes to produce the haploid offspring, which is then mutated at a rate of $\mu = 0.0001$ per locus (bit). Such a very low rate of mutation is used so that it will be clear that mixis is the primary source of novel gene combinations, while still providing a constant source of genetic diversity. In the hosts, the same recombination rate is used to perform crossovers between the modifier chromosomes; then mutations are performed at the same rate of $\mu = 0.0001$ per bit. The identical method of selection and mating is used within both the host and parasite populations.

As we noted above, parasites by definition reproduce more quickly than their hosts. To simulate this effect, we allow the parasites to reproduce $p \geqslant 1$ times during each host generation. The simulation proceeds like this:

1. Initialize the host's recombination modifier genes to 0.
2. Initialize all other loci in both the host and parasite to 0 (hosts arbitrarily begin with perfect defenses).
3. Determine fitness in the host population.
4. Select and reproduce in the host population.
5. Do p times
   (a) determine fitness in the parasite population;
   (b) select, and reproduce in the parasite population.
6. Go to 3.

All of the evolution parameters (e.g., mutation rate) are the same for both species, with the exception of generation time and recombination rate. The multiple parasite generations per host generation gives the evolutionary advantage to the parasites. On the other hand, we allow the hosts to compensate by increasing their rate of mixis via the recombination modifier gene.

Figure 15.2 summarizes our simulation results. Each simulation run lasts 2000 generations, allowing the modifier gene to reach its 'equilibrium' value. We vary the speed of the parasite evolution by varying both p (the number of parasite generations per host generation), and the parasite recombination rate (ρ_p). In all cases, there is strong selection for non-zero recombination rates. The equilibrium rate generally increases with faster parasite evolution.

The dynamics of these simulations are quite interesting (Figures 15.3 and 15.4). The runs begin with the arms race biased completely in favor of the hosts, and with no mixis in the host. It turns out that both of these initial conditions are far from their equilibrium values. Because the hosts begin so far ahead in the arms race, there is not strong selection pressure for higher recombination rates in the early generations.

In the run described by Figure 15.3, the reproductive rates of the host and parasite are equal, and the parasite is locked into a very low rate of recombination ($\rho_p = 0.0001$).

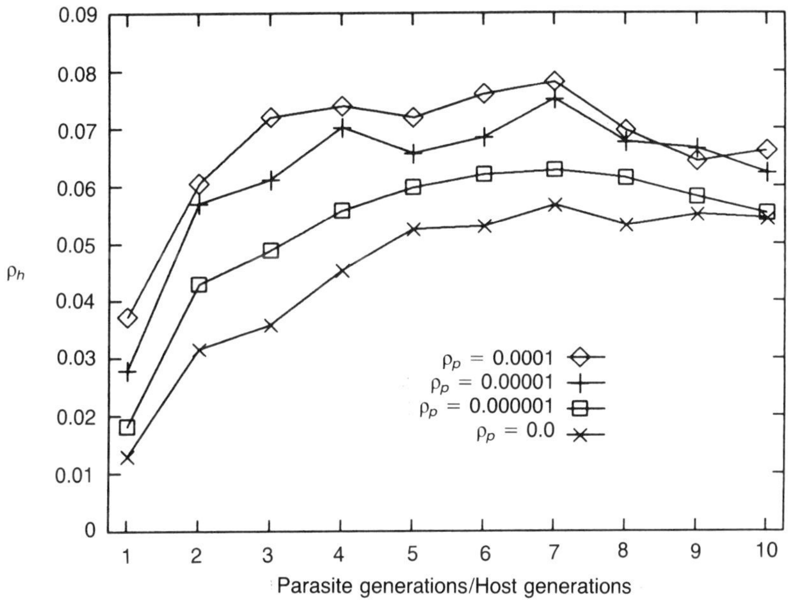

Figure 15.2 The equilibrium recombination rate in the host population (ρ_h) as a function of the number p of parasite generations per host generation. Each data point is the mean of nine runs at generation 2000.

While the parasite population begins the competition much less fit than the hosts, by around generation 400 the parasites are outcompeting their hosts, creating selection pressure for higher recombination rates. By generation 600, the mean host recombination rate has increased significantly, and in fact the hosts begin to get ahead of their parasites in the arms race. The parasite fitness curve exhibits a damped oscillation which appears to settle down by about generation 2000. The mean host recombination rate stabilizes around $\bar{\rho}_h = 0.04$, which translates into about four crossovers per individual in each generation.

Figure 15.4 presents another simulation, differing from the run in Figure 15.3 only in that it has faster parasite evolution due to $p = 5$ (versus $p = 1$) parasite generations per host generation. The overall behavior is qualitatively similar. The major differences due to faster parasite evolution are wider oscillations in the mean parasite fitness curve, a higher mean host recombination rate $\bar{\rho}_h$, and $\bar{\rho}_h$ seems to peak and then drop to a lower level over the course of many generations.

The system seems to have stabilized by generation 5000, although this is not to say that evolution has stopped. While the mean recombination rate and fitness for the population as a whole may have stabilized, rapid evolution continues at the host–parasite interaction genes throughout the simulations. This evolution represents the discovery exploitation of effective attacks/defenses. The frequency of the associated genes increases until the other species evolves a counteracting combination of genes, at which time selection begins removing it from the gene pool.

15.7 DISCUSSION

The empirical evidence from the Parasite simulations suggests that under fairly realistic evolutionary dynamics, parasites can cause selection for higher recombination rates.

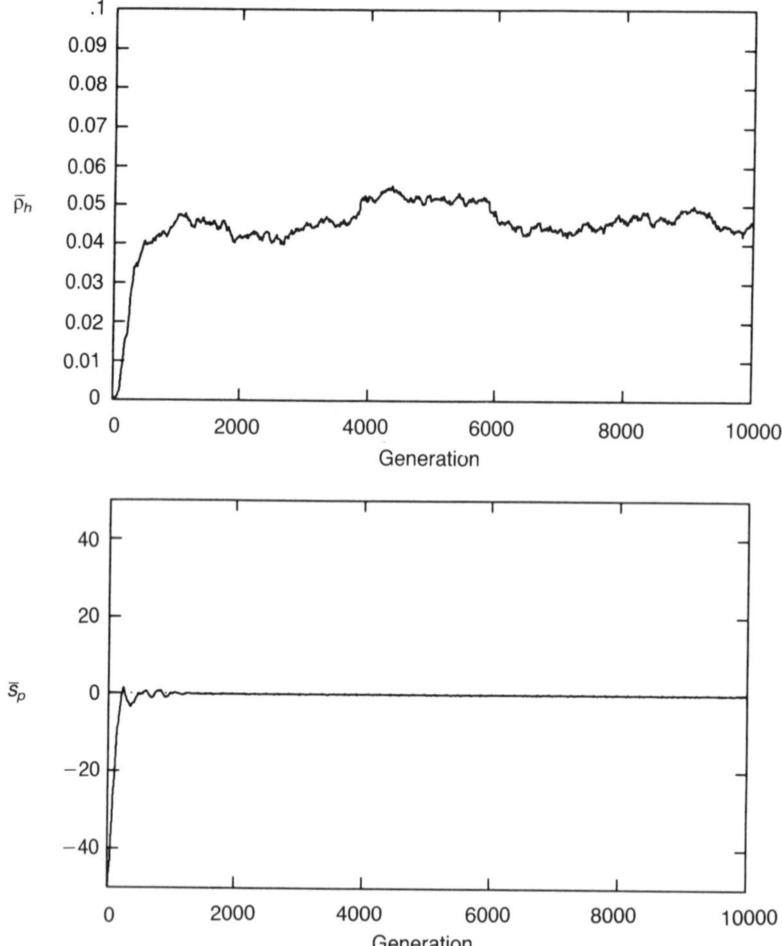

Figure 15.3 The dynamics of the host and parasite evolution with the parasite recombination rate $\rho_p = 0.0001$ and the number of parasite generations each host generation $p = 1$. The run up to host generation 10 000 is shown. The top graph tracks the evolution of the mean host recombination rate $\bar{\rho}_h$. The bottom graph shows the damped oscillation of the mean fitness of the parasite population. Where $\bar{s}_p < 0$, the hosts are more fit on average, where $\bar{s}_p > 0$, the parasites are more fit, and where $\bar{s}_p = 0$, the two populations are equally fit. The size of each population is $N = 65\,536$.

Although our model is based on very simple host–parasite interactions, it does have some connection to the real world. The one-to-one interactions between the host and parasite loci appear to be common in some groups of organisms (Flor 1956; Day 1974; Barrett 1983; Barrett 1985), but in many cases the interactions are more complex (Barrett 1985). This is really not a problem; more complex genetics are not difficult to implement in our artificial evolution paradigm, so the more complex situations can be simulated.

These are significant results, because they provide empirical evidence that host–parasite coevolution can result in strong selection for higher recombination rates (mixis) in a relatively realistic simulation. This selective advantage due to mixis bears directly on the

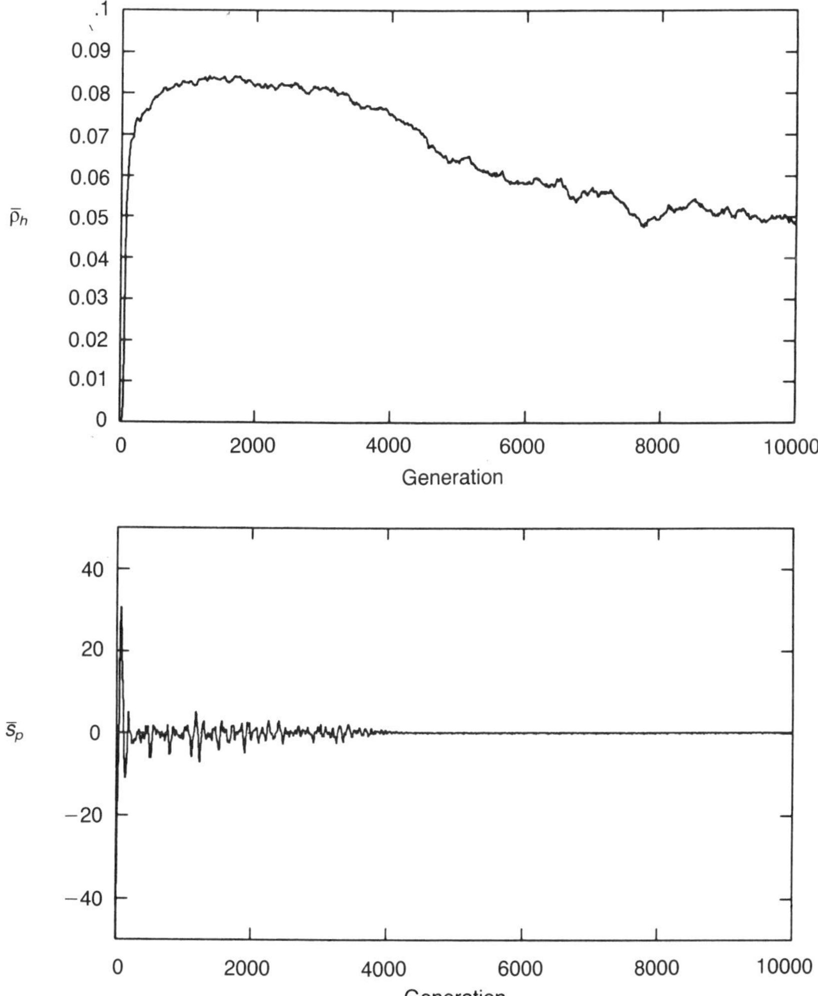

Figure 15.4 The dynamics of the host and parasite evolution with the parasite recombination rate $\rho_p = 0.0001$ and the number of parasite generations each host generation $p = 5$. The run up to host generation 10 000 is shown. The top graph tracks the evolution of the host recombination rate ρ_h. The bottom graph shows the damped oscillation of the mean fitness of the parasite population. Where $\bar{s}_p < 0$, the hosts are more fit on average, where $\bar{s}_p > 0$, the parasites are more fit, and where $s_p = 0$, the two populations are equally fit. The size of each population is $N = 65\,536$.

problem of the maintenance of sexual reproduction, and under what conditions there is selection for higher rates of mixis.

In addition, Parasite demonstrates very clearly the strong and unending selective pressure that host–parasite competition can create. A common problem in the application of genetic algorithms to optimization problems and to the evolution of artificial organisms is that the population usually moves quickly to a suboptimal peak in the adaptive landscape and stagnates. The addition of a competitive parasite species has the effect of constantly deforming the adaptive landscape, quickly turning a peak into a valley. This allows the gene pool of the host species to continue to evolve indefinitely. As long as stagnation (convergence) can be avoided, there is still hope that better solutions will be discovered. Hillis (1991) has applied this technique with success to the problem of evolving sorting algorithms.

A shortcoming of the Parasite study is that we have measured neither the actual strength of the selective advantage for or against higher recombination rates. This information would tell us how great an influence the parasite population must have on the host's fitness in order to provide selection for higher recombination rates (and thus sexual reproduction). Another shortcoming of this study is that we have directly measured the value of the recombination modifier gene. A more accurate measure of mixis would be to measure the effective recombination rate: the rate of crossover events that result in new gene combinations. For example, if two parents are genetically identical a crossover cannot produce an offspring with a different combination of genes from its parents. In this case, the effective recombination rate will be zero, no matter what the value of the recombination modifier gene might be. The local mating used in the Parasite simulation tends to create pockets of genetically similar individuals, so it is quite likely that $\bar{\rho}_h$ overstates the effective recombination rate. However, this is not a critical failing of this study because real populations exhibit local mating and thus pockets of genetically similar individuals. It is important to note that these shortcomings are features of the particular simulation, not of the paradigm of artificial evolution. In fact, we are currently refining the Parasite simulation to address these problems.

In this chapter, we have shown that low-level artificial evolution simulations generate realistic population dynamics, allowing us to use the simulations as a laboratory for studying natural evolution. While Parasite does not use the full power of the artificial evolution methodology (no environment, trivial organism behaviors, etc.), we have still produced exciting and important results.

ACKNOWLEDGMENTS

Special thanks to David Jefferson, who acted as my dissertation advisor. This work was supported in part by W. M. Keck Foundation grant number W880615, University of California Los Alamos National Laboratory award number CNLS/89-427, and University of California Los Alamos National Laboratory award number UC-90-4-A-88. The empirical data was gathered on a Connection Machine-2 computer at UCLA under the auspices of National Science Foundation Biological Facilities grant numbers BBS8714206 and DIR9024251.

REFERENCES

Anderson, R. M. and May, R. M. (1982). Coevolution of hosts and parasites. *Parasitology* **85**, 411–26.

Bäck, T., Hoffmeister, F. and Schwefel, H.-P. (1991). A survey of evolution strategies. In *Proceedings of the Fourth International Conference on Genetic Algorithms* (ed. R. K. Belew and L. B. Booker). Morgan Kaufmann, Los Altos, CA, pp. 2–9.

Barrett, J. A. (1983). Plant-fungus symbioses. In *Coevolution* (ed. D. J. Futuyama and M. Slatkin). Sinauer Associates Inc., Sunderland, MA.

Barrett, J. A. (1985). The gene-for-gene hypothesis: Parable or paradigm. In *Ecology and Genetics of Host–Parasite Interactions* (ed. D. Rollinson and R. M. Anderson). Academic Press.

Bell, Graham (1982). *The Masterpiece of Nature: The Evolution and Genetics of Sexuality*. University of California Press.

Bremermann, H. J. (1980). Sex and polymorphism as strategies in host–pathogen interactions. *Journal of Theoretical Biology* **87**, 671–702.

Bremermann, H. J. and Pickering, J. (1983). A game-theoretical model of parasite virulence. *Journal of Theoretical Biology* **100**, 411–26.

Brooks, Lisa D. (1987). The evolution of recombination rates. In *The Evolution of Sex: An Examination of Current Ideas* (ed. R. E. Michod and B.

R. Levin). Sinauer Associates Inc., Sunderland, MA, pp. 87–105.

Collins, R. J. (1992). Studies in Artificial Evolution. Ph.D. thesis, University of California, Los Angeles.

Collins, R. J. and Jefferson, D. R. (1991a). AntFarm: Towards simulated evolution. In *Artificial Life II*. Volume 10 of *Santa Fe Institute Studies in the Sciences of Complexity* (ed. C. G. Langton, C. Taylor, J. Doyne Farmer and S. Rasmussen). Addison-Wesley, Reading, MA, pp. 579–601.

Collins, R. J. and Jefferson, D. R. (1991b). An artificial neural network representation for artificial organisms. In *Proceedings of the First Workshop on Parallel Problem Solving* (ed. H.-P. Schwefel and R. Männer). Lecture Notes in Computer Science No. 496, Springer-Verlag, Berlin, pp. 259–63.

Collins, R. J. and Jefferson, D. R. (1991c). Representations for artificial organisms. In *Proceedings of the First International Conference on Simulation of Adaptive Behavior: From Animals to Animates* (ed. J.-A. Meyer and S. W. Wilson). MIT Press, Cambridge, MA, pp. 382–90.

Collins, R. J. and Jefferson, D. R. (1991d). Selection in massively parallel genetic algorithms. In *Proceedings of the Fourth International Conference on Genetic Algorithms* (ed. R.K. Belew and L. B. Booker). Morgan Kaufmann, pp. 249–56.

Collins, R. J. and Jefferson, D. R. (1992). The evolution of sexual selection and female choice. In *Proceedings of the First European Conference on Artificial Life* (ed. P. Bourgine and F. Varela). MIT Press, Cambridge, MA.

Coulson, R. N., Folse, J. and Loh, D. K. (1987). Artificial Intelligence and natural resource management. *Science* **237**, 262–7.

Crosby, J. L. (1963). Evolution by computer. *New Scientist* **327**, 415–17.

Darwin, Charles (1859). *On the Origin of Species by Means of Natural Selection*. Murray, London.

Day, P. R. (1974). *The Genetics of Host–Parasite Interactions*. W. H. Freeman, San Francisco.

Dobzhansky, T. (1956). What is an adaptive trait? *The American Naturalist* **190**, 337–47.

Feldman, M. W., Christiansen, F. B. and Brooks, Lisa D. (1986). Evolution of recombination in a constant environment. *Proceedings of the National Academy of Science, USA* **77**, 4838–41.

Felsenstein, J. (1987). Sex and the evolution of recombination. In *The Evolution of Sex: An Examination of Current Ideas* (ed. R. E. Michod and B. R. Levin). Sinauer Associates Inc., Sunderland, MA, pp. 74–86.

Felsenstein, J. and Yokoyama, S. (1976). The evolutionary advantage of recombination. *Genetics* **83**, 845–59.

Fisher, R. A. (1930). *The Genetical Theory of Natural Selection*. Dover Press, New York.

Flor, H. H. (1956). The complementary genic systems in flax and flax rust. *Advances in Genetics* **8**, 29–54.

Franklin, I. and Lewontin, R. C. (1970). Is the gene the unit of selection? *Genetics* **65**, 707–34.

Fry, J., Taylor, C. E. and Devgan, U. (1989). An expert system for mosquito control in Orange County California. *Bulletin of the Society of Vector Ecology* **14**(2), 237–46.

Ghiselin, M. T. (1987). The evolution of sex: A history of competing points of view. In *The Evolution of Sex: An Examination of Current Ideas* (ed. R. E. Michod and B. R. Levin). Sinauer Associates Inc., Sunderland, MA, pp. 7–23.

Glesener, R. R. and Tilman, D. (1978). Sexuality and the components of environmental uncertainty: Clues from geographic parthenogenesis in terrestrial animals. *American Naturalist* **112**, 659–73.

Goldberg, D. E. (1989). *Genetic Algorithms in Search, Optimization and Machine Learning*. Addison-Wesley, Reading, MA.

Haigh, J. (1978). The accumulation of deleterious genes in a population: Muller's ratchet. *Theoretical Population Biology* **14**, 251–7.

Hamilton, W. D. (1980). Sex versus non-sex versus parasite. *Oikos* **35**, 282–90.

Hamilton, W. D. (1982). Pathogens as causes of genetic diversity in their host populations. In *Population Biology of Infectious Diseases* (ed. R. M. Anderson and R. M. May). Springer-Verlag.

Hamilton, W. D. (1986). Instability and cycling of two competing hosts with two parasites. In *Evolutionary Process Theory* (ed. S. Karlin and E. Nevo). Academic Press, New York.

Hamilton, W. D. (1990). Sexual reproduction as an adaptation to resist parasites. *Proceedings of the National Academy of Science, USA* **87**, 3566–73.

Hamilton, W. D., Henderson, P. A. and Moran, N. A. (1981). Fluctuation of environment and coevolved antagonist polymorphisms as factors in the maintenance of sex. In *Natural Selection and Social Behavior* (ed. R. D. Alexander and D. W. Tinkle). Chiron Press.

Hartl, D. L. and Clark, A. G. (1989). *Principles of Population Genetics*. Sinauer Associates Inc., Sunderland, MA.

Hillis, W. D. (1985). *The Connection Machine*. MIT Press, Cambridge, MA.

Hillis, W. D. (1991). Co-evolving parasites improve simulated evolution as an optimization procedure. In *Artificial Life II*. Volume 10 of *Santa Fe Institute Studies in the Sciences of Complexity* (eds. C. G. Langton, C. Taylor, J. Doyne Farmer and S. Rasmussen). Addison-Wesley, Reading, MA, pp. 313–24.

Hillis, W. D. and Barnes, J. (1987). Programming a highly parallel computer. *Nature ACM* **326**(6108), 27–30.

Hillis, W. D. and Steele Jr, G. L. (1986). Data parallel algorithms. *Communications of the ACM* **29**(12), 1170–83.

Holland, J. H. (1975). *Adaptation in Natural and Artificial Systems*. The University of Michigan Press, Ann Arbor, MI.

Jaenike, J. (1978). An hypothesis to account for the maintenance of sex within population. *Evolutionary Theory* **3**, 191–4.

Jefferson, D., Collins, R., Cooper, C., Dyer, M., Flowers, M., Korf, R., Taylor, C. and Wang, A. (1991). The Genesys System: Evolution as a theme in artificial life. In *Artificial Life II*. Volume 10 of *Santa Fe Institute Studies in the Sciences of Complexity* (ed. C. G. Langton, C. Taylor, J. Doyne Farmer and S. Rasmussen). Addison-Wesley, Reading, MA, pp. 549–78.

Kauffman, S. and Levin, S. (1987). Towards a general theory of adaptive walks on rugged landscapes. *Journal of Theoretical Biology* **128**, 11–45.

Keightley, P. D. and Hill, W. G. (1989). Quantitative genetic variability maintained by mutation-stabilizing selection balance: Sampling variation and response to subsequent directional selection. *Genetical Research* **54**, 45–57.

Kimura, M. (1968). Evolutionary rate at the molecular level. *Nature* **217**, 624–6.

Langton, C. G. (ed.) (1989). *Artificial Life*. Volume 6 of *Santa Fe Institute Studies in the Sciences of Complexity*. Addison-Wesley, Reading, MA.

Langton, C. G., Taylor, C., Doyne Farmer, J. and Rasmussen, S. (eds.) (1991). *Artificial Life II*. Volume 10 of *Santa Fe Institute Studies in the Sciences ofComplexity*. Addison-Wesley, Reading, MA.

Levin, D. A. (1975). Pest pressure and recombination in plants. *American Naturalist*, **109**, 437–51.

Lewontin, R. C. (1964). The interaction of selection and linkage, I. General considerations; heterotic models. *Genetics* **49**, 49–67.

Martin, F. G. and Cockerham, C. C. (1960). High speed selection studies. In *Biometrical Genetics* (ed. O. Kempthorne). Pergamon.

May, R. M. (1983). Parasitic infections as regulators of animal populations. *American Scientist* **71**, 36–44.

Maynard Smith, J. (1987). The evolution of recombination. In *The Evolution of Sex: An Examination of Current Ideas* (ed. R. E. Michod and B. R. Levin). Sinauer Associates Inc., Sunderland, MA, pp. 106–25.

Mayr, Ernst (1983). How to carry out the adaptationist program? *The American Naturalist* **123**(3), 324–34.

Meyer, J.-A. and Wilson, S. W. (eds.) (1991). *Proceedings of the First International Conference on Simulation of Adaptive Behavior: From Animals to Animats*. MIT Press, Cambridge, MA.

Michod, R. E. and Levin, B. R. (eds.) (1987). *The Evolution of Sex: An Examination of Current Ideas*. Sinauer Associates Inc., Sunderland, MA.

Muller, H. J. (1964). The relation of recombination and mutational advance. *Mutation Research* **1**, 2–9.

Ohta, T. (1987). Simulating evolution by gene duplication. *Genetics* **115**, 207–13.

Ohta, T. (1989). Time for spreading of compensatory mutations under gene duplication. *Genetics* **123**, 579–84.

Ohta, T. and Tachida, H. (1990). Theoretical study of near neutrality. I. Heterozygosity and rate of mutant substitution. *Genetics* **126**, 219–29.

Price, M. V. and Waser, N. M. (1982). Population structure, frequency-dependent selection and the maintenance of sexual reproduction. *Evolution* **36**, 35–43.

Rennie, J. (1992). Trends in parasitology: Living together. *Scientific American* **266**(1), 122–33.

Rice, W. R. (1983). Parent-offspring pathogen transmission: A selective agent promoting sexual reproduction. *American Naturalist* **121**, 187–203.

Rosenberg, R. S. (1967). Simulation of Genetic Populations with Biochemical Properties. Ph.D. thesis, University of Michigan.

Schaffer, J. D. and Morishima, A. (1987). An adaptive crossover distribution mechanism for genetic algorithms. In *Genetic Algorithms and Their Applications: Proceedings of the Second International Conference on Genetic Algorithms* (ed. J. J. Grefenstette). Lawrence Erlbaum Associates, New Jersey, pp. 36–40.

Schull, W. J. and Levin, B. R. (1964). Monte Carlo simulations: Some uses in the genetic study of primitive man. In *Stochastic Models in Medicine and Biology* (ed. J. Gurland). The University of Wisconsin Press, Madison, pp. 179–96.

Seger, J. and Hamilton, W. D. (1987). Parasites and sex. In *The Evolution of Sex: An Examination of Current Ideas* (ed. R. E. Michod and B. R. Levin). Sinauer Associates Inc., Sunderland, MA, pp. 176–93.

Shields, W. M. (1987). Sex and adaptation. In *The Evolution of Sex: An Examination of Current Ideas* (ed. R. E. Michod and B. R. Levin). Sinauer Associates Inc., Sunderland, MA, pp. 253–69.

Swartzman, G. L. and Kaluzny, S. P. (1987). *Ecological Simulation Primer*. Macmillan.

Taylor, Charles E. (1983). Evolution of resistance to insecticides: The role of mathematical models and computer simulations. In *Pest Resistance to Pesticides* (eds. G. P. Georghiou and Tetsuo Saito). Plenum Press, New York, NY.

Taylor, C. E., Jefferson, D. R., Turner, S. R. and Goldman, S. R. (1989a). RAM: Artificial life for the exploration of complex biological systems. In *Artificial Life*. Volume 6 of *Santa Fe Institute Studies in the Sciences of Complexity* (ed. C. G. Langton). Addison-Wesley, Reading, MA, pp. 275–95.

Taylor, C. E., Muscatine, L. and Jefferson, D. R. (1989b). Maintenance and breakdown of the *hydra-chlorella* symbiosis: A computer model. *Proceedings of the Royal Society of London* **238**, 277–89.

Tooby, J. (1982). Pathogens, polymorphism, and the evolution of sex. *Journal of Theoretical Biology* **97**, 557–76.

Werner, G. M. and Dyer, M. G. (1991). Evolution of communication in artificial organisms. In *Artificial Life II*. Volume 10 of *Santa Fe Institute Studies in the Sciences of Complexity* (ed. C. G. Langton, C. Taylor, J. Doyne Farmer and S. Rasmussen). Addison-Wesley, Reading, MA, pp. 659–87.

Williams, G. C. (1980). Kin selection and the paradox of sexuality. In *Sociobiology: Beyond Nature/Nurture* (ed. G. W. Barlow and J. S. Iverberg). Westview.

Wright, S. (1931). Evolution in Mendelian populations. *Genetics* **16**, 97–159.

Wright, S. (1932). The roles of mutation, inbreeding, crossbreeding and selection in evolution. In *Proceedings of the Sixth International Congress of Genetics*. Volume 1, pp. 356–66.

BOTH WRIGHTIAN AND 'PARASITE' PEAK SHIFTS ENHANCE GENETIC ALGORITHM PERFORMANCE IN THE TRAVELLING SALESMAN PROBLEM

16

Brian H. Sumida and William D. Hamilton

Orestes: You don't see them, you don't – but I see them:
they are hunting me down, I must move on.
Choephoroi (from *Sweeney Agonistes* (T. S. Eliot))

16.1 BACKGROUND

A genetic algorithm (GA) is a heuristic optimization technique based on natural evolution (Holland 1975). Although GAs are executed with the aid of a computer, the GA itself is inspired by principles derived from both computer science and biology. Hence there are two potential sources from which to seek insights towards improving the performance of GAs. A particular GA implementation may be improved in the first instance by seeking more efficient algorithms (e.g., for sorting) and secondly, by deriving more appropriate evolutionary models (Sumida 1992). In this chapter we describe a GA based on a new evolutionary model which, by adding parasites and structured populations in the model, greatly improves the performance on a particular highly multi-peaked, adaptive landscape, that of the travelling salesman problem (TSP).

Parasites as potential stimulants of evolution have been the theme of recent theoretical (Hamilton 1990*a*; 1990*b*; 1990*c*) and computational (Hamilton *et al.* 1990; Hillis 1991; Collins, in this volume) studies. Parasites have also been credited as a key factor for the maintenance of sex (Hamilton 1980; Hamilton *et al.* 1990), and, as Holland realized much earlier, sex, apart from natural selection itself, is the most important process that GAs use.

Populations of organisms which inhabit the natural world are usually clustered into groups of individuals. For a population of a given size, having its members compete and mate largely within groups greatly alters its properties under natural selection. Sewall Wright (1969 and earlier) was the first to give serious attention to this fact and developed a body of theory describing what he called a 'shifting balance' referring to alternative adaptations and levels of adaptation among semi-isolated populations. In the important case for this theory local populations, or demes, interbreed freely within themselves but only occasionally transfer individuals from one deme to another. Such partial isolation is highly preservative of global population variation (Wright 1931) and from the point of view of evolution becomes most creatively so if the demes are small because isolation facilitates random genetic drift (Futuyma 1986). Since drift is a non-directional process, in the ensemble all directions of drift are experienced, including, at times, different allele and genotype fixations; hence the ensemble stores variability. Adding natural selection Wright saw that (a) different adaptive peaks would tend to be attained in different demes, and (b) by a combination of drift and migration, demes finding superior peaks could transform neighbouring demes to their discovery. Independently, Holland (1975) conceived of evolutionary 'building blocks' arising in GAs which, in a context of demes, means that one deme might discover one part of a superior adaptation (a strong opening in chess, say) while another deme might discover another complementary part (a strong end game). In the meta-population (i.e., the aggregate of demes) interdemic migration and subsequent intrademic recombination could bring the two parts together.

16.2 THE TRAVELLING SALESMAN PROBLEM (TSP)

The TSP, although difficult to solve, is easy to describe. A travelling salesman provided with a list of cities must visit each city once only and end the tour at the starting city. The cost of a journey between any two cities is known and the salesman's objective is to organize a tour permutation which incurs the least cost. Given a tour containing n cities, the number of possible permutations is $n!$. It is obvious that an exhaustive search for the optimal permutation is impossible for even moderately sized (~50) values of n. The problem is inherently extremely multi-peaked and at many levels, so eliminating classical optimization methods. Such 'ruggedness' of landscape arises because there are many radically different approaches to touring cities, both within any cluster and also for touring between the clusters. Thus, should Oxford, Banbury and Aylesbury be visited while touring up the east side of England or when touring down the west? Are they indeed best treated as a cluster, neighbours to be visited in some close sequence? If a salesman decides to change his western tour to include these, it is almost certain he will need to adjust other parts of both eastern and western limbs of his total tour to say nothing of changing order within the three mentioned cities themselves. Once a fairly efficient tour has been devised all simple 'cut and paste' operations of segments of tours are most likely to be harmful even though they may suggest, to an intelligent viewer, dramatically improved tours that could follow. More explicit examples of such inherent ruggedness peaks will appear later. Non-GA optimization methods that improve on blind exhaustive search by eradicating blocks of alternatives that have no chance to be successful are available. These methods, however, are still extremely tedious. Using them the current record for the longest solved TSP of 3038 cities (Sangalli 1992) was obtained. The unsolved and perhaps unsolvable difficulty of the TSP provides a challenging test bed for the GA to prove its effectiveness. A concise review of the application of GAs to the TSP can be found in Michalewicz (1992).

16.3 TOUR REPRESENTATION, HOST SEXUALITY AND HOST FITNESS

A novel chromosomal representation of travelling salesman tours is presented. A given tour's representation comprises two arrays: a next city array and a previous city array. In each the array index corresponds to the city name. Actually one of the two arrays is sufficient to define the tour but there is an advantage of clarity in the algorithm as so far developed to maintain both.

Both arrays are used in sexual reproduction to form hybridized offspring tours. In each deme, for each offspring to be produced, the individual with the best tour becomes one parent and a random deme member the other parent. If there are to be further offspring, this procedure is repeated exactly – the best tour becomes in effect polygamous. An offspring tour is created from a pair of parents by an act resembling bacterial conjugation: the best tour parent donates a subtour (a sequence of city visits) from itself to be inserted into a copy of the tour of the other parent. The transformed copy is the offspring.

The first step in the insertion procedure is to pick a random start city, SD, in the donor and a random subtour length, l (i.e., number of cities, $0 < l < n$ where n is total number of cities). The second step is to pick a random city, SR, in the recipient tour. SD is to be the first novel city visited in the transformed recipient tour, immediately following city SR. Thus the 'next' of SR in the recipient tour is changed to SD and the 'previous' of SD is changed to become SR. What about the next of SD itself? Soon it will be the next city, if there is to be one, from the donated segment; but in case there is not to be one, and to keep tours legal at every point of the insertion procedure, what is done immediately is to make the next of SD into what was the next of SR and to make the previous of that next into SD.

An example may help to clarify this procedure. Suppose that one route being tried in a particular deme, let us call it R (denoting 'recipient'), around British cities includes the sequence: → Manchester, Shrewsbury, Gloucester → . Another route, D (denoting 'donor'), in the same deme has a sequence: → Leicester, Wolverhampton, Birmingham, Nottingham, Sheffield → and let us suppose that this tour is the best (i.e., shortest) tour yet. The R and D tours happen to be selected as parents for a 'mating' and in the random city drawings for the mated pair SR chances to be Shrewsbury and SD to be Wolverhampton. The procedure described above declares that for the second tour R, in two steps, the next of Shrewsbury and previous of Gloucester are to become Wolverhampton (a change that happens to look rather good on the map). But the R tour, of course, was already visiting Wolverhampton at some other stage. Suppose it was in a sequence: → Coventry, Wolverhampton, Oxford → . Since R must not visit Wolverhampton twice, this section of its tour is now made to be simply: → Coventry, Oxford → . (This 'bypass' also happens to look good, but that is only because we have chosen it to show how improvements can occur.) Reverting to the language of the arrays, the final step involves the next of SD's former previous and the previous of SD's former next which must be changed to specify each other. Once all this is complete city SD has been 'inserted' to follow SR and the recipient still has a legal tour encompassing all the cities. The algorithm can now proceed in the same way to insertion of the second city of the donated subtour (in the example it would be Birmingham) and so on onward until all l cities have been inserted.

For clarity in programming the algorithm, it was found convenient to think in terms of operations that were called 'arrows', an arrow being the creation of a step from one city to another. An arrow may itself be thought of as comprising two 'half arrows', these being simply the outward bound 'next' of the arrow-tail city and a corresponding inward bound 'previous' of the arrow-head city. In these terms, the insertion of one novel city as described above can be accomplished (given also a few storage additions to make sure one is not using values already altered) by calling a function of the form 'arrow(A,B)' three times, which effects a step A → B. The three steps can be described as 'moving in', 'continuing' and 'bypassing old position' as above.

For simplicity the description so far has assumed that the inserted segment of donor tour keeps its original direction. We saw no biological reason why this order should be kept, however, and saw certain advantages for the TSP that could be gained if it were not. Therefore we arranged that, with $p = 0.5$, the original direction would be used, otherwise the segment of l cities would be read 'backwards' along the donor tour from the start at city SD. This is easily done by referring at a certain point of the program to elements of the 'previous' array instead of the elements of the 'next' array. The obvious advantage for the TSP is that inserting self-segments backwards can 'untwist' geometrical crossovers in the tour since the presence of such sequences can never be optimal. Insertion of sequences already present in the recipient (i.e., self-segments) of course occurs with increasing frequency as demes tend to monomorphism.

The set of values in, say, an offspring's next array is always a permutation of the cities. But to be a TSP tour it has to be a permutation of a restricted kind, otherwise detached subtours and immobile salesmen will occur. This however is easily arranged at the start of the GA by creating a set of differing full random tours and then ensuring, as in the three-step insertion procedure above, that every transformation effected on a tour creates a full tour.

A host's primary fitness is obtained by summing the distance travelled whilst traversing the tour specified by the chromosome – fitness increases with decreasing tour length. Primary fitnesses yield secondary fitnesses by increasing host tour length (i.e., decreasing fitness) by a quantity reflecting the success of parasitism. Secondary fitnesses are used to allocate reproductive parents and to determine who dies. As already stated the individual with the highest secondary fitness in the deme is always selected as one member of each mating pair. This mating advantage effects part of the Darwinian selection driving the evolutionary process. Mortality, the other common aspect of selection, is represented by overwriting the individuals with the lowest fitnesses with newly created offspring.

16.4 PARASITE TOURS AND PARASITE SEX

Parasite chromosomes are the same length as host chromosomes but a parasite tour differs in that not every city need be visited. Indeed it is common for a parasite tour to visit only a few cities. However, a single-city parasite would always match and we felt that even a two-city matching parasite would too easily evolve. The GA needs time to consolidate its discoveries and to export them from the demes making the discovery before being forced by parasitism to 'doubt' and so to change them. Small matching parasites may reduce the value of new features almost from the moment of their arrival. From these considerations we decided that a parasite tour could be any legal closed TSP tour or subtour subject to $n \geqslant 3$. Cities not visited are marked on the chromosome by having their next and previous array values set to a null value for which we used -1. Since these null parts of the parasite's chromosome do not need to be handled in most of the operations involving parasites their effective chromosomes are shorter and their replications are carried out speedily. Parasites can become small or large in the course of evolution. As it turns out most of the time they are on average small and we feel that this is life-like and potentially beneficial to the GA (see below). In contrast, antagonistic organisms of equal size to the hosts such as are used by Hillis (1991) and Collins (this volume) seem to us more aptly described as coevolving predators.

At the stage of applying parasites, random hosts and random parasites from their respective demic populations are paired and out of the pairing a parasite fitness is calculated which is then deducted from host fitness, as mentioned above. A parasite's fitness bears no relation to the geometrical length of its tour. Instead it depends upon the sequences of cities that the parasite manages to imitate in the tour of the host it is paired with. All 'arrows' of the host tour duplicated by the attached parasite are noted in the program and each such arrow is a potential contribution to fitness; but in the version we have used so far we count only the longest unbroken sequence of corresponding arrows as the parasite's primary fitness. The parasite's secondary fitness was taken to be this count divided by the total number of cities. In such a scoring system, 'short' parasites should find it relatively easy to mimic a host along most of their length; however, such parasites do not contain as much potential for pay-off as longer parasites have. We wanted short parasites to be present for speed in the early and very changeable stages of host evolution but also to allow for the possibility of a slow acquisition by the parasites of long mimetic sequences that reflected major elements of TSP strategy. These long sequences would correspond to the chromosomes of hosts perched on the well-stabilized adaptive hilltops. The essence of using parasites in the GA is that their effect should be gently to lower and flatten achieved peaks so that eventually the slopes of neighbouring peaks can be discovered (Wright 1977; Lenski 1988).

In order to allow parasite size to evolve we arranged that when parasites mated two offspring would result. The procedure of selecting parents, one best and the other random, was the same as for hosts and the procedure of donating a segment of tour from this donor to create offspring by modifying a copy of the recipient's tour was also the same except that the 'bypass' operation could often be skipped, an omission keyed in the algorithm by discovery of '−1' values in the arrays. Each time the bypass operation is omitted, the recipient subtour grows in length by one city. While this was going on, differently than with hosts, a copy of the donor tour was simultaneously modified by elimination of the cities it donated. 'Next' and 'previous' array elements of all the eliminated cities were set to −1 and the cities bounding the gap connected by an arrow. This is a 'bypass' operation similar to that already described for a recipient host tour but nevertheless is quite distinct from it. That former kind of bypass may be made in a recipient parasite tour also, as already mentioned, but (a) it never involves more than one city at a time, and (b) it does not shorten the recipient tour. The present operation on a parasite donor is an elimination rather than a bypass and has the result that every act of mating usually produces one augmented tour and always produces one diminished tour, as compared to what had existed before. Either of the new tours may turn out to be an improvement in the context of the current host population. Thus mean parasite length might evolve up or down. With this procedure we felt we provided both speed when variability was high in the host population and the parasite's target very changeable, and effective discovery and harassment of hosts when subpeak host adaptations were becoming stabilized: parasites as long as necessary to damage hosts seriously would evolve to discourage any spreading conviction that a best tour had been found.

An illustration of the parasite pursuit arising from our variable parasite length model is given in Figure 16.1. In this figure we can observe the evolving distribution of parasite chromosome lengths as the parasite attempts to mimic the host chromosome. The distribution of parasite chromosome lengths is initially flat (Figure 16.1(a)) due to the random start. As the hosts evolve, selection finds and preserves the fittest individuals, thus carrying the population towards the nearest peak in the fitness landscape. As the peak is approached the population becomes more uniform. A uniform population presents the parasites with a static target and, consequently, parasites quickly evolve to imitate the chromosome characterizing the peak and in doing so grow long (Figure 16.1(b)). Increased success by parasites is matched by decreasing host fitness which in turn eventually gives advantage to residual

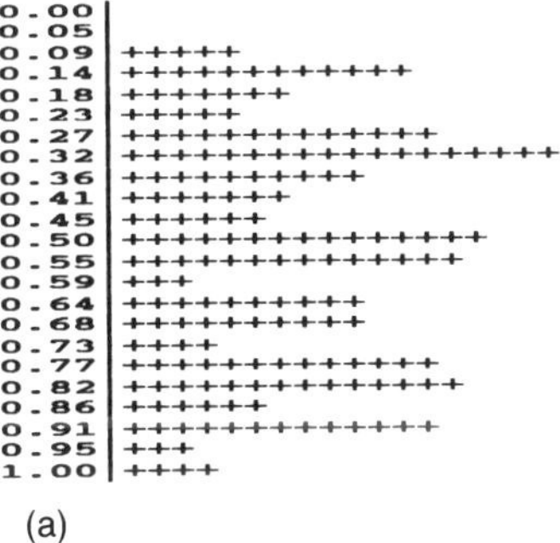

(a)

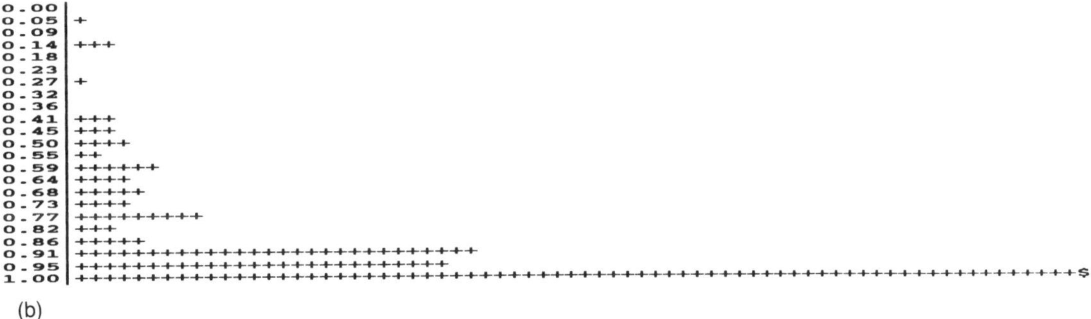

(b)

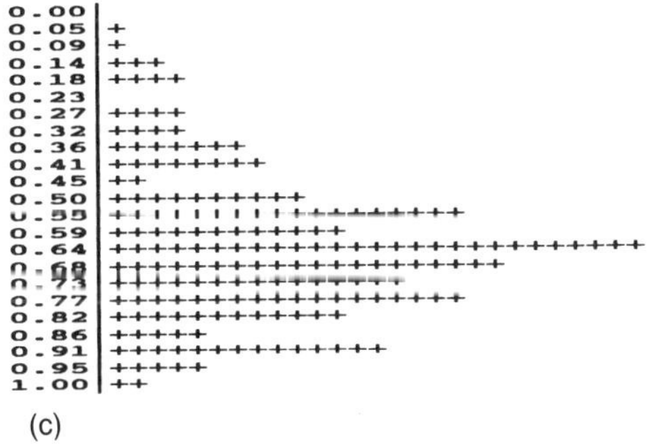

(c)

Figure 16.1 Distribution of parasite tour lengths at three stages of a run where parasites are present in a single deme. When a parasite tour length is equal to a host tour length it receives a value of '1.0'. The 0th bin represents parasite tours of lengths up to 5% of the host tour. Each '+' symbol represents an individual of the total population; when the limit of a bin is reached (72 individuals) an overflow symbol, '$', is printed (Figure 16.1(b)). Population size for both hosts and parasites was 195 individuals in a single deme. Host mortality and parasite mortality was 16 and 40 deaths per generation respectively. (a) gives distribution at the start of the simulation; (b) is the distribution after 500 generations; and (c) is the distribution after 1500 generations.

variants and new suboptimal mutants. These residuals, mutants, and their combinations with each other, along with the current peak genotypes, explore the valleys and saddles around the current peak. When the upward slope of another peak is discovered and that peak begins to be 'colonized', a peak shift follows accompanied by a large increase of both host and parasite variability. The latter manifests itself in part as a decrease in mean length of parasites and an increase in their length variance. Both result from the fact that during the peak shift long mimetic sequences in the parasite are fractured and there is therefore no longer an advantage in being long. Figure 16.1(c) illustrates a final state where the best peak has been flattened by successful parasitism but no next ascending slope has been found. In this terminal state considerable variability is again to be expected. Indeed to identify the final best tour/peak genotype it is necessary to switch off parasite action and continue the run in order to eliminate the suboptimal, opportunistically probing genotypes the parasites are still causing to be created. Such a clearance of parasites probably sometimes occurs in the real world when a host species colonizes an oceanic island or a new continent. Populations on islands are sometimes very innovative; an example of this might be the finches that uniquely use cactus spines as tools to extract insect larvae from dead trees in the Galapagos Islands (Grant 1986).

16.5 DEME STRUCTURE AND MIGRATION

As already implied, matings, selection and reproduction occur between individuals within demes. In our GA simulation the demes are spatially arranged as a toroidal world forming a continuous lattice with no barriers. Movement of individuals between demes is possible but constrained to create the condition of semi-isolation. In our model, individuals are dispersed through the toroidal world by two classes of migration. In nearby migration an individual may migrate into any one of the eight adjacent 'near' demes with equal probability. Far migration differs from this in that an individual is allowed to migrate to any deme in the toroidal world with equal probability as in Wright's 'Island' model (Wright 1931). As in the natural world, migration to and from nearby demes in the toroidal world was made much more likely than a distant migration. The frequency of interdemic migration strongly affects gene distribution in the population (Crow and Kimura 1970). If set too high or too low, we regress either to one large deme or to many isolated small ones. In both cases the benefit to the GA of having population structure is nullified (Tanese 1987 and Sumida *et al.* 1990).

The population structure of our model superficially resembles the models of Hillis (1991) and Collins (in this volume), although in the absence of precise detail of their population handling we cannot say how functionally or demographically equivalent they are. Our implementation discussed in this chapter is clearly distinguished from theirs by our use of much smaller population size (hundreds of individuals as compared to tens of thousands) and by the presence of far migration. Obviously GAs themselves should evolve to be efficient 'organisms' in their intended niche and if a certain degree of difficulty of TSP can be overcome by a program that is economical in its code, storage requirements and CPU time, then this bodes well for its success on bigger problems when more time and storage are available. Such extension of efficiency probably applies most strongly, however, if merely tours of a given standard of efficiency and not ultimate optimal tours are being sought. There seems to be no escape from the rapid increase of difficulty of the strict optimal tour problem as the number of cities increases. For the strict TSP applying to very many cities, some degree of 'concept formation' and of 'appreciation of the pattern of the whole' seem necessary. Our program is certainly not yet at a stage of such abilities.

16.6 OUTCOMES USING PARASITES AND POPULATION STRUCTURE

An experiment was run to determine whether adding parasites and partitioning the population into demes aided the GA. We tested four conditions: (1) no population structure and no parasites; (2) a structured population but no parasites; (3) no population structure but parasites present; and (4) a structured population with parasites.

A population of 192 individuals was divided into 16 demes each containing 12 individuals. Truncation selection (see Crow and Kimura 1970 for definition) at the lower end of the secondary fitness ordering was used with the mortality in each deme set to two deaths for the hosts and six deaths for the parasites in every generation. This implies that in each host's deme the two worst individuals are killed and the best host individual is twice made a donor parent whereas in each parasite deme the six worst individuals are killed with the best parasite individual made parent three times (recall that parasite matings give two offspring each). This different mortality level for the parasite population is one of two ways in which selection on parasites is made stronger. The other is that parasites are given a generation time half of that of the host. In detail the last assumption means that after a generation is run in which both parasites and hosts undergo selection, parasites are run again in a generation in which hosts do not change; then both change again, and so on throughout.

Nearby migration and distant migration probabilities for both hosts and parasites per individual per generation were set at 0.01 and 0.001 respectively. Ten replicates for each condition were run, each run lasting 5000 generations.

16.7 RESULTS

The evolution of the population towards the optimal tour is shown in a series of tour graphs (Figures 16.2 to 16.5). The best tour from each of the sixteen demes is plotted in a subdivided square, with the best overall tour outlined. Figure 16.2 shows the population at the start of the simulation. Evolution has not begun at generation zero so the best tour here is merely the best of the random initial tours. After 1500 generations, in Figure 16.3, we can see several strategies for visiting cities emerging. It can also be observed from the figure that tours in adjacent demes, due to the mainly local migration, are beginning to exhibit common subtour elements. At generation 4500, Figure 16.4, we can see that most of the population variance has been lost, but an example of an optimal tour has evolved in the upper left-hand corner of the figure. At the run's conclusion at 5000 generations, Figure 16.5, the optimal tour (the length of which is $(11 + 3\sqrt{10})1.5 = 30.7302$) is present in nearly every deme in the population. Notice, however, that even at this late stage three other variants remain. Given the presence of the best tour in the demes surrounding these variants, their existence must certainly be due to the influence of the parasites. In other words, in terms of secondary fitnesses, these host variants have become locally better in secondary fitness than the globally best TSP tour because they have broken the mimicking sequences of their local parasites.

The results of runs are shown, as means of the best evolved tour lengths over the ten replicates, in Table 16.1.

Table 16.1

	No parasites	*Parasites*
No demes	34.152	32.187
Demes	31.879	31.369

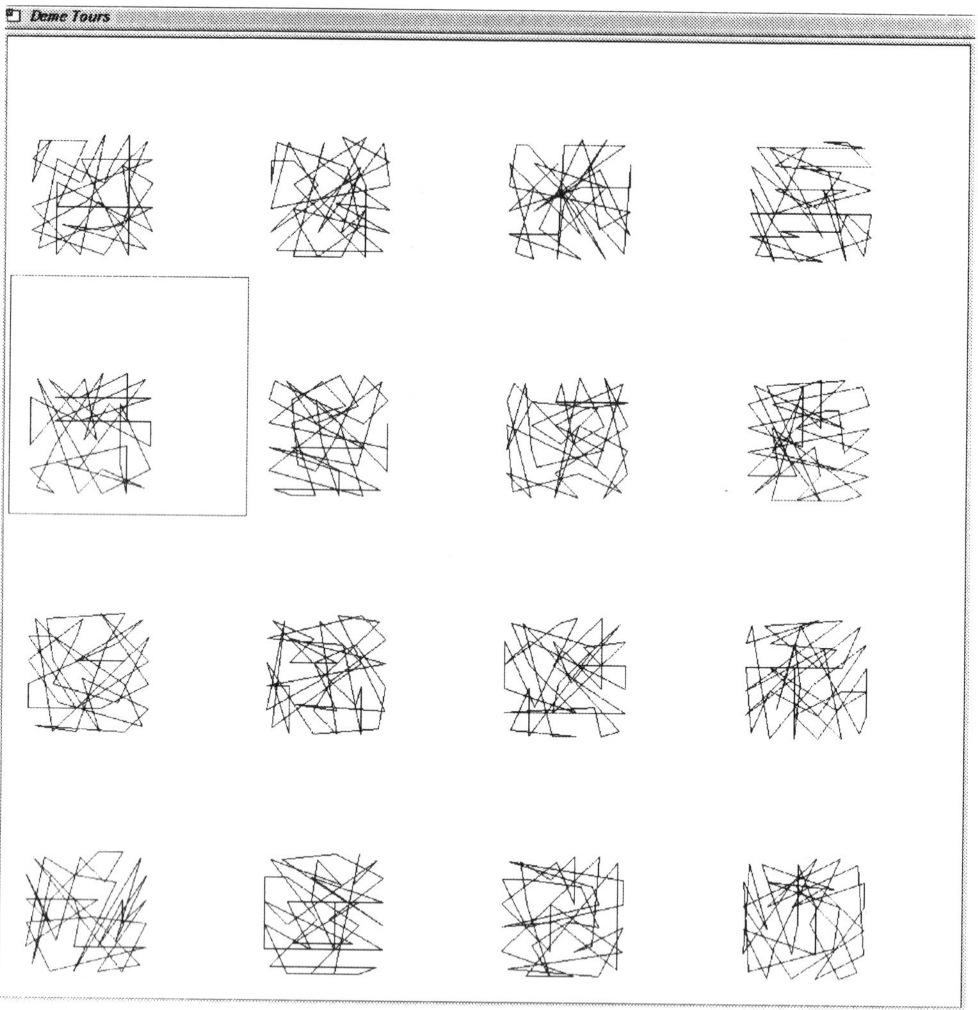

Figure 16.2 Best host tours from each of 16 demes. The overall best tour is surrounded by a white box. The population of 192 was arranged into 16 demes of 12 individuals and there were two parasite generations per host generation. This figure is for the 0th and consequently the tours here are random.

The full data set was analysed with a two-way analysis of variance (DF = 39). Subdividing the population into the toroidal arrangement of demes gave a highly significant improvement ($F = 20.407$, $p \leqslant 0.0002$). A presence of parasites also gave a highly significant improvement ($F = 13.09$, $p \leqslant 0.0012$), but this factor as shown in the means and significance levels was slightly less effective than deme structure. The interaction between deme structure and parasites was significant ($F = 4.524$, $p \leqslant 0.0381$) but comparatively weak.

16.8 DISCUSSION

Variation must come from somewhere if evolution is to be effective. Subdividing the population into small demes increases the power of random genetic drift and this counters

Figure 16.3 Best host tours from each of 16 demes after 1500 generations. Alternative general approaches can be seen. It can be seen from the figure that tours in adjacent demes share similar subtour elements due to the mainly local migration.

the loss of variation in the population. In addition, in our model it is likely, especially early on, that different demes will be attracted to different peaks. This, of course, also helps to maintain variation although it is not in obviously useful form. How it can be made useful depends on the effects of migration.

The role of migration is somewhat subtle. In one sense, in unifying the demes into a meta-population, even if only barely, a loss of variation may result because the meta-population now has a chance to fix all demes at the same peak. On the other hand when the attraction of a new and better peak is found in any deme, migration enables the discovery to disseminate to other demes and in the course of this dispersal, assisted by recombination, a new wave of variation becomes liberated for global experimentation (Figure 16.1(c)).

Figure 16.4 Best host tours from each of 16 demes after 4500 generations. An optimal tour has been discovered, seen in the upper left-hand corner. Much of the variability in the population has been lost.

We have not had time yet to examine the relative effects of distant versus local migration. It may have appeared that the evolutionary process of mutation (often treated as equivalent to far migration in population genetical accounts) was omitted from our simulation. Mutation-like effects, however, are implicit in the way that the subtours are inserted during reproduction and therefore explicit mutation was deemed unnecessary. It should be recalled that even with a uniform population, because SRs and

Figure 16.5 Best host tours from each of 16 demes after 5000 generations. The optimal tour discovered in generation 4500 (Figure 16.4) has been dispersed throughout the population. Only three of the 16 demes give a non-optimal tour, these deviations having almost certainly arisen secondarily as the result of the local advantage of tours which escape attacking parasites.

SDs are picked at random and tour parts are being inserted in normal or reversed orientation, sexual reproduction is continually causing changed tours to be tried. It is worth pointing out here that the potential of migration to prevent (almost) any local fixation was seen clearly by Wright (1943):

> Moreover, the short range means of dispersal that have been postulated are likely to be supplemented by occasional long range dispersal. All of these tend to prevent fixation of one type even locally.

The same passage goes on:

> On the other hand, selection may favor one allele in some places and others in other

places. This would tend to increase local differentiation.

In our model this is true in two ways, one of which Wright may have had in mind, while the other he probably did not. Firstly, differences of selection arise through different demes being approximated to different peaks. Secondly, hosts from different demes coevolve with different sets of parasites. In the real world both factors must occur along with the even more obvious one of selection differing locally due to heterogeneity of the physical environment.

Evolutionary pressures acting on our host population stem from the advantage of better tour permutations and, at the same time, from avoidance of parasites. The action of these pressures can be described geometrically using the adaptive landscape metaphor (Wright 1969). Mapping the possible tour permutations onto a fitness surface results in the TSP case in an adaptive landscape of innumerable peaks of varying heights, representing fitter tours, and valleys of various depths representing the less fit tours. As one example demonstrating the existence of peaks, note that our chosen map of cities actually has four equivalent optimal peaks (Figure 16.6). It so happens that the run we illustrate in Figures 16.2–16.5 finds the one we intended to be optimal but it could equally well have found one of the other three optimal tours of whose existence the GA quickly informed us. (To be more precise the observation informed us of the existence of at least two forms and further analysis suggested a third and fourth which were subsequently proven to complete the optimal set. The fourth (top right), which has not yet been found by our GA in any run, presumably has a smaller 'catchment' in the total set of tours. From frequent observations of all the alternative optimal tours, catchment sizes appear to decrease counter-clockwise from the top left (most common) tour.)

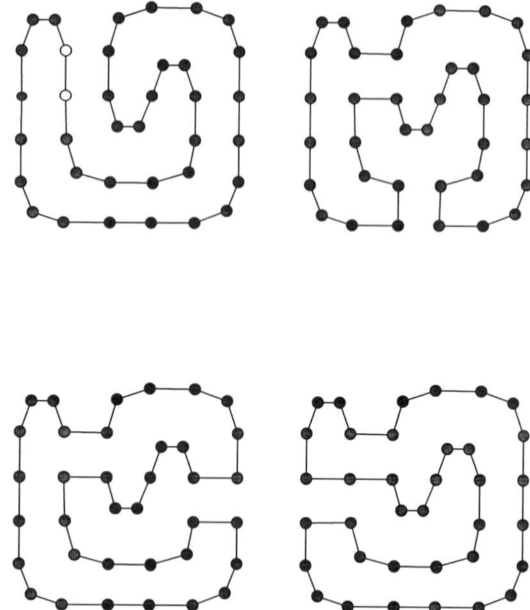

Figure 16.6 Optimal tour permutations for the 36 city TSP experiment. Each of the four illustrated tours have the same tour length. The chosen city pattern is based on a 6 × 6 square grid of points. Twelve points are moved inwards half-way towards the centres of lattice squares that they bound so as to suggest the 'worm' shape at the top left. This pattern could be made the only optimal tour (as originally intended) by moving the lower of the two lattice points represented by unfilled circles (in top-left tour) toward the one above it by any small distance, thus lengthening slightly the 'bridge' to the right that appears from the worm's neck in the three other tours.

It is obviously impossible to go from any one of the four optimal tours to any other by making any single insertion from the tour into an identical copy. It is even more obvious that no minimal modification (transfer of one city to another place) will effect the transition. Therefore it must involve steps down into a valley. Minor 'foothills' are also very easily illustrated. We are dealing with a landscape with four equal peaks and a very great number of foothills.

Over many generations populations of hosts move about on the landscape surface, blindly but effectively (Dawkins 1986), seeking the highest peak and the fittest tour. When the hosts are in a valley but within the basin of attraction of any given local peak, selection will drive the population towards the summit. This hill-climbing is a local optimization. The best or peak tour, once found by any individual, will spread by means of migration throughout the population, eventually resulting in the population becoming homogeneous. Selection in the absence of any other agency will never produce other than the local peak tour, implying that if the local population is isolated and large then, as Wright pointed out, the population is trapped on its peak. However, if there is (a) small deme size and/or (b) mutation/migration the population may be able to move away from its adapted peak towards other basins of attraction according to the Wrightian scheme.

The other evolutionary pressure moving populations away from current peaks is provided by parasites. This will work in opposition to the ultimate landscape selection even in large isolated populations. With or without the presence of (a) and/or (b) above, parasite pressure may be able to dislodge hosts and move them onward very much like random genetic drift in Wright's theory. As the best tour's genotype increases in frequency in the population, parasites, with their shorter generation time, quickly evolve tours which, by successfully matching a now common best host tour, drive the fitness of its parasitized host downwards from its peak onto adjacent saddles – or, as a more accurate image, flattens the subpeak itself. The best of the depressed hosts is temporarily liberated from parasites by having broken, by recombination, the matching parasite sequences. When such individuals attain the slopes of neighbouring peaks, selection becomes the stronger influence again: hill-climbing then recommences and the process begins anew. Moving from peak to peak in this manner (Wright 1939, see Figure 16.9) the population has a better chance to discover a global optimum.

In the present study a new evolutionary model of the GA incorporating both parasites and structured populations was tested with a 36 city travelling salesman tour using a desktop microcomputer (33 MHz 80486). Finding the optimal tour of this shortish problem cannot be considered trivial yet is not difficult either. As it turns out our GA finds the solution quite easily. What the experiment demonstrates is that significant improvements are gained both by adding parasites and by altering the population structure. Each factor alone was significant: two-sample T-test, population variances unknown and not assumed equal. Significance level was $p \leqslant 0.0015$ for the test of no population and no parasites versus no population structure but with parasites. For the test of no population structure and no parasites versus population structure but no parasites, the significance level was $p \leqslant 0.0008$. Effort and data are insufficient so far to assess the degree of interaction between the two factors, however. Probably the moderately significant anti-synergistic interaction observed in Table 16.1 is due to nothing more than that each factor by itself was proving so effective that there was not much left that a combination could achieve.

Of course our model need not be limited to short tours. To demonstrate this our GA was run with a much more difficult tour of 100 cities randomly placed within a square. To compensate for the increased complexity, population size was increased slightly to 384 (32 demes of 12 individuals) and the number of generations was increased from 5000 to 20 000. Mortality and migration for hosts and parasites were set to the parameter values used in the previously described experiment. Figure 16.7 shows an evolved best tour for this 100 city tour. As can be seen in more than one place the evolved tour is certainly not quite optimal. But it is, at least to the eye or to a not-too-fastidious salesman, a satisfyingly efficient tour beyond which probably only a small percentage improvement can be made.

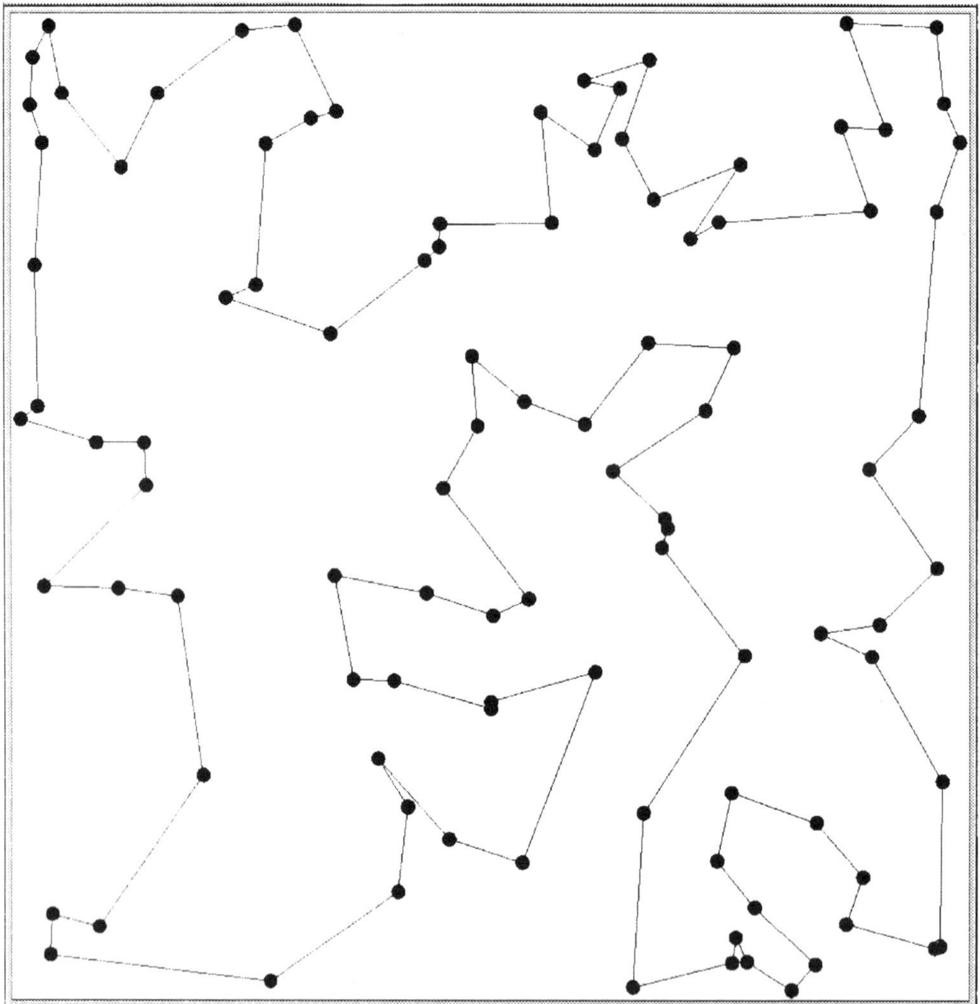

Figure 16.7 Evolved best tour for visiting 100 randomly placed cities. Population sizes for host and parasites were 384 individuals arranged into 32 demes of 12 individuals. Parasite and host mortalities were two and six deaths per generation respectively. Parasites were allowed two generations for every host generation. Total number of generations was 20 000. The optimal tour for this configuration of cities is not known but the evolved tour appears an efficient route unlikely to be far from optimal in terms of distance. Minor local imperfections, at least one of which is easily seen, are likely to be due to the continuing influence of parasites (see text).

Indeed some fairly obvious local algorithmic attention could polish this evolved tour into a form that would be quite difficult to improve at all. That such polishing has not been done by the GA is probably due to the presence of the parasites as indicated by the earlier discussion. If so, then their final slightly adverse effect can be eliminated, as already mentioned, by continuing the simulation in a terminal optimizing period with parasite influence switched off.

In this chapter we have demonstrated

how Nature's own software solutions can be transformed and transplanted into the language of the GA to provide results which can rival or even exceed the improvements that could be obtained by the use of larger populations and faster computers. By restricting ourselves to small GAs and, for now, small problems we hope we have shown that parasites as well as population structure have potential to help with far more substantial problems.

ACKNOWLEDGEMENTS

B. H. Sumida was supported by a SERC postdoctoral research grant. The authors wish to thank A. I. Houston for helpful comments.

REFERENCES

Crow, J. F. and Kimura, M. (1970). *An Introduction to Population Genetics Theory*. Harper & Row, New York.

Dawkins, R. (1986). *The Blind Watchmaker*. Longman, Essex.

Futuyma, D. J. (1986). *Evolutionary Biology*. Sinauer, Sunderland, MA.

Grant, P. R. (1986). *Ecology and Evolution of Darwin's Finches*. Princeton University Press, Princeton, NJ.

Hamilton, W. D. (1980). Sex versus non-sex versus parasite. *Oikos* **35**, 282–90.

Hamilton, W. D. (1990*a*). Seething genetics of health and the evolution of sex. In *The Proceedings of the International Symposium on the Evolution of Life*, March 25–28 1990. Springer, Tokyo, pp. 225–52.

Hamilton, W. D. (1990*b*). Mate choice near and far. *American Zool.* **30**, 341–52.

Hamilton, W. D. (1990*c*). Memes of Haldane and Jayakar in a theory of sex. *J. Genet.* **69**, 17–32.

Hamilton, W. D., Axelrod, R. and Tanese, R. (1990). Sexual reproduction as an adaptation to resist parasites (A review). *Proceedings of the National Academy of Sciences, USA* **87**, 3566–73.

Hillis, W. D. (1991). Co-evolving parasites improve simulated evolution as an optimization procedure. In *Artificial Life II* (ed. C. G. Langton, C. Taylor, J. D. Farmer and S. Rasmussen). Addison-Wesley, Reading, MA, pp. 313–24.

Holland, J. H. (1975). *Adaptation in Natural and Artificial Systems*. The University of Michigan Press, Ann Arbor, MI.

Lenski, R. E. (1988). Experimental studies of pleiotropy and epistasis in *Escherichia coli*. II. Compensation for maladaptive effects associated with resistance to virus T4. *Evolution* **42**, 443–40.

Michalewicz, Z. (1992). *Genetic Algorithms + Data Structures = Evolution Programs*. Springer-Verlag, Berlin.

Sangalli, A. (1992). Short circuiting the travelling salesman problem. *New Scientist* **134**, 16.

Sumida, B. H. (1992). Genetics for genetic algorithms. *SIGBIO Newsletter* **12**, 44–6.

Sumida, B. H., Houston, A. I., McNamara, J. M. and Hamilton, W. D. (1990). Genetic Algorithms and Evolution. *J. Theoret. Biol.* **147**, 59–84.

Tanese, R. (1987). Parallel genetic algorithms for a hypercube. In *Genetic Algorithms and Their Applications: Proceedings of the Second International Conference on Genetic Algorithms* (ed. J. J. Grefenstette). Lawrence Erlbaum, New Jersey.

Wright, S. (1931). Evolution in Mendelian populations. *Genetics* **16**, 97–159.

Wright, S. (1939). Statistical genetics in relation to evolution. In *Actualités scientifiques et industrielles, 802: Exposés de Biometrie et de la statistique biologique XIII*. Hermann & Cie, Paris.

Wright, S. (1943). Isolation by distance. *Genetics* **28**, 114–38.

Wright, S. (1969). *Evolution and Genetics of Populations*, Vol. 2. *The Theory of Gene Frequencies*. University of Chicago Press, Chicago.

Wright, S. (1977). *Evolution and Genetics of Populations*, Vol. 3. *Experimental Results and Evolutionary Deductions*. University of Chicago Press, Chicago.

EVOLUTION OF EMERGENT COOPERATIVE BEHAVIOR USING GENETIC PROGRAMMING

17

John R. Koza

17.1 INTRODUCTION AND OVERVIEW

In nature, populations of biological individuals compete for survival and the opportunity to reproduce on the basis of how well they grapple with their environment. Over many generations, the structure of the individuals in the population evolves as a result of the Darwinian process of natural selection and survival of the fittest. That is, complex structures arise from fitness. The question arises as to whether behaviors can also be evolved using the metaphor of natural selection. Another question is whether complex artificial structures, such as computer programs, can also be evolved using this metaphor. This chapter explores these questions in the context of a problem requiring the discovery of emergent cooperative behavior in the form of a computer program that controls a group of independently acting agents (e.g., robots or social insects such as ants).

17.2 THE PAINTED DESERT PROBLEM

The repetitive application of seemingly simple rules can lead to complex overall behavior (Steels 1990, 1991; Forrest 1991; Manderick and Moyson 1988; Moyson and Manderick 1990; Collins and Jefferson 1991). The study of such emergent behavior is one of the main themes of research in the field of artificial life (Langton *et al.* 1991; Varela and Bourgine 1992; Meyer and Wilson 1991).

The 'painted desert' problem was proposed by Mitchel Resnick and Uri Wilenski of the MIT Media Laboratory at the *Logo workshop at the Third Artificial Life Conference in Santa Fe in 1992. In the modified, low-dimensionality version of this problem discussed in this chapter, there are 10 ants and 10 grains of sand of each of three colors (black, gray and striped). The 10 ants and 30 grains start at random positions in a 10 by 10 toroidal grid as shown in Figure 17.1.

Each ant executes a common computer program and can only sense information about its immediate local position in the grid. There is no direct communication between the ants. There is no central controller directing the ants.

The objective is to find a common computer program for directing the behavior of the 10 ants which, when executed in parallel by all 10 ants, enables the ants to arrange the 30 grains of sand into three vertical bands of like color. Specifically, the 10 black grains are to be arranged into a vertical band immediately to the right of the *Y*-axis, the 10 gray grains are to be arranged into a second vertical band immediately to the right of the black band, and the 10 striped grains are to be arranged into a third vertical band immediately to the right of the gray band. No more than one

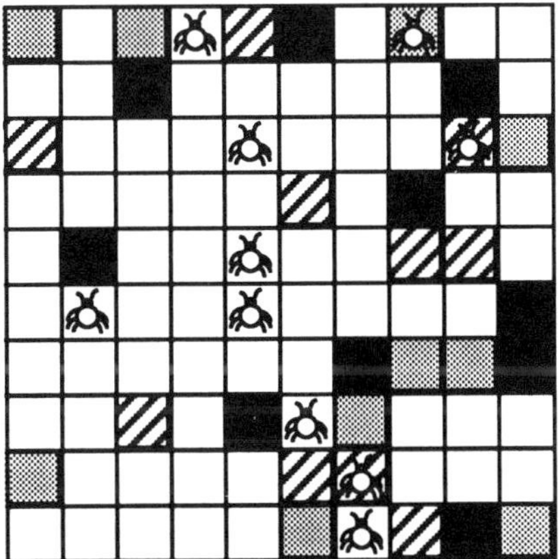

Figure 17.1 Starting arrangement of 10 ants, 10 black grains, 10 gray grains and 10 striped grains of sand.

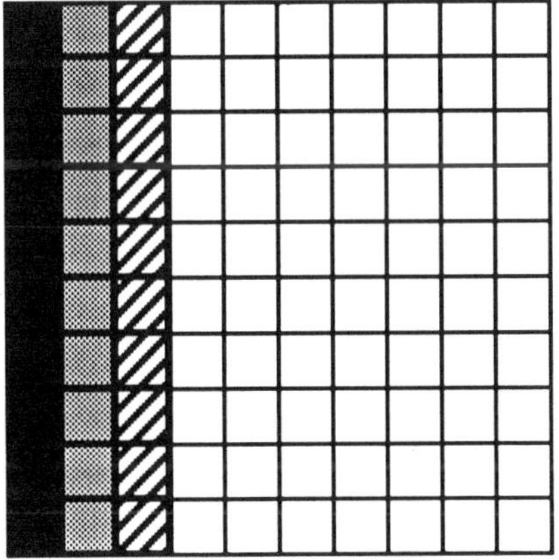

Figure 17.2 Desired final arrangement of the sand.

grain of sand is allowed at any position in the grid at any time.

Figure 17.2 shows the desired final arrangement of the 30 grains of sand.

It has been demonstrated that it is possible for higher level cooperative behavior to emerge via the purely local sensing and the parallel application by each ant of an identical (and relatively simple) program. Deneubourg *et al.* (1986, 1991) conceived and wrote an intriguing program involving only local interactions which, when simultaneously executed by a group of ants, enables the ants to consolidate widely dispersed pellets of food into one pile. Travers and Resnick (1991) conceived and wrote a parallel program and produced a videotape that showed how a group of ants could locate piles of food and transport the food to a certral place. Their central place foraging program permitted the ants to drop slowly dissipating chemicals, called pheromones, to recruit and guide other ants to the food (see also Resnick (1991)). These programs were written using considerable ingenuity and human intelligence.

In this chapter, the goal is to genetically breed a common computer program that, when simultaneously executed by a group of ants, causes the emergence of interesting higher level collective behavior. In particular, the goal is to evolve a common program that solves the painted desert problem as described above.

17.3 BACKGROUND ON GENETIC ALGORITHMS

John Holland's pioneering 1975 *Adaptation in Natural and Artificial Systems* described how the evolutionary process in nature can be applied to artificial systems using the genetic algorithm operating on fixed length character strings (Holland 1975). Holland demonstrated that a population of fixed length character strings (each representing a proposed solution to a problem) can be genetically bred using the Darwinian operation of fitness proportionate reproduction, the genetic operation of recombination, and occasional mutation. The recombination operation combines parts of two chromosome-like fixed length character strings, each selected on the basis of their fitness, to produce new offspring strings. Holland established, among other things, that the genetic algorithm is a mathematically near optimal approach to adaptation in that it maximizes expected overall average payoff when the adaptive process is viewed as a multi-armed slot machine problem requiring an optimal allocation of future trials given currently available information.

The genetic algorithm has proven successful at searching nonlinear multidimensional spaces in order to solve, or approximately solve, a wide variety of problems (Goldberg 1989; Davis 1987; Davis 1991; Michalewicz 1992; Belew and Booker 1991; Schwefel and Männer 1991; Davidor 1991; Whitley 1992; Männer and Manderick 1992).

17.4 BACKGROUND ON GENETIC PROGRAMMING

For many problems the most natural representation for the solution to the problem is a computer program (i.e., a composition of primitive functions and terminals), not merely a single point in a multidimensional numerical search space. Moreover, the exact size and shape of the composition needed to solve many problems is not known in advance.

Genetic programming provides a domain-independent way to search the space of computer programs composed of certain given terminals and primitive functions to find a program of unspecified size and shape which solves, or approximately solves, a problem. The book *Genetic Programming: On the Programming of Computers by Means of Natural Selection* (Koza 1992*a*) describes the genetic programming paradigm and demonstrates that populations of computer programs (i.e., compositions of primitive functions and terminals) can be genetically bred to solve a surprising variety of different problems in a wide variety of different fields. Specifically, genetic programming has been successfully applied to problems such as:

1. planning (e.g., navigating an artificial ant along a trial, developing a robotic action sequence that can stack an arbitrary initial configuration of blocks into a specified order) (Koza 1991);
2. discovering control strategies for backing up a tractor-trailer truck, centering a cart and balancing a broom on a moving cart;
3. discovering inverse kinematic equations to control the movement of a robot arm to a designated target point;
4. evolution of a subsumption architecture for robotic control (Koza 1992*c*; Koza and Rice 1992*b*);
5. emergent behavior (e.g., discovering a computer program which, when executed by all the ants in an ant colony, enables the ants to locate food, pick it up, carry it to the nest and drop pheromones along the way so as to produce cooperative emergent behavior) (Koza 1991);
6. classification and pattern recognition (e.g., distinguishing two intertwined spirals);
7. generation of maximal entropy random numbers;
8. induction of decision trees for classification;
9. optimization problems (e.g., finding an optimal food foraging strategy for a lizard) (Koza, Rice and Roughgarden 1992);
10. symbolic 'data to function' regression, symbolic integration, symbolic differentiation, symbolic solution to general functional equations (including differential equations with initial conditions, and integral equations) and sequence induction (e.g., inducing a recursive computational procedure for generating sequences such as the Fibonacci sequence);
11. empirical discovery (e.g., rediscovering Kepler's third law, rediscovering the well-known nonlinear econometric 'exchange equation' $MV = PQ$ from actual, noisy time series data for the money supply, the velocity of money, the price level, and the gross national product of an economy);
12. Boolean function learning (e.g., learning the Boolean 11-multiplexer function and 11-parity functions);
13. finding minimax strategies for games (e.g., differential pursuer-evader games,

discrete games in extensive form) by both evolution and coevolution;
14. simultaneous architectural design and training of neural networks.

A videotape visualization of 22 applications of genetic programming can be found in the *Genetic Programming: The Movie* (Koza and Rice 1992a). See also Koza (1992b).

In genetic programming, the individuals in the population are compositions of primitive functions and terminals appropriate to the particular problem domain. The set of primitive functions used typically includes arithmetic operations, mathematical functions, conditional logical operations and domain-specific functions. The set of terminals typically includes inputs appropriate to the problem domain and various constants.

The compositions of primitive functions and terminals described above correspond directly to the computer programs found in programming languages such as LISP (where they are called symbolic expressions or S-expressions). In fact, these compositions in turn correspond directly to the parse tree that is internally created by the compilers of most programming languages. Thus, genetic programming views the search for a solution to a problem as a search in the space of all possible functions that can be recursively composed of the available primitive functions and terminals.

17.5 STEPS REQUIRED TO EXECUTE GENETIC PROGRAMMING

Genetic programming is a domain independent method that proceeds by genetically breeding populations of compositions of the primitive functions and terminals (i.e., computer programs) to solve problems by executing the following three steps:

1. Generate an initial population of random computer programs composed of the primitive functions and terminals of the problem.
2. Iteratively perform the following substeps until the termination criterion for the run has been satisfied.
   (a) Execute each program in the population so that a fitness measure indicating how well the program solves the problem can be computed for the program;
   (b) Create a new population of programs by selecting program(s) in the population with a probability based on fitness (i.e., the fitter the program, the more likely it is to be selected) and then applying the following primary operations:
      (i) Reproduction: Copy an existing program to the new population.
      (ii) Crossover: Create two new offspring programs for the new population by generatically recombining randomly chosen parts of two existing programs.
3. The single best computer program produced during the run is designated as the result of the run of genetic programming. This result may be a solution (or approximate solution) to the problem.

17.5.1 CROSSOVER (RECOMBINATION) OPERATION

In genetic programming, the genetic crossover (sexual recombination) operation operates on two parental computer programs and produces two offspring programs using parts of each parent.

For example, consider the following computer program (LISP symbolic expression):

(+ (* 0.234 Z) (− X 0.789)),

which we would ordinarily write as

$$0.234z + x - 0.789.$$

This program takes two inputs (X and Z) and produces a floating point output. In the prefix notation used in LISP, the multiplication function * is first applied to the terminals 0.234 and Z to produce an intermediate

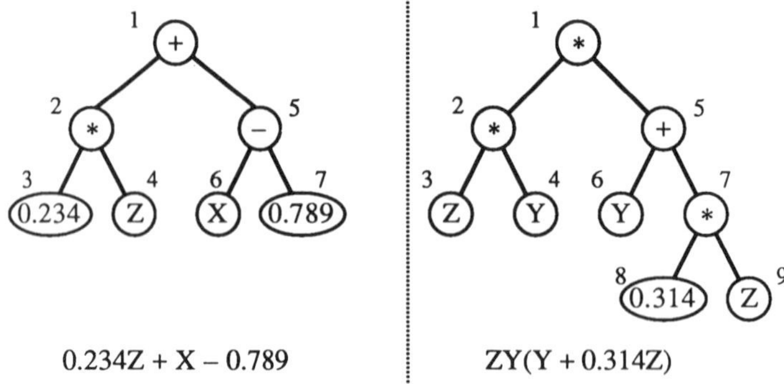

Figure 17.3 Two parental computer programs.

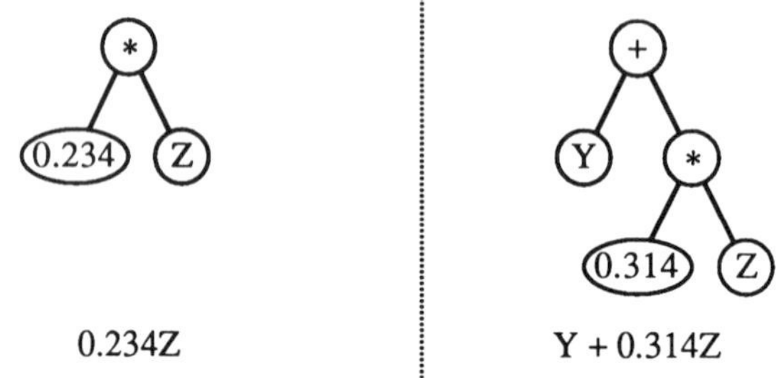

Figure 17.4 Two crossover fragments.

result. Then, the subtraction function − is applied to the terminals X and 0.789 to produce a second intermediate result. Finally, the addition function + is applied to the two intermediate results to produce the overall result.

Also, consider a second program:

(* (* Z Y) (+ Y (* 0.314 Z))),

which we would ordinarily write as

$$zy\,(y + 0.314z).$$

In Figure 17.3, these two parental programs are depicted as rooted, point-labeled trees with ordered branches. Internal points (i.e., nodes) of the tree correspond to functions (i.e., operations) and external points (i.e., leaves, endpoints) correspond to terminals (i.e., input data). The numbers beside the function and terminal points of the tree appear for reference only.

The crossover operation creates new offspring by exchanging subtrees (sublists, subroutines, subprocedures) between the two parents.

Assume that the points of both trees are numbered in a depth-first way starting at the left. Suppose that the point number 2 (out of the seven points of the first parent) is randomly selected as the crossover point for the first parent and that the point number 5 (out of the nine points of the second parent) is randomly selected as the crossover point of the second parent. The crossover points in the trees above are therefore the * in the first parent and the + in the second parent. The two crossover fragments are the two subtrees shown in Figure 17.4.

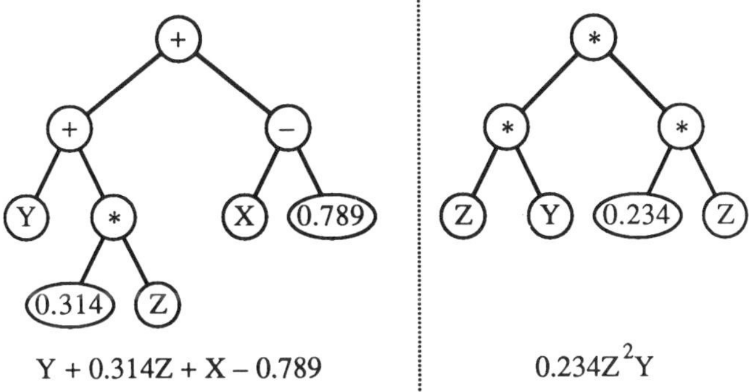

Figure 17.5 Two offspring.

These two crossover fragments correspond to the underlined subprograms (sublists) in the two parental computer programs. The two offspring resulting from crossover are:

(+ (+ Y (* 0.314 Z)) (− X 0.789))

and

(* (* Z Y) (* 0.234 Z)).

The two offspring are shown in Figure 17.5.

Thus, crossover creates new computer programs from parts of existing parental programs. Because entire subtrees are swapped if the set of functions and terminals are closed, this crossover operation always produces syntactically and semantically valid programs as offspring regardless of the choice of the two crossover points. Because programs are selected to participate in the crossover operation with a probability proportional to their fitness, crossover allocates future trials to parts of the search space whose programs contain parts from promising programs.

17.6 PREPARATORY STEPS FOR USING GENETIC PROGRAMMING

There are five major steps in preparing to use genetic programming, namely, determining:

1. the set of terminals;
2. the set of primitive functions;
3. the fitness measure;
4. the parameters for controlling the run; and
5. the method for designating a result and the criterion for terminating a run.

The individual programs in the population are compositions of the primitive functions from the function set and terminals from the terminal set. Since the set of ingredients from which the computer programs are composed must be sufficient to solve the problem, it seems reasonable that the programs would include (1) the inputs to the common computer program being evolved to govern the behavior of each ant; (2) functions enabling the ant to move, to pick up a grain of sand, to drop a grain of sand; and (3) functions enabling the ant to make decisions based on the limited information that it receives as input.

The first major step in preparing to use genetic programming is the identification of the set of terminals. In this problem, the inputs to each ant consist only of information concerning the ant's current location in the grid. This information consists of:

1. X, an indication of the ant's current horizontal location (from 0 to 9) in the grid;
2. Y, an indication of the ant's current vertical location (from 0 to 9) in the grid;
3. CARRYING, an indication of the color of the grain of sand (0 for black, 1 for gray

or 2 for striped) that the ant is currently carrying or −1 otherwise; and
4. COLOR, an indication of the color of the grain of sand, if any, on the square of the grid where the ant is currently located or −1 if there is no grain at that square.

The side-effecting functions GO-N, GO-E, GO-S and GO-W enable the ant to move one square north, east, south and west, respectively. The side-effecting function MOVE-RANDOM randomly selects one of the above four functions and executes it. The side-effecting function PICK-UP picks up the grain of sand (if any) at the ant's current position provided the ant is not already carrying a grain of sand. These side-effecting functions return the value (1, 2 or 3) of the color of sand at the ant's current position. Since these six side-effecting functions take no arguments, they are best viewed as terminals. That is, the terminal set represents the endpoints of the program tree.

Thus, the terminal set $\mathcal{T}$ for this problem is:

$\mathcal{T}$ = {X, Y, CARRYING, COLOR,
(GO-N), (GO-E), (GO-S), (GO-W),
(MOVE-RANDOM), (PICK-UP)}.

The second major step in preparing to use genetic programming is the identification of a function set for the problem.

The four-argument conditional branching function IFLTE evaluates and returns its third (then) argument if its first argument is less than or equal to its second argument and otherwise evaluates and returns its fourth (else) argument. For example, (IFLTE 2 3 A B) evaluates to the value of A. The three-argument conditional branching function IFLTZ evaluates and returns its second (then) argument if its first argument is less than zero and otherwise evaluates and returns its third (else) argument. If the ant is currently carrying a grain of sand and if the ant's current position does not have a grain of sand, the two-argument conditional branching function IF-DROP drops the sand that the ant is carrying and evaluates and returns its first (then) argument, otherwise IF-DROP evaluates and returns its second (else) argument. Testing functions such as these are commonly used in genetic programming and are implemented as macros.

Thus, the function set $\mathcal{F}$ for this problem consists of:

$\mathcal{F}$ = {IFLTE, IFLTZ, IF-DROP}

taking 4, 3 and 2 arguments, respectively.

Each computer program in the population is a composition of primitive functions from the function set $\mathcal{F}$ and terminals from the terminal set $\mathcal{T}$. Since genetic programming operates on an initial population of randomly generated compositions of the available functions and terminals and later performs genetic operations, such as crossover, on these individuals, each primitive function in the function set should be well defined for any combination of arguments from the range of values returned by every primitive function that it may encounter and the value of every terminal that it may encounter. The above choices of terminals and functions (specifically, the use of numerically valued logic and the return of a numerical value for the side-effecting functions) ensures this desired closure.

The third major step in preparing to use genetic programming is the identification of the fitness measure that evaluates the goodness of an individual program in the population. The fitness of a program in the population is measured by executing a simulation for 300 time steps with the 10 ants and the 30 grains of sand starting from the randomly selected positions shown in Figure 17.1.

The fitness of an individual program is the sum, over the 30 grains of sand, of the product of the color of the grain of sand (1, 2 or 3) and the distance between the grain and the *Y*-axis when execution of the particular program ceases. If all 30 grains of sand are located in their proper vertical band, the fitness will be 100. This fitness measure is higher and less favorable when the grains of sand are less perfectly positioned. If a program leaves all the grains of sand at their starting positions shown in Figure 17.1, it will score a fitness of 395.

This work was done on a serial computer, so the common computer program is, in practice, executed for each ant in sequence. Thus, the locations of the grains of sand are altered by the side-effecting actions of earlier ants in the sequence. In addition, certain special situations must also be considered in implementing this particular problem. To conserve computer time, an individual computer program is never allowed to execute more than 1000 side-effecting operations over the 300 occasions that each individual program is allowed to be executed for each ant. If this number is exceeded, further execution of that particular program ceases and fitness is then measured at that moment. If an ant happens to be carrying a grain of sand when fitness is being measured and the ant's current position does not have a grain of sand, the grain is deemed to be located at that position. If the ant's current position is occupied by a grain of sand, but the place from which it picked up its grain is unoccupied by a grain, the grain is deemed to be located at that place (i.e., it is rubber-banded there). If both these positions are occupied by a grain, the grain is deemed to be located at the first unoccupied position near the ant's current position (these positions being considered in a particular orderly way).

The fourth major step in preparing to use genetic programming is the selection of values for certain parameters. Our choice of values for the primary parameters (i.e., 500 for the population size, *M*, and 51 as the maximum number of generations, *G* to be run) reflects our judgment as to the likely complexity of the solution to this problem. These same two choices were made for the vast majority of problems presented in Koza (1992*a*). Our choice of values for the various secondary parameters that control the run of genetic programming are, with one expection, the same default values as we have used on numerous other problems (Koza 1992*a*). The exception is that we used tournament selection (with a group size of seven), rather than fitness proportionate selection. In tournament selection, a group of a specified size (seven here) is selected at random from the population and the single individual with the best value of fitness is selected from the group. For the crossover operation, two such selections are made to select the two parents to participate in crossover.

Finally, the fifth major step in preparing to use genetic programming is the selection of the criterion for terminating a run and the selection of the method for designating a result. We will always terminate a given run after 51 generations. We count each grain of sand that is in its proper vertical band as a 'hit'. We also terminate a run if 30 hits are attained at any time step. We designate the best individual obtained during the run (the best-so-far individual) as the result of the run.

17.7 RESULTS FOR ONE RUN

A review of one particular run will serve to illustrate how genetic programming discovers progressively better computer programs to solve the painted desert problem.

One would not expect any individual from the randomly generated initial population to be very good. In generation 0, the median individual in this population of 500 consists of four points (i.e., functions and terminals), and is:

```
(IFLTZ Y (GO-S) X).
```

Depending on the value of the first argument, Y, to the conditional branching function IFLTZ, this program either moves an ant to the south or pointlessly evaluates the third argument, X. This program does not move any grains of sand and therefore scores a

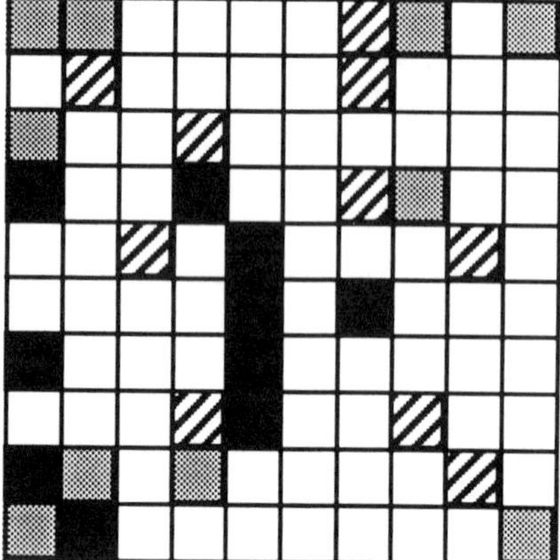

Figure 17.6 Sand arrangement after execution of the best-of-generation individual from generation 0.

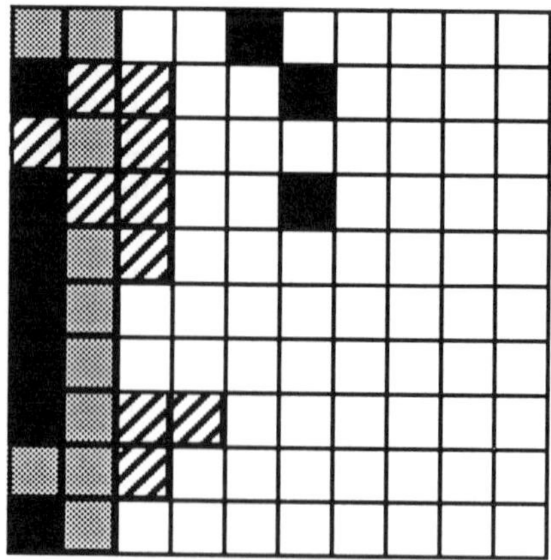

Figure 17.7 Result of executing the best-of-generation individual from generation 4.

fitness of 395. This median level of performance for generation 0 represents a baseline value for a random search of the space of possible compositions of terminals and primitive functions for this problem. Of the 500 programs in generation 0, 64% score 395.

The fitness of the worst individual program among the 500 individuals in the population is 436. This worst-of-generation program consists of eight points and actually moves some grains of sand to less desirable positions than their random starting positions.

However, even in a randomly created population, some individuals are better than others. The fitness of the best-of-generation individual program from generation 0 is 262. This program has 494 points and is not shown here.

Figure 17.6 shows the arrangement of the grains of sand after execution of the best-of-generation individual from generation 0. As can be seen, the three colors of sand are not arranged in anything like the desired final configuration. Nonetheless, 90% of the black grains, 60% of the gray grains, and 40% of the striped grains have been moved onto the left half of the grid. This compares favorably to the starting arrangement shown in Figure 17.1 where less than half of each color happened to start on the left half of the grid.

In succeeding generations, the fitness of the median individual and the best-of-generation individual all tend to progressively improve (i.e., drop). In addition, the average fitness of the population as a whole tends to improve. For example, the best-of-generation individual from generation 1 has a fitness value of 167 and 11 points. The best-of-generation individual from generation 2 is the same individual as generation 1. The best-of-generation individual from generation 3 has a fitness value of 151 and 11 points. Of course, the vast majority of individual computer programs in the population of 500 still perform poorly.

By generation 4, the fitness of the best-of-generation individual improves to 135. This individual has 17 poins and is shown below:

```
(IFLTE (GO-W) (IF-DROP (GO-N) (LFLTE
(MOVE-RANDOM) COLOR Y COLOR))
(IFLTE X COLOR COLOR (IF-DROP
CARRYING (PICK-UP))) (MOVE-
RANDOM)).
```

Figure 17.7 shows the result of executing the best-of-generation individual from generation 4. As can be seen, 70% of the black

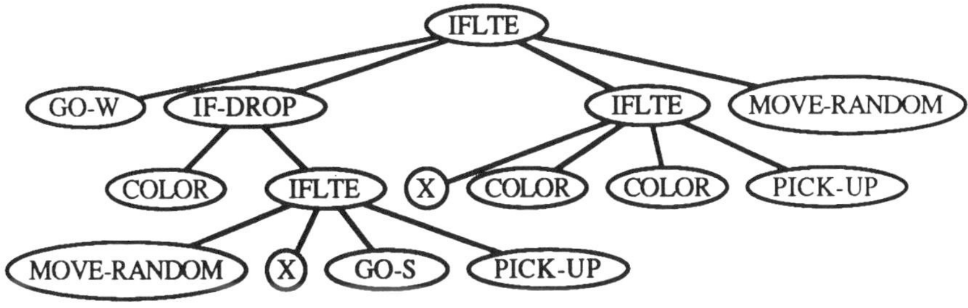

Figure 17.8 Best-of-run individual from generation 8.

grains are in their proper vertical band immediately adjacent to the *Y*-axis. Eighty per cent of the gray grains are in their proper vertical band. Sixty per cent of the striped grains are in their proper band. This represents a considerable improvement over earlier generations.

The fitness of the best-of-generation individual drops to 127 for generation 5, to 112 for generation 6, and then rises to 115 for generation 7.

By generation 8, the best-of-generation individual has 15 points and a fitness value of 100 (i.e., it scores 30 hits and is a perfect solution to this problem). This best-of-run individual is shown below:

```
(IFLTE (GO-W) (IF-DROP COLOR (IFLTE
(MOVE-RANDOM) X (GO-S) (PICK-UP)))
(IFLTE X COLOR COLOR (PICK-UP))
(MOVE-RANDOM)).
```

Figure 17.8 shows this best-of-run individual from generation 9 as a rooted, point-labeled tree with ordered branches.

As one watches an animated simulation of the execution of the above best-of-run program by all 10 ants, the GO-W and the GO-S functions cause the ants and sand to move in a generally south-westerly direction. The best-of-run program works by first moving to the left (west) and unconditionally dropping its sand if the square is free. If the grain of sand is to the right of its correct vertical column (as it usually will be), the ant then picks the sand up again. If the sand cannot be dropped at the ant's current location and the sand is in its correct vertical column or to the right of its correct vertical column, the ant moves south and again tries to drop the sand in a different position in the same vertical column. If the ant ever carries the sand to the left of its correct vertical column, the toroidal geometry of the grid returns the ant to the far right.

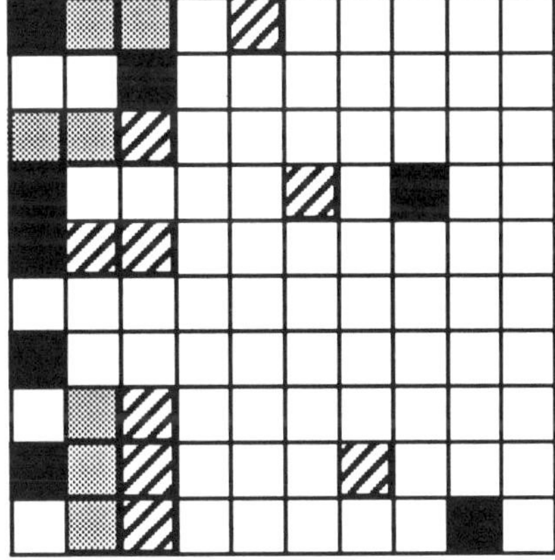

Figure 17.9 Intermediate result of executing the best-of-run individual from generation 8 after 30 time steps.

Figure 17.9 shows the result of executing the best-of-run individual from generation 8 after only 30 time steps. As can be seen, 25 of the 30 grains of sand are already located in the leftmost three vertical columns of the grid. Moreover, 50% of each color are already located in their correct vertical column.

This particular 100% correct program operates by filling each of the three vertical columns at the far left with grains of the correct color at approximately the same rate.

Note that we did not pre-specify the size and shape of the result. As we proceeded from generation to generation, the size and shape of the best-of-generation individuals changed as a result of the selective pressure exerted by Darwinian natural selection and crossover.

17.7.1 GENEALOGICAL AUDIT TRAIL

A genealogical audit trail can provide further insight into how genetic programming works. An audit trail consists of a record of the ancestors of a given individual and of each genetic operation that was performed on the ancestors in order to produce the individual. For the crossover operation, the audit trail includes the particular crossover points chosen within each parent.

The creative role of crossover is illustrated by an examination of the genealogical audit trail for the best-of-generation individual from generations 7 and 8 of this run.

Parent 1 is the 51st best individual from generation 7. It has a fitness value of 139, has 15 points, and scores 24 hits, and is shown below:

```
(IFLTE (GO-W)
       (IF-DROP COLOR (IFLTE (MOVE-RANDOM) J (GO-S) (GO-E)))
       (IFLTE J COLOR COLOR (PICK-UP)) (MOVE-RANDOM))
```

Parent 2 is the 27th best individual from generation 7. It has a fitness value of 131, has 31 points, and scores 20 hits, and is shown below:

```
(IFLTE (GO-W)
       (IF-DROP (GO-N)
                (IFLTE (MOVE-RANDOM)
                J
                (GO-S)
                (IFLTE (GO-W)
                       (IF-DROP (GO-N) (IFLTE (MOVE-RANDOM) J (GO-S) (GO-E)))
                       (IFLTE J COLOR COLOR (PICK-UP))
                       (MOVE-RANDOM))))
(IFLTE J COLOR COLOR (PICK-UP))
(IF-DROP (GO-N) COLOR))
```

Parent 1 is not 100% effective in solving this problem because it has no way to deal with a grain of sand that is located to the left (west) of its correct vertical column. This defect is relatively minor since it only involves the relatively small number of grains of sand that might be improperly located in the first and second vertical columns. The insertion of the PICK-UP function into parent 1 from generation 7, in lieu of its underlined GO-E function, corrects the defect in parent 1 by enabling the ant to pick up such a grain. The general westward movement of sand and the toroidal geometry of the grid then causes this grain to reappear at the far right so that it can eventually be located in its correct vertical column.

As it happens in this run, parent 1 from generation 7 and its grandparent 1 from generation 6 are identical.

Great-grandparent 1 from generation 5 is the 23rd best individual from its generation, has a fitness value of 157, has 15 points, and scores 17 hits, and is shown below:

```
(IFLTE (GO-W)
      (IF-DROP (GO-E) (IFLTE (MOVE-RAN) J (GO-S) (GO-E)))
  (IFLTE J COLOR COLOR (PICK-UP))
  (MOVE-RAN))
```

Great-grandparent 2 from generation 5 is the 42nd best individual from its generation, has a fitness value of 164, has 18 points, scores 9 hits, and is shown below:

```
(IFLTE (IFLTZ (GO-W) (MOVE-RAN) (GO-E)
      (IF-DROP (GO-N) (IFLTE (MOVE-RAN) COLOR I COLOR))
  (IFLTE J COLOR COLOR (PICK-UP))
  (MOVE-RAN))
```

Great-grandparent 1 from generation 5 is less effective than it might be, in part because it does not check the color of the sand it is currently carrying and because it moves back towards the east. The insertion of the COLOR function into great-grandparent 1 from generation 5, in lieu of its underlined GO-E function, enables the ant to check the color of the sand it is currently carrying and thereby improve its fitness from 157 to 139 and to increase its number of hits from 17 to 24.

Great-great-grandparent 1 from generation 4 of great-grandparent 1 from generation 5 has a GO-N function in lieu of the GO-E function found in great-grandparent 1 from generation 5. This GO-N function in generation 4 was actually superior to the GO-E function found in its offspring in generation 5 because it prevented the counter-productive and regressive movement back to the east. Consequently, great-great-grandparent 1 from generation 4 scored 20 hits, whereas its offspring in generation 5 scored 17 hits. This highlights the fact that genetic algorithms are not mere hill-climbers. Crossover most typically recombines parts of individuals that are not the current optimum to produce an improved offspring in the next generation.

17.7.2 FITNESS CURVE SHOWING PROGRESSIVE LEARNING

Figure 17.10 shows that, as we proceed from generation to generation, the fitness of the

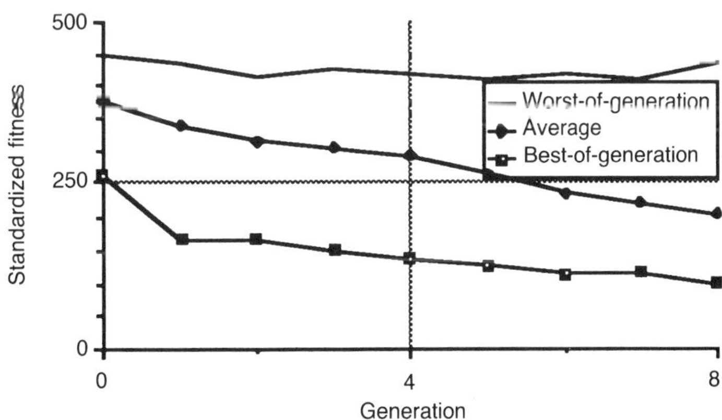

Figure 17.10 Fitness curves.

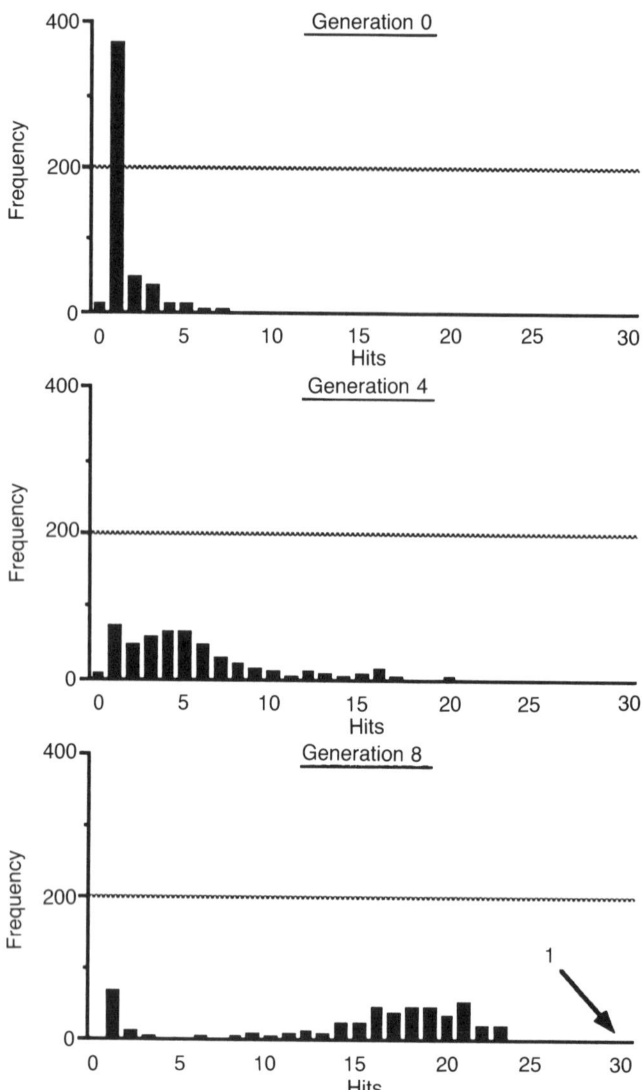

Figure 17.11 Hits histogram for generations 0, 4 and 8.

best-of-generation individual and the average fitness of the population as a whole tends to progressively improve (i.e., drop).

17.7.3 HITS HISTOGRAM SHOWING PROGRESSIVE LEARNING

The hits histogram is a useful monitoring tool for visualizing the progressive learning of the population as a whole during a run. The horizontal axis of the hits histogram represents the number of hits (0 to 30) while the vertical axis represents the number of individuals in the population (0 to 500) scoring that number of hits.

Figure 17.11 shows the hits histograms for generations 0, 4 and 8 of this particular run. Note in the progression the left-to-right undulating movement of both the high point and the center of mass of these three histograms. This 'slinky' movement reflects the improvement of the population as a whole.

17.7.4 RETESTING OF BEST-OF-RUN INDIVIDUAL

As it happens, the best-of-run individual from generation 8 from the particular run described above was sufficiently parsimonious and understandable that we were able to satisfy ourselves that it indeed is a solution to the general painted desert problem. However, programs that score 100% are generally not as parsimonious or as understandable as the above best-of-run individual. We envision a general painted desert problem and we think of the randomly selected starting arrangement for the ants and sand in Figure 17.1 as one instance of the possible starting positions for this general problem. However, there is no guarantee that a computer program that is evolved while being exposed to only one random starting arrangement of ants and sand (or any incomplete sampling of starting arrangements) will necessarily generalize so as to correctly handle all other possible starting arrangements. Genetic programming knows nothing about the more general problem or about the larger class of potential starting positions. There is a possibility that the best-of-run individual from generation 8 described above is idiosyncratically adapted (i.e., overfitted) to the one random starting arrangement (i.e., fitness case) shown in Figure 17.1. We can acquire evidence as to whether the solution evolved by genetic programming generalizes to the more general problem that we envision by retesting the above best-of-run individual scoring 100% using different random starting positions. When we retested this individual on five different random starting positions, we found that it correctly solved the general painted desert problem.

17.8 RESULTS OVER A SERIES OF RUNS

In the foregoing section, we illustrated genetic programming by describing a particular successful run. In this section, we discuss the performance of genetic programming over a series of 32 runs. The solutions evolved in different runs exhibit very different behavior from the solution previously described. Some solutions, for example, tend to move the ants and sand in north-westerly, south-easterly, or north-easterly directions while others exhibit no directional bias. Other solutions completely fill the third vertical column with 10 grains of striped sand before placing much sand in the first and second columns. Moreover, all of the other solutions are considerably less parsimonious and less easily understood than the particular run described above. The number of points in the nine solutions found in the 32 runs averaged 92.6 points, as compared to the 15 points found in the unusually parsimonious solution discussed in detail in the previous section. The crossover operations in those other runs almost always recombined larger subtrees, rather than individual terminals.

17.8.1 PERFORMANCE CURVES

Figure 17.12 presents two curves, called the performance curves, for the painted desert problem over a series of 32 runs. The curves are based on a population size, M, of 500 and a maximum number of generations to be run, G, of 51.

The rising curve in this figure shows, by generation, the experimentally observed cumulative probability of success, $P(M, i)$, of yielding an acceptable result for this problem by generation i. We define an acceptable result for this problem to be at least one S-expression in the population which scores 30 hits (this being equivalent to scoring a value of fitness of 100). As can be seen, the experimentally observed value of the cumulative probability of success, $P(M, i)$, is 25% by

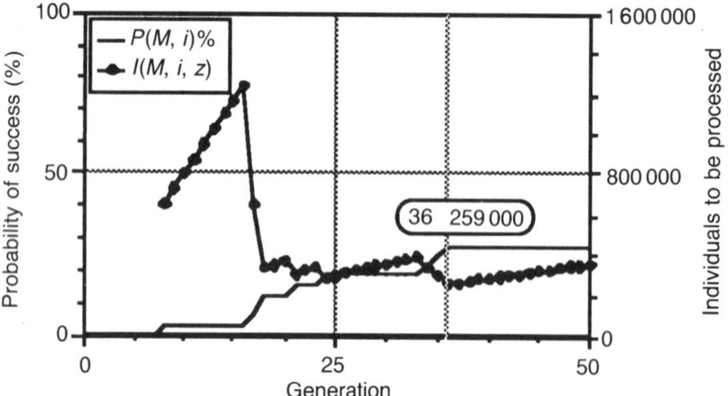

Figure 17.12 Performance curves showing that it is sufficient to process 259 000 individuals to yield a solution with 99% probability.

generation 36 and 28% by generation 50 over the 32 runs.

The second curve (which first falls and then rises) in this figure shows, by generation, the number of individuals that must be processed, $I(M, i, z)$, to yield, with probability z, a solution to the problem by generation i. $I(M, i, z)$ is derived from the experimentally observed values of $P(M, i)$. Specifically, $I(M, i, z)$ is the product of the population size M, the generation number i, and the number of independent runs, $R(z)$, necessary to yield a solution to the problem with probability z by generation i. In turn, the number of runs $R(z)$ is given by

$$R(z) = \left\lceil \frac{\log(1 - z)}{\log(1 - P(M, i))} \right\rceil,$$

where the brackets indicate the ceiling function for rounding up to the next highest integer. The probability z is 99% herein.

As can be seen, the $I(M, i, z)$ curve reaches a minimum value at generation 36 (highlighted by the light dotted vertical line). For a value of $P(M, i)$ of 25%, the number of independent runs $R(z)$ necessary to yield a solution to the problem with a 99% probability by generation i is 14. The two summary numbers (i.e., 36 and 259 000) in the oval in this figure indicate that if this problem is run through to generation 36 (the initial random generation being counted as generation 0), processing a total of 259 000 individuals (i.e., 500 × 37 generations × 14 runs) is sufficient to yield a solution to this problem with 99% probability. This number, 259 000, is a measure of the computational effort necessary to yield a solution to this problem with 99% probability.

17.8.2 FITNESS CURVES OVER A SERIES OF RUNS

We have already seen, for the single run of this problem presented above, that the fitness of the best-of-generation individual and the average fitness of the population tended to progressively improve via small, evolutionary improvements from generation to generation. The same is true when fitness is tracked over a series of runs.

Figure 17.13 shows the incremental learning behavior of genetic programming by depicting, by generation, the average, over the 32 runs, of the fitness of the best-of-generation individual and the average of the average fitness of the population as a whole (the 'average average' fitness).

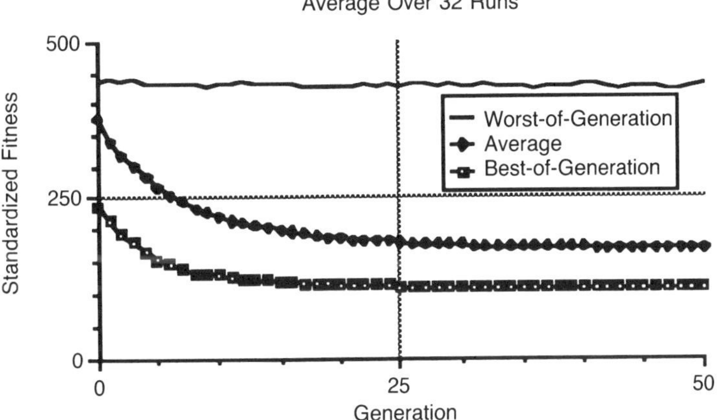

Figure 17.13 Fitness curves over a series of 32 runs.

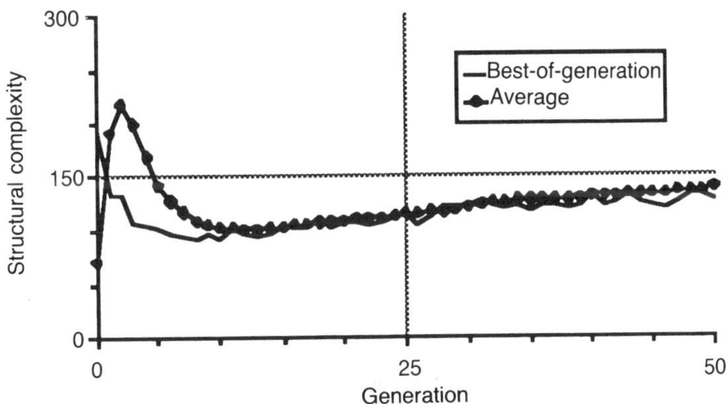

Figure 17.14 Structural complexity curves over a series of 32 runs.

17.8.3 STRUCTURAL COMPLEXITY CURVES OVER A SERIES OF RUNS

Figure 17.14 shows, by generation, the average, over the 32 runs, of the structural complexity of the best-of-generation individual and the average of the average structural complexity of the population as a whole (the 'average average' structural complexity). We can see that as fitness improves over the generations, the structural complexity changes.

17.8.4 VARIETY CURVE OVER A SERIES OF RUNS

Figure 17.15 shows, by generation, the average, over the 32 runs, of the variety of the population (i.e., the percentage of unique individuals). As can be seen, average variety starts at 100% for generation (because we specifically remove duplicates from the initial random population) and remains near 100% over the generations with genetic programming (i.e., there is no loss of genetic diversity or premature convergence).

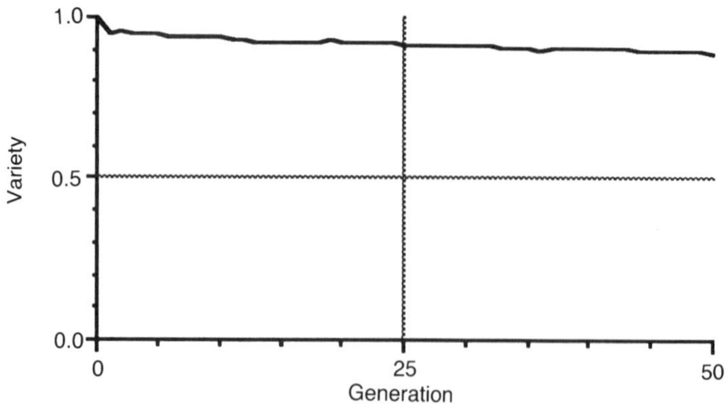

Figure 17.15 Variety curve over a series of 32 runs.

17.9 CONCLUSIONS

We have shown that a computer program to solve a version of the painted desert problem can be evolved using genetic programming. The computer program arises from the selective pressure exerted by a Darwinian reproduction operation, a genetic crossover (recombination) operation, and a fitness measure appropriate to the problem.

Past research has indicated that surprisingly complex behavior can emerge from the repetitive application of seemingly simple, locally applied rules. We have shown here that complex overall behavior, such as the behavior required to solve the painted desert problem, can be evolved using a domain-independent genetic method based on the Darwinian principle of survival of the fittest and genetic operations. The fact that such emergent behavior can be evolved for this problem is suggestive of the fact that a wide variety of other complex behaviors can be evolved.

ACKNOWLEDGEMENTS

James P. Rice of the Knowledge Systems Laboratory at Standford University made numerous contributions in connection with the computer programming of the above. Simon Handley of the Computer Science Department at Stanford University made helpful comments on this chapter.

REFERENCES

Belew, R. and Booker, L. (eds.) (1991). *Proceedings of the Fourth International Conference on Genetic Algorithms*. Morgan Kaufmann, San Mateo, CA.

Collins, R. and Jefferson, D. (1991). Ant Farm: Toward simulated evolution. In *Artificial Life II*, Vol. 10 of *SFI Studies in the Sciences of Complexity* (ed. C. Langton, C. Taylor, J. D. Farmer and S. Rasmussen). Addison-Wesley, Redwood City, CA, pp. 579–601.

Davidor, Y. (1991). *Genetic Algorithms and Robotics*. World Scientific, Singapore.

Davis, L. (ed.) (1987). *Genetic Algorithms and Simulated Annealing*. Pitman, London.

Davis, L. (1991). *Handbook of Genetic Algorithms*. Van Nostrand Reinhold, New York.

Deneubourg, J. L., Aron, S., Goss, S., Pasteels, J. M. and Duerinck, G. (1986). Random behavior, amplification processes and number of participants: How they contribute to the foraging properties of ants. In *Evolution, Games, and Learning* (ed. D. Farmer, A. Lapedes, N. Packard and B. Wendroff). North-Holland, Amsterdam, pp. 176–86.

Deneubourg, J. L., Goss, S., Franks, N., Sendova-Franks, A., Detrain, C. and Chretien, L. (1991). The dynamics of collective sorting robot-like ants and ant-like robots. In *From Animals to Animats: Proceedings of the First International Conference on Simulation of Adaptive Behavior* (ed. J.-A.

Meyer and S.W. Wilson). MIT Press, Cambridge, MA, pp. 356–63.

Forrest, S. (ed.) (1991). *Emergent Computation: Self-Organizing, Collective, and Cooperative Computing Networks*. MIT Press, Cambridge, MA.

Goldberg, D. E. (1989). *Genetic Algorithms in Search, Optimization, and Machine Learning*. Addison-Wesley, Reading, MA.

Holland, J. H. (1975). *Adaptation in Natural and Artificial Systems*. University of Michigan Press, Ann Arbot, MI.

Koza, J. R. (1991). Genetic evolution and co-evolution of computer programs. In *Artificial Life II*, Vol. 10 of *SFI Studies in the Sciences of Complexity* (ed. C. Langton, C. Taylor, J. D. Farmer and S. Rasmussen). Addison-Wesley, Redwood City, CA, pp. 603–29.

Koza, J. R. (1992a). *Genetic Programming: On the Programming of Computers by Means of Natural Selection*. MIT Press, Cambridge, MA.

Koza, J. R. (1992b). The genetic programming paradigm: Genetically breeding populations of computer programs to solve problems. In *Dynamic, Genetic, and Chaotic Programming* (ed. B. Soucek and the IRIS Group). John Wiley, New York.

Koza, J. R. (1992c). Evolution of subsumption using genetic programming. In *Toward a Practice of Autonomous Systems: Proceedings of the First European Conference on Artificial Life* (ed. F. J. Varela and P. Bourgine). MIT Press, Cambridge, MA, pp. 110–19.

Koza, J. R. and Rice, J. P. (1992a). *Genetic Programming: The Movie*. MIT Press, Cambridge, MA.

Koza, J. R. and Rice, J. P. (1992b). Automatic programming of robots using genetic programming. In *Proceedings of Tenth National Conference on Artificial Intelligence*, Menlo Park, CA. AAAI Press/MIT Press, Cambridge, MA, pp. 194–201

Koza, J. R., Rice, J. P. and Roughgarden, J. (1992). Evolution of food foraging strategies for the Caribbean *Anolis* lizard using genetic programming. *Simulation of Adaptive Behavior* **1**(2), 47–74

Langton, C., Taylor, C., Farmer, J. D. and Rasmussen, S. (eds.) (1991). *Artificial Life II*, Vol. 10 of *SFI Studies in the Sciences of Complexity*. Addison-Wesley, Redwood City, CA.

Manderick, B. and Moyson, F. (1988). The collective behavior of ants: An example of self-organization in massive parallelism. In *Proceedings of the AAAI Spring Symposium on Parallel Models of Intelligence*. American Association of Artificial Intelligence, Stanford, CA.

Männer, R. and Manderick, B. (1992). *Proceedings of the Second International Conference on Parallel Problem Solving from Nature*. North-Holland, Amsterdam.

Meyer, J.-A. and Wilson, S. W. (1991). *From Animals to Animats: Proceedings of the First International Conference on Simulation of Adaptive Behavior*. MIT Press, Cambridge, MA.

Michalewicz, Z. (1992). *Genetic Algorithms + Data Structures = Evolution Programs*. Springer-Verlag, Berlin.

Moyson, F. and Manderick, B. (1990). Self-organization versus programming in massively parallel systems: A case study. In *Proceedings of the International Conference on Parallel Processing in Neural Systems and Computers* (ICNC-90) (ed. R. Eckmiller).

Resnick, M. (1991). Animal simulations with *Logo: Massive parallelism for the masses. In *From Animals: Proceedings of the First International Conference on Simulation of Adaptive Behavior* (ed. J.-A. Arcady and S. W. Wilson). MIT Press, Cambridge, MA, pp. 534–9.

Schwefel, H.-P. and Männer, R. (eds.) (1991). *Parallel Problem Solving from Nature*. Springer-Verlag, Berlin.

Steels, L. (1990). Cooperation between distributed agents using self-organization. In *Decentralized AI* (ed. Y. Demazeau and J.-P. Muller). North-Holland, Amsterdam.

Steels, L. (1991). Toward a theory of emergent functionality. In *From Animals to Animats: Proceedings of the First International Conference on Simulation of Adaptive Behavior* (ed. J.-A. Meyer and S. W. Wilson). MIT Press, Cambridge, MA, pp. 451–61.

Travers, M. and Resnick, M. (1991). Behavioral dynamics of an ant colony: Views from three levels. In *Artificial Life II Video Proceedings* (ed. C. G. Langton). Addison-Wesley, Reading, MA.

Varela, F. J. and Bourgine, P. (eds.) (1992). *Toward a Practice of Autonomous Systems: Proceedings of the First European Conference on Artificial Life*. MIT Press, Cambridge, MA.

Whitley, D. (ed.) (1992) *Proceedings of Workshop on the Foundations of Genetic Algorithms and Classifier Systems*, Vail, Colorado. Morgan Kaufmann, San Mateo, CA.

AN EVOLUTIONARY APPROACH TO DESIGNING NEURAL NETWORKS 18

Aviv Bergman

18.1 OBJECTIVES

One of the most interesting properties of neural networks is their ability to learn appropriate behavior by being trained on examples. Established learning algorithms, which typically work by minimizing error through backpropagation in weight space, tend to get stuck in local optima – a tendency typical of gradient-descent methods applied to nonconvex objective functions. Therefore, for problems of nontrivial complexity these systems must be handcrafted to a significant degree, but the distributed nature of neural network representations make this handcrafting difficult.

My goal is to develop learning and adaptation mechanisms capable of coping with complex and dynamic problem domains. Once we obtain a machine that performs a certain task well, we want to understand *why* its structure leads to good performance, and thereby help a network designer to create even more successful designs. More specifically, the aim of this program is to design a system that can learn to recognize signals adaptively. That is, the system should learn to respond in a distinctive, repeatable way to those signals to which it has been exposed; should track changes to its signal environment (including possibly the introduction of entirely new classes of signals); and should do these things spontaneously, with no instruction. Adaptive signal recognition should be the result of a self-reorganization of the system in the face of a changing environment.

18.2 APPROACH

We are investigating an evolutionary approach to learning with two levels of representation: genotypic and phenotypic. In this approach a genotype is a highly structured encoding of a class of neural networks, which play the role of phenotypes. The genotype specifies general properties of the networks, such as initial patterns of connectivity, distributions of weights, thresholds or gains, and so on. A phenotypic network can then be further modified to respond appropriately to experienced stimuli (in particular, to classify stimuli).

A fundamental hypothesis motivating this approach is that the principles of biological evolution and population genetics provide the basis for such behavior. The processes of variation and selection, operating at both levels of representation, are known to produce in natural populations the kind of emergent behavior we seek to emulate. By simulating these processes on the computer, we observe similar kinds of behavior in artificial systems.

Genotypic variation is caused by random mutation and recombination of the network descriptions; genotypic selection is caused by differential reproduction governed by the performance of networks as measured by an explicit or implicit fitness function. These processes operate over a comparatively long time scale and produce networks with comparatively general adaptations.

My purpose is not to model biological processes explicitly, but rather to explore a genetic and ecological metaphor of computation. I am interested in investigating this metaphor for two reasons. First, adaptive behavior may lead to very general methods of dealing with difficult and ill-defined problems in signal understanding. A system that can learn from experience without explicit training by examples, that can exploit contextual information, and that can modify itself to adapt to possibly radical changes in its input could be useful for difficult problems such as speaker-independent speech recognition. In addition, the inherent parallelism of the evolutionary metaphor, with its emphasis on populations, can lead to effective methods for exploiting the power of parallel computer systems.

18.3 DEFINITION OF THE PROBLEM

Suppose that we have a system, for the time being regarded as a 'black box', that receives as input a signal vector of length n, $x = \langle x_0, \ldots, x_{n-1} \rangle$. These signals could be, for example, speech waveforms. The components of x are real numbers within some limited dynamic range. In practice, since any measurement of a real signal will be uncertain to some degree, we can represent the signal vector with nonnegative integers to some precision b bits. Each possible signal is a point in the n-dimensional metric signal space.

Now suppose the system is stimulated only by a much smaller, structured ensemble of signals generated by a few unknown, relatively low-dimensional physical processes, possibly corrupted by noise. They are called sources. They could be, for example, a few speakers of English. There may be considerable variation within a single source, so we should imagine a source to be represented by a subset of the signal space: its attractor. The task of the system is to respond distinctively to each source. From looking at a macroscopic feature of the system, we should be able to tell when it has been presented with a source and which source it is.

In the simplified problem I restrict the components of the input vector to binary values ($b = 1$) and restrict the sources to single values (point attractors). Under these assumptions, the system will be learning a subset of the numbers $\{0, \ldots, 2^n - 1\}$. The signal vector can be visualized as the corners of an n-dimensional hypercube, and the response of the system will be to select one of these corners.

18.4 ENCODER POPULATIONS

Each subsystem is an instantiation of a simple neural network called an encoder (Ackley *et al.* 1985; Rumelhart *et al.* 1986) as shown in Figure 18.1. An n_1-n_2-n_3 encoder has n_1 inputs that feed into n_2 hidden units, which in turn feed into n_3 output units. Each unit computes a weighted sum of the inputs and compares the result with a threshold. If the sum exceeds the threshold, the unit is activated and outputs a one; otherwise, it produces a zero.

Originally, these networks were used to attack the encoding problem (Rumelhart *et al.*

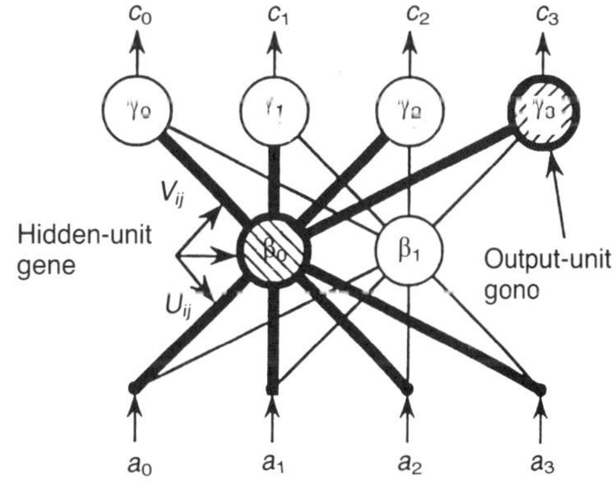

Figure 18.1 A 4-2-4 encoder.

1986). Assume that $n_1 = n_3$ and $n_2 = \log_2 n_1$, and that the inputs consist of a single one bit, with all the rest zeros. The position of this bit then represents one of the first n natural numbers. The encoding problem is to learn to encode these numbers into a pattern of $\log n$ bits, and also to learn to decode this $\log n$ bits pattern into an output pattern, usually identical to the input pattern. We, however, are using the population of encoders in quite a different way. Instead of finding a single network that solves the encoding problem for all sources, I want to construct subpopulations of networks that are specialized for encoding different sources.

In general, an encoder is a tuple

$$\xi = [\beta, \gamma, U, V]$$

where $\beta = \langle \beta_0, \ldots, \beta_{n_2-1} \rangle$ and $\gamma = \langle \gamma 0, \ldots, \gamma_{n_3-1} \rangle$ are thresholds for the hidden units and the output units, respectively, and $U = \{u_{ij} | 0 \leqslant i < n_2, 0 \leqslant j < n_1\}$ and $V = \{v_{ij} | 0 \leqslant i < n_2, 0 \leqslant j < n_3\}$ are weight matrices.

An encoder accepts an n_1-bit input vector $\boldsymbol{a}$, produces an n_2-bit hidden vector $\boldsymbol{b}$, and then produces an n_3-bit output vector $\boldsymbol{c}$. Each unit applies a threshold function

$$\Theta(\phi, s) = \begin{cases} 1 & \text{if } s \geqslant \phi \\ 0 & \text{otherwise} \end{cases}$$

to the sum of its weighted inputs:

$$b_i = \Theta(\beta_i, \sum_{0 \leqslant j < n_1} u_{ij} a_j)$$

$$c_j = \Theta(\gamma_j, \sum_{0 \leqslant i < n_3} v_{ij} b_j).$$

It is essential to the genetic algorithm described below that a description of an encoder may be decomposed into parts, called genes, in such a way that a new encoder (a child) can be constructed with parts from two others (the parents) (Holland 1975; Goldberg 1989). In part, I have chosen the encoder network for this work because it can be decomposed in a fairly natural way. The genetic structure of an encoder is illustrated in Figure 18.1. Each encoder has n_2 hidden-unit genes and n_3 output-unit genes. The hidden-unit genes are the more complex of the two types. The ith hidden-unit gene of an encoder ξ consists of the hidden-unit threshold β_i, a vector of input weights $\langle u_{ij} | 0 \leqslant j < n_1 \rangle$, and a vector of hidden-unit weights $\langle v_{ij} | \ 0 \leqslant j < n_3 \rangle$. The jth output-unit gene consists simply of the output-unit threshold γ_j.

The system consists of a population of N encoders

$$\Xi = \{\xi_k, 0 \leqslant k < N\}$$

with, in general, different thresholds and weights. I always have $n_1 = n_3$ and typically, but not necessarily, $n_2 = \log_2 n_1$. Every encoder in the population is presented simultaneously with the same input vector, and tries to reconstruct the input. Success is measured by a fitness function (Bergman and Kerszberg 1987; Kerszberg and Bergman 1988)

$$f_k(\boldsymbol{a}) = - \sum_{0 \leqslant j < n} |a_i - c_i|.$$

Note that fitness is simply the negative of the Hamming distance between the input and the output vectors. The idea behind the genetic algorithm described below is to increase the frequency of genes and combinations of genes in Ξ by selection, thereby causing the population to learn to encode the inputs it sees most frequently.

18.5 A GENETIC ALGORITHM

Genetic algorithms can be effective for exploring large design spaces (Goldberg 1989; Holland 1975). The essential idea is to simulate many generations of populations of individual subsystems, with each generation produced from previous generations by selection and differential reproduction (Bergman and Kerszberg 1987; Ewens 1979; Hartl and Clark 1989; Roughgarden 1979). Each individual is graded by a fitness function that is intended to measure its performance on one or more instances of a problem. Those individuals that are most fit are selected and then a set of new subsystems is created by applying genetic operators to the descriptions of the selected individuals. Commonly used genetic operators are called crossover and mutation, modeled after similar processes that drive biological evolution (Bergman and Feldman 1990; Goldberg 1989; Holland 1975). Although the concepts behind genetic algorithms are very general, there are inevitably a wide variety of parameters, reproduction schemes, representations and so on that could be used. Part of the aim of this preliminary work is to understand the consequences of and interactions among these choices.

My genetic algorithm consists of an initialization,

$$\Xi \leftarrow \Xi^0$$

followed by an iteration of the generation operator, $\mathcal{G}$:

$$\Xi \leftarrow \mathcal{G}(\Xi, \boldsymbol{a}^t), \quad t = 0, 1, \ldots$$

In the initialization step, a population of at least $N = 4096$* encoders with n inputs and m hidden units is created. All thresholds and weights are chosen from a uniform random distribution over the interval $[-1, 1]$. Initially, all of the members of Ξ are marked as alive and are assigned an age chosen from a random distribution of integers in the range $[0, \ldots, \text{age}_{max} - 1]$. Only those encoders marked as alive, denoted by Ξ_a, are active and available for input, selection and reproduction. All encoders that are not alive are treated as available space for the next generation. The age of ξ is an integer indicating the number of generations for which ξ has been continuously alive.

The generation function $\mathcal{G}$ is defined as the following sequence of steps:

$$\begin{aligned}
\Omega &\leftarrow \text{select}\ (f, \Xi, a) \\
\Omega^* &\leftarrow \text{reproduce}\ (\Omega) \\
\Xi &\leftarrow \text{insert}\ (\Omega^*, \Xi) \\
\Xi &\leftarrow \text{age}\ (\Xi) \\
\Xi &\leftarrow \text{kill}\ (\Xi)
\end{aligned}$$

These steps can be performed in several ways, but each step has the basic characteristics outlined below, in Section 18.6.

Selection: $\Omega \leftarrow \text{select}\ (f, \Xi, \boldsymbol{a})$

An input bit vector, $\boldsymbol{a}$, is chosen and presented to the system. The input can be selected in a variety of ways. The simplest is to select the vector from a set of sources according to some prior probability distribution. Input vectors can be degraded with noise by inverting bits with some probability. Inputs can also be chosen randomly from the set of 2^n possible inputs with some specified frequency. All living encoders are ranked by fitness and a subset Ω of the most fit is selected. The size of Ω could be determined dynamically by a threshold on fitness. Instead, in this preliminary investigation, I set the size of Ω as a fixed proportion of the size of Ξ (usually $\frac{1}{16}$).

Reproduction: $\Omega^* \leftarrow \text{reproduce}\ (\Omega)$

Every member of Ω is paired at random with another member of Ω (possibly itself), which is called its mate. The pairs are combined to produce a fixed number of children. The combination is performed by applying two genetic operators, crossover and mutation. In the crossover operation, every child's gene is selected from one or the other parent with probability $\frac{1}{2}$, a process called free recombination (Hartl and Clark 1989; Packard 1989). In the mutation operation, every gene constituent, whether a weight or a threshold, is replaced by a random value with some probability of mutation μ, which is usually quite low.

*I use a Connection Machine with 4096 processors for my simulations. N can be larger than 4096, but must be a power of 2.

Insertion: $\Xi \leftarrow \text{insert}(\Omega^*, \Xi)$

A random number $k \in \{0, \ldots, N-1\}$ is generated for every child in Ω^*. If ξ_k is not alive, the child is inserted into Ξ at that location, is marked as alive, and is assigned an age of zero. If more than one child tries to occupy the same location, one child is chosen at random.

Aging: $\Xi \leftarrow \text{age}(\Xi)$

The ages of all living encoders are increased by 1.

Death: $\Xi \leftarrow \text{kill}(\Xi)$

Every encoder whose age is greater than age_{max} is marked as not alive. Its space in Ξ then becomes available for the children in the next generation.

18.6 RESULTS

When interpreting the performance of the system, I consider only those encoders that can reconstruct their outputs perfectly. These are said to respond to the input; that is, $r_k(\boldsymbol{a}) = 1$, where

$$r_k(\boldsymbol{a}) = \max(0, 1 + f_k(\boldsymbol{a})).$$

I want many networks to respond to the sources, few or none to respond to nonsource signals, and different subpopulations to respond to each different source.

Two measures of the effectiveness of the system depend on computing the probability distribution $P(\boldsymbol{a}|r)$, which is the probability that the signal is $\boldsymbol{a}$, given that a randomly chosen encoder is responding. This distribution is computed assuming no prior knowledge of the frequency of occurrence of the source. Therefore, using a uniform (maximum entropy) distribution of priors

$$P(\boldsymbol{a}) = \frac{1}{2^n}$$

and writing the probability of an encoder responding to $\boldsymbol{a}$ as

$$P(r|\boldsymbol{a}) = \frac{\sum_k r_k(\boldsymbol{a})}{N},$$

and the probability of an encoder responding to any signal as

$$P(r) = \frac{\sum_x \sum_k r_k(\boldsymbol{x})}{N2^n},$$

I use Bayes's rule to determine the desired distribution:

$$P(\boldsymbol{a}|r) = \frac{P(r|\boldsymbol{a})P(\boldsymbol{a})}{P(r)},$$

or

$$P(\boldsymbol{a}|r) = \frac{N\sum_k r_k(\boldsymbol{x})}{\sum_x \sum_k r_k(\boldsymbol{x})}$$

Ideally, this distribution should be identical to the prior probability $P(\text{a})$ after many generations.

I can compute the entropy of $P(\boldsymbol{a}|r)$

$$S = -\sum_x p(\boldsymbol{x}|r)\log^2 P(\boldsymbol{x}|r)$$

to summarize the degree of oganization of the system in terms of the uncertainty associated with its response. I can also compute the correlation between $P(\boldsymbol{a}|r)$ and some prior model distribution $P_M(\boldsymbol{a})$ from which the sources were chosen:

$$C = \frac{\sum_x (P(\boldsymbol{x}|r) - \overline{P(\boldsymbol{x}|r)})}{\sqrt{\sum_x (P(\boldsymbol{x}|r) - \overline{P(\boldsymbol{x}|r)})^2}} \times \frac{(P_M(\boldsymbol{x}) - \overline{P_M(\boldsymbol{x})})}{\sqrt{\sum_x (P_M(\boldsymbol{x}) - \overline{P_M(\boldsymbol{x})})^2}}$$

The first three experiments described below use entropy and correlation to examine the evolution of the system under different

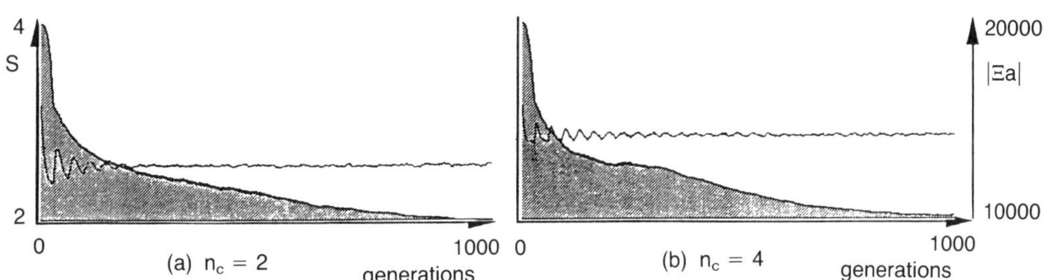

Figure 18.2 Typical behaviour (no mutation).

conditions. Because the time required to compute $P(\boldsymbol{a}|r)$ grows exponentially with the length of the input vector, n, these experiments were done only on small 4-2-4 encoders. The fourth experiment examines the behavior of the system when n is larger and, in particular, when the number of possible inputs greatly exceed the size of the population. Finally, the fifth experiment examines whether the population becomes specialized to the sources.

18.6.1 EXPERIMENT 1: TYPICAL BEHAVIOR (NO MUTATION)

The first experiment examines the typical behavior of a population of 16K 4-2-4 encoders with no mutation ($\mu = 0$). The inputs were chosen at random with equal frequency from a set of four sources. Figure 18.2 shows the entropy of $P(\boldsymbol{a}|r)$ over 1000 generations when the maximum number of children n_c is 2 and 4 ((a) and (b), respectively). Also shown is the size of the population that is living.

In both cases the entropy eventually drops to the ideal value of $\log_2 4 = 2$, which is the entropy of the model distribution. The correlation with the model distribution (not shown) is very nearly 1 after only about 20 generations. The fraction of the population that is living fluctuates at first, but eventually approaches some limit, which is greater for the $n_c = 4$ case.

18.6.2 EXPERIMENT 2: CHANGING ENVIRONMENT

The previous simple experiment illustrated that adaptation can occur without mutation, relying only on the crossover operation. This experiment shows that mutation is essential in a more challenging problem. Figure 18.3 shows the entropy and the correlation measures when the system is successively stimulated with two different sets of four signals, L_1 and L_2. Two cases are shown: $\mu = 0$ and $\mu = 0.01$. The interesting feature of this experiment is that in the first case, $\mu = 0$, the system 'collapses' into an irreversible condition of total insensitivity on the third presentation of the set L_1. The entropy drops to zero, indicating that the system can respond to no signals (or possibly to only one), and the correlation with the model distribution drops effectively to zero. Apparently, the successive presentations and epochs of selection have eliminated variation in Ξ. Selection for L_1 eliminates genes effective for L_2, selection for L_2 eliminates genes effective for L_1, and so on, until by the third presentation of L_1, Ξ has been so depleted that it cannot adapt.

In the case of $\mu = 0.01$ this does not happen. Even this low rate of mutation is sufficient to maintain adequate variation in Ξ. The crossover operation is effective for making large jumps through the space of genotypes, while mutation is effective as a continual source of variation.

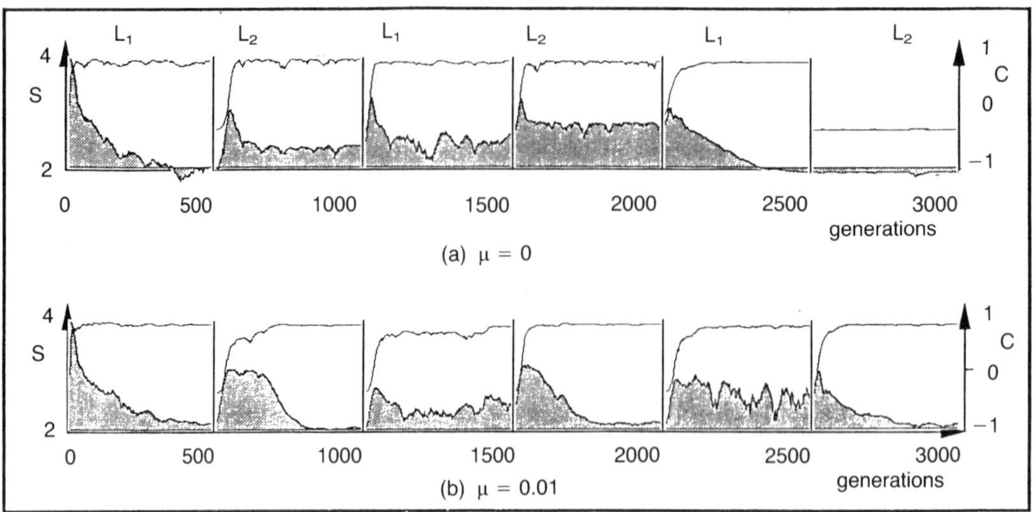

Figure 18.3 Changing environment.

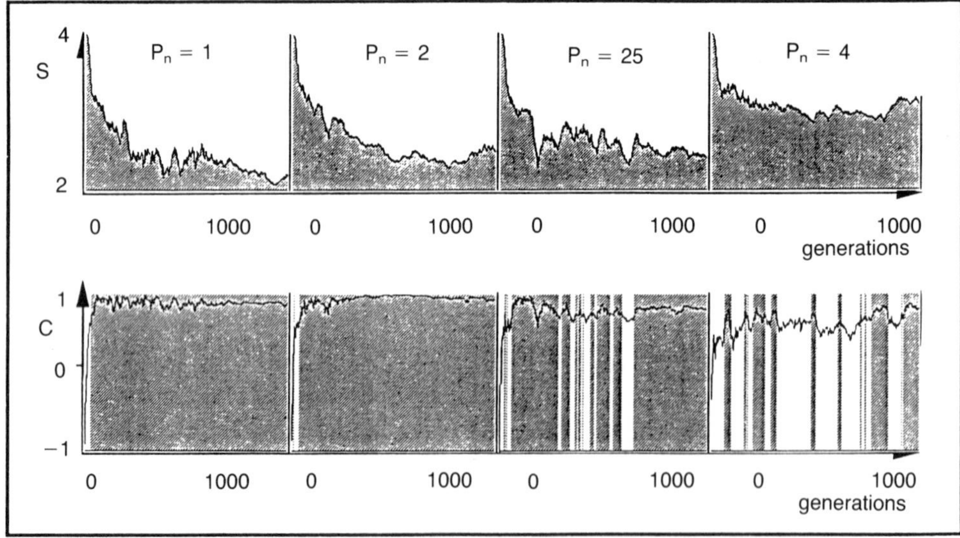

Figure 18.4 Effects of noise.

18.6.3 EXPERIMENT 3: EFFECTS OF NOISE

Experiment 3 examines the effects of noise in the input. The population size is 4K, the encoders are 4-2-4, four different sources are used with equal probability, $\mu = 0.01$, $n_c = 4$ and $age_{max} = 30$. Each encoder is presented with an input vector, selected from the four sources, but each vector has a probability P_n of having (at least) one bit changed at random. All encoders receive input from the same source, but the inputs are corrupted by noise independently, so that any two encoders may see different signals. Figure 18.4 shows four cases: $P_n = 0.1, 0.2, 0.25, 0.4$. Entropy is shown above and correlation below. The shaded portions of the correlation graphs indicate when the system is working, in the sense that the four signals of highest probability are identical to the sources. The system performs well up to $P_n = 0.2$ but degrades quickly for higher noise levels.

18.6.4 EXPERIMENT 4: LARGE n

To test the system on a larger problem, and in particular on a problem in which the number of possible signals greatly exceeds the size of Ξ, I performed a simulation with 16-4-16 encoders and eight sources. As in the previous simulation the population size is 4K, $\mu = 0.01$, $n_c = 4$ and $\text{age}_{\max} = 30$. Because the number of possible inputs is $2^{16} = 64\text{K}$ it is not practical to compute the complete distribution $P(\boldsymbol{a}|r)$, especially not for every generation. Instead, I let the system run for 4000 generations and then counted the number of encoders that responded averaged over all eight sources, which was 488.5, and the average number of encoders that responded averaged over 1000 randomly chosen signals, which was 0.13.

18.6.5 EXPERIMENT 5: SPECIALIZATION

The last experiment examines whether the population divides into disjoint subpopulations specialized for the sources. Suppose we have s sources with R_i being the subpopulation of encoders that respond to source i. The following equation gives a normalized measure of the overlap between two subpopulations:

$$O_{ij} = \frac{|R_i \cap R_j|}{|R_i \cup R_j|} \qquad 0 \leq i, j < s.$$

Ideally, O_{ij} should be one if $i = j$ and zero otherwise for complete specialization. Figure 18.5 shows matrices of overlap measures for four cases. When I adapt 4-2-4 encoders to only two sources, shown in Figure 18.5(a), no specialization occurs at all: nearly every encoder that responds to one source also responds to the other. When I adapt the same system to four sources (b) or seven sources (c), there is some specialization, with relatively more specialization occurring when there are more sources. Finally, when I adapt a system of 16-2-16 encoders to ten sources, Figure 18.5(d), the specialization is nearly perfect, with only two subpopulations having a significant degree of overlap.

18.7 COEVOLUTION

Thus far in my research I have dealt with individual populations in isolation and having no interaction with other populations. However, in the natural world populations do not exist in isolation. The interactions (between populations) are intrinsically interesting because they produce perhaps the most intricate and fascinating patterns in biology. In this section I will introduce the notion and implication of the evolution of population of processes in the presence of 'parasites'. A parasite can be considered a low-level process which depends on its host for survival and reproduction. The host provides the environment for the parasite, and as long as the parasite can exploit the host, it can survive while causing harm to its host (a synergetic behavior can also be included, but for the arguments below I will not consider that option). Consider the following interaction between a host and its environment and a parasite and its environment, that is, the host itself. The host is selected such that its fitness is maximized. The parasite reproduces, and is therefore considered successful, if it can exploit its host, say, by recognizing its genetic makeup. Once recognition is achieved, the host is no longer operational and the parasite spreads its offspring, copies of itself, to neighboring hosts. In the case where the neighboring hosts have similar genetic makeup to that of the original host, in the next generation they will be nonfunctional

(a) 2 sources overlap matrix (for 4-2-4 encoders)

	S1	S2
S1	1.0	0.99
S2	0.99	1.0

(b) 4 sources overlap matrix (for 4-2-4 encoders)

	S1	S2	S3	S4
S1	1.0	0.99	0.99	0.0
S2	0.99	1.0	0.99	0.0
S3	0.99	0.99	1.0	0.0
S4	0.0	1.0	0.0	1.0

(c) 7 sources overlap matrix (for 4-2-4 encoders)

	S1	S2	S3	S4	S5	S6	S7
S1	1.0	0.22	0.62	0.53	0.0	0.0	0.0
S2	0.22	1.0	0.43	0.33	0.0	0.0	0.0
S3	0.62	0.43	1.0	0.69	0.0	0.0	0.0
S4	0.53	0.33	0.69	1.0	0.0	0.0	0.0
S5	0.0	0.0	0.0	0.0	1.0	0.99	0.99
S6	0.0	0.0	0.0	0.0	0.99	1.0	0.99
S7	0.0	0.0	0.0	0.0	0.99	0.9	1.0

(d) 10 sources overlap matrix (for 16-2-16 encoders)

	S1	S2	S3	S4	S5	S6	S7	S8	S9	S10
S1	1.0	0.0	0.0	0.0	0.0	0.0	0.0	0.0	0.0	0.0
S2	0.0	1.0	0.0	0.0	0.0	0.0	0.0	0.0	0.0	0.0
S3	0.0	0.0	1.0	0.0	0.0	0.0	0.0	0.0	0.0	0.0
S4	0.0	0.0	0.0	1.0	0.0	0.0	0.0	0.0	0.0	0.66
S5	0.0	0.0	0.0	0.0	1.0	0.0	0.0	0.0	0.0	0.0
S6	0.0	0.0	0.0	0.0	0.0	1.0	0.0	0.0	0.0	0.0
S7	0.0	0.0	0.0	0.0	0.0	0.0	1.0	0.0	0.0	0.0
S8	0.0	0.0	0.0	0.0	0.0	0.0	0.0	1.0	0.0	0.0
S9	0.0	0.0	0.0	0.0	0.0	0.0	0.0	0.0	1.0	0.0
S10	0.0	0.0	0.0	0.66	0.0	0.0	0.0	0.0	0.0	1.0

Figure 18.5 Specialization.

and will no longer produce offspring. The hosts that survive are those that have enough variability in their genetic makeup to avoid the parasite. Since the host is subjected to its environment and the process of selection causes the elimination of the processes that responds poorly to the environment, the processes that survive are the processes that are successful in responding to the environment and simultaneously avoiding the parasite. Such a behavior could be achieved if the variation in the host is such that it occurs in places that are not critical for the selection process that occurs at the phenotypic level. For example, consider a process which is the conversion of a binary bit string to its integer representation. The parasite, a binary bit string of the same length, looks only at the binary bit string and measures its Hamming distance to it regardless of its integer representation. In case the selection is based on the highest integer representation for the host, the variability that will be maintained in the host that will have the minimal effect on its phenotypic fitness and still maintain high distance from the parasite will be at the least significant bits. Such a behavior of controlled variability is better than random mutation since its effect on the phenotypic level is minimal while random mutation has no bias to maintain high fitness at the phenotypic level.

I have shown (in preparation) that, in the presence of a changing environment, the coevolved population in the presence of a parasite can evolve to fit the new environment, while maintaining a memory about the past environment, longer than when variation is maintained by random mutation.

18.8 CONCLUSIONS

For the encoding problem, the evolutionary algorithm exhibits effective adaptation. Differential reproduction amplifies the frequency of selected genes and leads to the emergence of a population that is progressively more fit. In my model, free recombination (crossover) seems to be the primary means of adaptation. Two relatively fit parents clearly have a better-than-average chance of producing more fit offspring. Mutation, on the other hand, has only an average chance of producing an offspring that is more fit, regardless of the parents' fitness. However, by itself free recombination causes a progressive loss of information: those genes that are amplified replace others that are lost forever. This loss of diversity in the gene pool is disastrous if the ensemble of sources changes, as demonstrated in Experiment 2. The mutation operator continuously injects diversity into the gene pool, thereby preventing the system from becoming trapped in a low-diversity dead end.

My approach differs from some genetic-algorithm and neural-network approaches in a fundamental way. I do not seek an individual encoder that is 'most fit' overall; instead, I seek subpopulations of networks that have specialized their responses to particular sources. The response of the system is an aggregate, macroscopic feature of the individual responses of a large population of individual, interacting subsystems. I view fitness as a very general concept: simply a measure of the similarity between the input and the output. Rather than being built into the fitness function, the evolutionary trend toward specialization is instead an emergent property of the population as a whole, and a consequence of the informational bottleneck in the encoders. Unlike the more standard optimization methods for designing systems, this method results in subpopulations that resemble species adapted to different ecological niches that are determined by the sources.

Variability is one of the important driving forces that causes a population to evolve. One way of maintaining variability in population is by mutation, but mutation is a random process that causes a reduction in the population performance and may lead, together with drift, to an unfit population. Are there more sophisticated mechanisms by which nature chooses to operate? Is coevolution a process that can be artificially reproduced and generate populations that will be able to adapt to a changing environment while memorizing the important features of the history? My experiments indicate that such a mechanism and behavior can be mimicked and rather interesting dynamical behavior can be observed. The question that can be asked here is the relation of such systems to dissipative dynamical systems, where the environment acts as an energy source and the parasites ('viruses') act as the dissipative part of the system.

REFERENCES

Ackley, D. H., Hinton, G. E. and Sejnowski, T. J. (1985). A learning machine for Boltzman machines. *Cognitive Science* **9**, 147–69.

Bergman, A. and Feldman, M. W. (1990). More on selection for and against recombination. *Journal for Theoretical Population Biology*.

Bergman, A. and Kerszberg, M. (1987). Breeding intelligent automata. *IEEE First Annual Conference on Neural Networks*, San Diego.

Ewens, W. J. (1979). *Mathematical Population Genetics*. Springer-Verlag.

Goldberg, D. (1989). *Genetic Algorithms*. Addison-Wesley.

Hartl, D. L. and Clark, A. G. (1989). *Principles of Population Genetics*, 2nd edn. Sinauer.

Holland, J. H. (1975). *Adaptation in Natural and Artificial Systems*. University of Michigan Press, Ann Arbor, MI.

Kerszberg, M. and Bergman, A. (1988). The evolution of data processing abilities in competing automata. In *Computer Simulation in Brain Science* (ed. R. M. J. Cotterill). Cambridge University Press, pp. 249–59.

Packard, N. (1989). Evolving bugs in a simulated ecosystem. In *Artificial Life* (ed. C. G. Langton). Addison Wesley.

Roughgarden, J. (1979). *Theory of Population Genetics and Evolutionary Ecology: An Introduction*. Macmillan.

Rumelhart, D. E., Hinton, G. E. and Williams, R. J. (1986). Learning internal representations by error propagation. In *Parallel Distributed Processing*, Vol. 1 (ed. D. E. Rumelhart and J. L. McClelland). MIT Press, Cambridge, MA.

PART FOUR
ECOLOGY

FREE THE SPIRIT OF EVOLUTIONARY COMPUTING: THE ECOLOGICAL GENETIC ALGORITHM PARADIGM

19

Yuval Davidor

19.1 BACKGROUND

The ECO GA paradigm (ECOlogical Genetic Algorithm) was originally proposed in order to improve the robustness of function optimization GAs by introducing emergent speciation, thus reducing the risk of premature convergence (Davidor 1991). Serendipitously, the ECO framework exhibited some fundamental robustness properties which makes it an attractive generic framework for many evolutionary computing models designed to solve problems effectively.

The original motivation in developing the ECO framework resulted from the fact that robustness was an illusive goal in standard GAs practice. Researchers in numerous publications have attempted to suggest that there are additional operators for restricting mating and selection which should be applied, or specific activation profiles for the genetic operators (such as varying the mutation rate), if controlled convergence is desired. The ECO framework suggests that it is possible to increase robustness with virtually no additional computation overhead by simply rearranging the genetic interactions.

In Section 19.1.1 I briefly review the operators suggested in the past to solve (or at least improve) the robustness and controlled convergence problems in GAs. This will help to explain why the ECO framework is an attractive framework to apply GAs to optimization problems. Simplified concepts of ecology, which form the basis of the ECO framework, are discussed in the following section (Section 19.1.2), and are followed by a short section on cellular automata concepts, which also bear some resemblance to the ECO framework.

19.1.1 GA OPERATORS TO CONTROL CONVERGENCE

Since the early days of the 1970s two important aspects of GAs have been realized; first, that controlled convergence is an illusive goal, difficult to achieve with global operators; second, that fine-tuning of the GA operators is not the key to robustness and controlled convergence. In other words, it was realized that the operators themselves, through mutual interaction, should provide an adaptive, and hopefully optimal, balance between exploration and exploitation – exploration and exploitation in the sense of global information processing and not in the sense of resource allocation and similar partial mechanisms of evolution.

Mechanisms to provide a desirable controlled convergence property were suggested as soon as the first simulations of GAs started to be experimented with, with varying degrees of success. The main idea is that there should be a mechanism that will control and prevent an unbalanced proliferation of population members (or segments of genetic material) at the expense of global search. It is clear that such an operator can be imposed through an explicit testing of genotype diversity, but it is equally clear that such a mechanism is computationally prohibitive. The 'battle' for controlled convergence focused on the attempt to obtain the desired controlled property indirectly in order to minimize computational overheads. The following is a brief history of the mechanisms that were suggested as an answer to the control of convergence issue.

As early as 1970, Cavicchio suggested a preselection mechanism, a mechanism which replaces parent members in the population with their offspring (Cavicchio 1970). The idea is that since offspring contain the same bit diversity as their parents, replacing parents with offspring reduces premature loss of bit diversity. A similar mechanism was suggested by De Jong – the crowding scheme in which an offspring replaces the most similar string in bit terms (Hamming distance) from a randomly drawn subpopulation of size CF (crowding factor) from the main population (De Jong 1975). Mauldin's uniqueness operator helped to maintain diversity by incorporating a 'censorship' operator with which the insertion of an offspring into the population is possible only if the offspring is genotypically different from all members of the population at a specified number of loci (Hamming distance) (Mauldin 1984). Related ideas in a classifier system environment include Booker's sharing (Booker 1982) and Schaffer's VEGA model (Schaffer 1985).

Until the mid 1980s globally monitoring genotypic diversity was regarded as the proper way to solve the instability exhibited by GAs, and the only concern was how to reduce the computation cost of such operations while increasing their effectiveness. Goldberg and Richardson incorporated some of the ideas mentioned above into a mechanism they called a **sharing function** (Goldberg and Richardson 1987; Deb and Goldberg 1989). This mechanism modifies the reproduction probability of a population member according to how many population members occupy a similar 'niche' of the solution space (usually measured in the genotypic space).

The idea of sharing functions helps to maintain diversity and multiple peaks in the solution space, but introduces a few problems, the main ones being computational overheads, the need to specify *a priori* the niche count (unknown for arbitrary functions), and the fact that it restricts mating within high fitness species.

Eshelman and Schaffer introduced a more direct mating restriction with an operator they called **incest prevention** (Eshelman and Schaffer 1991). This operator prevented mating between two randomly paired population members if they are too similar to each other.

Although the above mechanisms partly improve the convergence, they suffer from the fact that they employ additional operators, sometimes at heavy additional computing costs. Also, they are not applicable to all problem domains and evolutionary models, representations and so forth. More importantly, most of the above mechanisms are applicable to binary representations only and limited in their applicability to high-dimensional problems.

Muhlenbein and Schleuter explored a particular hardware-dependent simulation they called ASPARAGOS (Muhlenbein 1989; Schleuter 1989). The underlying idea of ASPARAGOS is that multiple populations are held on a network of transputers, and acts as interacting subpopulations. There are other relevant works involving parallel implementations of GAs, but mostly they just recapitulate previous approaches (Cohoon *et al.* 1987; Jog and VanGucht 1987; Pettey and Leuze 1989; Tanese 1989).

There are two recent models (Collins and

Jefferson 1991; Spiessens and Manderick 1991) which attempt to parallelize GAs and which were published about the same time as the earlier publication of the model discussed in this chapter (Davidor 1991). They contain some of the essence of the latter model.

In summarizing the state of the art of controlled convergence operators, it must be acknowledged that there is no model which is hardware-independent, capable of handling arbitrary representation formats (variable length, real value genes, etc.), and yet allows controlled convergence without substantially increasing computation overheads.

19.1.2 POPULATION GENETICS AND ECOLOGY

Since Holland's central genetic algorithm theorem, many new 'genetic' operators have been introduced and used (such as the ones mentioned in the previous section). Usually the introduction of such new or modified operators resulted from *ad hoc* experience of experimental results on limited problem domains, and not as a result of theoretical concepts. The ECO model was originally suggested on the basis of observations of evolutionary models in the more biologically oriented field of ecology and population dynamics. Simplified concepts of population genetics and ecology are discussed here in order to introduce intuitive understanding, which, in the author's opinion, was missing in conventional GA models. This led to the development of a generic ECO model for evolutionary computations. It is neither suggested that the ECO framework simulates ecosystem dynamics, nor that accurate simulation of ecosystems must necessarily become a generic problem solver. It is suggested, however, that evolutionary computation models should be based, like ecosystem dynamics, on topology as a framework for genetic interaction.

There are certain assumptions regarding ecological systems interaction and population dynamics which affect the interactions between the genetic operators. The following are mentioned as those most relevant to the evolutionary computation model suggested here:

1. An organism's life can be characterized by three fundamental activities: foraging (looking for food), reproduction (mating) and survival (winning conflicts).
2. Mating partners are selected from the organism's local environment, and proportional to fitness.
3. Offspring remain in the geographical vicinity of their parents.
4. The frequency of aggressive conflicts is inversely proportional to resource availability.
5. Aggresive conflicts are resolved probabilistically, but proportional to the relative strengths of the opponents.

It is clear from the above list that the behaviour of the individual is a result of mechanisms which rely on local information only. It can be argued whether this aspect of evolution has or has not contributed to quicker convergence and to better genetic patterns than other possible interactions. It is, however, clear that a strong locality in the global genetic interaction contributed to the basic control of the process.

19.1.3 CELLULAR AUTOMATA

Cellular automata concepts, though not directly related to the discussion about controlled convergence in GAs and evolutionary computation, are based on similar topological local interactions as the ECO interactions. As such, cellular automata models are a good tool to explain the ECO rules of genetic interactions. In cellular automata, a set of fixed rules is applied locally and uniformly to all the system elements (Toffoli and Margolus 1987). Cellular automata models are more interested in representing geometrical or physical phenomena such as fluid and thermodynamics, but they share this topology dependence with the dynamics of population genetics and ecosystems.

In cellular automata modelling, the topology dependence is all that is being described. Hence, to achieve a certain state dynamics, an appropriate local interaction rule must be applied. In population genetics it serves a different purpose, the purpose of controlled convergence. So, though the ECO framework is not about drawing time variant pictures, or simulating time invertible flows around a foil, it is helpful to realize the similar dynamic separability in the two paradigms.

19.2 THE ECO GENETIC ALGORITHM PARADIGM

The most fundamental difference between common GA models and the ECO model is the mechanism of manipulating population members. By and large, in common GAs population members are processed globally. Such a centralized simulation results in a situation in which the simulation considers, and is directed by, global information. The implications of global interactions in respect to convergence are well understood by now, and if one wants to reduce the genetic material mixing, one has to either unrealistically increase the population size, or modify the basic genetic interactions with the above mentioned problems. In the ECO framework, however, population members are held in a n-dimensional grid and are manipulated locally, primarily according to grid position. For reasons discussed later, the dimensionality of the grid is set to 2, resulting in a two-dimensional topology. For isomorphism considerations, the opposite edges of the grid are connected together so that each grid element has eight adjacent nodes as shown in Figure 19.1.

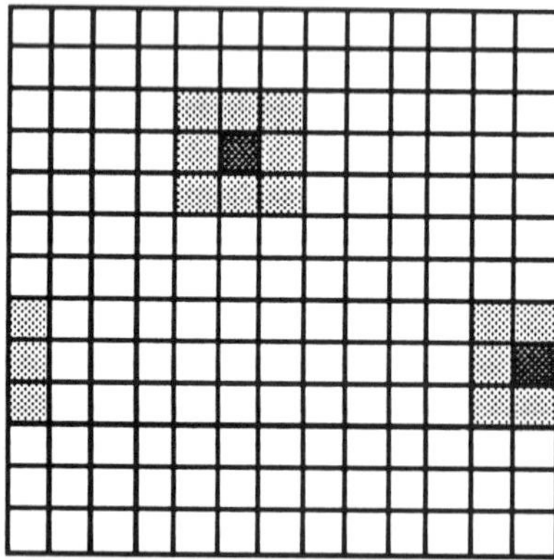

Figure 19.1 The two-dimensional ECO topology environment shows two grid elements, each containing one population member (or a solution) and their eight-member local environments respectively. (The opposite edges of the grid are connected creating a torus.)

19.2.1 THE GENETIC OPERATORS TURNED LOCAL

The ECO model requires local genetic interactions. Therefore, reproduction, recombination and selection have to be changed from global to local mechanisms. This section discusses those changes necessary to turn the genetic operators local. Figure 19.2 presents a flow diagram of a fairly generic ECO GA.

Reproduction

The ECO reproduction is a type of steady state reproduction (Syswerda 1989; Syswerda 1991; Whitley 1988). At each reproduction cycle, a grid element i,j is selected at random. The population member which occupies this i,jth node is called the first parent, and together with the eight population members adjacent to this i,jth member forms the local and temporarily active subpopulation (such as in Figure 19.1).

The average fitness of this nine-member subpopulation is calculated (first parent + eight neighbours). The relative fitness of the first parent of the i,jth node relative to the average fitness of the nine-member subpopulation determines how many offspring are going to be produced in the current repro-

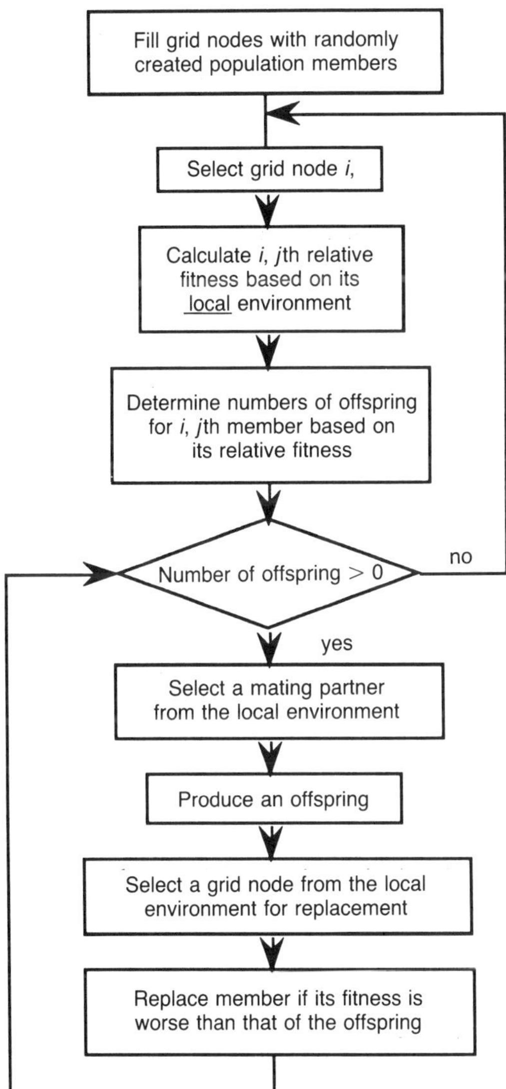

Figure 19.2 A fairly generic ECO simulation for a genetic algorithm.

duction cycle in the following way. The number of offspring is the fitness of the first parent relative to its nine-member subpopulation rounded probabilistically. For example, if the relative fitness of a given first parent is 1.4, then this first parent will produce at least one offspring, and another one with probability 0.4, while a relative fitness of 0.5 means probability 0.5 for no offspring and 0.5 for one offspring.

Mating

After determining how many offspring the first parent will help produce, a second parent is selected from the subpopulation with replacement probabilistically relative to fitness excluding the first parent for each offspring produced.

Please note that the type of crossover mechanism, whether it be a two-point, uniform, PMX or other problem-specific crossover, is not mentioned and nor is the type of mutation because it is not relevant to the ECO model. By thus defining the choice of parents one is free to use whatever crossover, mutation or other mechanisms one wishes. The mechanisms used in this work are those suitable for continuous representations as discussed in Grossman and Davidor (1992).

Selection

After producing the offspring and calculating their fitness, each offspring is introduced into the grid by selecting at random a grid element from the nine grid elements of the *i*,*j*th environment. The population member in the selected element and the offspring are in conflict because of limited habitat resources (as only one population member is allowed at each grid element). The conflict is resolved probabilistically according to the relative fitnesses of the two opponents. For example, if the offspring's absolute fitness is 12 and the invaded member's fitness is 8, then the offspring has a 0.6 probability of invading the population and replacing the occupant of the said grid element.

This selection process is repeated for all offspring sequentially so there is the possibility that a second or later offspring will replace an earlier invading offspring. This selection pressure bears some similarity to Cavicchio's preselection mechanism.

19.2.2 ECO EFFECTS ON CONVERGENCE

In this section the effects of the ECO local interactions on global behaviour are analysed. Resulting from the locality of operators, some strong granularity is expected in the global dynamics of the ECO GA simulation. Since the 'effective' population size for each of the population members is small,* much smaller than the total number of population members included in the simulation or used in traditional GAs, it is reasonable to expect fast local (in grid terms) convergence.

Reproduction and sampling are proportionate to a very small number of population members. Members of superior fitness (superior relative to local environment) have a much more pronounced sampling bias (a similar effect to that achieved by scaling operators used in global GAs, but without the problems of premature global loss of diversity). The fast local convergence resulting from the small effective population size is augmented by the ECO selection pressure leaving offspring in the same local environment as their parents, thus depositing the biased genetic material in its origin environment, and enhancing local convergence.

Thus, if one monitors the local environment of a given population member, one can observe the growth of genotypic isomorphism in its immediate vicinity. The local speedy convergence occurs in many parts of the grid, and produces a speciation-like phenomenon as the genetic diversity in many local neighbourhoods is reduced. The genotypic granularity of the grid becomes more pronounced, and small 'islands' of genetically isomorphic colonies begin to emerge. This emerging granularity formation phenomenon is graphically described in Figure 19.3.

Figure 19.3 Islands of isomorphic genotype material emerge as a result of the ECO reproduction, mating and selection pressures.

As the degree of genotypic isomorphism increases inside a region, a process of specialization takes place. The sampling bias is reduced as the relative fitness approaches unity. Crossover becomes less disruptive as there are many shared schemata in the local neighbourhood. Mutations, if active, supply a greedy optimization. The overall effect is that of local optimization around the phenotypic value of the dominating schema.

On the perimeter of high genetic isomorphic regions a different process takes place. Since reproduction is biased towards highly fit members, if the average fitness of an isomorphic region is higher than its surroundings, it will expand (Figures 19.4 and 19.6(a)–(j)). If the region's average fitness is lower, it will shrink. The mating on the perimeter of a genetic highly isomorphic island is relatively dramatic compared to the effects of mating inside islands. Since the genetic material that surrounds an island may be very different from that within it, mating results in offspring of diverse genetic makeup.

Therefore, at initial stages one can observe the formation of small islands of similar genetic material. As the simulation progresses

*Though not strictly the size of the subpopulation, as genetic material can propagate across the grid, the effective population size in this discussion is considered as the size of the mating neighbourhood.

Figure 19.4 The growth and expansion of highly fit genetically isomorphic islands.

the islands of relative high fitness grow and take over grid space that was previously occupied by low fitness members (Plate 3). Also, within islands there will be specialization and the occasional discovery of locally (in phenotypic sense) optimized members, and on the perimeter of islands the creation and test of new genotypic combinations. Once such a new genotype of high relative fitness (to its local grid environment) is created, it will start to reproduce and expand.

In summary, the ECO simulation is characterized by the formation of genetically isomorphic islands, the growth of islands of relative high fitness, the continued convergence and local optimization within islands, and discovery of new rewarding genetic formations on the borders of islands. In the next section a chronicle of an ECO simulation is presented which is intended to demonstrate the unique convergence dynamics described in this section.

19.3 DEMONSTRATION OF THE LOCAL SPECIATION

The local speciation through the genetic isomorphic dynamics is demonstrated in this section. A floating point GA model (Grossman and Davidor 1992) running under the ECO framework is applied to the following problem:

$$\min\left(y_0^2 + \sum_{i=1}^{n}(y_i^2 + x_i^2)\right) \qquad (19.1)$$

subject to

$$y_i = y_{i-1} + x_i \qquad (19.2)$$

where $x_0 = 100$ is a given initial state, $n = 45$, and $x_i \in \leqslant \langle -200.0, 200.0 \rangle$.

Plate 3 shows a graphic display of an ECO simulation environment designed for solving control problems such as the one in equation (19.1). In this environment a 64×64 grid is used, as well as two time plots to study the online performance. In both the lower and right graphs the ordinate is the online performance (or error obtained by equation (19.1)), and the abscissa marks the simulation time. In the lower graph the online performance is in logarithmic scale, and the right graph is a linear magnification of a small range around the global optimum. At the top left part of the display, the simulation grid of 64×64 size is shown with a coloured index based on the marked different error values. At each plot interval the grid is scanned, grid node by grid node, and the size of the error of the population member located at each node is indicated with the appropriate colour from the colour map immediately to the left of the grid.

Plate 4(a)–(j) are a time series of the 64×64 grid presenting the error improvement of equation (19.1) as a function of the simulation time. Note the formation of error isomorphic island dynamics throughout the grid, and that new peaks (represented by colour elements of lower error value) appear in different regions of the grid suggesting a simultaneous exploration of different local minima in the solution space. Also, new peaks often appear within lower fitness, but isomorphic regions.

Plate 4(a)

The state of the grid 'occupants' shortly after the random initialization of the simulation process. After a few hundred local interactions most of the population members have reduced their error from over 1 000 000 to less than 600 000. There are also few population members with an even smaller error which will form the centres of future growing isomorphic regions as shown in the following figures.

Plate 4(b)

A further few thousand local interactions, and the isomorphic regions start to solidify around the population members that are of a lower error value. It is already clear that the distribution of the quality of population members is not uniform, but arranged in clusters according to grid metric.

Plate 4(c)

About half the grid already has in error of less than 300 000. Two grid occupants in the centre right of the grid have an error less than 100 000. In the following figure these two population members form a focal point for the creation of other low error population members.

Plate 4(d)

Most of the grid now contains solutions with an error less than 100 000. The concentration of good quality solutions in the vicinity of the two population members mentioned in Plate 4(c) has grown, and a new 'best' solution with an error less than 70 000 is found in the middle of one of the two centres mentioned.

Plate 4(e)

Further speciation. Several other solutions with an error less than 70 000 have been found in different regions of the grid. As these regions are separated and cannot interact directly, there is ample time for specialization and local optimization.

Plate 4(f)

The first discovery of solutions with an error less than 50 000 and further development of all other isomorphic regions. All of these 'currently best solutions' are in the vicinity of each other and thus form one region, but not for too long.

Plate 4(g)

As the local speciation continues throughout the grid, many more small isomorphic regions with an error less than 40 000 are discovered. The regions with an error less than 50 000 continue to expand at the expense of higher error population members in their vicinity.

Plate 4(h)

Local speciation continues and there are few regions with an error less than 32 000. At this stage, the isomorphic regions become more pronounced and further improved solutions are formed and found in those isomorphic regions which protect the good genetic material from overdisruptive crossovers.

Plate 4(i)

The discovery of solutions with an error less than 26 000 in two distinct grid regions. Note that these regions are not those in which new best solutions were discovered earlier. Like the growing circles around drops of rain on a pond, local specialization produces the outward motion of low error solutions.

Plate 4(j)

Further improvements throughout the grid.

Plate 5

After about 25% of the simulation time (100% designates 1 200 000 function evaluations used as the benchmark limit) solutions with an error less than 17 500 are found. In the two time plots, the green line designates the best solution in the grid, and the blue line designates the average of the whole population. The red line marks the position of the global optimum.

19.4 THE PROOF OF THE PUDDING: EXPERIMENTAL RESULTS

After introducing the ECO framework, discussing the effect of its local interactions on the global genetic process, and demonstrating this dynamics through a simulation, it is now appropriate to show how all this translates into improved GA performance. Towards this end three problems in two different domains are discussed. A high-dimensional quadratic assignment problem (Section 19.4.1), the 10 × 10 job shop scheduling (JSS) problem, and the 5 × 20 JSS problem.

19.4.1 A HIGH-DIMENSIONAL QUADRATIC ASSIGNMENT PROBLEM

The following problem is a quadratic minimization problem which has many applications in dynamics. Given a structure which is not dynamically balanced around its rotational axis Z (center of gravity is not on the Z axis), and n locations in the structure in which mass can be added or subtracted, find that combination of mass changes which minimizes the yaw and pitch moments and the total mass added.

$$\min(C\Delta W + (1 - C)\,\| M \|),$$

ΔW is the net weight added to the structure according to

$$\Delta W = \sum_{i=1}^{n} w_i.$$

$\| M \|$ is the resulting Z right angle moment given by

$$\| M \|^2 = (M_{x1})^2 + (M_{y1})^2 + (M_{x2})^2 + (M_{y2})^2$$

where, in cylindrical coordinates, z_i, r_i and φ_i designate distance along the Z axis, radius and angle from some reference radius, respectively,

$$M_{x1} = M^0_{x1} + \sum_{i=1}^{n} w_i r_i \cos\varphi_i$$

$$M_{x2} = M^0_{x2} + \sum_{i=1}^{n} z_i w_i r_i \cos\varphi_i$$

$$M_{y1} = M^0_{y1} + \sum_{i=1}^{n} w_i r_i \sin\varphi_i$$

$$M_{y2} = M^0_{y2} + \sum_{i=1}^{n} z_i w_i r_i \sin\varphi_i$$

(the superscript '0' designates the initial moment in the said direction). We further require that

$$w_i^{\min} \leqslant W_i \leqslant w_i^{\max}, \quad \forall i = 1, 2, \ldots, n$$

$$W_{\min} \leqslant \left(W_0 + \sum_{i=1}^{n} w_i\right) \leqslant W_{\max},$$

given n locations (z_i, r_i and φ_i, $i = 1, 2, \ldots, n$) in which weight w_i can be added or subtracted.

This type of problem was traditionally solved by a simplex method which theoretically guarantees optimality. However, numerical problems encountered when the problem is of high dimensionality, and the notorious slow convergence of simplex

Table 19.1 A summary of 200 experiments comparing a traditional GA which was optimized to solve the dynamic balancing problem and the performance of the same GA model running under the ECO simulation framework

	Average best	*SD*
Traditional GA	0.1999	0.1268
Under ECO	0.0450	0.0003

Table 19.2 A summary of 200 experiments of a 10 × 10 job shop scheduling problem with population size of 2025 (45 × 45 grid size for the ECO simulation)

	Average best	*SD*
Traditional GA	976	16
Under ECO	963	14

Table 19.3 A summary of 200 experiments of a 5 × 20 job shop scheduling problem with population size of 5041 (71 × 71 grid size for the ECO simulation)

	Average best	*SD*
Traditional GA	1236	19
Under ECO	1214	16

methods make the use of a GA attractive. A GA model was developed and optimized to solve this problem. Its performance superseded that of the available simplex packages.

The performance of the optimized GA developed for this problem is compared in Table 19.1 with that of the same GA running under the ECO framework. The number of balancing points is 14, leading to a dimensionality of 14 (the rest of the problem parameters can be obtained from the author upon request). The lower bound for this problem is LB* ≅ 0.0115 and the best known solution is 0.0287.

19.4.2 JOB SHOP SCHEDULING PROBLEMS RUNNING UNDER ECO

Job shop scheduling (JSS) is a proven NP-hard problem. It is shown in this section that the ECO framework can further improve the performance of an already very good GA model if it is used in the JSS domain by applying the ECO framework. The GA model used for the test is that of Yamada and Nakano called by them GT/GA simulation and its description can be found in Yamada and Nakano (1992). This algorithm exhibits excellent performance in comparison to other GAs for JSS on several benchmarks.

Two difficult JSS problems are investigated here, the 10 × 10 and 5 × 20 benchmarks.

10 × 10 JSS Problem

The GT/GA simulation optimal population size for the 10 × 10 JSS problem (Muth and Thompson 1963) is around 2000 (Nakano *et al.*, submitted). Accordingly, the population size selected to suit both simulations is 2025 resulting in a 45 × 45 grid. Table 19.2 summarizes the results of 200 independent experiments running the GT/GA simulation alone and under the ECO framework. The optimal make-span for this JSS benchmark is 930.

5 × 20 JSS Problem

The 5 × 20 JSS problem (Muth and Thompson 1963) is a more difficult one than the 10 × 10 problem discussed in the previous section, and its optimal population size for the GT/GA simulation is around 5000 (Nakano *et al.*, submitted). Table 19.3 summarizes the results of 200 independent experiments running the GT/GA simulation alone and under the ECO framework. The optimal make-span for this JSS benchmark is 1180.

19.5 SUMMARY AND FURTHER DEVELOPMENT

The ECO GA model discussed in this paper presents a new synthesis of the conventional genetic operators. In this model all operators are based on local interaction in a 2D grid topology. The choice of a 2D topology rather than a grid with higher dimensionality is not arbitrary and neither is the range for the local genetic interaction. The number of dimensions in the topology and the range for the genetic interaction determines the size of the subpopulation. The 2D configuration and first immediate neighbour were chosen to achieve a balance between the overall number of subpopulations and their size. This new population interaction arrangement results in a rapid local convergence while maintaining both global diversity and speciation as an emergent property of the simulation. A too large local subpopulation, whether the result of high dimensionality of the grid or a long genetic interaction range, and the ECO improved control over convergence property becomes diminished. Of course the other extreme of a very small subpopulation and the genetic process becomes too weak.

The ECO GA model as presented, does not claim to maintain niche and species *ad infinitum*. It does allow, however, the emergence of subpopulation niches in order to enhance the efficiency and robustness of a GA search. The achievement of a complete niche and species equilibrium would advance the GA technology a giant step forward, and the ECO paradigm is a step toward this goal. Depending also on the space topology of the solution and the core GA model used, the speed of local convergence is linked to the size of the grid. The ECO framework is not usually sensitive to the size of the grid over a large range of simulation conditions, but nevertheless, is affected by it. For example, if local convergence is slow then the grid dimension should be big enough to allow sufficient time for local maturation before the genetic material of high fitness members propagates through the grid.

A disadvantage of the ECO framework compared with traditional GA models already discussed is that at the final stages of the search it converges more slowly than the traditional global interaction based models. This phenomenon is clearly the result of the mixing ability of the local interaction model and can be easily overcome by turning the ECO grid population into a global population model. Experiments to that end are under way.

The author is currently involved in extending the research on ecological models of GAs in two directions; first, gathering additional experimental data on the behaviour of this framework; and second, extending the basic model to a model with extended control over convergence, and the ability to maintain several niches at an equilibrium.

ACKNOWLEDGEMENTS

The author wishes to thank for their warm support Dr Seishi Nishikawa, Dr Ryohei Nakano and NTT Communication Science Labs, Kyoto, Japan where early versions of this chapter were written while the author spent time as a visiting scientist. A special thanks to Dr Ueda for his help in developing the Open Windows environment used for the simulations presented in Section 19.3, and to Mr Takeshi Yamada for running the JSS experiments.

REFERENCES

Booker, L. B. (1982). Intelligent behavior as an adaption to the task environment (doctoral dissertation, University of Michigan). *Dissertation Abstracts International* **43**(2), 469B.

Cavicchio, D. J. (1970). Adaptive search using simulated evolution. Doctoral dissertation, University of Michigan.

Cohoon, J. P. *et al.* (1987). Punctuated equilibria: A parallel genetic algorithm. *Second International Conference on Genetic Algorithms* (ed. J. J. Grefenstette). Lawrence Erlbaum, NJ, pp. 148–54.

Collins, R. J and Jefferson, D. R. (1991). Selection in massively parallel genetic algorithms. *Fourth International Conference on Genetic Algorithms* (ed. R. K. Belew and L. B. Booker). Morgan Kaufmann, Los Altos, CA, pp. 249–56.

Davidor, Y. (1991). A naturally occuring niche and species phenomenon: The model and first results. *Fourth International Conference on Genetic Algorithms* (ed. R. K. Belew and L. B. Booker). Morgan Kaufmann, Los Altos, CA, pp. 257–63.

Deb, K. and Goldberg, D. E. (1989). An investigation of niche and species formation in genetic function optimization. *Third International Conference on Genetic Algorithms* (ed. J. D. Schaffer). Morgan Kaufmann, San Mateo, CA, pp. 42–50.

De Jong, K. (1975). An analysis of the behavior of a class of genetic adaptive systems (doctoral dissertation, University of Michigan). *Dissertation Abstracts International* **36**(10), 5140B.

Eshelman, L. J. and Schaffer, J. D. (1991). Preventing premature convergence in genetic algorithms by preventing incest. *Fourth International Conference on Genetic Algorithms* (ed. R. K. Belew and L. B. Booker). Morgan Kaufmann, Los Altos, CA, pp. 115–22.

Goldberg, D. E. and Richardson, J. (1987). Genetic algorithms with sharing for multimodal function optimization. *Second International Conference on Genetic Algorithms* (ed. J. J. Grefenstette). Lawence Erlbaum, NJ, pp. 41–9.

Grossman, T. and Davidor, Y. (1992). *An Investigation of a Genetic Algorithm in Continuous Parameter Space* (CS92-20). The Weizmann Institute, Israel.

Jog, P. and VanGucht, D. (1987). Parallelisation of probabilistic sequential search algorithms. *Second International Conference on Genetic Algorithms* (ed. J. J. Greffenstette). Lawrence Erlbaum, NJ, pp. 170–6.

Mauldin, M. L. (1984). Maintaining diversity in genetic search. *National Conference on Artificial Intelligence*, pp. 247–50.

Muhlenbein, H. (1989). Parallel genetic algorithms, population genetics and combinatorial optimization. *Third International Conference on Genetic Algorithms* (ed. J. D. Schaffer). Morgan Kaufmann, San Mateo, CA, pp. 416–21.

Muth, J. F. and Thompson G. L. (1963). *Industrial Scheduling*. Prentice Hall, New York.

Nakano, R., Davidor, Y. and Yamada, T. (1993). Optimal population size under the constant cost criterion. *Fifth International Conference on Genetic Algorithms* (ed. S. Forrest). Morgan Kaufmann, San Mateo, CA.

Pettey, C. C. and Leuze, M. R. (1989). A theoretical investigation of a parallel genetic algorithm. *Third International Conference on Genetic Algorithms* (ed. J. D. Schaffer). Morgan Kaufmann, San Mateo, CA, pp. 398–405.

Schaffer, J. D. (1985). Multiple objective optimization with vector evaluated genetic algorithms. *First International Conference on Genetic Algorithms* (ed. J. J. Grefenstette). Lawrence Erlbaum, NJ, pp. 93–100.

Schleuter, M. G. (1989). ASPARAGOS, An asynchronous parallel genetic optimization strategy. *Third International Conference on Genetic Algorithms* (ed. J. D. Schaffer). Morgan Kaufmann, San Mateo, CA, pp. 422–7.

Spiessens, P. and Manderick, B. (1991). A massively parallel genetic algorithm: Implementation and first analysis. *Fourth International Conference on Genetic Algorithms* (ed. R. K. Belew and L. B. Booker). Morgan Kaufmann, Los Altos, CA, pp. 279–86.

Syswerda, G. (1989). Uniform crossover in genetic algorithms. *Third International Conference on Genetic Algorithms* (ed. J. D. Schaffer). Morgan Kaufmann, San Mateo, CA, pp. 2–9.

Syswerda, G. (1991). A study of reproduction in generational and steady state genetic algorithms, In *International Workshop on the Foundations of Genetic Algorithms* (ed. G. J. E. Rawlins). Morgan Kaufmann, Bloomington.

Tanese, R. (1989). Distributed genetic algorithms. *Third International Conference on Genetic Algorithms* (ed. J. D. Schaffer). Morgan Kaufmann, San Mateo, CA, pp. 434–9.

Toffoli, C. and Margolus, N. (1987). Cellular automata machines. MIT Press, Cambridge, MA.

Whitley, D. (1988). GENITOR: a different genetic algorithm. *Rocky Mountain Conference on Artificial Intelligence*, pp. 118–30.

Yamada, T. and Nakano, R. (1992). A genetic algorithm applicable to large-scale job-shop problems. *Parallel Problem Solving from Nature* (ed. R. Männer and B. Manderick). North-Holland, Amsterdam, pp. 281–90.

THE ECOLOGY OF COMPUTATION 20

Bernardo A. Huberman

20.1 INTRODUCTION

Distributed computational systems can be thought of as a community of concurrent processes which in their interactions and lack of perfect knowledge resemble ecosystems. A theory that incorporates many of the features endemic to such systems, including distributed control, asynchrony, resource contention and extensive communication among agents is presented and its results discussed. It is also shown how global stability through local controls can be achieved by using fitness mechanisms similar to those found in nature.

Propelled by advances in software design and increasing connectivity, distributed computational systems are starting to spread throughout offices, laboratories, countries and continents. Unlike the more familiar stand-alone computers, these growing networks seldom offer centralized scheduling and resources allocation. Instead, computational processes consisting of the active execution of programs migrate from work stations to printers, servers and other machines of the network as the need arises, without *a priori* knowledge of their state or their availability at run time.

In these systems, computational processes are born, spawn new ones in remote machines of the network, die when finished, and often collaborate in the solution of problems while competing for resources contested by other processes. They thus become a community of concurrent processes which, in their interactions, strategies and lack of perfect knowledge, behave like whole ecologies. This analogy between distributed intelligence and computational ecosystems suggests the possibility of using the latter for studying issues of distributed problem solving.

In contrast to biological systems, however, computational agents are programmed to complete their tasks as soon as possible, which in turn implies a desirability for their earliest death. This task completion may also involve terminating other processes spawned to work on different aspects of the same problem, as in parallel search, where the first process to find a solution terminates the others.

Another interesting difference between biological and computational ecologies lies in the fact that for the latter the local rules (or programs for the processes) can be arbitrarily defined, whereas in biology those rules are quite fixed. Moreover, in distributed computational systems the interactions are not constrained by a Euclidean metric, so that processes separated by large physical distances can strongly affect each other by passing messages of arbitrary complexity between them. And last but not least, in computational ecologies the rationality assumption of game theory can be explicitly imposed on its agents, thereby making these systems amenable to game dynamic analyses, suitably adjusted for their intrinsic characteristics.

The appearance of computational ecosystems creates a number of challenging problems in terms of their design and implementation. At the operational level, the lack of global perspectives for determining resource allocation gives rise to a whole different approach to system-level programming and the creation of suitable languages. Just to implement procedures whereby processes can manage to compute in machines with diverse characteristics is a challenging task with no optimally known solution. Biological organizations have evolved to deal successfully with the problem of asynchronous operation with imperfect knowledge and delays, but the implementation of a computational analog is far from obvious. Nevertheless, pieces of such systems are already in place, and a serious effort at designing open computational networks is under way at a number of laboratories.

There are by now a number of distributed computational systems which exhibit many of the above characteristics and that offer increased performance when compared with traditional operating systems. **Enterprise** is a market-like scheduler where independent processes or agents are allocated at run time among remote idle workstations through a bidding mechanism. A more evolved system, **Spawn**, is organized as a market economy composed of interacting buyers and sellers. The commodities in this economy are computer processing resources; specifically, slices of CPU time on various types of computer workstations in a distributed computational environment. The system has been shown to provide substantial improvements over more conventional systems, while addressing the problems of resource contention, fair dynamic load sharing, resource management for concurrent computations, and the notion of priority in distributed systems.

From a scientific point of view, the analogy between open systems and natural ecologies brings to mind the spontaneous appearance of organized behavior in biological and social systems, where agents can engage in cooperating strategies while working on the solution of particular problems. In some cases, the strategy mix used by these agents evolves towards an asymptotic ratio which is constant in time and stable against perturbations. This phenomenon sometimes goes under the name of evolutionarily stable strategy (ESS). Recently, it has been shown that spontaneous organization can also exist in open computational systems when agents can choose among many possible strategies while collaborating in the solution of computational tasks. In this case, however, imperfect knowledge and delays in information introduce asymptotic oscillatory and chaotic states which exclude the existence of simple ESSs. This is an important finding in light of studies which resort to notions of evolutionarily stable strategies in the design and prediction of the performance of open systems.

In what follows I will describe a theory of distributed computation which we developed and which explicitly takes into account incomplete knowledge and delayed information on the part of its agents. The theory describes the collective dynamics of computational agents, while incorporating many of the features endemic to such systems, including distributed control, asynchrony, resource contention and extensive communication among agents. When processes can choose among many possible strategies while collaborating in the solution of computational tasks, the dynamics leads to asymptotic regimes characterized by complex attractors. Detailed experiments have confirmed many of the theoretical predictions, while uncovering new phenomena, such as chaos induced by overly clever decision-making procedures.

Next, I will deal with the problem of controlling chaos in such systems, for we have discovered ways of achieving global stability through local controls inspired by fitness mechanisms found in nature. Furthermore, I will show how diversity enters into the picture and the minimal amount of such diversity that is required to achieve stable behavior in a distributed computational system.

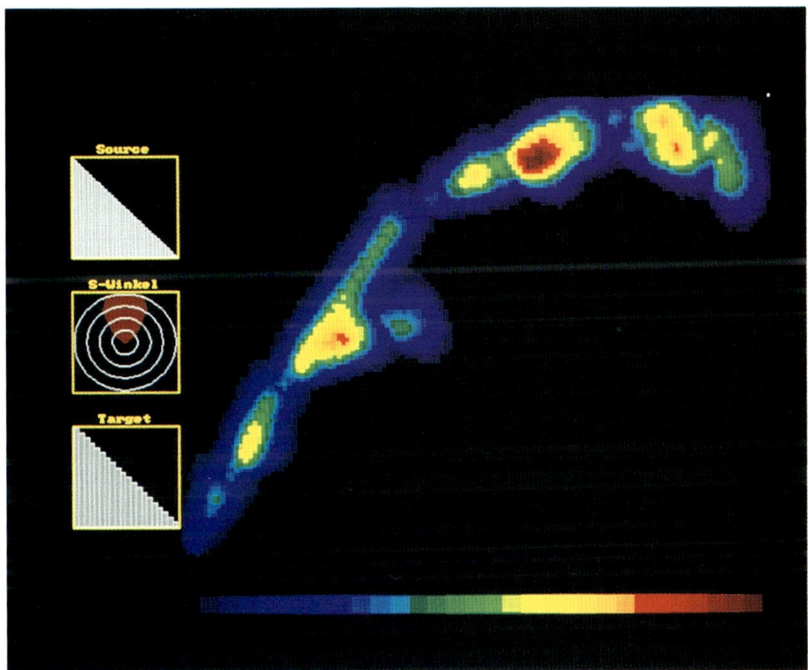

Plate 1. An example of individual chemogradient field in false colour coding. **(See Chapter 7).**

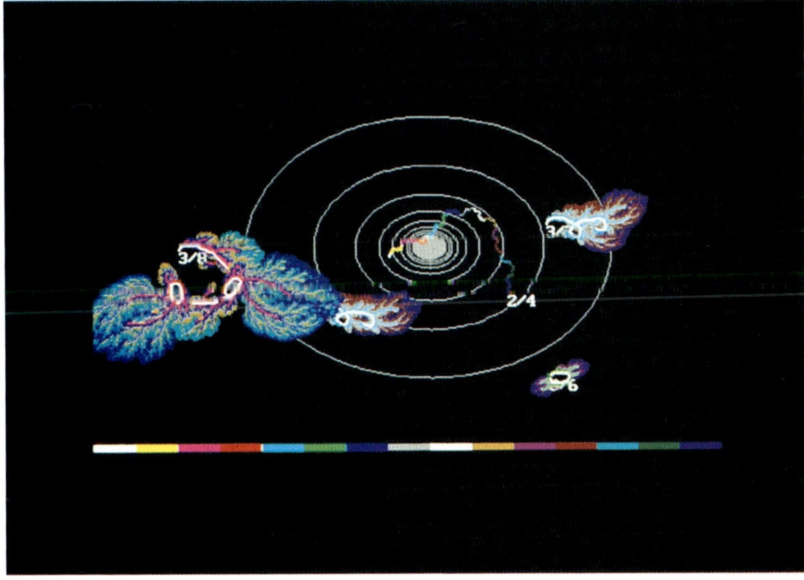

Plate 2. An example of individual growing objects. (See Chapter 7).

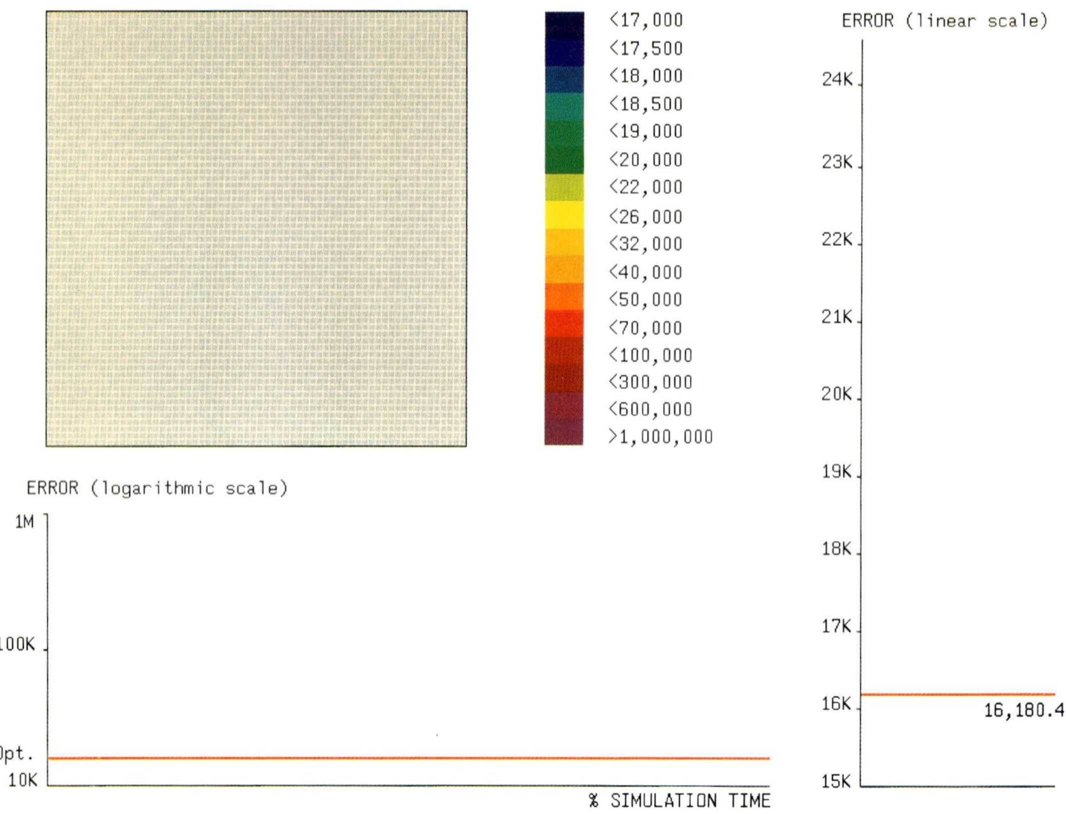

Plate 3. Graphic display of an ECO simulation environment designed for solving control problems such as the one in equation (19.1).

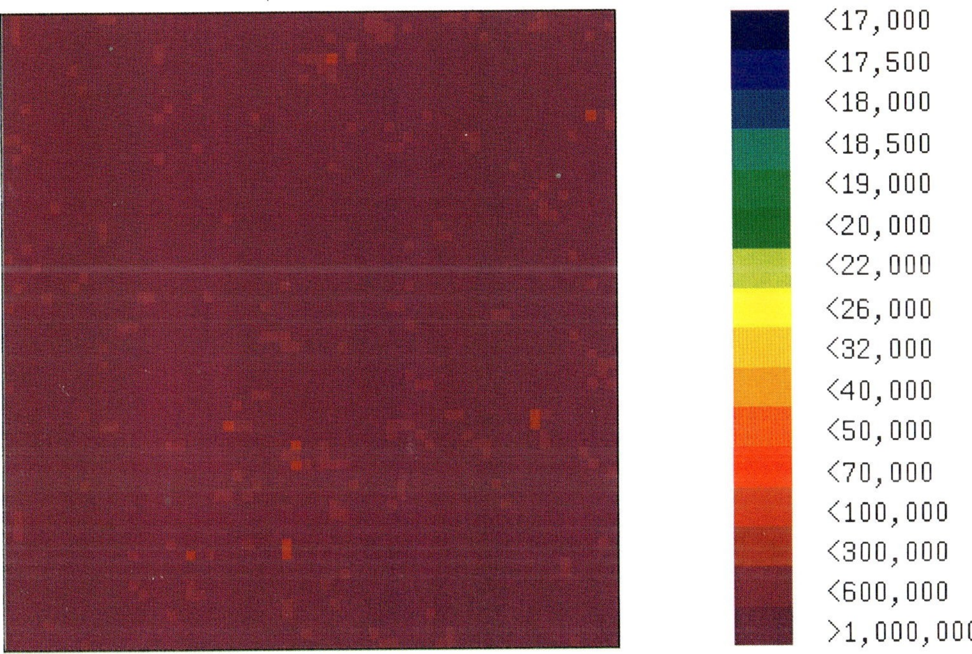

Plate 4. Time series presenting the error improvement of equation (19.1) as a function of the simulation time. (a) The state of the grid's occupants shortly after the random initialization of the simulation process.

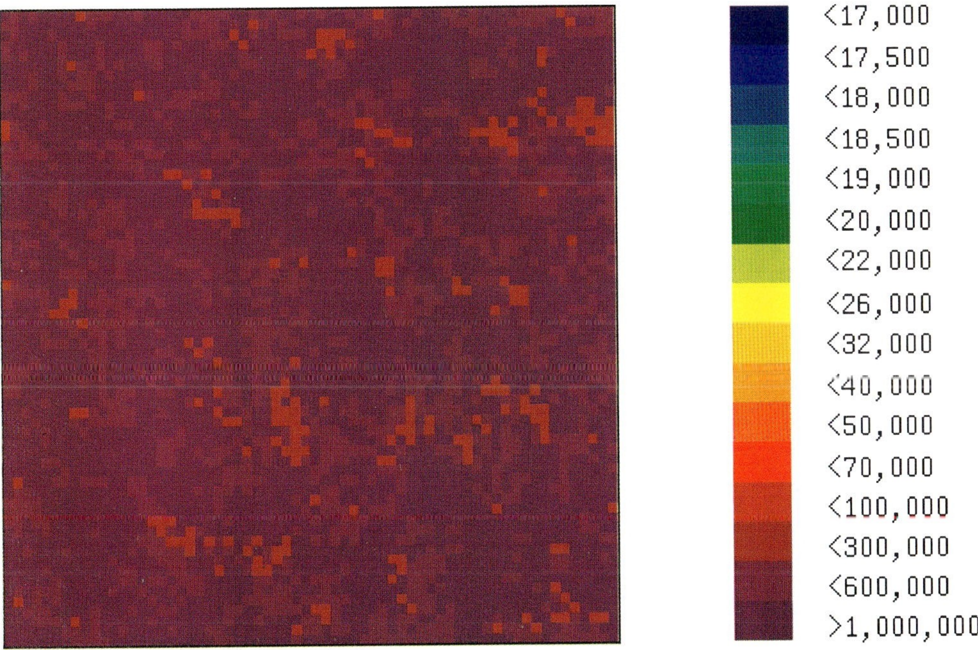

(b) A further few thousands of local interactions, and the isomorphic regions start to solidify around the population members there are of lower error value.

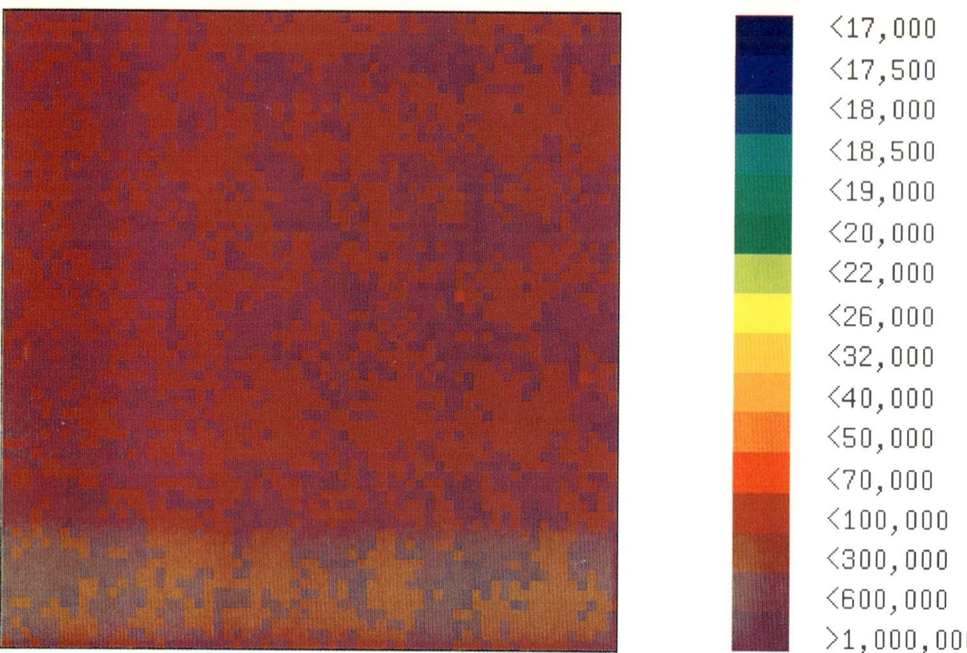

(c) About half the grid already has an error of less than 300 000.

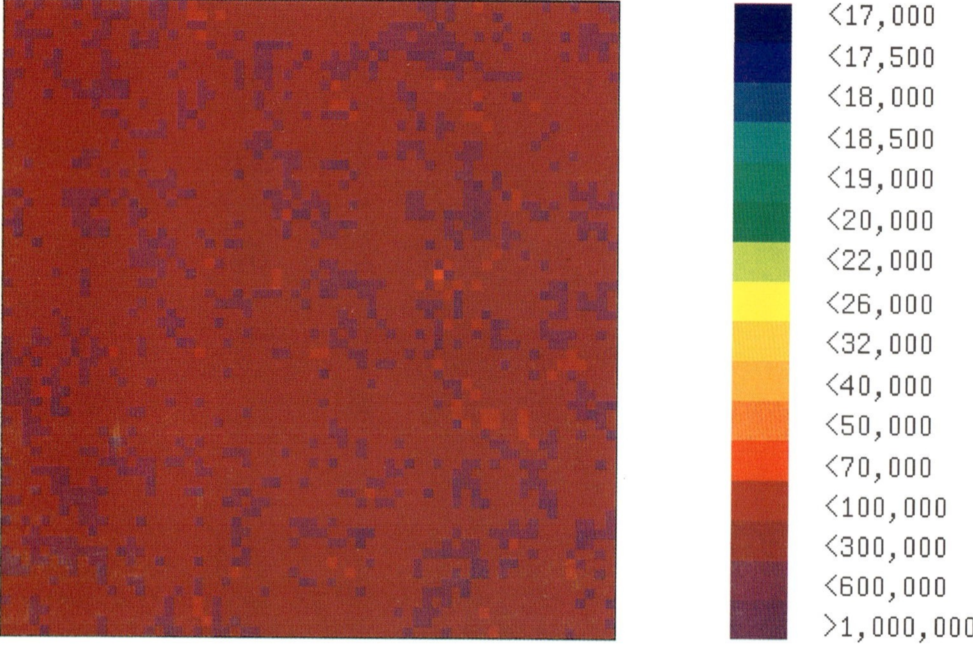

(d) Most of the grid now contains solutions with an error less than 100 000.

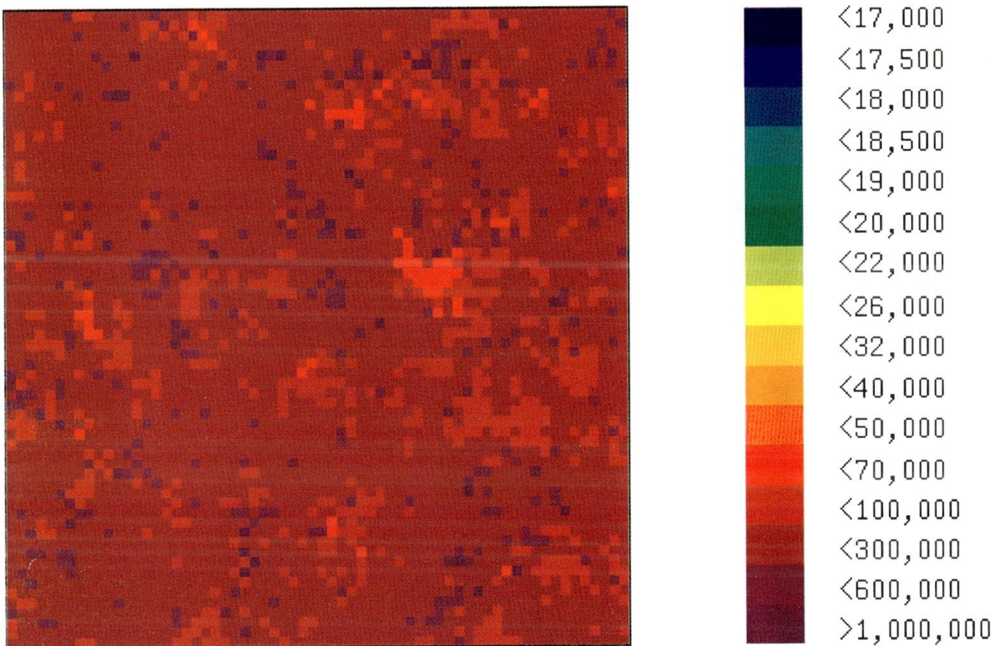

e) Further speciation, several other solutions with an error less than 70 000 have been found in different regions of the grid.

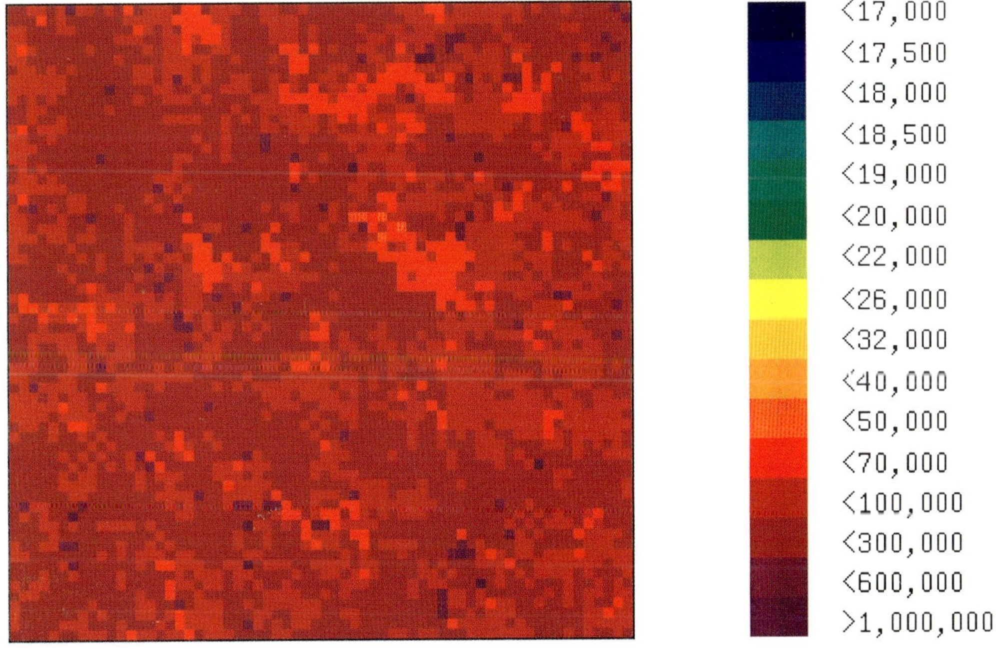

(f) The first discovery of solutions with an error less than 50 000 and further development of all other isomorphic regions.

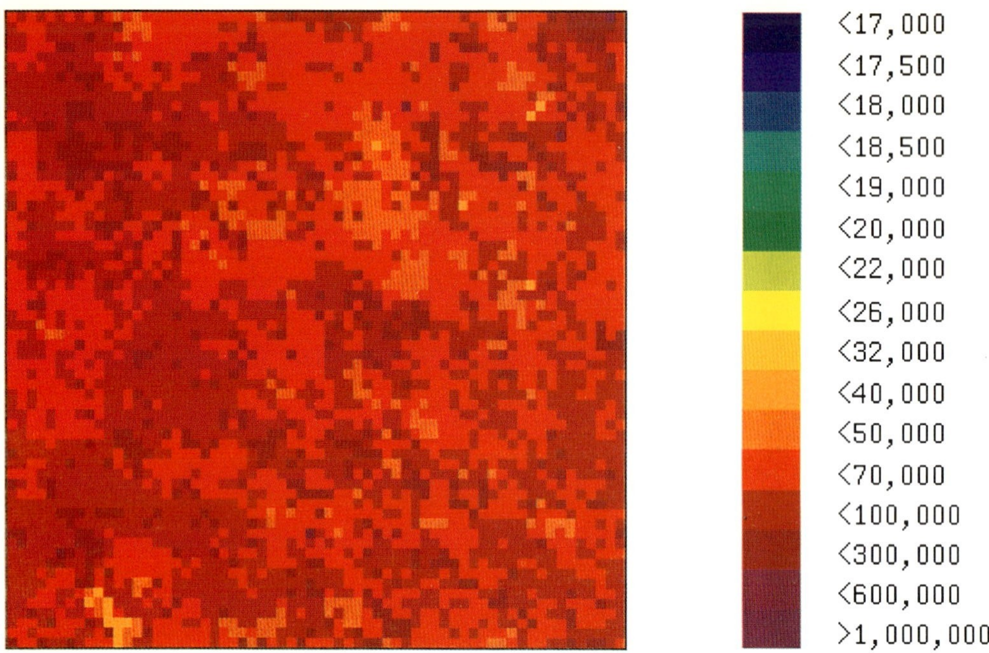

(g) As the local speciation continues, many more small isomorphic regions with an error less than 40 000 are discovered.

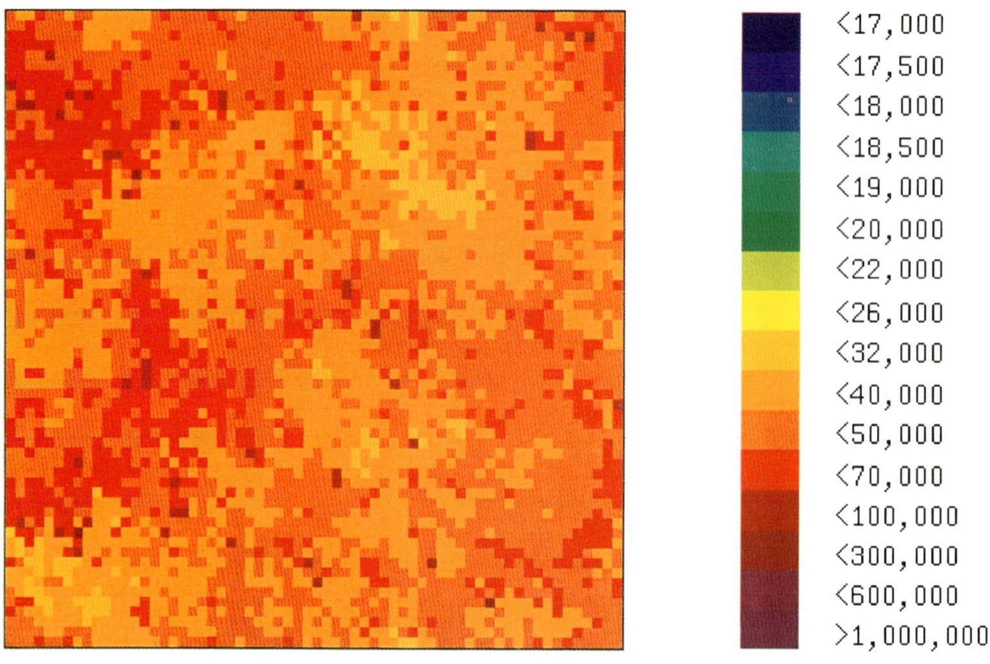

(h) Local speciation continues and there are a few regions with an error less than 32 000.

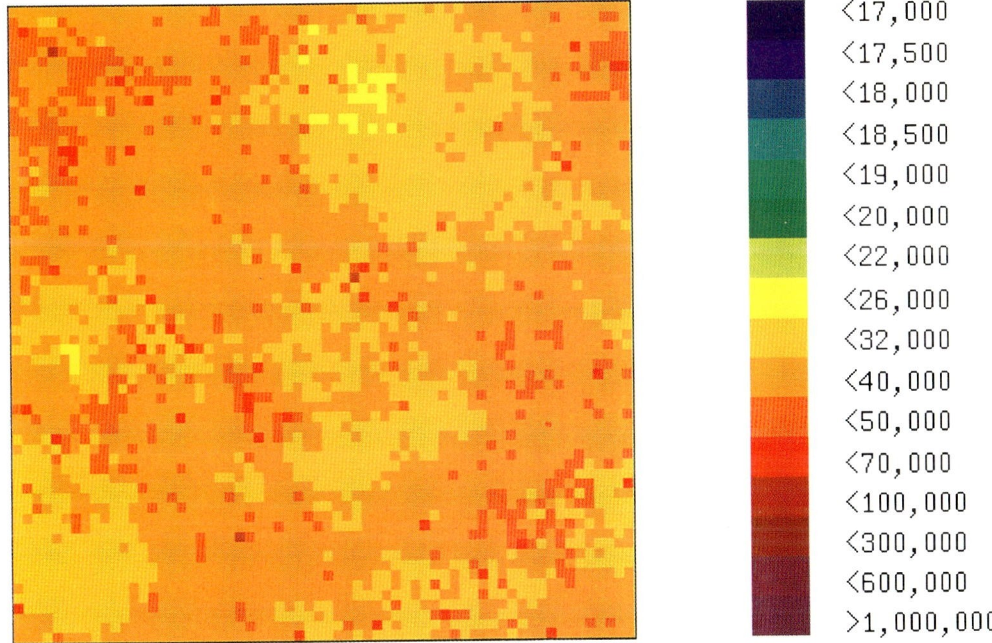

(i) The discovery of solutions with an error less than 26 000 in two distinct grid regions.

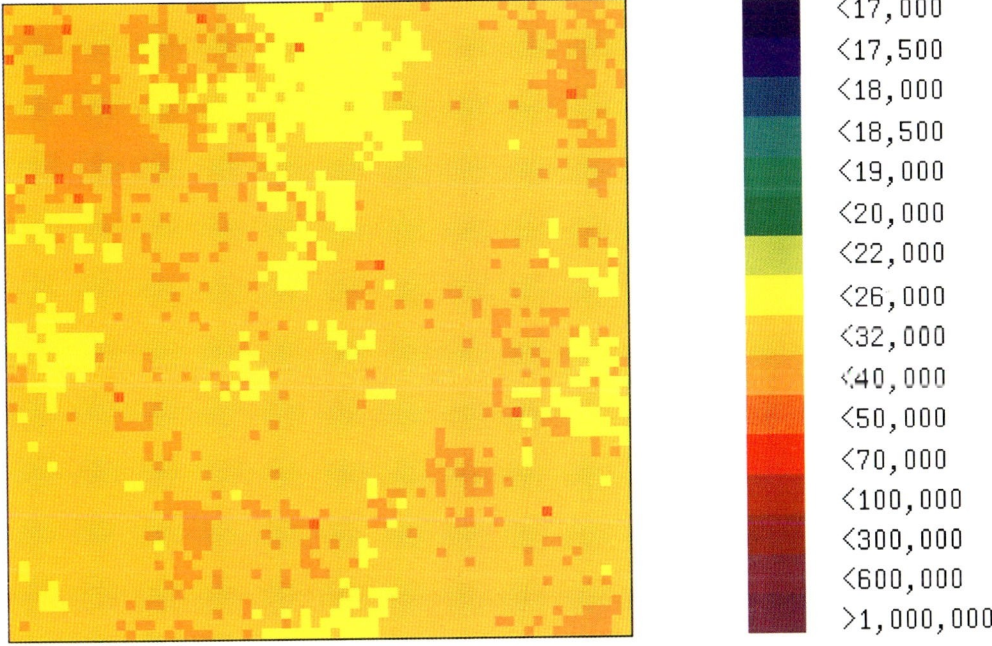

(j) Further improvements throughout the grid.

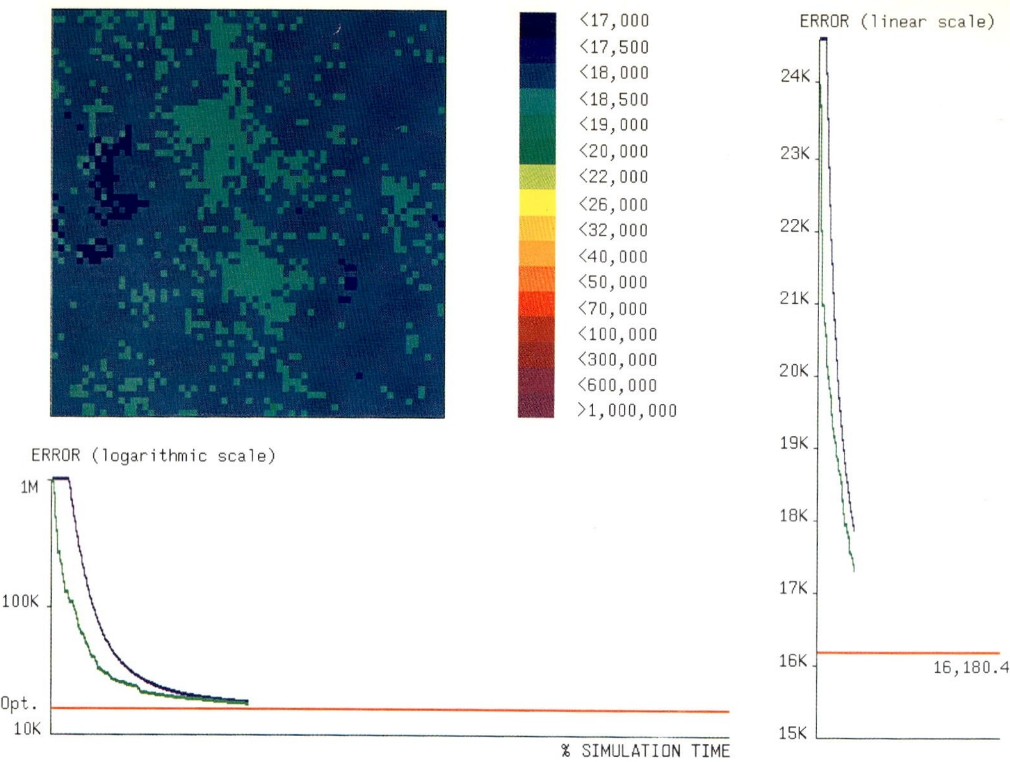

Plate 5. After about 25% of the simulation time, solutions with an error less than 17 500 are found. In the two time plots, the green line designates the best solution in the grid, and the blue line designates the average of the whole population.

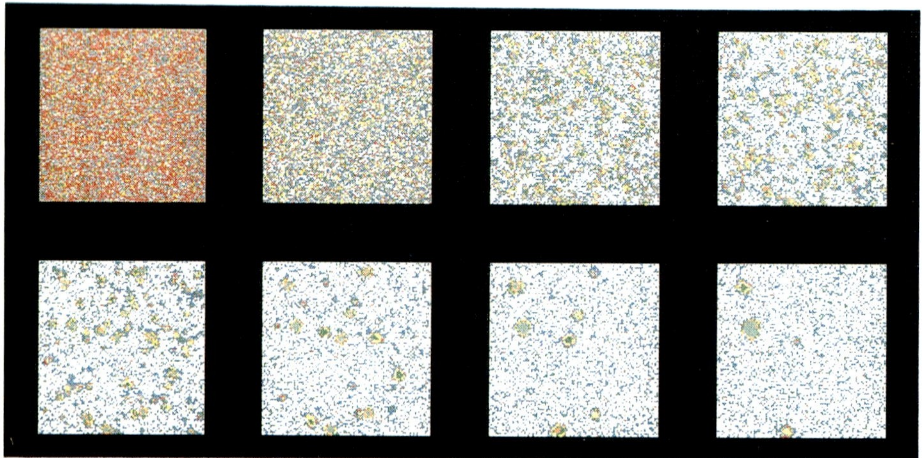

Plate 6. Life-like structures emerge in a system of abstract chemical catalysts. Components are assigned random initial properties that can change with the simulated chemical environment. Without such a context-dependent information production, there is not enough flexibility in the remaining network to develop self-sustaining cycles. This illustrates the principle of self-modification. Colour represent numbers of the respective components; each component occupies one grid point. The control parameters are:
p_1- probability of inital catalytic activity, identical for every component
p_1- per time step probability of the development of a new catalytic activity

20.2 CHAOS IN COMPUTATIONAL ECOSYSTEMS

A useful view of distributed computer systems considers them as a collection of processes or agents that choose between various resources with which to accomplish their tasks (Huberman and Hogg 1988). Since decisions are not centrally controlled, the agents independently and asynchronously select among the available choices based on their perceived payoff. These payoffs are actual computational measures of performance, such as the time required to complete a task, accuracy of the solution, amount of memory required, and so on. In general, the payoff G_r for using resource r depends on the number of agents already using it. In a purely competitive environment, the payoff for using a particular resource tends to decrease as more agents make use of it. Alternatively, the agents using a resource could assist one another in their computations, as might be the case if the overall task could be decomposed into a number of subtasks. If these subtasks communicate extensively to share partial results, the agents will be better off using the same computer rather than running more rapidly on separate machines and then being limited by slow communications. As another example, agents using a particular database could leave index links that are useful to others. In such cooperative situations, the payoff of a resource would then increase as more agents use it, until it became sufficiently crowded.

Imperfect information about the state of the system causes each agent's perceived payoff to differ from the actual value, with the difference increasing when there is more uncertainty in the information available to the agents. This type of uncertainty concisely captures the effect of many sources of errors such as some program bugs, heuristics incorrectly evaluating choices, errors in communicating the load on various machines and mistakes in interpreting sensory data. Specifically, the perceived payoffs are taken to be normally distributed, with standard deviation σ, around their correct values. In addition, information delays cause each agent's knowledge of the state of the system to be somewhat out of date. Although for simplicity we will consider the case in which all agents have the same effective delay, uncertainty and preferences for resource use, we should mention that the same range of behaviors is also found in more general situations (Miller and Drexler 1988).

As a specific illustration of this approach, we consider the case of two resources, so the system can be described by the fraction f of agents which are using resource 1 at any given time. Its dynamics is then governed by (Huberman and Hogg 1988; Kephart *et al.* 1989):

$$\frac{df}{dt} = \alpha(\rho - f) \tag{20.1}$$

where α is the rate at which agents reevaluate their resource choice and ρ is the probability that an agent will prefer resource 1 over 2 when it makes a choice. Generally, ρ is a function of f through the density-dependent payoffs. In terms of the payoffs and uncertainty, we have

$$\rho = \frac{1}{2}\left(1 + \text{erf}\left(\frac{G_1(f) - G_2(f)}{2\sigma}\right)\right) \tag{20.2}$$

where σ quantifies the uncertainty and erf is an error function. Notice that this definition captures the simple requirement that an agent is more likely to prefer a resource when its payoff is relatively large. Finally, delays in information are modeled by supposing that the payoffs that enter into ρ at time t are the values they had at a delayed time $t - \tau$.

For a typical system of many agents with a mixture of cooperative and competitive payoffs, the kinds of dynamical behaviors exhibited by the model are shown in Figure 20.1. When the delays and uncertainty are fairly small, the system converges to an

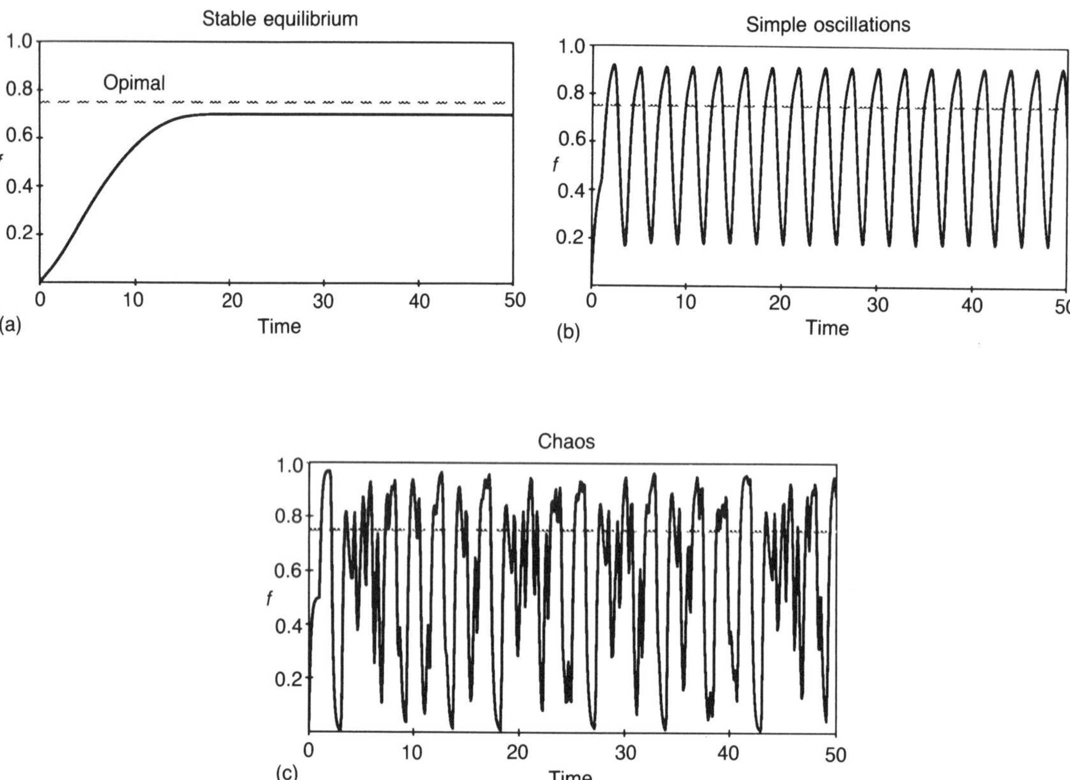

Figure 20.1 Typical behaviors for the fraction f of agents using resource 1 as a function of time for successively longer delays: (a) relaxation toward stable equilibrium; (b) simple persistent oscillations; and (c) chaotic oscillations. The payoffs are $G_1 = 4 + 7f - 5.333f^2$ for resource 1 and $G_2 = 4 + 3f$ for resource 2. The time scale is in units of the delay time r, $\sigma = 1/4$ and the dashed line shows the optimal allocation for these payoffs.

equilibrium point close to the optimal obtainable by an omniscient, central controller. As the information available to the agents becomes more corrupted, the equilibrium point moves further from the optimal value. With increasing delays, the equilibrium eventually becomes unstable, leading to the oscillatory and chaotic behavior shown in the figure. In these cases, the number of agents using particular resources continues to vary so that the system spends relatively little time near the optimal value, with the consequent drop in its overall performance.

20.3 THE USES OF FITNESS

I will now describe an effective procedure for controlling chaos in distributed systems (Hogg and Huberman 1991). It is based on a mechanism that rewards agents according to their actual performance. As we shall see, such an algorithm leads to the emergence of a diverse community of agents out of an essentially homogenous one. This diversity in turn eliminates chaotic behavior through a series of dynamical bifurcations which render chaos a transient phenomenon.

The actual performance of computational processes can be rewarded in a number of ways. A particularly appealing one is to mimic the mechanism found in biological evolution, where fitness determines the number of survivors of a given species in a changing environment. This mechanism is used in computation under the name of **genetic algorithms** (Goldberg 1989). Another example is provided by computational systems modelled on ideal economic markets (Miller and Drexler 1988; Waldspurger *et al.* 1992), which reward good performance in terms of profits. In this case, agents pay for the use of resources, and they in turn are paid for completing their tasks. Those making the best choices collect the most currency and are able to outbid others for the use of resources. Consequently they come to dominate the system.

While there is a range of possible reward mechanisms, their net effect is to increase the proportion of agents that are performing successfully, thereby decreasing the number of those who do not do as well. It is with this insight in mind that we developed a general theory of effective reward mechanisms without resorting to the details of their implementations. Since this change in agent mix will in turn change the choices made by every agent and their payoffs, those that were initially most successful need not be so in the future. This leads to an evolving diversity whose eventual stability is by no means obvious.

Before proceeding with the theory we point out that the resource payoffs that we will consider are instantaneous ones (i.e., shorter than the delays in the system), work actually done by a machine, currency actually received, and so on. Other reward mechanisms, such as those based on averaged past performance, could lead to very different behavior from the one exhibited in this chapter.

In order to investigate the effects of rewarding actual performance we generalize the previous model of computational ecosystems by allowing agents to be of different types, a fact which gives them different performance characteristics. Recall that the agents need to estimate the current state of the system based on imperfect and delayed information in order to make good choices. This can be done in a number of ways, ranging from extremely simple extrapolations from previous data to complex forecasting techniques. The different types of agents then correspond to the various ways in which they can make these extrapolations.

Within this context, a computational ecosystem can be described by specifying the fraction of agents, f_{rs} of a given type s using a given resource r at a particular time. We will also define the total fraction of agents using a resource of a particular type as

$$f_r^{\mathrm{res}} = \sum_s f_{rs}$$
$$f_s^{\mathrm{type}} = \sum_r f_{rs} \qquad (20.3)$$

respectively.

As mentioned previously, the net effect of rewarding performance is to increase the fraction of highly performing agents. If γ is the rate at which performance is rewarded, then equation (20.1) is enhanced with an extra term which corresponds to this reward mechanism. This gives

$$\frac{\mathrm{d}f_{rs}}{\mathrm{d}t} = \alpha(f_s^{\mathrm{type}}\rho_{rs} - f_{rs}) + \gamma(f_r^{\mathrm{res}}\eta_s - f_{rs}) \qquad (20.4)$$

where the first term is analogous to that of the previous theory, and the second term incorporates the effect of rewards on the population. In this equation ρ_{rs} is the probability that an agent of type s will prefer resource r when it makes a choice, and η_s is the probability that new agents will be of type s, which we take to be proportional to the actual payoff associated with agents of type s. As before, α denotes the rate at which agents make resource choices and the detailed interpretation of γ depends on the particular reward mechanism involved. For example, if they are replaced on the basis of their fitness it is the rate at which this happens. In a market system, on the other hand, γ corresponds to the rate at which agents are paid. Notice that in this case, the fraction of each type is proportional to the wealth of agents of that type.

Since the total fraction of agents of all types must be 1, a simple form of the normalization condition can be obtained if one considers the relative payoff, which is given by

$$\eta_s = \frac{\sum_r f_{rs} G_r}{\sum_r f_r^{\text{res}} G_r} \tag{20.5}$$

(This form assumes positive payoffs; for example, they could be growth rates. If the payoffs can be negative (e.g., they are currency changes in an economic system), one can use instead the difference between the actual payoffs and their minimum value m. Since the η_s must sum to 1, this will give

$$\eta_s = \frac{\sum_r f_{rs} G_r - m}{\sum_r f_r^{\text{res}} G_r - Sm} \tag{20.6}$$

which reduces to the previous case when $m = 0$.)

Note that the numerator is the actual payoff received by agents of type s given their current resource usage, and the denominator is the total payoff for all agents in the system, both normalized to the total number of agents in the system.

Summing equation (20.4) over all resources and types gives

$$\begin{aligned} \frac{df_r^{\text{res}}}{dt} &= \alpha\left(\sum_s f_s^{\text{type}} \rho_{rs} - f_r^{\text{res}}\right) \\ \frac{df_s^{\text{type}}}{dt} &= \gamma(\eta_s - f_s^{\text{type}}) \end{aligned} \tag{20.7}$$

which describes the dynamics of overall resource use and the distribution of agent types, respectively. Note that this implies that those agent types which receive greater than average payoff (i.e., types for which $\eta_s > f_s^{\text{type}}$) will increase in the system at the expense of the low-performing types.

Note that the actual payoffs can only reward existing types of agents. Thus in order to introduce new variations into the population an additional mechanism is needed (e.g., corresponding to mutation in genetic algorithms or learning).

20.4 RESULTS

In order to illustrate the effectiveness of rewarding actual payoffs in controlling chaos, we examine the dynamics generated by equation (20.4) for the case in which agents choose between two resources with cooperative payoffs, a case which we have shown to generate chaotic behavior in the absence of rewards (Kephart *et al.* 1989). As in the particular example of Figure 20.1(c), we use $\tau = 10$; $G_1 = 4 + 7f_1 - 5.333f_1^2$; $G_2 = 7 - 3f_2$; $\sigma = \frac{1}{4}$ and an initial condition in which all agents start by using resource 2.

One kind of diversity among agents is motivated by the simple case in which the system oscillates with a fixed period. In this case, those agents that are able to discover the period of the oscillation can then use this knowledge to reliably estimate the current system state in spite of delays in information. Notice that this estimate does not necessarily guarantee that they will keep performing well in the future, for their choice can change the basic frequency of oscillation of the system.

In what follows, we take the diversity of agent types to correspond to the different past horizons, or extra delays, that they use to extrapolate to the current state of the system. These differences in estimation could be due to having a variety of procedures for analyzing the system's behavior. Specifically, we identify different agent types with the different assumed periods which range over a given interval. Thus, we take agents of type s to use an effective delay of $\tau + s$ while evaluating their choices.

The resulting behavior is shown in Figure 20.2 which should be contrasted with Figure 20.1(c). We used an interval of extra delays

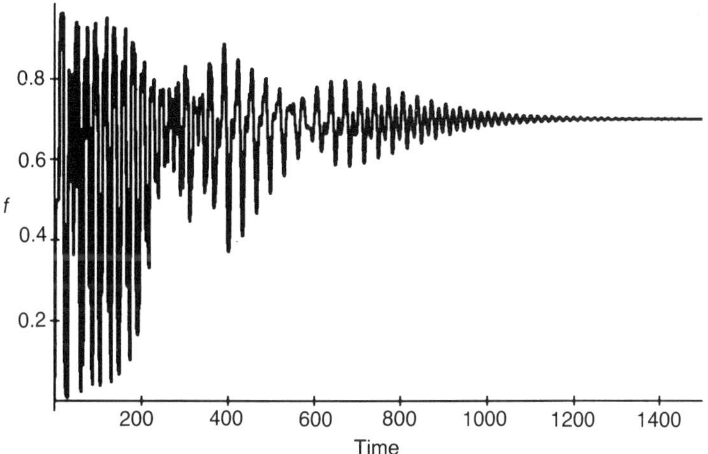

Figure 20.2 Fraction of agents using resource 1 as a function of time with adjustment based on actual payoff. These parameters correspond to Figure 20.1(c), so without the adjustment the system would remain chaotic.

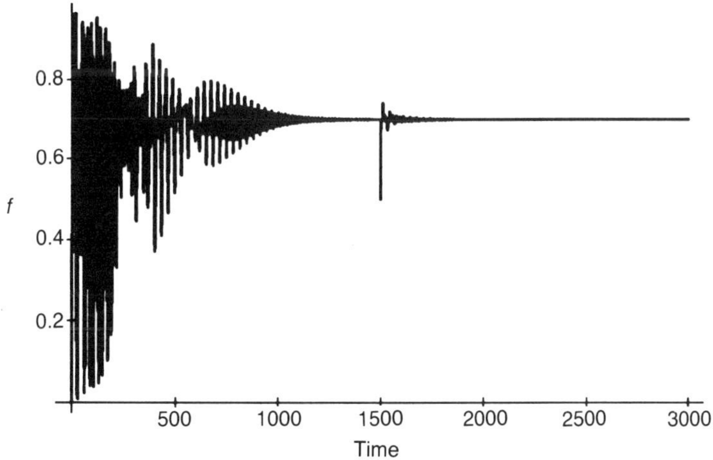

Figure 20.3 Behavior of the system shown in Figure 20.2 with a perturbation introduced at time 1500.

ranging from 0 to 40. As shown, the introduction of actual payoffs induces a chaotic transient which, after a series of dynamical bifurcations, settles into a fixed point that signals stable behavior. Furthermore, this fixed point is exactly that obtained in the case of no delays. This equilibrium is stable against perturbations can be seen by the fact that if the system were perturbed again (as shown in Figure 20.3), it rapidly returns to its previous value. In additional experiments, with a smaller range of delays, we found that the system continued to oscillate without achieving the fixed point.

This transient chaos and its eventual stability can be understood from the distribution of agents with extra delays as a function of time. As can be seen in Figure 20.4 actual

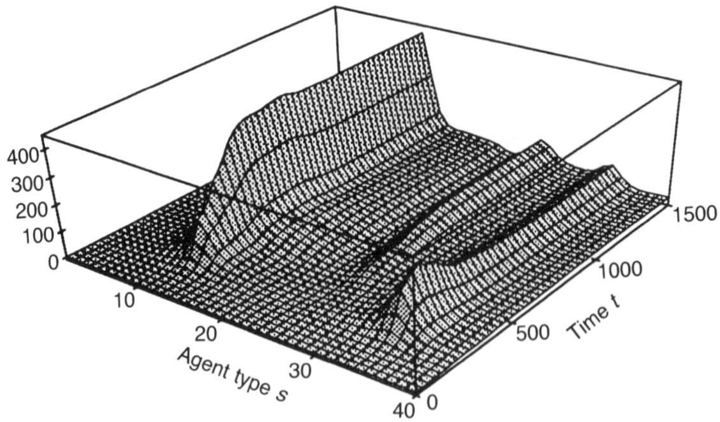

Figure 20.4 Ratio $f_s^{\text{type}}(t)/f_s^{\text{type}}(0)$ of the fraction of agents of each type, normalized to their initial values, as a function of time. Note there are several peaks, which correspond to agents with extra delays of 12, 26 and 34 time units. Since $r = 10$, these match periods of length 22, 36 and 44 respectively.

payoffs lead to a highly heterogeneous system, characterized by a diverse population of agents of different types. It also shows that the fraction of agents with certain extra delays increases greatly. These delays correspond to the major periodicities in the system.

20.5 STABILITY AND MINIMAL DIVERSITY

As we showed in the previous section, rewarding the performance of large collections of agents engaging in resource choices leads to a highly diverse mix of agents that stabilize the system. This suggests that the real cause of stability in a distributed system is that provided by sufficient diversity, and that the reward mechanisms are an efficient way of automatically finding a good mix.

This raises the interesting question of the minimal amount of diversity needed in order to have a stable system. The stability of a system is determined by the behavior of a perturbation around equilibrium, which can be found from the linearized version of equation (20.4).

In our case, the diversity is related to the range of different delays that agents can have. For a continuous distribution of extra delays, the characteristic equation is obtained by assuming a solution of the type $e^{\lambda t}$ in the linearized equation, giving

$$\lambda + \alpha - \alpha\rho' \int ds\, f(s)\, e^{-\lambda(s+\tau)} = 0. \quad (20.8)$$

Stability requires that all the values of λ have negative real parts, so that perturbations will relax back to equilibrium. As an example, suppose agent types are uniformly distributed in $(0, S)$. Then $f(s) = 1/S$, and the characteristic equation becomes.

$$\lambda + \alpha - \alpha\rho' \frac{1 - e^{-\lambda S}}{\lambda S} e^{-\lambda\tau} = 0. \quad (20.9)$$

Defining a normalized measure of the diversity of the system for this case by $\eta \equiv S/\tau$, introducing the new variable $z \equiv \lambda\tau(1 + \eta)$, and multiplying equation (20.9) by $\tau(1 + \eta)ze^{z}$ introduces an extra root at $z = 0$ and gives

$$(z^2 + az)e^{z} - b + be^{rz} = 0 \quad (20.10)$$

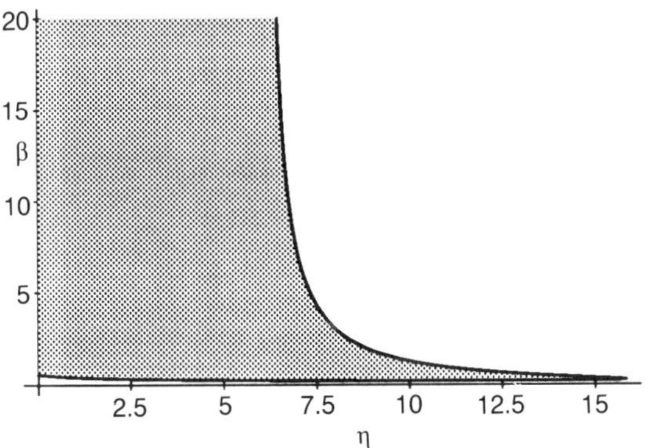

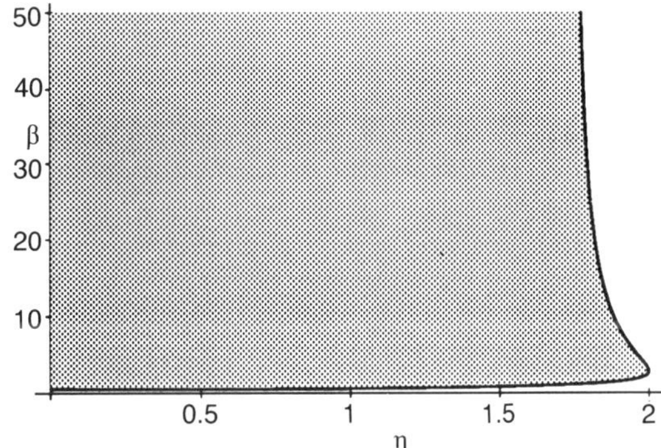

Figure 20.5 Stability as a function of $\beta = \alpha r$ and $\eta = S/r$ for two possible distributions of agent types: (a) $f(s) = 1/S$ in $(0, S)$; and (b) $f(s) = (1/S)e^{-s/S}$. The system is unstable in the shaded regions and stable to the right and below the curves.

where

$$a = \alpha\tau(1 + \eta) > 0$$

$$b = -\rho'\frac{\alpha\tau(1 + \eta)^2}{\eta} > 0 \qquad (20.11)$$

$$r = \frac{\eta}{1 + \eta} \in (0, 1).$$

The stability of the system with uniform distribution of agents with extra delays thus reduces to finding the condition under which all roots of equation (20.10) other than $z = 0$, have negative real parts. This equation is a particular instance of an exponential polynomial, whose terms consist of powers multiplied by exponentials. Unlike regular polynomials, these objects generally have an infinite number of roots, and are important in the study of the stability properties of differential-delay equations (Bellman and Cooke 1963). Established methods can then be used to determine when they have roots with positive real parts. This in turn defines the stability boundary of the equation. The result for the particular case in which $\rho' = -3.41044$, corresponding to the parameters used in Section 20.4, is shown in the left half of Figure 20.5.

Similarly, if we choose an exponential distribution of delays, i.e., $f(s) = \left(\frac{1}{S}\right)e^{-s/S}$ with positive S, the characteristic equation acquires the form

$$(z^2 + pz + q)e^z + r = 0 \qquad (20.12)$$

where

$$p = \alpha\tau + 1/\eta > 0$$

$$q = \alpha\tau/\eta > 0 \qquad (20.13)$$

$$r = -\alpha\tau\rho'/\eta > 0$$

and $z \equiv \lambda\tau$. An analysis similar to that for the uniform distribution case leads to the stability diagram shown on the right-hand side of the figure.

Although the actual distributions of agent types can differ from these two cases, the similarity between the stability diagrams suggests that regardless of the magnitude of β one can always find an appropriate mix that will make the system stable. This property follows from the vertical asymptote of the stability boundary. It also illustrates the need for a minimum diversity in the system in order to make it stable when the delays are not too small.

Having established the right mix that produces stability one may wonder whether a static assignment of agent types at an initial time would not constitute a simpler and more direct procedure to stabilize the system without resorting to a dynamic reward mechanism. While this is indeed the case in a nonfluctuating environment, such a static mechanism cannot cope with changes in both the nature of the system (e.g., machines crashing) and the arrival of new tasks or fluctuating loads. It is precisely to avoid this vulnerability by keeping the system adaptive that a dynamic procedure is needed.

Having seen how sufficient diversity stabilizes a distributed system, we now turn to the mechanisms that can generate such heterogeneity, as well as the time that it takes for the system to stabilize.

In particular, the details of the reward procedures determine whether the system can even find a stable mix of agents. In the cases described above, reward was proportional to actual performance, as measured by the payoffs associated with the resources used. One might also wonder whether stability would be achieved more rapidly by giving greater (than their fair share) increases to the top performers.

We have examined two such cases: (a) rewards proportional to the square of their actual performance; and (b) giving all the rewards to top performers (e.g., those performing at the 90th percentile or better in the population). In the former case we observed stability with a shorter transient, whereas in the latter case the mix of changes continued to change through time, thus preventing stable behavior. This can be understood in terms of our earlier observation that whereas a small percentage of agents can identify oscillation periods and thereby reduce their amplitude, a large number of them can no longer perform well.

Note that the time to reach equilibrium is determined by two parameters of the system. The first is the time that it takes to find a stable mix of agent types, which is governed by γ, and the second the rate at which perturbations relax, given the stable mix. The latter is determined by the largest real part of any of the roots, λ, of the characteristic equation.

20.6 DISCUSSION

In this chapter I have presented a case for treating distributed computation as an ecosystem, an analogy that turns out to be quite fruitful in the analysis, design and control of such systems. In spite of the many differences between computational processes and organisms, resource contention, complex dynamics and reward mechanims seem to be ubiquitous in distributed computation, making it also a tool for the study of natural ecosystems.

Since chaotic behavior seems to be the natural result of interacting processes with imperfect and delayed information, the problem of controlling such systems is of paramount importance. We discovered that rewards based on the actual performance of agents in a distributed computational system can stabilize an otherwise chaotic or oscillatory system. This leads in turn to greatly improved system performance.

In all these cases, stability is achieved by making chaos a transient phenomenon. In the case of distributed systems, the addition of the reward mechanism has the effect of dynamically changing the control parameters of the resource allocation dynamics in such a way that a global fixed point of the system is achieved. This brings the issue of the length of the chaotic transient as compared to the time needed for most agents to complete their tasks. Even when the transients are long, the results of this study show that the range gradually decreases, thereby improving performance even before the fixed point is achieved.

A particularly relevant question for distributed systems is the extent to which these results generalize beyond the mechanism that we studied. Since we only considered the specific situation of a collection of agents with different delays in their appraisal of the system evolution, it is of interest to enquire whether using rewards to increase diversity works more generally than in the case of extra delays.

Since we only considered agents choosing between only two resources, it is important to understand what happens when there are many resources the agents can choose from. One may argue that since diversity is the key to stability, a plurality of resources provides enough channels to develop the necessary heterogeneity, which is what we observed in situations with three resources. Another note of caution has to do with the effect of fluctuations on a finite population of agent types. While we have shown that sufficient diversity can, on average, stabilize the system, in practice a fluctuation could wipe out those agent types that would otherwise be successful in stabilizing the system. Thus, we need either a large number of each kind of agent or a mechanism, such as mutation, to create new kinds of agents.

A final issue concerns the time scales over which rewards are assigned to agents. In our treatment, we assumed the rewards were always based on the performance at the time they were given. Since in many cases this procedure is delayed, there is the question of the extent to which rewards based on past performance are also able to stabilize chaotic distributed systems.

REFERENCES

Bellman, R. and Cooke, K. L. (1963). *Differential-Difference Equations*. Academic Press, New York.

Goldberg, D. E. (1989). *Genetic Algorithms in Search, Optimization and Machine Learning*. Addison-Wesley, Reading, MA.

Hogg, T. and Huberman, B. A. (1991). Achieving global stability through local controls. *Proceedings of the 1991 IEEE International Symposium on Intelligent Control*, p. 67.

Huberman, B. A. and Hogg, T. (1988). The behavior of computational ecosystems. In *The Ecology of Computation* (ed. B. A. Huberman). North-Holland, Amsterdam, pp. 77–115.

Kephart, J. O., Hogg, T. and Huberman, B. A. (1989). Dynamics of computational ecosystems. *Physical Review A* **40** 404–21.

Miller, M. S. and Drexler, K. E. (1988). Markets and computation: agoric open systems. In *The Ecology of Computation* (ed. B. A. Huberman). North-Holland, Amsterdam, pp. 133–76.

Waldspurger, C. A., Hogg, T., Huberman, B. A., Kephart, J. O. and Stornetta, S. (1992). Spawn: A distributed computational economy. *IEEE Trans. Software Engineering* **18**, 103.

AN ECOLOGICAL ANALYSIS OF A SYSTEM FOR DETECTING NODS OF THE HEAD

21

Ian Horswill

21.1 INTRODUCTION

In this chapter, I will discuss how we can analyze and understand the control and perceptual systems of agents that are implicitly adapted to their environments, and analyze a simple perceptual system that detects nods of the head. Our goal will be to tease apart the ways in which an agent is specialized to its environment and to make that specialization explicit. I will show how we can do this by finding a series of formal transformations, each representing a distinct specialization, that will produce the specialized agent from a more general agent. Throughout the chapter, I wil approach the problem with a definite engineering orientation: my principal goal is to build and understand artifacts, not to reverse-engineer biological organisms. There is, of course, an important dialog between these two pursuits, as this volume will attest to. Many of the ideas in this work are drawn from the notions of biology. The reader should not assume however that biology can be considered to be a form of engineering, either forward or reverse. As an engineer, I can usually assume that I have a specific task that I must make my agent carry out. Such assumptions are much more speculative in biology.

Science and mathematics each seek general principles. Engineering often seeks to build general mechanisms: single mechanisms that can perform a broad range of tasks. In engineering however, generality often comes at a high price. General mechanisms may be less efficient, more fragile, or simply more expensive to construct than more specialized ones. Engineers are thus forced to build specialized mechanisms in many cases, and must content themselves with searching for general principles for building specialized systems.

In the case of systems that interact with their environments, this specialization can take the form of specialization to a particular task or tasks, or to a particular form of environment. Computer scientists routinely construct systems that are specialized to a given task, such as payroll or data analysis. The relationship between tasks and the structure of computer programs that are specialized to them is relatively well understood. Often the structure of the task is directly reflected in the division of the program into modules and subroutines.

Specialization to an environment is a very different matter. Whereas the designer of an artificial agent often has an explicit, or even formally defined, task for the agent, the designer rarely has an explicit definition of the environment in which the agent is to operate (an exception being those who create simulated worlds specifically for their artificial agents). When I build a robot to operate in my office building, I have no formal specification of 'office-buildingness' from which to proceed. Whereas traditional computer science is thought of as being a branch of applied mathematics, an axiomatizable domain within which absolutely true conclusions can be deduced from absolutely true premises by way of universal rules, the study of environments is, of necessity, a form of natural science: hypotheses can be made and tested, but the domain is never completely specified by a set of *a priori* premises.

This empirical character complicates work on specialized agents in two ways. It complicates synthesis (design) because the designer never has a fully worked out theory of the environment from which possible specializations can be deduced. It complicates analysis because explanations of the agent's behavior are contingent on the validity of the domain theory upon which the explanation was based. I will focus on the latter here.

21.1.1 PREVIOUS WORK

Relatively little attention has been devoted to environmental specialization in computer science, mostly likely because it is only recently that we have begun to construct computational systems that are closely coupled to outside environments.

In biology, a great deal of attention has been given to the specialization of complete agents to their environments. Cybernetics, the progenitor of artificial intelligence, also focused on the agent/environment interactions, although not necessarily on the properties of specific, complex environments (Wiener 1961). Ideas from these areas are now being applied to artificial intelligence and robotics (McFarland 1991; Paton *et al.* 1992; Meyer and Guillot 1991).

In perceptual psychology, Gibson proposed an 'ecological' theory of perception that stressed the role of the environment in forming an agent's perceptions. Gibson argued that the structure of an environment determines a number set of invariants in the energy flowing through the environment and that these invariants can be directly picked up by the perceptual apparatus of the organism via a process akin to resonance.

Marr (1982) argued that in order to properly understand the operation of a perceptual system (or more generally, of any AI system), we must understand the problem it solves at the level of a computational theory. (Note: Marr's actual story is more complicated than this, and used three levels of explanation, not two. See Marr (1982), and Section 21.6.) The computational theory defines the desired input–output behavior of the perceptual system, along with a set of constraints on the possible interpretations of a given input. The constraints were necessary because a single stimulus can usually be generated by an infinite number of different external events. The value of the computational theory was that it was a high level theory that abstracted away from the details of an individual mechanism. A single computational theory could be used to explain and unify an infinite number of mechanisms that instantiated it. The role of the constraints within the computational theory was to show how the structure of the environment made interpretation possible. I will build upon this approach below. Marr believed that the human visual system was a general mechanism for constructing three-dimensional descriptions of the environment, and so was relatively unconcerned with understanding how a system could be specialized to take advantage of fortuitous properties of the environment.

Rosenschein and Kaelbing have formalized activity in terms of interacting finite state machines (FSMs) representing the agent and environment (Rosenschein 1987; Rosenschein and Kaelbling 1986). The formalization allows specialized mechanisms to be directly synthesized from descriptions of desired behavior and a formalization of the behavior of the environment. The formalization was powerful enough to form the basis of a programming language used to program a real robot. Later, Rosenschein developed a method for synthesizing automata whose internal states had provable correlations to the state of the environment given a set of temporal logic assertions about the dynamics of the environment. Donald and Jennings (1992) use a geometric, but similar, approach for constructing virtual sensors.

Wilson (1991) has specifically proposed the classification of simulated environments based on the types of mechanisms which can operate successfully within them. Like Rosenschein, Wilson modeled environments as finite state machines. He divided environments into three classes based on properties such as determinacy. Todd and Wilson (1993) have used the FSM approach to begin to build a taxonomy of grid worlds for a class of artificial agents created by a genetic algorithm. Littman (1993) has also used the FSM approach to classify environments for reinforcement learning (RL) algorithms. Littman parameterized the complexity of RL agents in terms of the amount of local storage they use and how far into the future the RL algorithm looks. He then classified environments in terms of the parameters of the simplest agent that could be found to learn the optimal control policy.

21.1.2 A TRANSFORMATIONAL APPROACH

We need theoretical tools that can (1) make explicit an agent's dependence on its task and environment, (2) make explicit the manner in which that dependence simplifies the agent's computational problems, (3) predict the agent's performance in novel domains and (4) allow us to recycle insights used in the design of the agent for use in other agents. In this chapter, I will discuss some first steps toward these goals. These steps are based on two ideas. The first is to reinterpret Marr's notion of a constraint to be a constraint on possible environments. Rather than viewing a constraint as restricting the possible interpretations of a scene, we will view it as restricting the possible environments in which a perceptual system will provide the correct interpretation or, more generally, the possible environments in which an agent as a whole will perform its task.

The second idea is to treat specialization as a kind of optimization. By an 'optimization' I mean a discrete operation which transforms one system (an agent or component of an agent) into another that is somehow better but not necessarily optimal. This is the sense used in compiler optimization in computer science (Aho *et al.* 1986), where mechanically deriving an optimum is often either intractable or uncomputable. It is to be distinguished from the more common sense of an extended process of finding a truly optimal solution. This latter sense is used in evolutionary theory (McFarland 1991; or Houston 1991).

We can view a specialized system as a conditionally optimized version of an idealized general system, one which performs better than the general system in some way, but which only works in environments that satisfy some constraint. Roughly, we will view the optimizations as the form of the specialization, and the constraints as the reasons for their validity. By finding a series of small optimizations, each with its own constraint, that collectively transform the general system into the specialized system, we can tease apart the additional assumptions made by the specialized system over the general system and bring out their computational ramifications.

In the next two sections, I will describe the approach somewhat more formally. Then in Section 21.4, I will apply the technique to the analysis of a nod-of-the-head detector. Then, in Section 21.5, I will discuss a complex vision-based robot based on these ideas. Finally, I will compare this approach to the approach of Marr in more depth and suggest general conclusions about the nature of AI.

21.2 DEFINING AGENTS AND ENVIRONMENTS

We will treat the world as a dynamic system and assume that it can be roughly separated into two interacting actors, the agent and the agent's environment. The world here need not mean the entire physical universe, only that portion of it which is relevant to our analysis. If we build a robot to operate in an office building, then the robot's environment would be the office building. The planet Pluto would be largely irrelevant. If the agent and environment are roughly separable, then we can imagine different possible agents and environments interacting. As agents and environments interact, they will give rise to different behaviors of the complete dynamic system (the world). In some cases, we will be able to replace an agent with another agent, or an environment with another environment, without affecting the qualitative behavior of the world.

Beyond this, I will not try to define what agents, environments, or behavioral equivalence are. I will simply assume a set $\mathcal{A}$ of agents and a set $\mathcal{E}$ of environments. The possible worlds are agent/environment pairs, that is, elements of the set $\mathcal{A} \times \mathcal{E}$. We will treat constraints as sets of environments (subsets of $\mathcal{E}$). We will represent behavioral equivalence with the $\equiv$ relation. For two agent/environment pairs (a_1, e_1) and (a_2, e_2), we will write $(a_1, e_1) \equiv (a_2, e_2)$ when their behavior is equivalent.

We can define the equivalence of agents in terms of the equivalence of worlds (agent/environment pairs). We will say that two agents are behaviorally equivalent if their interactions with all possible environments are equivalent, that is

$$a_1 \equiv a_2 \quad \text{iff} \quad \forall e_1, e_2\ (a_1, e_1) \equiv (a_2, e_2).$$

We can also define conditional equivalence (equivalence given a domain constraint) in terms of equivalence of worlds. We will say that two agents are equivalent given a constraint $C \subseteq \mathcal{E}$ if they give rise to equivalent behavior in all environments satisfying that constraint. We will write this $a_1 \overset{C}{\equiv} a_2$. Thus

$$a_1 \overset{C}{\equiv} a_2 \quad \text{iff} \quad \forall\, a_1, e_2 \in C\ [(a_1, e_1) \ldots] \equiv (a_2, e_2)$$

In the discussion that follows, we will assume that the designer has a particular class of behavior that they want the agent to achieve. Then, the only useful behavioral distinction is whether the agent 'works' or not, and so the $\equiv$ relation will divide the possible behaviors into only two classes: the working ones and the non-working ones. This is a reasonable assumption in engineering, although it may not be reasonable in biology. For a given agent, we can distinguish two types of environments – those with which the agent 'works' and those with which it does not. We will call the set of environments with which the agent works the agent's **habitat**. Thus, a habitat is formally like a constraint – it is a subset of $\mathcal{E}$. Moreover, a constraint on environments is also a constraint on habitats: if an agent only works in environments that satisfy a constraint C, then its habitat is restricted to be a subset of C. For this reason, I will sometimes refer to constraints as habitat constraints.

These definitions ignore many important points. Instantiating them with actual formal definitions of $\mathcal{A}$, $\mathcal{E}$ and $\equiv$ can involve a great deal of work. For examples of attempts to define large classes of agents and environments, see Rosenschein and Kaelbling (1986), Wilson (1991), Smithers (1992), Beer (1992) or Horswill (1993).

It is important to note that the dividing line between agent and environment is largely arbitrary. For a human, we would be perfectly free to draw it at the boundary of the body, at the outputs of the motor neurons, the outputs of the motor cortex, or even at the boundary of his/her automobile, if they happened to be driving at the time. There is no compelling ontological reason to place the dividing line at the body. One consequence of this is that we can analyze the specialization of individual components of agents simply by placing the dividing line around the component rather than the agent as a whole. The rest of the agent then becomes a part of the environment of the component.

21.3 SPECIALIZATION AS OPTIMIZATION

Suppose we want to understand an agent S that is somehow specialized to its environment. Since S is specialized, it is efficient, but does not work in all environments. Rather, it works in some restricted habitat, $H_s \subseteq \mathscr{E}$. We would like to determine both what H_s is and also why S is restricted to it.

Suppose there is some more general system G (i.e., a system whose habitat H_G is larger than H_s but which may be less efficient) that solves the task we are interested in engineering. If we can find a sequence of mechanisms S_i and domain constraints C_i, such that

$$G \overset{C_1}{\equiv} S_1 \overset{C_2}{\equiv} S_2 \ldots \overset{C_n}{\equiv} S$$

then we have that $G \overset{C_1 \cap \ldots \cap C_n}{\equiv} S$. We can phrase this latter statement in English as: within the environments that satisfy $C_1, \ldots, C_n$, G and S are behaviorally equivalent – they will work in exactly the same cases. This lets us express the habitat of S in terms of the habitat of G:

$$H_S \supseteq H_G \cap C_1 \cap \ldots \cap C_n.$$

Note that the left- and right-hand sides are not necessarily equal because there may be situations where S works but G does not. One of the constraints on the right-hand side might also be overly strong. Thus the derivation only provides a lower bound on the habitat of S.

I will call such a sequence of equivalences, in which G is gradually transformed into S, a derivation of S from G, in analogy to the derivations of equations. As the derivation proceeds from start to finish, the agent becomes progressively more specialized. With each step, the agent specialized a little and its performance is optimized a little. The specialization and the optimization go together: the specialization (the constraint) allows the transformation from one mechanism to the next to preserve behavioral equivalence. The benefits of performing the derivation are that it (1) breaks S's specialization into smaller pieces that are easier to understand and (2) places constraints and optimizations in correspondence, thus making the computational value of each constraint explicit.

By teasing apart the different assumptions made by the mechanism and placing them in correspondence with optimizations, we can predict the performance of a mechanism in novel environments. If the environment satisfies all the mechanism's constraints, the mechanism will work. If it does not, then we know which optimizations will fail, and consequently, what parts of the design of the mechanism will need to be modified.

Derivations are primarily a tool for *post hoc* analysis of specialized agents. They have an additional advantage for synthesis (design) of new systems, however. Often, we can view a derivation step, such as $G \overset{C_1}{\equiv} S_1$, as the application of a general optimizing transformation. When this is the case, we can write a general lemma stating that any agent A satisfying certain properties will be conditionally equivalent given C_1 to some new agent $f(A)$, where f is a transformation over possible agents or agent components. The fact that $G \overset{C_1}{\equiv} S_1$ would then be a consequence of the fact that G satisfies the right properties and $S_1 = f(G)$. Phrasing the optimization as a general lemma allows the optimization to be applied to other systems by other designers.

It is important to remember that there is not a one-to-one correspondence between mechanisms and derivations; the derivation of one mechanism from another need not be unique. As with the derivation of one equation from another, it will sometimes be possible to merge the steps of a derivation, to

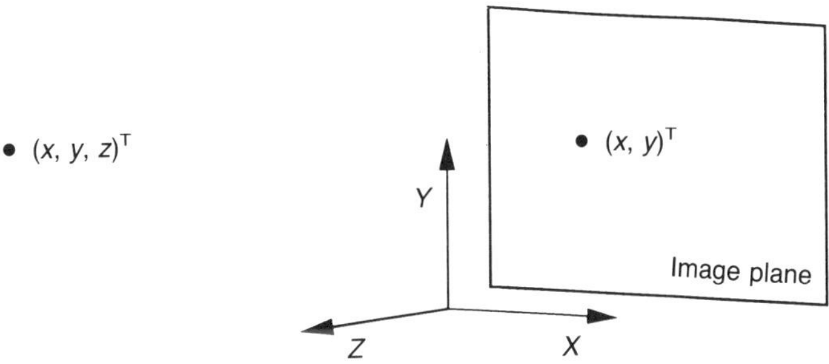

Figure 21.1 A point imaged under orthographic projection.

break them down further, or to reorder them. It may even be possible to derive S from G in several completely different ways. Thus the final derivation is as much a matter of taste as of objective reality. In particular, if a derivation shows that $G \overset{C_1 \cap C_2 \cap \cdots \cap C_n}{\equiv} S$, then we cannot interpret the C_i as an absolute specification of S's tacit knowledge of its habitat, or even of its additional tacit knowledge over that of G. It is simply a useful characterization of the latter, and perhaps one of many.

21.4 DETECTING NODS OF THE HEAD

Consider the problem of deciding whether a person facing the camera is nodding their head. We will assume that the camera and the person are facing one another. For simplicity, we will assume orthographic projection with the X axis horizontal, the Y axis vertical, and the Z axis aligned with the axis of the camera lens. Thus a point $(x, y, z)^T$ in the world is projected to a point $(x, y)^T$ in the image, and points higher in the image have larger y coordinates (see Figure 21.1). (We use the transpose notation $(\cdot)^T$ because coordinates are conventionally considered to be column vectors, but are more easily typeset as row vectors.) Actual camera projections are more complicated, but these considerations are unimportant for our purposes (see Ballard and Brown (1982) or Horn (1986) for introductions to the mathematics of camera projection). Finally, we will assume that the objects in view are rigid or at least piecewise rigid. We will treat the motion of a given object at a given point in time as a combination of a translation $\boldsymbol{T}$ and a rotation ω about a point $\boldsymbol{P}$. (Note: this is a gross oversimplification of the kinematics of jointed objects, but it will suffice for our purposes.) A point $\boldsymbol{R}$ on the object thus moves with an instantaneous velocity of

$$\boldsymbol{T} + (\boldsymbol{R} - \boldsymbol{P}) \times \omega.$$

Since a nod is an alternating rotation of the head, a nod is then a motion of the head in which ω maintains a constant direction, but a varying velocity, $\boldsymbol{R}$ remains constant, and $\boldsymbol{T}$ remains zero. The conceptually simplest approach to detecting nods would be to first find the head in the image, then determine its pose (position and orientation in 3-space), to track the pose through time to recover the motion parameters $\boldsymbol{T}$, $\boldsymbol{R}$, $\boldsymbol{P}$ and ω, and then test these parameters for rotational oscillation. We will write this schematically as follows:

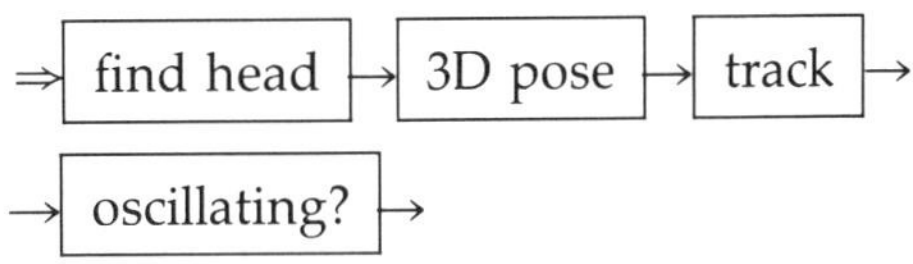

Here the double arrow on the left is meant to express input from sensors, and the single arrows represent signals connecting the computations.

The system just described would be expensive and difficult to implement, although it is certainly conceptually possible. The main problems are that finding faces can be difficult, and that determining the pose of the face is also very difficult. A particular problem with the latter is that it is easy to confuse translation with a rotation. In particular, it is very difficult to distinguish the cases of my bending (rotating) my head downward, and my head moving down 3 cm, but still pointing forward. Both would involve the points of my face moving downward in the image, along the Y axis, while staying at the same point along the X axis. Of course, the translation would require my neck to suddenly grow 3 cm shorter.

21.4.1 A SIMPLER SYSTEM ADAPTED TO ITS ENVIRONMENT

An alternative system, which is much simpler, is the following:

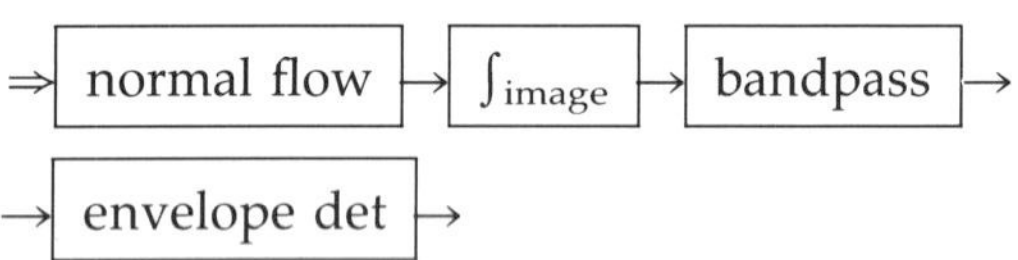

First we compute the vertical component of normal flow field (see below) of the image, then we integrate the field over the entire image to obtain a single net vertical motion, and then we look for oscillations in the net vertical motion. The optic flow of an image is the apparent two-dimensional motion of the texture in the image. It is the projection into the image plane of the motion vectors of the actual objects in the world. Since the 3-space motion of a point $\boldsymbol{R}$ is simply

$$\boldsymbol{T} + (\boldsymbol{R} - \boldsymbol{P}) \times \omega$$

the optic flow of $\boldsymbol{R}$'s projection is simply the x and y coordinates of this vector, or

$$f_{\boldsymbol{R}} = \begin{bmatrix} T_x + (R_y - P_y)\omega_z - (r_z - p_z)\omega_y \\ T_y + (R_z - P_z)\omega_x - (R_x - P_x)\omega_z \end{bmatrix}.$$

The normal flow field is the component of the optic flow field in the direction of the image's intensity gradient.

The useful property of the normal flow field is that it is very easy to compute. If we let $I(x, y, t)$ represent the brightness of a point (x, y) in the image at time t, then the vertical component of the normal flow field is given by

$$-\frac{\frac{\partial I}{\partial t}(x, y, t)}{\frac{\partial I}{\partial y}(x, y, t)}$$

Since the derivatives can be approximated by subtracting adjacent points in time or space, respectively, we can compute this value very quickly.

21.4.2 DERIVING THE ADAPTED SYSTEM

While it seems plausible that the simplified system would work, we would like an explanation of exactly why it would work, and more importantly, under what conditions it would work. We can do this by performing a derivation. We start from the original system:

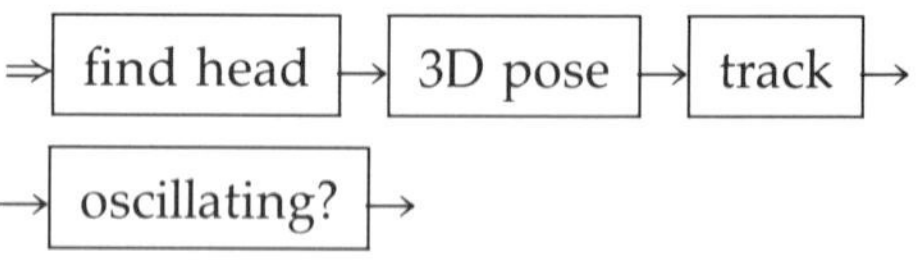

Recall that computing the 3D pose is difficult, since translations and rotations both generate vertical motions in the image. However, since the head is not physically capable of translating, any up/down motion of the head in the image must be due to a rotation, unless the person is bending their knees, jumping up and down, bowing rapidly, or standing on an oscillating platform. If we assume that people do not do these things, then we may safely interpret any vertical oscillation of a head as a nod. Let's call this the **head kinematics constraint**: that the translation vector $\boldsymbol{T}$ of the head is nearly always zero, and that it never oscillates. The head kinematics constraint allows us to use the 2D position of the head (i.e., the position of the head in the image itself), instead of its 3D pose. An oscillation of the 2D position will indicate a nod. Thus we can replace the '3D pose' module with a module which finds the centroid of the image region occupied by the head:

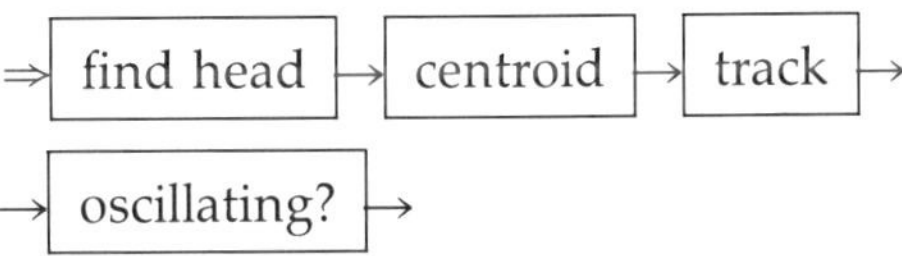

Thus we have reduced the problem to finding 2D oscillations of the head. Finding the head can be difficult in itself however. Fortunately, we are not looking for arbitrary heads, only oscillating ones. Thus we need not bother looking in parts of the image which are not moving. Indeed, if we assume that vertical oscillations are generally uncommon, that is, people do not nod their hands or play with yo-yos, then we can use the oscillatory motion itself to find the head. We shall call this the **motion salience constraint**: no motion parameter of any object is allowed to oscillate, save for the ω parameter of a head. The motion salience constraint removes the need for a full recognition engine and allows us to use a system like:

First we find the pixels in the image which are oscillating, and then we look for a head-shaped region of oscillating pixels. We can find the oscillating points by first computing the 2D motion (optic flow) at each point in the image, and then testing the motion for vertical oscillation. The test can be done, in turn, by applying a bandpass filter and an envelope detector at each point. The resulting system is thus:

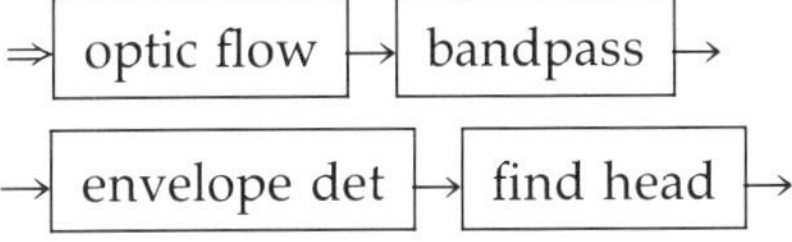

Computing optical flow can still be fairly expensive however. The normal flow field is much easier to compute and, fortunately, its vertical component will always have the same sign as the vertical component of the optical flow, provided that the normal flow is non-zero. The normal flow will be zero when the actual motion is non-zero only if there is no texture in the image at that point, or if the y (vertical) derivative of the image intensity is zero. Fortunately, faces have considerable vertical intensity variation and so this is not a problem. Thus we can reduce the system to

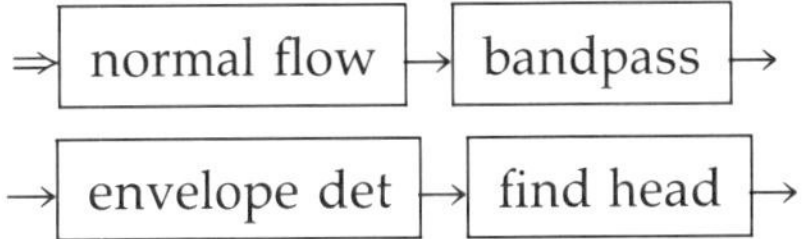

Even looking for oscillations at each point might be too expensive, however. If we assume that the motion of the head will dominate any other motion in the image, then we can look for oscillations in the net vertical motion of the entire image.

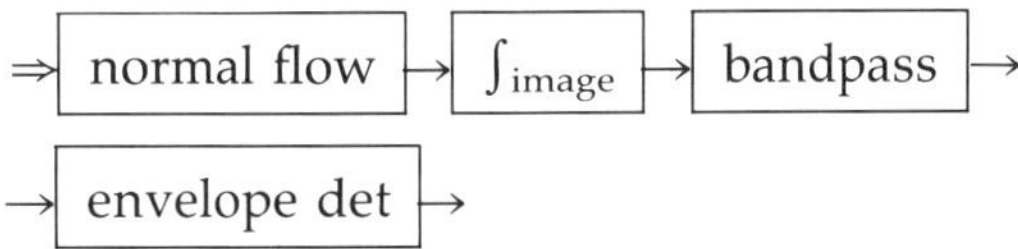

which is the system we were trying to derive. The assumption that the head motion dominates the motion in the rest of the image is needed to rule out the case of motions in different parts of the image canceling with one another to generate the appearance of oscillatory motion when, in fact, no single part of the image was oscillating.

A summary of the constraints used in the derivation of this system is given in Table 21.1. I have implemented this system on a Macintosh personal computer. The system uses 64 × 48 grey-scale images at a rate of approximately 5 frames per second. The system performs reliably provided that the subject nods their head several times to give the bandpass filter a chance to respond.

On the whole, this system is not adequate

Table 21.1 Constraints used in the derivation of the nod detector, and the problems they were used to simplify

Constraint	*Problem*	*Optimization*
Head kinematics	Motion disambiguation	Substitution of 2D motion for 3D
Motion salience	Head detection	Use oscillation to guide head detection
Horizontal lines	Flow computation	Substitute normal flow for optic flow
Nod dominance	Head detection	Use oscillation of net flow

for general use because the bandpass filter requires more oscillations of the head than people generally make. A non-linear oscillation detector such as a zero-crossing detector would probably do better. The result would be something like this:

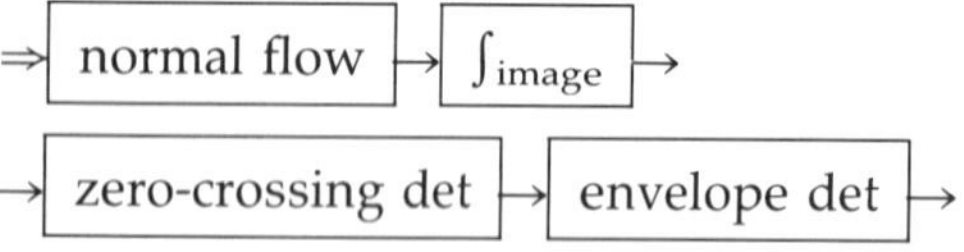

However we would expect the net flow to have a relatively large number of random zero crossings, even when the head is not moving. We can deal with this problem by backing off from the net flow optimization and returning to computing the oscillations on a per-pixel basis:

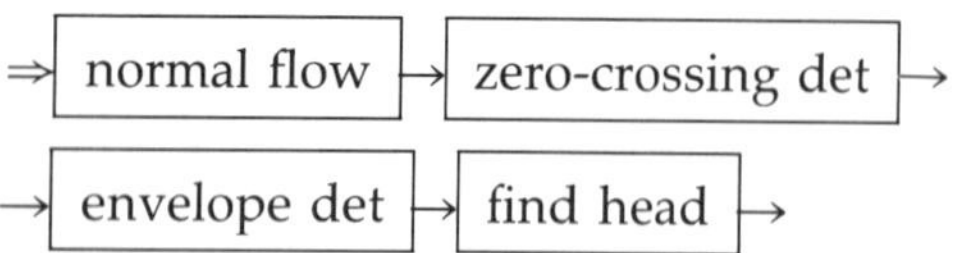

The flow and zero-crossing detectors have been implemented on the Polly robot, described below. Unfortunately, the robot is so short, and the viewing angle of the camera so wide, that it can only see the bottom of a person's chin, so it has been difficult to get the flow detector to respond to any head movements, much less to detect nods.

The point of this analysis is not that this is the right way to build a nod detector (this remains to be seen), or that I went through these steps while designing the system (I did not, although I did go through some of the steps), but that viewing the final system as an optimized form of the more general system is a useful way of articulating the implicit assumptions that underlie the system's operation, the ways in which they facilitate the computation, and which assumptions are independent of one another.

21.5 A ROBOT ADAPTED TO ITS ENVIRONMENT

I have developed a simple vision-based robot, Polly, designed to give primitive 'tours' of the MIT AI lab using a number of simple visual systems such as the nod detectors discussed above. The nod detectors were originally developed for the robot, but were not used in the end because the robot was so short that it could only reasonably see the bottom of a person's chin.

Polly drives about the lab looking for visitors to whom it can offer tours. In a typical run, the robot patrols the hallways until a visitor takes note of it (as signaled by their walking up to the robot). It then introduces itself, offers the visitor a tour, and tells the visitor to wave her foot around if she wants a tour. If the visitor answers yes by waving her foot, then Polly will lead her from point to point, giving a canned speech each time it recognizes a landmark.

The robot is surprisingly simple and economical considering its level of functionality. The robot is built principally from a Real World Interface B-12 robot base (which houses the batteries, motors and motor controllers), a C30-based digital signal processor, and a frame grabber (see Figures 21.2 and 21.3).

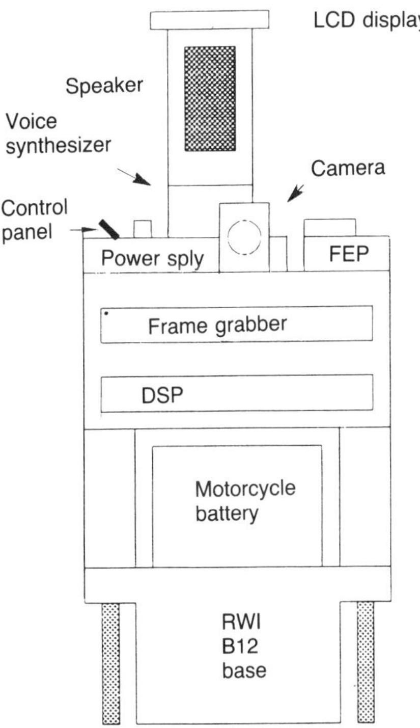

Figure 21.2 The Polly robot.

The robot also has various I/O peripherals and a programmable micro-controller to service them. The complete robot was built from off-the-shelf boards for less than US$20 000, but could be built for half that today. We are in the process of porting parts of the vision system to run on a new hardware platform costing less than US$1000.

Polly's vision system consists of a number of simple subsystems, each of which computes a simple signal describing the world. The systems compute the direction of the corridor the robot is currently in, the distance to the nearest obstacle in each direction, whether there is a left turn or a right turn in view, whether there is a person's leg in view (the robot can only look down, so it can only see people's legs), whether there is a wall ahead, whether the floor is dark, and whether there is motion in the visual field. These systems can be extremely simple because of their tacit understanding of their habitat. This understanding can be made explicit in each case by performing a derivation, although in some cases, the derivations are not sufficiently interesting to formalize. I have also used the derivation method to compare the habitat constraints of the robot's navigation algorithms to those of a general path planning system. For a detailed discussion of the structure and analysis of Polly, see Horswill (1993).

21.6 DISCUSSION

The reader may have interpreted the foregoing as an argument for some sort of in-

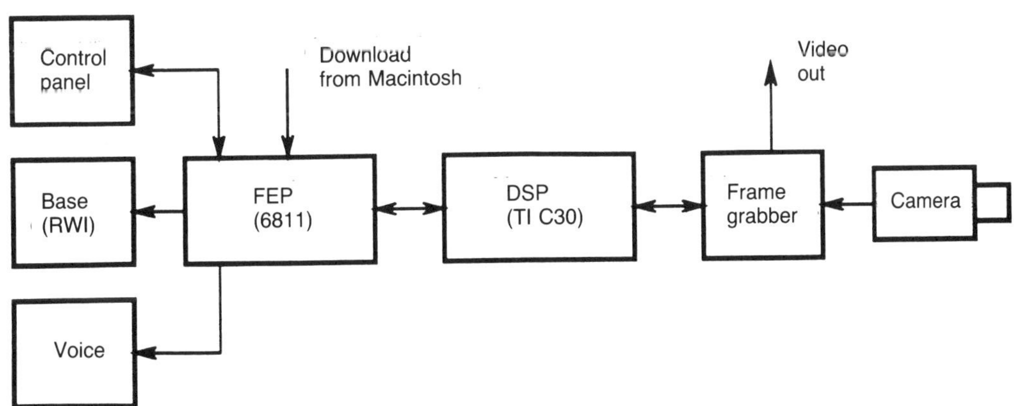

Figure 21.3 Hardware architecture of Polly.

natism. This is not my intention, although the most obvious application of this framework is to the analysis of agents that have 'innate knowledge' of their habitats. I have tried to be epistemologically agnostic. The same theoretical tools I have used to analyze carefully hand-coded agents could also be applied to the end result of Littman's learning systems (1993), of Wilson's genetic algorithms (1991), or of the selectionist systems of Edelman (1987) or Manderick (1992) and this book.

Fundamentally, this chapter is about explanation. For one reason or another, we are often faced with agents or other mechanisms that operate properly in one type of environment but not in another. In such cases, we want to explain the agent's performance in different environments.

Marr argued that computational mechanisms should be explained at three levels (1982):

1. a computational theory that specifies the desired input/output behavior of a computational module, along with a set of constraints that force a unique output for a given input;
2. an analysis of the representations used by the mechanism to encode information, the algorithms used to transform between representations, and the way in which the algorithm/representation pair realize the computational theory;
3. an analysis of how the mechanism implements the representation and algorithm.

The advantage of Marr's methodology is that each level is more general than the next. The barest level, the computational theory, describes only what is computed, without describing how it is computed. A given computational theory can then be implemented by many different algorithm/representation pairs and a given algorithm/representation pair can be implemented by many different mechanisms. This strict division of what and how is both a strength and a weakness. Since the computational theory specifies input/output behavior in an arbitrary environment, it acts as a formal program specification for mechanisms intended to operate in arbitrary environments. The specification adjudicates whether a given mechanism truly solves the computational problem or not. The structure of the specification, its constraints in particular, are often manifest in the structure of the mechanism, thus allowing connections to be made between very different mechanisms by showing how they each implement the same computational theory. For example, very different optimization mechanisms in early vision can be seen as using the same constraint, surface smoothness.

While Marr's methodology takes a strict stance on the division between what and how, it makes little provision for where, that is, for the agent's habitat. This makes it difficult to apply to agents with restricted habitats. For example, Polly avoids obstacles by avoiding texture in the image. Because the floor of its environment happens to be textureless, we can view Polly's texture detector as a not-floor detector, that is, as an obstacle detector. What are we to take as the computational theory for this problem? Is it to be a broad theory, such as a computational theory of obstacle detection in general, or a narrow theory, such as a theory of obstacle detection in rooms with textureless carpets? It seems to be a matter of taste, and worse, there is a dilemma. The use of narrow theories leads to an explosion of computational theories, one for each task/habitat pair. Indeed there threatens to be a computational theory for each mechanism, since slightly different mechanisms will have slightly different habitats. On the other hand, if one chooses a broad theory, then it can no longer be viewed as a formal specification since the agent no longer instantiates it. The theory specifies proper behavior for all types of paths but the agent only instantiates the theory within its restricted habitat of rooms with textureless floors. More importantly, the computational theory is not nearly so valuable a resource for interpreting the mechanism, since the most important constraint that is manifest in the mechanism, the texturelessness of carpets, is no longer a part of the theory.

Computational theories, then, no longer provide a unifying view of different mechanisms unless they happen to share precisely the same habitat. Broad or narrow, the computational theory will say little about the relationship between the general mechanism and the specialized mechanism. In fact, the methodology cannot speak to this issue because it maintains a strict division of explanation between specification and implementation. In so far as a specialization is a kind of optimization, an explanation of it must necessarily combine specification of habitat with possibilities for implementation.

Finally, I would like to stress the fact that AI is and must be a natural science in the sense that it must continually make and test hypotheses about the nature of the external world. Our algorithms, representations and formalizations of the world must eventually be compared with the structure of external reality, and doing so early reduces the risk of wasted effort. If we wish to understand intelligence and agency, then we must study not only ourselves but the world in which we live.

ACKNOWLEDGEMENTS

Phil Agre, Rod Brooks, Eric Grimson, Maja Mataric, David Michael, Ray Paton and Lynn Stein all provided much needed feedback during the development of these ideas and provided comments on drafts. Support for this research was provided in part by the University Research Initiative under Office of Naval Research contract N00014-86-K-0685, and in part by the Advanced Research Projects Agency under Office of Naval Research contract N00014-85-K-0124.

REFERENCES

Aho, A. V., Sethi, R. and Ullman, J. D. (1986). *Compilers: Principles, Techniques, and Tools*. Addison Wesley, Reading, MA.

Ballard, D. H. and Brown, C. M. (1982). *Computer Vision*. Prentice-Hall, Englewood Cliffs, NJ.

Beer, R. (1992). A dynamical systems perspective on autonomous agents. CES 92-11, Case Western Reserve University, Cleveland, OH.

Donald, B. R. and Jennings, J. (1992). Constructive recognizability for task-directed robot programming. *Robotics and Autonomous Systems* **9**, 41–74.

Edelman, G. M. (1987). *Neural Darwinism: The Theory of Neuronal Group Selection*. Basic Books, New York.

Horn, B. K. P. (1986). *Robot Vision*. MIT Press, Cambridge, MA.

Horswill, I. (1993). Specialization of perceptual processes. Ph.D. thesis, Massachusetts Institute of Technology, Cambridge, MA.

Houston, A. I. (1991). Matching, maximizing, and melioration as alternative descriptions of behavior. In *Proceedings of the First International Conference on Simulation of Adaptive Behavior* (ed. J.-A. Meyer and S. W. Wilson). MIT Press, Cambridge, MA, pp. 498–509.

Littman, M. L. (1993). An optimization-based categorization of reinforcement learning environments. In *Proceedings of the Second International Conference on Simulation of Adaptive Behavior* (ed. J.-A. Meyer and S. W. Wilson). MIT Press, Cambridge, MA, pp. 262–70.

Manderick, B. (1992). Selectionist systems. In *The Proceedings of the First European Conference on Artificial Life* (ed. F. J. Varela and P. Bourgine). MIT Press, Cambridge, MA.

Marr, D. (1982). *Vision*. W. H. Freeman, San Francisco, CA.

McFarland, D. (1991). What it means for robot behavior to be adaptive. In *Proceedings of the First International Conference on Simulation of Adaptive Behavior* (ed. J.-A. Meyer and S. W. Wilson). MIT Press, Cambridge, MA, pp. 22–8.

Meyer, J. A. and Guillot, A. (1991). Simulation of adaptive behavior in animals: Review and prospect. In *Proceedings of the First International Conference on Simulation of Adaptive Behavior* (ed. J.-A. Meyer and S. W. Wilson). MIT Press, Cambridge, MA, pp. 2–14.

Meyer, J. A. and Wilson, S. W. (eds.) (1991). *From Animals to Animats: Proceedings of the First Inter-*

national Conference on Simulation of Adaptive Behavior. MIT Press, Cambridge, MA.

Meyer, J. A. and Wilson, S. W. (eds.) (1993). *From Animals to Animats: Proceedings of the Second International Conference on Simulation of Adaptive Behavior*. MIT Press, Cambridge, MA.

Paton, R. C., Nwana, H. S., Shave, M. J. R. and Bench-Capon, T. J. M. (1992). Computing at the tissue/organ level (with particular reference to the liver). In *The Proceedings of the First European Conference on Artificial Life* (ed. F. J. Varela and P. Bourgine). MIT Press, Cambridge, MA, pp. 411–20.

Rosenschein, S. J. (1987). Formal theories of knowledge in AI and robotics. Report CSLI-87-84, Center for the Study of Language and Information, Stanford, CA.

Rosenschein, S. J. and Kaelbling, L. P. (1986). The synthesis of machines with provable epistemic properties. In *Proceedings of the Conference on Theoretical Aspects of Reasoning about Knowledge* (ed. J. Halpern). Morgan Kaufmann, San Mateo, CA, pp. 83–98.

Smithers, T. (1992). Taking eliminative materialism seriously: A methodology for autonomous systems research. In *The Proceedings of the First European Conference on Artificial Life* (ed. F. J. Varela and P. Bourgine). MIT Press, Cambridge, MA, pp. 31–40.

Todd, P. M. and Wilson, S. W. (1993). Environment structure and adaptive behavior from the ground up. In *Proceedings of the Second International Conference on Simulation of Adaptive Behavior* (ed. J.-A. Meyer and S. W. Wilson). MIT Press, Cambridge, MA, pp. 11–20.

Varela, F. J. and Bourgine, P. (eds.) (1992). *Toward a Practice of Autonomous Systems: the Proceedings of the First European Conference on Artificial Life*. MIT Press, Cambridge, MA.

Wiener, N. (1961). *Cybernetics*. MIT Press, Cambridge, MA.

Wilson, S. W. (1991). The animat path to AI. In *The Proceedings of the First International Conference on Simulation of Adaptive Behavior* (ed. J.-A. Meyer and S. W. Wilson). MIT Press, Cambridge, MA, pp. 15–21.

SOCIO-ECOLOGICAL METAPHORS AND AUTONOMOUS AGENTS IN COMPUTER SUPPORTED COOPERATIVE AUTHORSHIP

22

Geof Staniford

22.1 INTRODUCTION

Metaphor is pervasive throughout the languages of developed societies and Aristotle, in his *Poetics*, is generally adjudged to have given the first extended analysis in western culture; this analysis firmly placed metaphorical association at the level of words. Through the centuries it has become apparent that metaphor extends to whole concepts, made plain by Richards (1936):

> fundamentally it is a borrowing between and intercourse of thoughts, a transaction between concepts.

Richards went further and proposed that human cognition is metaphoric in its essential nature and that the metaphors of our language actually derive from an interaction of thoughts. It seems reasonable that if Richards is even partly correct it often can be seen that one tool applied too exclusively to any area of human complexity falls short in some respect with time – we can extend the idea of metaphorical thought to models as well as language. Models are metaphorical in the sense that they represent one thing in terms of another.

We do of course use language, of more or less formality, to communicate and record ideas about our models. Consider the following examples. Throughout the *Society of Mind*, Minsky (1988) uses the notion of an agent; nowhere does he develop a definition, a formal statement, describing the essential properties of agentness; on the contrary he explains that the whole subject of the book 'revolves around' an exploration of the question, 'What could the agents be?' Similarly in *Logic and Information* Devlin (1991), in his search for a new direction in which to build upon first-order predicate logic with the overall goal of providing a mathematical framework that will form the backbone of a science of information, has not, as yet, provided a formal description of an agent, although he also uses the notion of an agent throughout the work. Both authors use the word 'agent' when considering entities that have very different degrees of internal complexity and very different types of external function; it is our contention that whilst the reliance upon implicit metaphors may be helpful in the development and understanding of some exploratory scientific model building and in the transmission of ideas across scientific (and other) boundaries, a more overt, more explicit approach is required when using metaphor as an aid while building models that are intended to be both understandable and computationally feasible.

A number of different metaphors are

considered in this chapter that were used to aid the design of a model for the study of cooperation between both autonomous (and semi-autonomous) computational agents on the one hand and human author agents on the other. The computational context of the design is a collaborative authorship support system. In such a model, when considering communication channels between cooperating agents, we find automaton/automaton communication, automaton/human communication and human/human communication. The main discussion in this chapter is confined to some automaton/automaton issues; discussion of the other issues is available in depth elsewhere (Staniford and Dunne 1994).

Paton (1992) defines two basic metaphor types that are important in the analysis of biosystems, systemic and spatial. Systemic metaphors help us to transfer information about the structure of biological systems, and spatial metaphors provide a means of transference of behavioural information. Taken together they point in the direction of ethology as a source of useful metaphors to transfer contexts into computing. The approach developed here to the application of metaphors from particular fields is heavily influenced by the work of Hinde (1982, 1987) who argues that principles from ethology are necessary for understanding some aspects of behaviour at lower levels of social complexity but have severe limitations at higher ones. Hinde argues that principles from sociology need to be considered when considering such higher levels and that the key to successful progress in understanding such behaviours lies in recognizing that there are crucial distinctions between levels of social complexity whilst at the same time that all are processes in dialectical relations with each other. Within the model, the processes of perception by, and communications between, automata represent areas in which biological metaphors are used to provide a rich source of ideas. At a higher level of abstraction the structured layering of the interaction between agents of differing architectural complexity makes use of sociological metaphors.

When humanity – certainly in the Euro-American cultural tradition traced to classical Greece and Rome – first began to produce documents, those documents have been primarily viewed, by classical scholars (McKnight *et al.* 1991) as a record of the spoken word. Over time, as authors and readers have become increasingly sophisticated in literary ability and sensibility, conventions have developed for the structural classification of meaningful documents. The understanding of and adherence to such conventions is a fruitful framework for the process of communication between author(s) and reader(s). In recent times many classes of document, large technical publications, joint reports, complex company plans and scientific papers, for example, are increasingly being written by groups of authors working in cooperative groupings of one sort or another on the creation of the final work. During the planning and creation phases of such documents, teams of authors need support facilities to record and report on joint decisions and inter-author communication. Latterly, with the inception and development of powerful networked computing facilities, it has become possible to explore paradigms for computer supported writing systems; such systems may provide an enabling technology for increasing authoring efficiency in the production of large cooperatively written documents. Such systems are large, loosely coupled, and contain a considerable degree of inherent concurrency. Chandrasakaran has delineated the need for the inclusion of ideas from many fields as a way of controlling complexity in distributed problem solving (Chandrasakaran 1981).

In both the biological and social sciences many different approaches have been taken to the study of agents. Latterly this research has been drawn on to provide support for work in both artificial intelligence and artificial life. The view of agents adopted in this work can best be summarized by describing agents as physical processes interacting in an environment with other agents and with the environment. Agents can be considered to fall into complexity classes, based upon their abstract architectural design, that define their functional abilities (Kiss 1991). Social grouping of agents may be formed within and between these complexity classes based upon an agent's language capabilities (Staniford 1993). The research described in this chapter can be considered to be the application of research in distributed artificial intelligence (DAI) to the field of computer supported cooperative work (CSCW). Both fields involve much interdisciplinary work which presents both a challenge and an opportunity. The opportunity is to approach things from a multifaceted perspective, and the challenge is to keep this approach within reasonable bounds. It is necessary from a practical point of view to limit the choice of perspectives and to leave consideration of some very important issues for other times and places.

Section 22.2 first provides the context for the following discussion in the form of a model of cooperative authorship and a formally defined document graph environment that defines the universe in which all agent operations take place. In the context of an authorship support system, agents are formally defined at the architectural level, whilst discussing a number of informal but explicit biological, sociological and ecological metaphors: sight, hearing, tropism, hierarchy, network and environment. Such metaphors while providing a rich set of resources for use during the design process, produce many consequences that must be taken into account during problem solving processes. In Section 22.3 two practical examples of an agent's sensory functions are discussed coupled with an example of their use to illustrate some of the advantages that metaphorical thinking can bring to programming. Finally the work is reviewed and consideration is given to possible directions for future research.

22.2 MODELS AND METAPHORS

22.2.1 A MODEL OF COOPERATIVE AUTHORSHIP

A cognitive model of communication which explains writing as the process of organizing a loosely structured network of internal ideas and external sources into a suitable hierarchy followed by mapping the hierarchy into a linear form; of sentence, paragraph, subsection, and so on is used by Smith *et al.* (1987). For a detailed account of the history and use of such networks and hierarchies see, for example, Way (1991). In Smith's model the author is viewed as an independent process, creating text or complete documents without cooperation with and communication between other processes. Barlow and Dunne (1991) model authors as cooperating processes, jointly working with other authors on the same document. The author processes (agents) communicate at the level of parallel processes within a single unified system.

In this section the ideas contained within the cognitive writing model are extended to include and support the notion of communication between the authors at the conceptual level, in order that they may cooperate on a jointly written document (Figure 22.1). This extended notion, coupled with the agent model, allows a description of writing that is flexible at the conceptual level and highly structured at the document level. The extended model includes the concepts of **exosystem communication** and **endosystem communication** in order to model the authors as independent, communicating processes. By exosystem communication we mean any channel, oral or written, by which authors communicate with each other. These communications may be exchanges of ideas, critiques of existing sections of the document, administrative matters and so on. The system will allow the importation of an electronic record of these communications and their storage, separate from the document that is

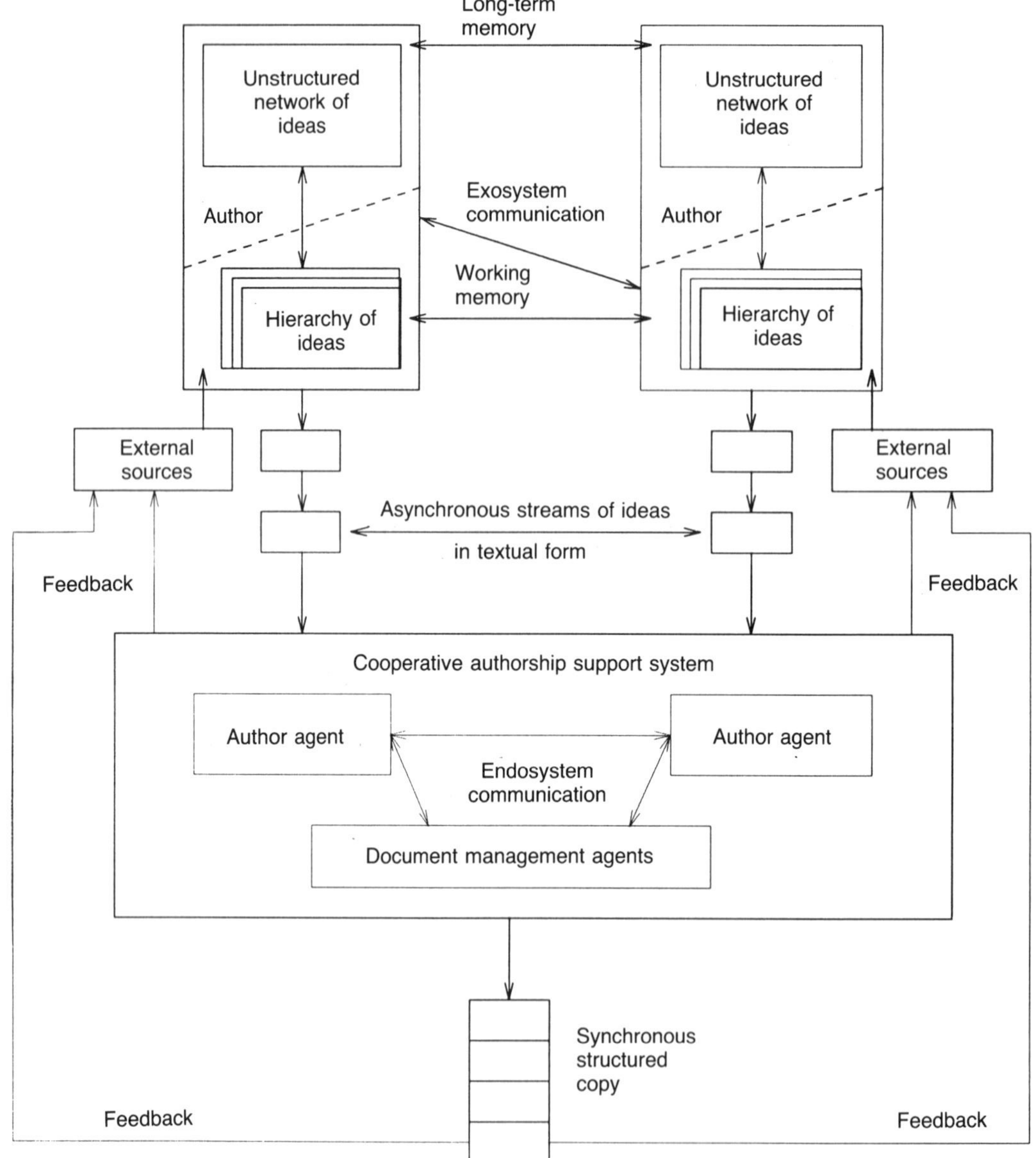

Figure 22.1 A model of collaborative writing using a cooperative authorship support system. (Note: for clarity only two authors have been shown in the diagram.)

currently being written. Endosystem communications are the communications that take place between the computational agents within the cooperative support system. Such communications do not use natural language but are composed into well-defined messages in a formal language. These facilities have been described in some detail in Staniford (1993).

Figure 22.1 gives a pictorial representation of the cooperative writing model just described and the reader is referred to this figure for an indication of the relationships existing between different parts of the model. Further discussion of this model may be found in Dunne and Staniford (1992). Note that, whilst the model presented here is primarily aimed at showing information flows in a concurrent communicating system, many workers use a functional model of authorship as a basis for discussion. Such functional models can be viewed as being implicitly present within the concurrent information flow model and the provision of an explicit functional model has not been considered necessary.

The environment in which the agents that we will discuss have their existence is the mathematical space of document graphs. This does not mean that we intend to imply that agents of this architectural design are only suitable for this type of environment. We hold the view that such agents have a great deal more generality than that and could be used in many models where the notion of agents of change in a dynamic environment would be a suitable paradigm. However, it is much easier to set up examples of actual agents to enable meaningful discussion of their properties to take place when a uniformly specified environment is available. In order to set the notion of a document graph in context we first consider ideas of document structure.

Structured documents form by far the largest class of documents, and can be found in applications covering every scientific, commercial, legal, technical and educational field. From early times conventions for different subclasses of structured documents have developed; see the *Canterbury Psalter* reproduced in McKnight *et al.* (1991). This book itself provides an example of a structured document, which would fall into a different subclass than say a textbook. Normally the reader is expected to be aware of the structural conventions and to read the document in some particular order. This in turn means that the author, irrespective of the semantic content of the work, must be aware of and adhere to the structural conventions expected by the reader.

Increasingly, in recent years, computers have been used to produce and store documents in electronic form, sometimes for the purpose of creating a paper document as the final version and sometimes purely to create and manage an electronic document. We are primarily concerned with the former class of document production; in such a system any electronic version of a document is a means to an end, rather than an end in itself. However, many of the problems to be solved and techniques that have been developed are entirely relevant to documents such as hypertext that have their existence solely within electronically readable form.

Clearly, electronic documents have a different set of structural conventions from the intended finished product and authors at present must be aware of both sets of conventions whilst engaging in the creative, semantic aspects of document writing. In addition, with current electronic systems, authors need to consider the layout and appearance of the printed page; a task undertaken by the editorial staff of the document publisher in more traditional systems. The difficulties caused by this additional cognitive loading has had some preliminary investigation in studies reported by Hansen and Haas (1988). Early results indicate that both speed and quality of writing may suffer. These results apply to single-author documents and we would hypothesize that scaling up to multiple, cooperative author documents would tend to multiply the problems in a very rapid way. An important objective of this research is to produce suitable representations of electronic documents that may be maintained and manipulated to some extent by well-specified automata to enable a reduction in this extra burden upon authors.

Furuta (1989) defines a taxonomy of document structures in some considerable detail. However, we content ourselves with noting that the document's structure specifies the form of the document, describing how the atomic objects are composed into higher level structures. We separate the structures that describe the relationships between objects in the document into constituent structures and define them separately. Each constituent structure is defined over some partitioning of the document's content into atomic objects, but we note that the granularity of partitioning of one constituent structure may not be at an appropriate level for use in another constituent structure. It follows then that we are, when describing the structure of an atomic object, specifying the minimum addressable unit of the constituent structure and therefore such descriptions are an essential component of a structure specification.

We divide constituent structures into three main types: primary, secondary and auxiliary. Primary structures are the structures that define the composition of atomic components into higher level components; secondary structures describe interrelationships among document objects and auxiliary structures describe relationships between document objects and objects that are outside the document.

Koo (1989) proposed a model which is used for representing and modifying electronic documents which employs simple graph grammars (Nagl 1978) as a means of translating changes in the document structure into modifications to the abstract representation. Bench-Capon and Dunne (1989) described the consistency checking problem and showed that in general this problem cannot be solved. They went on to consider certain special cases for which there is a solution. Building from this work Dunne and Staniford (1991) introduced the concept of a general document representation the definitions of which are reproduced below in Definitions 1–4, in order that we may consider them in our discussions later. The notation in these first four definitions is deliberately abstract to reflect the formal language approach to the manipulation of mathematical structures.

Definition 1

A document graph is a directed acyclic graph, $G\ (V,\ E)$, with a vertex labelling relation λ_V and an edge labelling relation λ_E. The vertices in V denote objects in the document and the edges in E depict logical connections between objects. Each object has an associated object type. This consists of two parts: a data type which specifies the domain of possible data values for the object and an attribute type which indicates the domain of possible properties that the object may possess. Similarly each edge has an associated edge type. This consists of two parts: a data type which specifies the domain of possible data values for the edge and an attribute type which indicates the domain of possible properties that the edge may possess. The labelled directed graph structure $\langle G\ (V,\ E);\ \lambda_V;\ \lambda_E\rangle$ is called the document graph of D.

Document graphs provide the primary resource used in specifying the structural relationships between different parts of a document (Figure 22.2). These structural relationships are abstract concepts and do not form part of the document's textual content but rather form a meta-level description of connections between different parts of the document.

Definition 2

A document specification consists of a pair DS = (C, Init). Here C is a finite set of constraints, $C = \{C_1, C_2, \ldots, C_k\}$ where each C_i is a (computable) predicate on document graphs. Init is a set of initial document graphs such that for each Init_j, $j \in \{1, 2, \ldots, n\}$, $C_i(\text{Init}_j)$ is **true** for each constraint C_i. Given a document specification DS and a document graph G, G is said to meet the specification

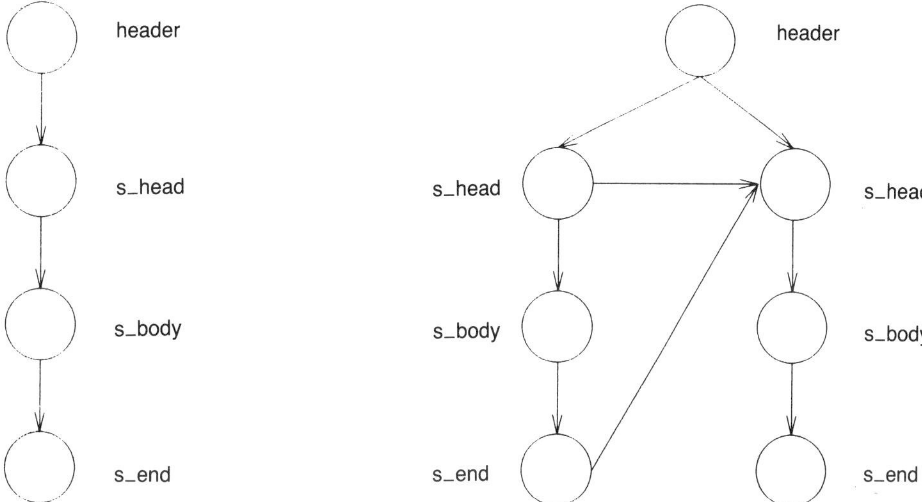

Figure 22.2 Document sections.

DS if and only if $G \in$ Init or $G_i(G)$ is **true** for each constraint C_i.

Document specifications allow the definition of precise, computable constraints that will enable autonomous agents to build complex document graph structures commencing with an initial graph and then adding vertices and edges in a way that is predetermined by the specification. In Figure 22.2, the right-hand document graph has been built from the left-hand document graph by the addition of three vertices and five edges.

Definition 3

A general document representation, D, is a seventuple

$$D = \langle \Sigma, V, E, \lambda_V, \lambda_E, \xi_V, \xi_E \rangle$$

where Σ is a finite alphabet of symbols, V is a finite set of vertices, $E \subseteq V \times V$ is a set of directed edges, λ_V is an vertex labelling relation, λ_E is an edge labelling relation, ξ_V is a vertex content relation and ξ_E is an edge content relation.

The general document representation extends the document graph definition with the addition of an alphabet of symbols and vertex and edge content relations. These additions provide all that is necessary to allow computational agents to operate within and upon a document graph environment. The vertex contents will typically contain the textual matter that is readable by the human reader and the edge contents will typically contain information that is to be used by computational agents in navigating and processing the environment.

Examples of specific representation classes are given in the following definition.

Definition 4

Below $D = \langle \Sigma, V, E, \lambda_V, \lambda_E, \xi_V, \xi_E \rangle$ is a general document representation, where $E \subseteq V \times V$ and given $\exists c_i$ such that $c_i = (v_m, v_n)$, $m \neq n$ then $\exists e_j$ such that $e_j = (v_n, v_m)$, $m \neq n \Leftrightarrow i = j$, that is, $G(V, E)$ is always a directed graph. $D \in$ Univ where the set Univ is the set of all possible document representations and is an uncountable set.

1. The linear representation (Λ): $D \in \Lambda \Leftrightarrow$
   (a) $G(V, E)$ is a set of chains, that is, every vertex in V has at most one incoming edge and at most one outgoing edge.
   (b) $\lambda_V = \lambda_E = \emptyset$.
   (c) $\xi_V \subseteq V \times \Sigma$ and is a mapping.
   (d) $\xi_E \subseteq E \times \Sigma$ and is a mapping.
   The linear representation can be used to represent documents that are handled as a flat file within a normal editor or word processing program and to represent a document version that has been stored within a higher level document graph structure and then been linearized as a precursor to output on a printer device.
2. The tree representation (T): $D \in T \Leftrightarrow$
   (a) $G(V, E)$ is a tree.
   (b) λ_V, λ_E, ξ_V and ξ_E are all mappings.
   The tree representation is commonly used in single author tools that provide an ideas outliner. This sort of tool is expressly designed to provide assistance with the process of organizing a loosely structured network of ideas into a hierarchy (Figure 22.1).
3. The acyclic representation (A): $D \in A \Leftrightarrow$
   (a) $G(V, E)$ is acyclic.
   (b) λ_V, λ_E, ξ_V and ξ_E are all mappings.
   The acyclic representation is the work horse of these four representations as far as autonomous agents are concerned; an example can be seen in Figure 22.2. It can be readily seen that in addition to the hierarchical edges of a tree structure there are cross edges to facilitate agent navigation at one level and the logical linking of semantically related elements of text, by authors, at another level.
4. The hypertext representation (H): $D \in H \Leftrightarrow$
   (a) λ_V is a one-to-one mapping.
   (b) ξ_V and ξ_E are both mappings.
   (c) λ_E is a mapping which satisfies for each $\alpha \in \Sigma^*$

   $$((u, v), \alpha) \in \lambda_E$$
   $$\text{and} \quad ((u, w), \alpha) \in \lambda_E \Rightarrow v \equiv w$$

   Thus one-to-many edge naming relations are not permitted.

The hypertext representation is primarily intended for documents that are intended to be created, viewed and manipulated solely in electronic form. This representation is only included here for the sake of completeness.

An extensive discussion of these definitions can be found in Staniford (1993) which presents an algebraic specification method for document graphs. Specifications in this algebraic form may readily be translated into first-order predicate calculus sentences that can be used by a logic programming language such as Prolog and this in fact has been done in writing the programs upon which the practical example, discussed later, is based. Taken together we have a dynamic environment that can be occupied and changed by computational agents.

Genesereth and Nilsson (1987) define four types of agent – tropistic, hysteretic, knowledge level and stepped knowledge level – in order of increasing sophistication and complexity. They then introduce the notion of a deliberate agent which is a stepped knowledge level agent with certain restrictions on its set of actions. These agents are designed to exist in an environment in which they have the sole occupancy. There is no need for them to be aware of nor communicate with others of their kind. In this chapter an extended architecture is presented that has been designed to enable agents to coexist with, communicate with and cooperate with other agents. For the sake of brevity we cannot summarize the work of Genesereth and Nilsson, but the interested reader is referred to the reference for a full account.

So that an agent may receive external stimuli we require two functions. Firstly

$$\textit{see}: R \rightarrow S$$

where the set R is the set of states that characterizes the agent's world. In order to characterize the agent's sensory capabilities we partition the agent's perception of the set R (of external states) into a set S, of disjoint subsets (which are internal states that represent the external environment) such that the agent is able to distinguish states in different partitions but is unable to distinguish states in the same partition. It follows that agents can never have a completely global view of their environment. The detailed specification of *see* makes use of the labelling relations ξ from Definitions 1, 3 and 4. The implementation of *see* is domain dependent and is discussed in Staniford (1993). Secondly

$$hear\colon T \rightarrow Q$$

The set T is a set of words that characterize communications that the agent may receive from other agents. Similar to the foregoing, in order to further characterize the agent's sensory capabilities, we partition the set T of words into a set Q of disjoint subsets such that the agent is able to distinguish words in different partitions but is unable to distinguish words in the same partition. Two worked examples, in IC-Prolog II, of *hear* functions are discussed in Section 22.3.

Uexküll (1921) has been greatly influential in showing experimentally that an organism can only perceive a limited part of its environment with its sense organs, so we use these sensory functions to characterize the way in which an agent perceives stimuli external to itself. *See* enables an agent to determine the local state of its environment (i.e., the state at a given graph node) and *hear* enables the agent to receive communications from other agents.

Milner (1989) argues that a valuable simplification in modelling communicating agents with respect to the act of communication is to consider a communication channel to be a one-way system that provides a medium for acts of communication to take place. An act of communication is considered to be an indivisible event, taking place instantaneously between two or more agents. We adopt this simplification for our models and we consider that if a message has been 'sent' (see the discussion below) by an agent then it has been received by the agent or agents to whom it was sent. Similarly we consider that any local attributes of a state in an environment that an agent can 'see' will have been perceived by that agent as soon as they become available as a result of some state change within the environment. It will often be necessary to define agents that carry out an action purely as a result of the influence of one or other of the external stimuli; consequently it seems reasonable to keep the two functions separate, particularly as a corollary to the foregoing we realize that agents are extremely myopic with reference to the *see* function but can *hear* from all parts of the environment.

Next we need to turn our attention to the manner in which the state of an environment is changed. We assume that agents make local changes to an environment and consider that a state S_n changes to $S_{(n+1)}$ the instant that an agent makes a local change. Agents do not have the power to make global changes in an environment. In order that an agent may make changes we define two more functions, firstly:

$$action\colon S \times Q \rightarrow A,$$

a discriminatory function which maps each disjoint subset of states and inputs onto a particular action to provide an agent with the ability to choose which action to perform according to its perceived stimuli. Secondly:

$$do\colon A \times R \times T \rightarrow R \times W$$

is an executory function which maps an action, state and input onto the new state and an output, providing the agent with the ability to change the local state of its environment (which includes the ability to move via some suitable arc from one node to an adjacent node) and to communicate messages to other agents. The set W is a set of words that characterize the communications that an agent may send to other agents. We assume that an agent can distinguish between all of its responses so there is no need to partition W.

Unlike *see* and *hear* we encapsulate acting and communicating in one function because although there will be occasions when we wish to communicate without changing a state, the converse is not the case. We do not allow an act which changes the state of the environment to take place without there also being a corresponding act of communication; hence the use of the word 'loquacious' in our name for these agents. We wish to indicate that these two operations are closely bound together in order that we may simplify the coordination of the knowledge – between autonomous agents – that environmental state changes have taken place. The ability to communicate without state changes is in fact crucial to the notions of cooperation that we will develop later. Although we do not explicitly map W into a set of disjoint subsets, outputs implicitly fall into one of three categories; successful completion of a state change, failure to complete a state change and communications regarding cooperation. In order to avoid ambiguity we require that these categories do form disjoint subsets within W and in fact that there is an implicit bijection between the success and failure messages.

For our purposes, then, acts of observation both visual and oral must take place in the internal state of an agent, between changes in state of the external environment. In general there is no simple linear correlation between the internal state changes within an agent and the external state changes of the environment. This presents problems for analysis and design, but we contend that these problems are more amenable to solution with the use of communicating agents than with simple non-communicating agents.

Definition 5

A loquacious tropistic agent in an environment is a 10-tuple of the form

$$\langle R, S, T, Q, W, A, see, do, hear, action \rangle$$

where the sets and functions are all as previously described.

A tropism is the tendency of biological life forms to react in response to an external stimulus (Loeb 1913). This term is particularly applied to plant forms and a good example is the opening and closing of the flower head, in response to the presence or absence of sunlight, by the popular garden plant dimorphotheca (the African daisy). The simplest agent – tropistic – is modelled with this metaphor in mind and we note two consequences of this:

1. Tropistic agents respond to external stimuli.
2. Tropistic agents do not have memory capacity regarding previous states.

It follows from (1) and (2) that tropistic agents cannot instigate actions without an external stimulus, nor can either previous actions or previous states have an influence upon current decisions.

Definition 6

A loquacious hysteretic agent in an environment is a 12-tuple of the form

$$\langle I, R, S, T, Q, W, A, see, do, hear, internal, action \rangle$$

where the sets and functions are all as previously described except that the set I is an arbitrary set of internal states and we assume that the agent can distinguish between all its internal states so that there is no need to partition I. The function *internal*: $I \times S \times Q \rightarrow I$ maps an internal state and both types of observation into the next internal state. Finally, the function *action*: $I \times S \times Q \rightarrow A$ maps each internal state, external state partition and input partition into the action that the agent is to perform whenever it finds itself in a particular combination of internal states, inputs and external states.

Hysteresis carries with it the notion of a simple form of internal memory. A previous internal state in conjunction with an external stimulus has an effect on the process of change to the next state; this notion is used to increase the complexity of an agent.

Definition 7

A loquacious knowledge level agent in an environment is a 12-tuple of the form

$$\langle D, S, T, A, R, Q, W, \mathit{see}, \mathit{do}, \mathit{hear}, \mathit{database}, \mathit{action}\rangle$$

where the sets and functions are all as previously described except that the set D is an arbitrary set of predicate calculus databases. The function *database*: $D \times T \times Q \rightarrow D$ maps a database and both types of observation into the new internal database. Finally, the function *action*: $D \times T \times Q \rightarrow A$ maps each internal database, external state partition and input partition into the action that the agent is to perform whenever it finds itself with a particular combination of internal databases, inputs and external states.

Moving on from the ability to memorize internal states, we allow that it will be necessary to consider relations between facts that represent entities in the external states in influencing the actions of an agent in its environment. In order to cope with this we introduce a database that can store relations as sets of tuples and the action function is amended to allow an agent to retrieve the tuples from the database.

Definition 8

A loquacious stepped knowledge level agent in an environment is a 12-tuple of the form

$$\langle D, S, T, A, R, Q, W, \mathit{see}, \mathit{do}, \mathit{hear}, \mathit{database}, \mathit{action}\rangle$$

where the sets and functions are all as previously described except that the function *database*: $D \times N \times T \times Q \rightarrow D$ maps a database, cycle number and both types of observation into the new internal database. Finally, the function *action*: $D \times N \times T \times Q \rightarrow A$ maps each internal database, cycle number, external state partition and input partition into the action that the agent is to perform whenever it finds itself with a particular combination of internal databases, inputs and external states.

When a sequence of different action patterns is proposed, interactions occur not only between feedback and the central control of the actions themselves, but also between feedback and the central control of groups of actions. Notice that the difference between a stepped knowledge level agent and an ordinary knowledge level agent is the dependence of the database and action functions on the agent's cycle number. This gives an agent the ability to make decisions based upon temporal considerations.

Definition 9

A loquacious deliberate stepped knowledge level agent in an environment is a 12-tuple of the form

$$\langle D, S, T, A, R, Q, W, \mathit{see}, \mathit{do}, \mathit{hear}, \mathit{database}, \mathit{action}\rangle$$

where the sets and functions are defined identically to those in Definition 8. The difference in defining agents in this class arises from the use of an automated inference method like resolution when defining the operation of the action function. The authors are currently actively investigating the use of argument schemas (Toulmin 1958) in deriving a sentence that indicates the required action on each cycle. An agent of this sort is deliberate in that it deliberates on every cycle about which external action to perform.

In order that agents may reason about their own sequences of actions we introduce the idea of deliberation. Deliberation allows us to design some sequences of actions based on classical decision theory (see McCleery (1978) for a discussion) or on the notion of satisficing (Simon 1956).

The names of the agents just defined, while giving a good descriptive feel, are too long to be comfortably used in practice. We will refer to the class of all such agents as L-Agents (i.e., loquacious agents) and will abbreviate each individual agent class from the set {Lt-Agent, Lh-Agent, Lk-Agent, Ls-Agent, Ld-Agent}.

Two important definitions of properties that L-Agents share with other autonomous agents, weak or observational equivalence and strong equivalence, are complex and protracted, see Milner (1980, 1989). For reasons of space we content ourselves with a simple explanation tailored to L-Agents. Two L-Agents are observationally equivalent if a third external agent cannot distinguish between them. This means that for every possible set of inputs the two agents produce the same set of actions/outputs. There is no requirement that the two agents should pass through the same internal states. Strong equivalence, on the other hand, requires that the inputs, internal states and outputs/actions are indistinguishable.

22.2.2 THE FOUR-LAYER MODEL OF AGENT COMMUNICATION

The description of the structure of a social grouping involves description of the constituent relationships within it and also descriptions of the meta-relationships that exist. Hinde (1982) divides the structural analyses of groups into two levels, surface structures and more general structure. Surface structure is taken to be a description of the pattern of relationships that describe interactions between individuals and may show idiosyncracity between individuals. The more general structure is viewed as an abstraction from the surface structures which shows the patterns or regularities of more general relationships across social groupings and neglects the peculiarities of individuals. It is to this more general structure that we look for a metaphor to build autonomous agents into societal units. Throughout the literature on the structure of social units there are a number of pervasive models, and workers in the field of distributed autonomous agents are actively engaged in investigating the application and suitability of such models in various projects. For a survey the reader may like to consult Bond and Gasser (1988). Two of the most pervasive models are the hierarchy and network (Thompson *et al.* 1991) and in the four-layer model of agent communication we make use of a combination of both these models to form a complex social structure.

The model that we use as a metaphor is based upon an analysis of the traditional management structure in the British construction industry (CIOB 1982). Hillebrandt (1984) observes that the typical construction process involves a custom-built product produced to a specification by a complex set of interacting organizations arguably having more environmental impact than any other industry. In Figure 22.3, it is apparent that the contractual relationships and main communication paths are complex and have evolved over a considerable period of time in response for the need to manage complexity in a very dynamic environment – British weather is well known for its changeability, to mention but one factor. We observe that this sort of structure, while in practice not being perfect, has a successful record of completing some large and complex projects, and can also be seen in many other human organizations. It should be apparent that there are many interesting parallels to be drawn between this model and the control and production of a complex virtual structure in a mathematical space by a group of interacting agents.

We are interested in particular in networks of cooperating equals and hierarchies of the

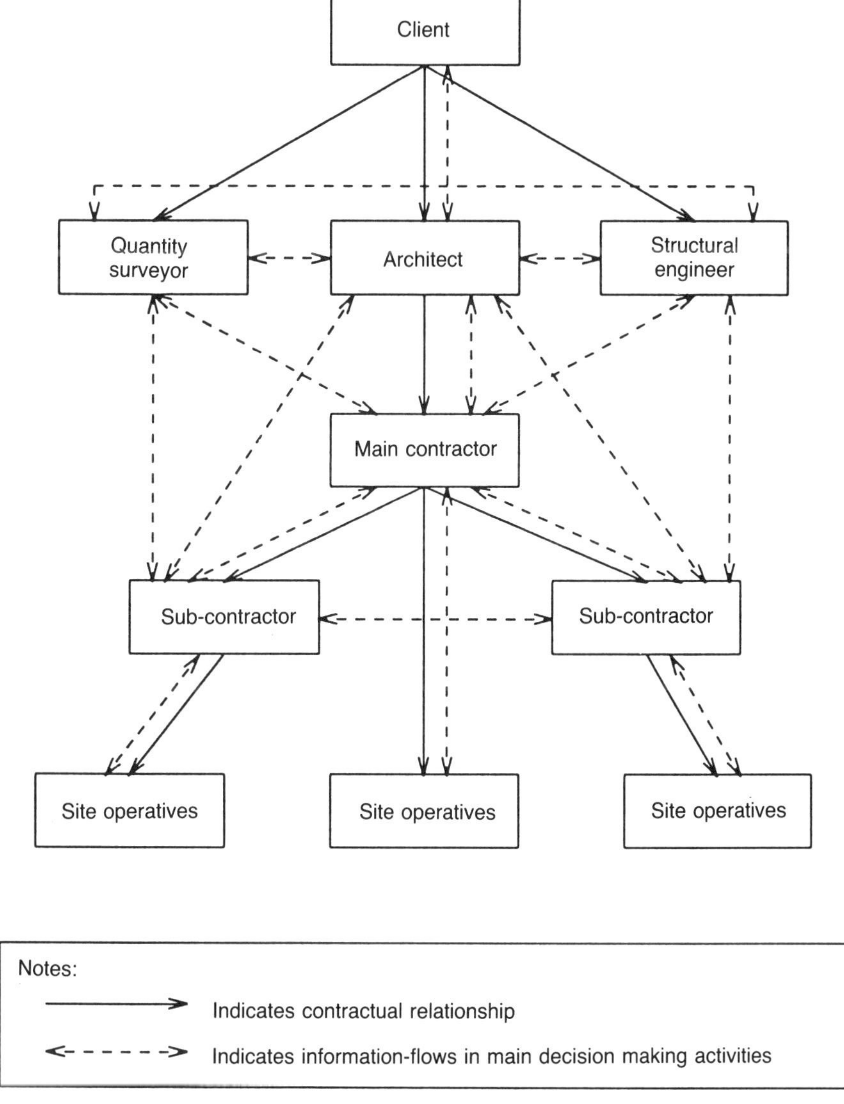

Figure 22.3 Traditional construction industry management structure. (Note: for simplicity only two subcontractors have been shown.)

division of responsibilities, and there are two primary mechanisms that may be used to impose a social structure upon autonomous agents. The vocabulary of the agent may be used to prescribe which other agents are allowed to meaningfully communicate with it and we can use the computational ability of an agent to implement cooperation strategies that enforce a hierarchy of control.

The L-Agents are partitioned into the four levels according to their inherent abilities (Figure 22.4). We have included both Ls-Agents and Ld-Agents in level zero, bearing in mind that Ld-Agents are a subset of

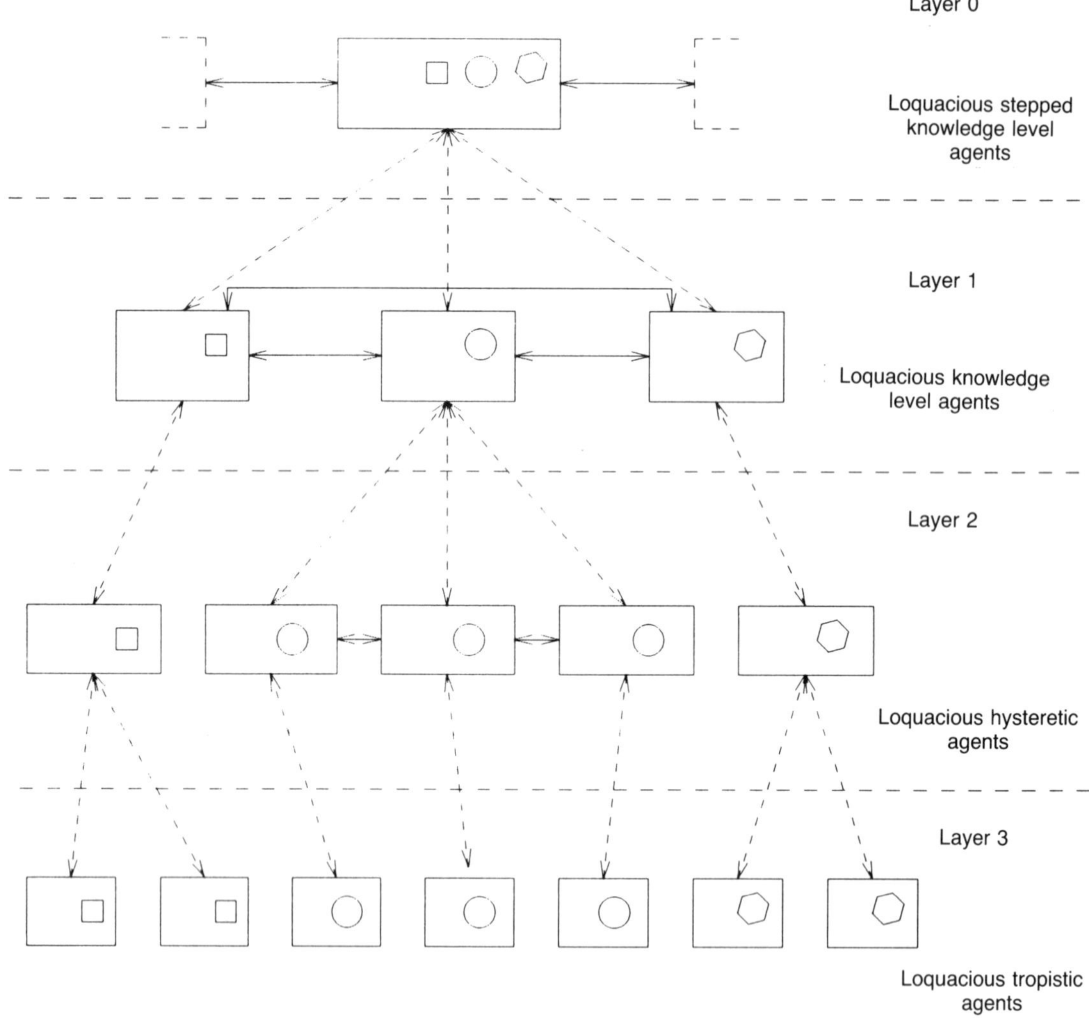

Figure 22.4 The four-layer model of autonomous agent communication.

Ls-Agents that are designed with particular internal functional methods in mind. Atomic agents are, informally, agents of any type that cannot be subdivided into simpler agents. Complex agents, on the other hand, are agents that are composed using at least two atomic agents from more than one level in the four-layer model. In a complex agent atomic agents will use preemptive cooperation in a downward direction which leads to the principle that a complex agent embodies within it the notion of a hierarchical control structure. We use the term 'preemptive cooperation' to allow for the situation in which lower level agents may refuse to accept a commission if they are already engaged upon an action commanded by some other higher level agent. This approach is necessary so that high level agents may share resources.

We do not view a number of agents on the same level cooperating upon a common task as a complex agent. Such a grouping is a set of autonomous communicating agents, that we might describe with collective nouns such as team, system, subsystem, society, and so on. From the foregoing it follows that whichever level the highest level agent in a complex agent is drawn from, there will be only one agent from that level in the complex agent. It also follows that we cannot combine groups of lower level agents into complex agents without the formation of a higher level agent; therefore Ls-agents can only be grouped together into systems and so on, not into agents.

Agents may be grouped together in systems of cooperating equals by means of intra-layer cooperation strategies. Precisely which form of strategies to choose is currently an open question which will form a major part of our forthcoming research (see the discussion later). We content ourselves at this point by stating that we do not allow Lt-Agents to have any intra-layer communication. Lh-Agents may communicate with observationally equivalent agents but have not the ability to cooperate with them. Lk-Agents may communicate with agents that are not observationally equivalent but may only cooperate with observationally equivalent agents. Finally Ls-Agents may communicate and cooperate with both observationally equivalent and non-observationally equivalent agents.

It is interesting to note that in Figure 22.3, which represents two graphs, that is, two sets of edges but only one set of nodes, the two graphs are not isomorphic. This feature has been lost with the rather more simplified structure of Figure 22.4 where the inter- and intra-layer cooperative structure is isomorphic to the communication channels. What has changed between the figures is the complete loss of an analogue for the client, of Figure 22.3, in Figure 22.4.

We see then that we have networks of communicating agents being grouped into hierarchies to form more complex agents in a manner that is both rich and precisely specifiable. These notions, we contend, when taken with the agent architectures described, allow the design of systems that fit in very well with the ideas contained in the top-down design philosophy first mooted by Dijkstra (1968) but enable the extension of those ideas to include the notion of cooperating equals in a system.

22.3 FROM METAPHORS TO PROGRAMS

22.3.1 A PLATFORM FOR PROTOTYPING AGENTS

Until recently researchers working with multiagent systems have been forced to either build individually tailored solutions to their research problems, in languages without specific constructs for distributed programming coupled with high level knowledge representational abilities, or to confine themselves to the use of formal specification languages and the theoretical presentation of their ideas. An attractive proposition for workers in such a position would be the enhancement of an existing language that is already in widespread use for programming knowledge-intensive applications. The benefits gained by such workers for their own work from a feedback loop via practical trials may be considerable, as would similarly be the case with an enhanced ability for explaining those practical trials through a well-understood high level programming paradigm. Groups throughout the world have been working at this problem since the late 1970s (see Newmarch (1990) for a discussion of some approaches and early problems). We consider one such approach and then present some short examples.

The group from Imperial College have been involved in this field since the early days with IC-Prolog (Clark and McCabe 1979), Parlog (Clark and Gregory 1983) and now IC-Prolog II (Cosmadopoulos and Chu 1992). IC-Prolog II (ICP) has had a number of features added over first-generation Prologs in order that workers may program distributed systems at a high level. In particular we note that ICP is multithreaded. Each thread is a process with its own execution and variable storage area and there are well-defined primitives by which threads may communicate with each other. Threads in one main ICP process do however share the one Prolog database, which does mean that considerable discipline is required when programming. Processes may communicate by one of three methods; pipes, TCP/IP primitives and mailboxes. Pipes have been specifically designed as a unidirectional channel of communication between threads running under the same ICP process and are supported by a set of primitives. Separate ICP processes running on the same or different machines may communicate via ICPs built in TCP/IP primitives; those who wish to program at this level of abstraction should refer to Cosmadopoulos and Chu (1992) and Stevens (1990).

Mailboxes also provide a means of process/process communication that has been designed both as a means of enabling workers unfamiliar with low level communications programming to use very high level constructs in building experimental communications prototypes and as a transparent method of minimizing network resource consumption. Each individual TCP/IP connection – used transparently by the mailbox program – supports up to 64 virtual channels and each virtual channel appears as a FIFO queue at the mailbox to communicating processes. A full set of primitives is provided for use with mailboxes (Chu 1992).

The author, using this platform and the architectures and environment given previously, has developed a number of communicating agents that are currently under testing and further development. In the next subsection we look at some of the functions that have been implemented.

22.3.2 SOME SHORT EXAMPLES USING AGENT SENSORY FUNCTIONS

Any autonomous or semi-autonomous communicating agent that displays even a rudimentary form of intelligence will of necessity be complex in its structure, as we have seen in the definitions, and contain considerable lengths of program code. We do not have space here to develop a complete agent but will content ourselves with considering the sensory functions of possible agents and how our choices of metaphor affect the design and coding of the functions. Readers interested in examining these agents more fully are referred to Staniford *et al.* (1993).

Example 1: sensory function of higher level agents

```
/*
hear/2, hear/4 – Agents level 0/1
   +Comm_path : must be instantiated to an atom that is
              : a channel identifier
   ?Sender    : Creator of incoming message
   +Recipient : Id of listening agent must be instantiated to
              : an atom

   Notes
              : message is a structure
              : msg( Recipient, Sender, Time, Msg_body)
   -Time             : Time of creation of message
   -Msg_body :       : Contents of message
*/
```

```
hear( Comm_path, Sender ) :-
  my_identifier( Recipient ),
  mbx_getid( Comm_path, Id, ok ),
  mbx_look(Id, Message, ok ), !,
  hear( Recipient, Sender, Message, Id ).

hear( Recipient, Sender, msg( Recipient, Sender, Time, Msg_body ), Id ):-
  mbx_commit( Id, ok ),
  assertz( short_mem( Sender, Time, Msg_body )).

hear( Recipient, _, msg( Rec_nt, _, _, _ ), Id ):-
  not compare( =, Recipient, Rec_nt ),
  mbx_discard( Id, ok ),
  reject_action( recipient ),
  fail.

hear( _, Sender, msg( _, Sndr, _, _ ), Id ):-
  not compare( =, Sender, Sndr ),
  mbx_discard( Id, ok ),
  reject_action( sender ),
  fail.

hear( _, _, Message, Id ):-
  not Message = msg( _, _, _, _ ),
  mbx_discard( Id, ok ),
  reject_action( message ),
  fail.
```

Example 2: sensory function of lower level agents

```
/*
hear/1, hear/3 – Agents level 2/3
  ?Msg_body      : Contents of message

  Notes
  +Comm_path : must be instantiated to an atom that is
             : a channel identifier
  +Recipient     : Id of listening agent must be instantiated to
             : an atom
             : message is a structure
             : msg( Recipient, Msg_body)
*/

hear( Msg_body ):-
  my_identifier( Recipient ),
  my_comm_path( Recipient, Comm_path ),
  mbx_getid( Comm_path, Id, ok ),
  mbx_look( Id, Message, ok ), !,
  hear( Recipient, Msg_body, Id ).
```

```
hear( Recipient, msg( Recipient, Msg_body ), Id ):-
  atomic( Msg_body ),
  mbx_commit( Id, ok ).

hear( Recipient, msg( Rec_nt, _ ), Id ):-
  not compare( =, Recipient, Rec_nt ),
  mbx_discard( Id, ok ),
  reject_action( recipient ),
  fail.

hear( Recipient, msg( Rec_nt, _ ), Id ) :-
  not atomic( Msg_body ),
  mbx_discard( Id, ok ),
  reject_action( message_contents ),
  fail.

hear( _, Message, Id ):-
  not Message = msg( _, _ ),
  mbx_discard( Id, ok ),
  reject_action( message_structure ),
  fail.
```

Consider first Example 1. This is a sensory function suitable for incorporation into agents in the highest two levels of the four-layer model (Figure 22.4 and Definitions 7, 8 and 9). As is common in Prolog programming, we implement the function as a set of clauses with the same name but different arities. In a full size system hear/2 would be public, to be called from anywhere in an agent's main thread, but hear/4 would be private and only called by hear/2. Note: both hear functions can exist in the same ICP process without confusion, due to their clauses being implemented with differing arities. This sensory function has considerable discriminatory properties. Hear/2 firstly recalls the identity of the agent via my_identifier/1 $\in D$ (see Definition 7) and secondly obtains the channel identifier, for the particular communication path that is being listened to, via a call to the built-in mbx_getid/3 and then calls built-in mbx_look/3 which is a blocking primitive that inspects the specified channel and returns a Message if one is available or waits for a Message to become available if the channel is empty. So, when a message is available in the communications channel hear/2 it passes to hear/4 which either accepts or rejects the message.

Clause 1 of hear/4 is the clause that accepts correctly specified messages using the inbuilt pattern matching of Prolog via unification in the head of the clause, that is, Recipient unifies with Recipient, Sender unifies with Sender and the message structure is correct. When all these conditions are satisfied Clause 1 fires and commits hear to reading the message, removing it from the communicating channel and unlocking the channel for the next incoming message to be made available. It then saves the message in short_mem/3 where short_mem/3 $\in D$ (see Definition 7). Clauses 2, 3 and 4 of hear/4 are the clauses that deal with all possible reasons for rejecting a message; mbx_discard/2 releases the communication channel without removing the message and we assume that reject_action/1 is suitably defined to deal with the three possible rejection situations – receiver does not match, sender does not match and incorrect message format. Hear/4 then is a completely declarative set of clauses and together with hear/2 provides a function that will receive messages for a specific agent, sent from a specific agent and that in a specific form. Note that the form shown in the example is, for the sake of simplicity, a named tuple of arity 4; it could be differently structured in other examples and in general could be any valid Prolog term, including being a section of executable code. In Example 1 there is no specification for Msg_body; therefore Msg_body may take the form of any Prolog term and so it is possible for agents to pass methods as well as factual knowledge to other agents.

In Example 2 (see also Figure 22.4 and Definitions 5 and 6) we see that the hear function is much simplified. It still consists of a set of clauses with the same name but different arities as in Example 1, but we now have hear/1 and hear/3. Note: as before both hear functions can exist in the same ICP process.

Similar to the above, hear/1 has one clause and hear/3 has four clauses. We see, however, that hear/1 can have two distinct purposes: one, to return a message in Msg_body and two, to be a decision function – by having Msg_body instantiated at the time of the call – returning true if some particular message has been received, otherwise returning false. Agents at this level have only one communication path that they can use. Again, similarly, hear/3 has four clauses, an acceptor and three rejectors, but now we only accept atoms, integers and floating point numbers as messages – atomic/1. These simpler agents cannot receive the identity of the sending agent except as part of the message body because they do not have databases (see Definitions 5 and 6) in which to store the information.

In order to appreciate the sort of messages that agents may communicate and the sort of functions that agents may be used for in a document graph environment we look at a simple example of a tropistic agent. It will be readily apparent from Examples 1 and 2 that the complete program code for even the simplest of agents is likely to be extensive. In Example 3 a specification of an agent's vocabulary and grammar is presented on which to focus the discussion.

Tropistic agents, by definition, can only react to external stimuli in proceeding from one internal state to the next; this means that it is not possible for such an agent to hold facts relating to its input messages and local environment while it accesses non-local information that is not immediately available. For example, a tropistic agent cannot carry out an operation requiring checking a higher level agent's access permission because the access permission is in general held non-locally. With few exceptions, because of the inherent concurrency within a distributed system, an agent operating within a document graph environment must explicitly open a graphical structure before it executes any operations upon the structure or upon any content or attributes belonging to members of the structure. The exceptions are limited to meta-level operations upon graphical structures such as comparison and status. Determining the status of a portion of the environment is an ideal task for a tropistic agent because the local status of a section of graph may be perceived from the attribute types of the local object and edges (see Definition 1). Non-local information and historical information are not required.

Example 3: structure status

LT_Status_Struct(Port_in, Port_out);

Input Messages:
STATUS •
agent_ID • sub_graph_ID •
{ Open | Read | Write }

Output Messages:
SUCCESS •
agent_ID • sub_graph_ID •
{
IsOpen | NotOpen | IsReadable |
NotReadable | IsWriteable |

NotWriteable
} |
FAIL •
agent_ID • sub_graph_ID • not_found

In Example 3, the LT_Status_Struct agent determines, as can be seen from the input/output messages, a wide variety of status information about the access status of a document graph structure.

The notation for messages uses • as the concatenation operator and | as the alternative operator. Using this notation we can see that there are only three legal input messages that the agent can receive as external stimuli from other agents:

1. STATUS agent_ID sub_graph_ID Open
2. STATUS agent_ID sub_graph_ID Read
3. STATUS agent_ID sub_graph_ID Write

For the sake of simplicity we ignore what actions and messages the LT_Status_Struct agent produces upon receipt of an incorrectly formulated or incomplete input message.

LT_Status_Struct is a good example of a tropistic agent that has a fairly complex degree of functionality while maintaining a relatively simple internal structure. It is trivial to see that each input message can be seen as a decision problem for successful outputs, there being a direct one-to-two relationship between input and successful output, for example,

$$\text{Open} \rightarrow \{ \text{IsOpen, NotOpen} \}$$

while the only possible reasons for failure, assuming a correctly formulated input message, leaves any of the originating questions unanswered. If a structure cannot be found, either because it does not exist or because it cannot be reached due to some system failure, then the question of its status cannot be resolved and the failure message reflects this state of affairs.

Staniford (1993) discusses in detail the specification of actual document graphs (Definition 1), document specifications (Definition 2), document representations (Definitions 3 and 4) and the mechanisms by which agents move around, communicate and co-operate in the document graph environment.

To complete this section we discuss briefly the remaining sensory function – *see*. Considering Definitions 1–4 then, we see that each vertex has a unique identifier and a vertex content and a set of outgoing edges associated with it. Each edge has a pair of identifiers, a label and an edge content associated with it. Similarly to *hear* we developed two identically named functions *see*. The simpler function can return one only of {'an identifier', 'vertex content', 'member of the set of outgoing edges', 'edge label', 'edge content' }. Considering the LT_Status_Struct agent of Example 3, it is readily apparent that the information required, that is, one of { IsOpen, NotOpen, IsReadable, NotReadable, IsWriteable, NotWriteable } can be seen by a tropistic agent from the object' attribute-type information. If the agent cannot *see* such information for any reason it replies with a message containing not_found.

The more complex function returns all the foregoing information in a Prolog structure, placed in short-term memory as before. Both sets of sensory functions are robust in operation and enable agents to perceive information in a form that suits their internal abilities within the agent architecture and their external roles within the larger societal structure.

22.4 DISCUSSION AND CONCLUSION

An architectural model for autonomous communicating agents has been discussed. This design depends on a combination of hierarchical and network features that make use of systemic metaphors from biology and sociology. The environment upon which the agents act, a spatial metaphor, was specified as a directed acyclic graph which we view as being in a dynamic state by the use of graph grammars.

The essential property of agentness extracted from our metaphors is the ability of a clearly defined agent to influence or change either some local part of its environment or the behaviour of some other agent or agents; it follows directly that the agent's own behaviour may respond to and change as a consequence of changes in its environment or communications from other agents. Agents influenced by other agents may be, but are not necessarily, situated in the same environment. A convenient device for use in reasoning about and proving the formal properties of agents is the ability of agents to have a null action in response to certain stimuli (Milner 1989). The architecture described provides for such actions to be specified by placing suitable restrictions upon particular agent's sets and functions.

L-Agents differ from the actor model of concurrent computation in distributed systems (Agha 1986), in which no distinction is made between actors and environment, and all entities within the model are viewed as dynamic actors, because we clearly separate L-agents and environment. L-agents may not change structurally over time, although they can change their behaviour in response to external influences; the environment, however, is viewed as a constantly changing structure over time, the changes being effected by the L-agents that operate within it.

Building upon actors and reactors in an environment we drew upon the metaphor of a complex human social agency to guide our design for the model of communicating agents working together to effect environmental change. The hierarchical element in the social model was produced by taking account of four levels of agent ability in conjunction with a very prescribed form of cooperation; the orthogonal network element was obtained using intra-layer communication to form networks of cooperating agents, the networks again being partitioned on the basis of agent ability.

We contend that this architecture provides a rich, specifiable model on which to build, predict and study the behaviour of autonomous communicating agents. The work is founded upon the explicit use of ideas from other scientific disciplines, ideas that have undergone a transformation and become metaphors, providing a rich source of interdisciplinary links in the chains of complex problem solving. There may be no quantified evidence for the benefits of such an approach but the anecdotal evidence that has been building up over recent decades and the arguments provided by such workers as MacCormac (1985), Way (1991) and Paton (1992) provide a basis for the belief that the practical consequences arising from the use of interdisciplinary metaphors will be of considerable help in enhancing our ability to design and build the ever more complex systems being investigated by researchers in computer science.

The next phase of the work with the agents described in this chapter will be the development of reasoning abilities in order that the agents may cooperate and coordinate joint actions on solving problems in their environment. Areas actively under consideration for suitable metaphors to include in the approach to this task are the legal profession, the philosophy of argument and the study of linguistics. It is hoped that they will provide a rich source of contexts for idea transference to the domain of reasoning automata.

ACKNOWLEDGEMENTS

The support provided by SERC, under research grant No. 90309983, and the helpful comments of the anonymous reviewers are gratefully acknowledged as are the many stimulating ideas and general help received from Trevor Bench-Capon, Paul Dunne and Ray Paton. Any inaccuracies remaining are, of course, my sole responsibility.

REFERENCES

Agha, G. A. (1986). *ACTORS: A Model of Concurrent Computation in Distributed Systems*. MIT Press, Cambridge, MA.

Barlow, J. and Dunne, P. E. S. (1991). *Applying a model of concurrency to computer supported cooperative work*. Proc. Computers in Writing IV, University of Sussex, pp. 175–83.

Bench-Capon, T. J. M. and Dunne, P. E. S. (1989). Some computational properties of a model for electronic documents. *Electronic Publishing–Origination, Dissemination and Design* **2**(4),231–56.

Bench-Capon, T. J. M., Dunne, P. E. S. and Staniford, G. (1991) RAPPORTUER: From dialogue to document, Proc. Computers in Writing IV, University of Sussex, pp 175–183.

Bond, A. H. and Gasser, L. (eds.) (1988). *Readings in Distributed Artificial Intelligence*. Morgan Kaufmann.

Chandrasakaran, B. (1981). Natural and social system metaphors for distributed problem solving: introduction to the issue. *IEEE Transactions on Systems, Man, and Cybernetics* **SMC-11**(1), pp. 1–5.

CIOB (1982). *Project Management in Building*. Chartered Institute of Building, p. 12.

Chu, D. A. (1992). A Beginner's Guide to Mailboxes. Draft, Logic Programming Section, Department of Computing, Imperial College, London.

Clark, K. L. and Gregory, S. (1983). *Parlog: A Parallel Logic Programming Language*. Technical Report 83/5, Logic Programming Section, Department of Computing, Imperial College, London.

Clark, K. L. and McCabe, F. G. (1979). The control facilities of IC-Prolog. In *Expert Systems in the Micro Electronic Age*, Michie D., Edinburgh University Press.

Cosmadopoulos, Y. and Chu D. A. (1992). *IC-Prolog II Version 0.94 Reference Manual*. Logic Programming Section, Department of Computing, Imperial College, London.

Devlin, K. (1991). *Logic and Information*. Cambridge University Press.

Dijkstra, E. W. (1968). A constructive approach to the problem of program correctness. *BIT* **8**, 174–86.

Dunne, P. E. S. and Staniford, G. (1991). A formal language basis for studying computational properties of graph-theoretic document models, *Bulletin of the EATCS* **44**, 292.

Dunne, P. E. S. and Staniford, G. (1992). CASS: a cooperative authorship support system. *Proc. JISI'92*, Tunis 1992, pp. 129–40.

Furuta, R. (1989). An Object-based Taxonomy for Abstract Structure in Document Models. Internal Report, Computer Science Department, University of Maryland, College Park, MD.

Genesereth, M. R. and Nilsson, N. J. (1987). *Logical Foundations of Artificial Intelligence*. Morgan Kaufman, Ch 13, pp. 307–27.

Hansen, W. J. and Haas, C. (1988). Reading and writing with computers: a framework for explaining differences in performance. *Communications of the ACM* **31**(9), 1080–9.

Hillebrandt, P. M. (1984). *Analysis of the British Construction Industry*. Macmillan, London.

Hinde, R. A. (1982). *Ethology, Its Nature and Relations with Other Sciences*. Fontana, Glasgow.

Hinde, R. A. (1987). *Individuals, Relationships and Culture: Links between Ethology the Social Sciences*. Cambridge University Press.

Kiss, G. (1992). *Variable coupling of agents to their environment: combining situated and symbolic automata*. Proc. Decentralized A.I. 3 (ed. E. Werner and Y. Demazeau). Elsevier, Amsterdam, pp. 231–48.

Koo, R. (1989). A Model for Electronic Documents. *ACM SIGOIS Bulletin* **10**(1), 23–33.

Loeb, J. (1913). Die Tropismen., Handb. Vergleich. Physiol., 4.

MacCormac, E. R. (1985). *A Cognitive Theory of Metaphor*. MIT Press, Cambridge, MA.

McCleery, R. H. (1978). Optimal behaviour sequences and decision making. In *Behavioural Ecology: An Evolutionary Approach* (ed. N. B. Davies and J. R. Krebs). Blackwell, Oxford.

McKnight, C., Dillon, A. and Richardson, J. (1991). *Hypertext in Context*. Cambridge University Press.

Milner, R. (1980). *A Calculus of Communicating Systems*. Lecture Notes in Computer Science **92**, Springer-Verlag, Berlin.

Milner, R. (1989). *Communication and Concurrency*. Prentice Hall International, London.

Minsky, M. L. (1972). *Computation: Finite and Infinite Machines*. Prentice Hall International, London.

Minsky, M. L. (1988). *The Society of Mind*. Simon & Schuster, New York.

Nagl, M. (1978). *A Tutorial and Bibliographic Survey on Graph Grammars*. Lecture Notes in Computer Science, Springer Verlag, New York.

Newmarch, J. D. (1990). *Logic Programming: Prolog and Stream Parallel Languages*. Prentice Hall of

Australia Ltd.

Paton, R. C. (1992). Towards a Metaphorical Biology. *Biol. and Phil.* **7**, 279–94.

Richards, I. A. (1936). *The Philosophy of Rhetoric*. Oxford University Press.

Simon, H. A. (1956). Rational choice and the structure of the environment. *Psych. Rev.* **63**, 129–38.

Smith, J. B., Weiss, S. F., Ferguson, G. F., Bolter, J. D., Lansman, M. and Beard, D. V. (1987). WE: A Writing Environment for Professionals. *Proceedings of the National Computer Conference '87*, Reston, VA. AFIPS Press, pp. 725–36.

Staniford, G. (1993). Multi-Agent Systems in Support of Cooperative Authorship. Ph.D. thesis, University of Liverpool, UK.

Staniford, G. and Dunne, P. E. S. (1994). *Autonomous Agents in the Support of Cooperative Authorship in CSCW and Artificial Intelligence* (ed. E. Edmonds and J. H. Connolly). Springer-Verlag, London, in press.

Staniford, G., Bench-Capon, T. J. M. and Dunne, P. E. S. (1993). *Cooperative dialogues with the support of autonomous agents*. Proc. of ICICIS'93, Rotterdam (ed. M. Huhns, M. P. Papazogolou and G. Schlageter). IEEE Computer Society Press, Los Alamitos, CA, pp. 144–52.

Stevens, W. R. (1990). *UNIX network programming*. Prentice Hall International.

Thompson, G., Frances, J., Levacic, R. and Mitchell, J. (1991). *Markets, Hierarchies and Networks*. Sage, London.

Toulmin, S. (1958) *The Uses of Argument*. Cambridge University Press.

Uexküll, J. V. (1921). *Umweltund Innenwelt Tiere*, 2nd. Edition, Berlin, Umweltsforchung. Z. Tierpsychol. **1**, 33–4.

Way, E. C. (1991). *Knowledge Representation and Metaphor*. Kluwer, Dordrecht.

PART FIVE

THEORETICAL AND CONCEPTUAL ISSUES

THE IMPORTANCE OF SELECTIONIST SYSTEMS FOR COGNITION 23

Bernard Manderick

23.1 INTRODUCTION

In order to cope with an uncertain and noisy environment, any autonomous cognitive creature – this could be a biological organism or an autonomous robot – has to solve a number of basic cognitive problems. For instance, a creature has to categorize the inputs coming from the environment or adapt its behaviour to the environment.

Moreover, we will suppose here that the creature has to solve these problems in the following circumstances. First of all, the creature has to do this without the help of a teacher. And second, the creature does not have initial variable environment-specific knowledge which could be used to solve these cognitive problems. The creature is connected with its environment in a feedback loop and the only information that it gets from the environment is via its sensors. At the same time, the creature affects the environment via its effectors. From the creature's point of view, the environment is essentially a black box.

So, an autonomous cognitive creature has to face the following **bootstrap problem**. How can an autonomous creature acquire initial knowledge about the environment and how can it behave adaptively (1) without having built-in environment-specific knowledge and (2) without being supervised while acquiring that initial knowledge or behaving adaptively?

This bootstrap problem resembles the circumstances in which biological systems like a species, an immune system and a brain have to adapt. Natural evolution produces species that are increasingly better adapted to their niche. The immune system of a vertebrate organism builds up an immune response to any foreign substance that might invade that organism. The brain categorizes stimuli coming from its environment and supports adaptive behaviour. Yet, these systems have not been designed with the characteristics of the environment in which they have to operate in mind. Still, they are able to adapt successfully to their environment of which they have no prior variable environment-specific knowledge.

In biology, theories have been proposed which explain the adaptive capabilities of these biological systems. Selectionist theories have proven to be the most successful ones: **neo-Darwinism**, the synthesis of Darwin's (1859) original theory of natural selection and modern genetics explains the evolution of species. The **clonal selection theory** for antibody formation (Jerne 1955; Burnet 1959) explains the working of the immune system. Edelman (1987) has proposed the theory of neuronal group selection to explain the working of the brain.

Scientific progress may result from the application of ideas in one domain to another one. We want to investigate how selectionism which underlies the adaptability of the above biological systems can be applied to an understanding of autonomous cognitive creatures.

In this chapter, we discuss three adaptive biological systems. For two of them, the natural evolution of species and the immune system, it is nowadays widely accepted that their adaptiveness is best explained by selectionist principles: the natural evolution of species is explained by neo-Darwinism which is the synthesis of Darwin's original theory and modern genetics, and the immune system is explained by the clonal selection theory. For the third system, the brain, a selectionist theory called **neuronal group selection theory** or **neural Darwinism** has been proposed recently to explain higher brain functions like perception, categorization and memory (Edelman 1987).

This chapter is organized as follows. First, we briefly discuss the natural evolution of species. Second, the immune system is described in somewhat more detail than the two other systems since it is an adaptive system with cognitive properties and a major source of inspiration for the selectionist automaton. Third, the brain according to the neuronal group selection theory is described. Fourth, the three biological systems are compared to abstract their common principles. Finally, the selectionist automaton is introduced, design decisions to be taken are discussed and their domain of application is defined.

The selectionist automaton has to be seen in the broader context of evolutionary computation which also uses natural evolution as a metaphor. Evolutionary computation is reviewed in Bäck's chapter in this book.

23.2 NATURAL EVOLUTION OF SPECIES

Darwin (1859) was the first to explain the natural evolution of species without having to resort to explicit design. The mechanisms driving natural evolution are known. A species evolves because all individuals in a population are different from each other (diversity). Individuals survive and reproduce according to their fitness (selection of the fitter individuals) and the produced offspring is similar but not necessarily identical to the parents (diversity is maintained). In this way species emerged and evolved which are highly adapted to their niche in morphology, structure and behaviour. We mention just three examples. The morphology of the albatross allows it to glide through the air for long periods of time. The structure of the human eye allows it to focus on objects in the external world. The complex social behaviour in ant colonies is responsible for the very efficient exploitation of food sources. Yet, the morphology of the albatross, the structure of the human eye and the behaviour of the ants were not designed in any way. Instead they are products of natural evolution which, according to Dawkins (1986), is a 'blind watchmaker'.

The neo-Darwinian theory of today synthesizes Darwin's original theory of natural evolution with the findings of modern genetics and is now common knowledge. The DNA in the fertilized cell contains vital information concerning the morphology, the structure and the innate behaviour of the new organism. This information is called the **genotype** of the organism and is expressed through a development process called **epigenesis**. The resulting organism with its morphology, structure and behaviour is called the **phenotype**. The combination of morphology, structure and behaviour determines how well the organism is adapted to its environment. The better an organism is adapted the higher its chance to reproduce itself. The reproductive success is called the fitness of the organism. Due to mutation and crossover, the genotype of the offspring differs from the genotype of the parents. Each piece of information in the genotype has a probability to be replaced by another piece. This is mutation. For sexually reproducing species, the genotype of the offspring is the recombination of the genotypes from its two parents. This recombination process includes

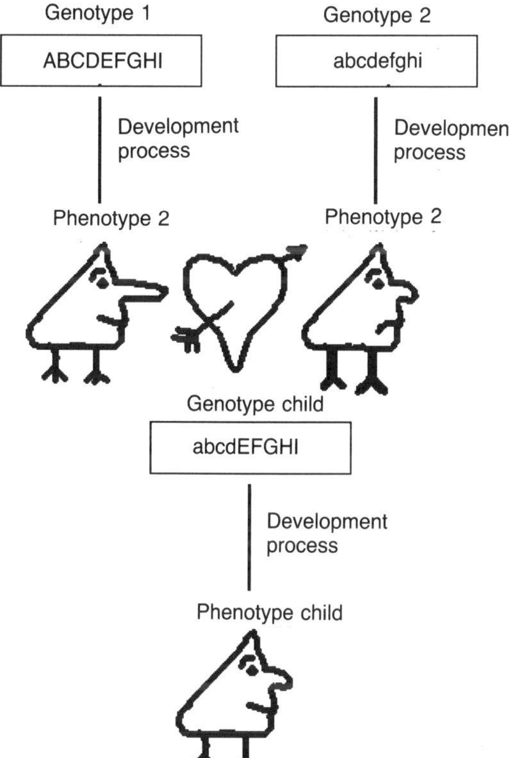

Figure 23.1 The genotype of an organism is a blueprint for the phenotype. The corresponding phenotype is the organism's morphology, structure and behaviour. This phenotype is the result of a development process called epigenesis and expresses the information present in the genotype. The combination of the organism's morphology, structure and behaviour determines how well the organism is adapted to the environment, that is, the organism's fitness. The fittest organisms survive and produce offspring. The genotype of the offspring differs from the genotypes of the parents as a result of mutation and crossover.

crossover. The above concepts are illustrated in Figure 23.1.

Differences in the phenotype and fitness result from differences in the genotype. For each genotype one can associate a corresponding fitness. This way, a fitness landscape is defined on the space of genotypes. A population of organisms can be mapped on the corresponding pool of genotypes. The evolution of species results from the combined action of (1) the recombination of the genotypes in the population, and (2) the selection of the better adapted organisms generation after generation. In this way, the evolutionary process makes the pool of genotypes move towards the regions of the genotype space with highest fitness. This is shown schematically in Figure 23.2.

Note that the above description is a simplification. For instance, the concept of adaptation is difficult to define. It runs the risk of tautology and is very difficult to measure. For more background information we refer the reader to Paton (1992).

23.2.1 DISCUSSION OF NATURAL EVOLUTION

Darwin's theory of natural evolution explained the origin of highly adapted species without having to resort to an intelligent creator who has explicitly designed the adapted organisms. The adaptation of species to their niche is the result of random diversity and selection of the fitter individuals. Individuals of one species are different from each other: there is diversity in the population. Differences in genotype result in differences in phenotype and fitness: the better adapted individuals reproduce more than the less adapted ones. The offspring is similar to but never identical to the parents. This way, the diversity is maintained. The generation and maintenance of diversity results from crossover and mutation of the genotypes present in the population. This recombination process is an autonomous process not influenced by the environment in any way. Diversity is first and selection by the environment of the fitter individuals is second.

Before Darwin, the adaptation of species was explained by Lamarck by an instructive theory. In instructive theories, the environment comes first and causes variants which

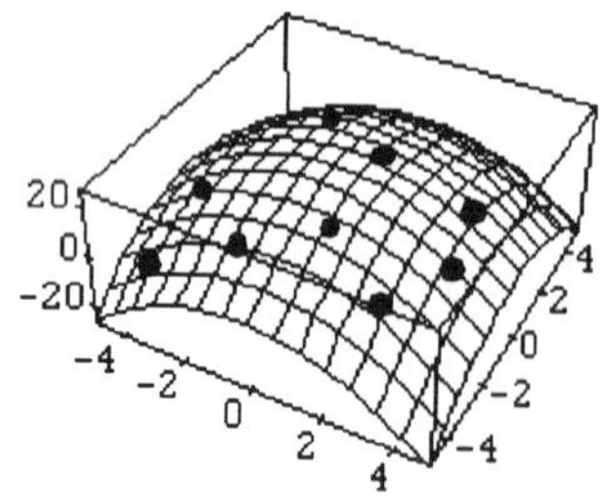

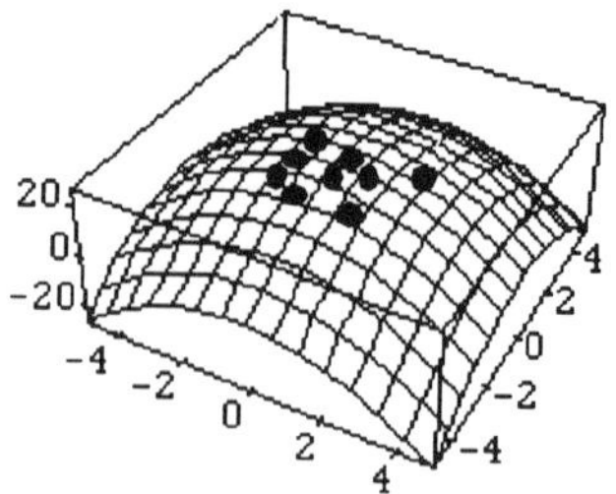

Figure 23.2 Shown is a fitness landscape defined on the genotype space. To each genotype in the genotype space a value is assigned; the fitness of the corresponding phenotype. To the left, we have the initial population of organisms (represented by dots) which is distributed over the whole genotype space. To the right, the population is shown several generations later. This population has moved towards regions with the highest fitness. This way, evolution is viewed as an optimization process.

are better adapted. Lamarck's theory makes the following claims:

1. Changes in the environment bring about changes in the needs of the organisms living in that environment.
2. Changes in needs trigger changes in behaviour to satisfy these new needs.
3. In the new behaviours, certain parts of the organism are used more frequently and others less frequently than before. What is used is strengthened and what is not used is weakened.
4. The characteristics acquired by the process described above are inherited.

Lamarck's theory, which consists of the combination of the above arguments, was the first instructive theory of adaptation that was refuted – but see Jablonka *et al.* (1992) for a recent discussion on the inheritance of acquired characteristics.

23.3 THE IMMUNE SYSTEM

The immune system (IS) is a network of interacting cells (lymphocytes) and molecules (antibodies) that protects an organism against foreign substances called antigens. It does this by producing antibodies that play a role in eliminating these antigens. So, the IS adapts to an unpredictable environment of antigens called the antigenic environment. It is nowadays accepted that the adaptability of the IS is best explained by selectionist principles. This is largely due to the work of Jerne and Burnet (Jerne 1955, 1975; Burnet 1959). Still, there exist different interpretations. The exposition of the IS given below is based on the interpretation of Varela, Coutinho and coworkers.

We discuss the IS for three reasons. First, the IS is a network with cognitive properties like pattern recognition, memory and learning (Varela *et al.* 1988):

1. Pattern recognition: the IS recognizes molecular shapes, that is, its cognitive domain is the space of all possible molecular shapes.
2. Memory: the IS retains a memory of the substances that it has encountered in the past.
3. Learning: the future immune response against a type of antigen already encountered will be both faster and larger in amplitude.

Second, the IS is an adaptive system whose knowledge about its cognitive domain is distributed and the system is fault-tolerant. Third, the IS is a source of inspiration for selectionist automata (Section 23.6).

The rest of this section is organized as

follows. In the first three subsections we describe the components of the IS, how these components interact, and how the whole adapts to a changing environment. We restrict our description of the IS to those features which are relevant for selectionist automata. In Section 23.4, we discuss the 'cognitive' properties of the IS, we show that the IS is an adaptive and fault-tolerant system whose knowledge is distributed, and we say a few words about the instructive theories proposed for the working of the IS.

23.3.1 COMPONENTS OF THE IS

The IS is a network of interacting cells and molecules which recognizes foreign substances invading the body of an organism and marks them for later destruction. These foreign substances are called **antigens**. Examples of antigens are bacteria, viruses, pollen grains, incompatible blood cells and man-made molecules.

The molecules of the IS that recognize antigens are called **antibodies**. An antibody does not recognize an antigen as a whole object. Instead, it recognizes small regions called **epitopes**. An antibody recognizes an antigen if it binds to one of its epitopes. The binding region of an antibody is called the **paratope**. We can think of a paratope as a key that has to fit one of the locks (epitopes).

The strength and the specificity of the interaction between antibody and antigen is measured by the affinity of the interaction. The affinity depends on the degree of complementarity in shape between the interacting regions of the antibody and the antigen. Each region is characterized by features like length, charge and molecular shape. The affinity can be expressed as a function of these chemical features. A given antibody can typically recognize a range of different epitopes. And a given epitope can be recognized by different antibody types. The latter property is called **degeneracy**.

Antibodies are produced by cells called **B-lymphocytes**. B-lymphocytes differ in the antibodies that they produce. Each type of antibody is produced by a corresponding B-lymphocyte which produces only this type of antibody. A B-lymphocyte has about 10^5 antibodies attached to its surface. They serve as sensors to detect the presence of complementary epitopes. An organism typically has approximately 10^7 different types of B-lymphocytes. This number is sufficient to recognize any possible antigen if we suppose that these types are distributed uniformly over the set of all possible B-lymphocyte types and if we take into account that each antibody recognizes a number of different epitopes.

Not only do antibodies recognize antigens but they also recognize other antibodies if they have the right epitope. An epitope characteristic for a given antibody type is called an **idiotope**. If an antibody i recognizes an antibody j (i.e., the paratope of i is complementary to the idiotope of j) then j is called the **anti-idiotypic** antibody of i (the anti-idiotypic antibody j of i can be regarded as the internal image within the IS of the external antigen recognized by i (Jerne 1975)). As a result of the interactions between different antibodies the IS exhibits a complicated dynamical behaviour.

23.3.2 THE DYNAMICS OF THE IS

When an antibody on the surface of a B-lymphocyte binds to another molecule (antigen or another antibody), the B-lymphocyte is stimulated to reproduce (i.e., to clone) and to secrete free antibodies and the lymphocyte is said to be activated. The amount of secretion is proportional to the affinity m_{ij} of the binding between molecules of type i and j (the index i refers to the paratope while the index j to the epitope). The molecule to which an antibody is bound is marked for destruction. The destruction is done by other cells in the IS. In contrast, B-lymphocytes that are not stimulated die after a few days. Thus, a selec-

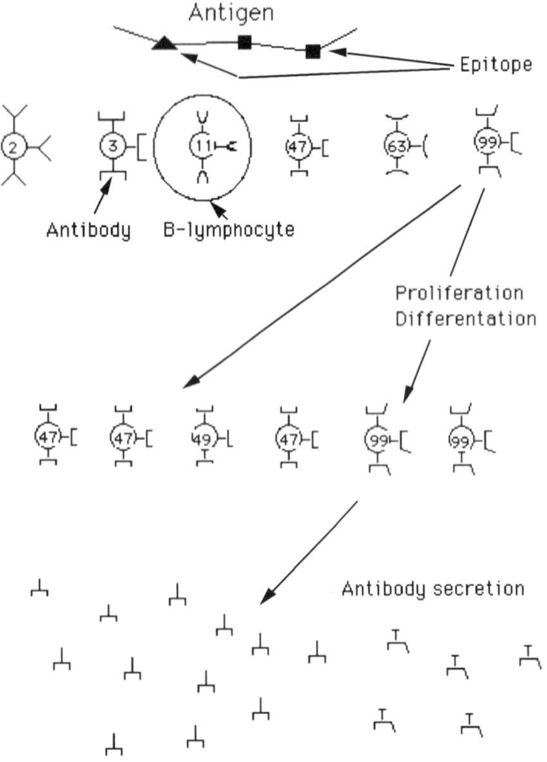

Figure 23.3 A schematic view of clonal selection. On top, an antigen with its epitopes (in black) is shown. Underneath, we have the current repertoire of B-lymphocytes. Each B-lymphocyte is represented by a circle labelled by a number and the corresponding antibodies are shown attached to the circle. Some of the B-lymphocytes (47 and 99) are activated by the antigen and are proliferated (47 and 99 start to clone) and differentiated (49 is a mutant of 47). This is shown in the middle. At the bottom, activated B-lymphocytes secrete free antibodies. Adapted from Farmer *et al.* (1986).

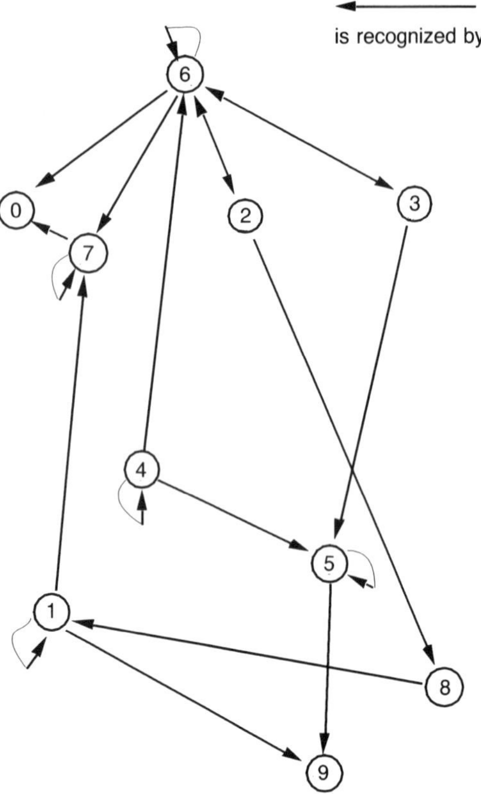

Figure 23.4 An idiotypic network of interacting antibody types. Each antibody type is represented by a circle labelled by a number. If antibody type *A* recognizes a type *B* then an arrow is drawn from *A* to *B*. Sometimes the recognition is bi-directional; for example, type 2 recognizes type 6 and vice versa. Some types also recognize themselves, for example, type 7. Adapted from Perelson (1990).

tion process is at work here where those antibodies that are stimulated by antigens or antibodies are amplified, while the other antibodies die out. This process is called **clonal selection** and is summarized in Figure 23.3.

A B-lymphocyte not only becomes activated by antigens but also by anti-idiotypic antibodies in the IS. In turn, these anti-idiotypic antibodies elicit the activation of still other antibodies. This way, we get a whole network of interacting antibodies called the **idiotypic network**. An example is displayed in Figure 23.4.

The IS can be mathematically modelled as follows. Let i and j be species or types of antibodies and let m_{ij} be the affinity between antibodies of type i and j. Let b_i and f_i be the concentration of B-lymphocytes and the free antibodies of type i, respectively. Let N be the number of different types.

Define the sensitivity $\sigma_j(t)$ of the current entire immune network as (Varela *et al.* 1988):

$$\sigma_j(t) = \sum_{i=1}^{N} m_{ij} f_i(t) + \sum_{k=1}^{M} m_{kj} a_k(t), \quad j \in S, \quad (23.1)$$

where the $f_i(t)$, $i = 1, \ldots, N$, are the concentrations of the free antibodies of type i at time t and the $a_k(t)$, $k = 1, \ldots, M$, are the concentrations of the M antigens in the antigenic environment at time t.

The network has a high sensitivity σ_j for a type j if type j will be recognized rapidly, that is, if there are a sufficient number of antibodies of types i that have a high affinity m_{ij} with j. The dynamics of the IS can then be modelled by the set of $2N$ equations (23.2) and (23.3) (Varela *et al.* 1988):

$$\frac{df_i}{dt} = -k_{i,1}\sigma_i f_i + k_{i,2}\mathrm{Mat}(\sigma_i)b_i - k_{i,3}\sum_{j=1}^{M} m_{ij} f_i a_j \quad (23.2)$$

$$\frac{db_i}{dt} = -k_{i,4}b_i + k_{i,5}\mathrm{Prol}(\sigma_i)b_i \quad (23.3)$$

for $i = 1, \ldots, N$. $\mathrm{Mat}(\sigma_i)$ and $\mathrm{Prol}(\sigma_i)$ are threshold functions of the sensitivity σ_i of the network and $k_{i,1}$, $k_{i,2}$, $k_{i,3}$, $k_{i,4}$ and $k_{i,5}$ are positive constants. The concentrations of the M antigens in the antigenic environment are represented by a_j, $j = 1, \ldots, M$.

The change of concentration df_i/dt of free antibodies of type i in equation (23.2) is the result of three processes: first, antibodies of other types attach to these antibodies and neutralize them resulting in a decrease of f_i (first term); second, matured B-lymphocytes of type i release the antibodies on their surface resulting in an increase of f_i (second term). The maturation Mat is a function of the sensitivity $\sigma_i(t)$ of the entire network for that type at time t. A typical example of Mat is shown in Figure 23.5. And third, free antibodies of type i attach to antigens of the antigenic environment resulting in a further decrease of f_i (third term).

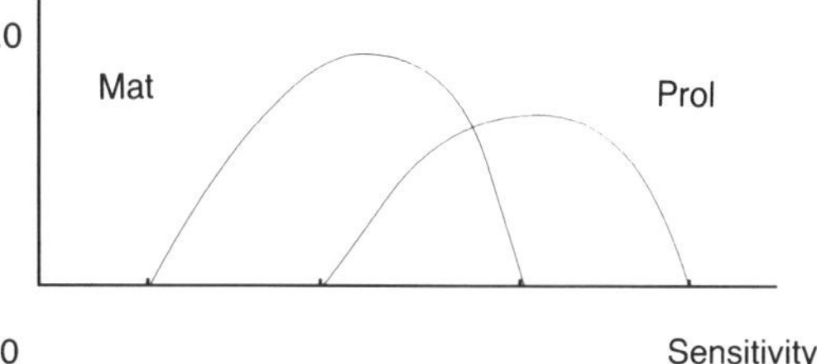

Figure 23.5 Typical examples of the maturation $\mathrm{Mat}(\sigma_i)$ and proliferation $\mathrm{Prol}(\sigma_i)$ of antibody type i. Both are functions of the sensitivity $\sigma_i(t)$ of the current network for type i. Adapted from Varela *et al.* (1988).

The change of concentration db_i/dt of B-lymphocytes of type i in equation (23.3) is the result of two processes: first, the death of B-lymphocytes of type i resulting in a decrease of b_i (first term); and second, an increase of B-lymphocytes of type i since B-lymphocytes are proliferated due to activation by antibodies of other types and/or activation by antigens (second term). The proliferation Prol is also a function of the sensitivity $\sigma_i(t)$ of the entire network for that type at time t. A typical example of Prol is shown in Figure 23.5.

Note that even without external agents the IS exhibits complicated dynamics (cf. equations (23.2) and (23.3)).

23.3.3 METADYNAMICS OF THE IS: RECRUITMENT OF NEW SPECIES

The dynamics of the IS describes the change of concentration of the existing free antibodies and the corresponding B-lymphocytes. In contrast, the metadynamics describes the incorporation of new types in the immune network and the removal of old types. The autoregulation of the IS keeps the total number of different types roughly constant. The adaptability of the IS results from the metadynamics and is an example of adaptation as innovation (Section 23.5).

The B-lymphocytes are produced in the bone marrow. They differ in the genes that code for the antibody type secreted by them.

Through reshuffling of these genes, new types of antibodies are produced continuously. This random process replaces part of the B-lymphocytes every day. This way, the diversity in antibody types of the IS is maintained. Furthermore, when B-lymphocytes are activated the genes that code for the corresponding antibody type start to mutate at a higher rate than the genes of the non-activated B-lymphocytes. This way, additional diversity is created but now biased towards the activated antibody types.

Although new types are created continuously, the probability that a newly created type is activated and is incorporated in the IS dynamics is different for different types. Let S be the space of all possible antibody types. The probability that a new species $s \in S$ will be incorporated in the IS dynamics depends on the sensitivity $\sigma_s(t)$ of the network for that type s at time t. Note that according to equation (23.1), the sensitivity $\sigma_s(t)$ of the network for type s is high if s has high affinity m_{sj} with a sufficient number of activated antibody types j, that is, types j with high concentrations f_j of free antibodies. These high concentrations f_j reflect the usefulness of the types j in the IS. As a consequence, the creation of new types s is biased towards the highly fit types j. This biased incorporation into the immune network of randomly created species is called the **recruitment strategy**.

In natural evolution, the creation of new individuals results from the mutation and crossover of the individuals in the population. In the immune system, new antibody types are created preferably in the neighbourhood of the existing high-fit antibody types. Why the IS has adopted this strategy is still an open question. According to Varela (1991) this has to do with the ruggedness of the fitness landscape defined over the space S. Given the antigenic environment, each antibody type and also the genes that code for that type can be assigned a fitness as follows. An antibody type has a high fitness if it has high affinity with some of the antigenes in the antigenic environment or with some of the existing antibody types. Two sets of genes can be very similar and still code for antibody types with very different molecular shapes. These antibody types then have very different fitnesses and the fitness landscape is rugged. In Manderick *et al.* (1991), we have shown that if the landscape is rugged, strong focus towards the high fitness regions works better than sustained exploration of the space of all conceivable antibody types. In the IS, additional diversity is created in the neighbourhood of the activated types and the search is focused towards the high fitness regions.

23.3.4 DISCUSSION OF THE IS

The IS is a system with cognitive properties like pattern recognition, memory and learning. Moreover, the IS is fault-tolerant and its knowledge is distributed. Just as for natural evolution, instructive theories have been proposed for the IS but all these theories have now been refuted. Although some mathematical results concerning the dynamics of the IS exist (Varela and Stewart 1990; Stewart and Varela 1990), the complexity of the IS dynamics and metadynamics make a complete mathematical understanding impossible. These topics are discussed below.

The IS is a cognitive network. Its cognitive domain is the space of molecular shapes. Also, the IS retains a memory of past encounters. This memory is a property emerging from the dynamics described in the equations (23.2) and (23.3). Once an antigen Ag is recognized by an antibody type i_1, the concentration of i_1 will start to increase. Another antibody type i_2 will respond to i_1 and its concentration will also increase. Eventually, an antibody type i_j will respond to i_{j-1} and i_j is recognized by i_1, the antibody that initially responded to the antigen Ag. Thus, the loop is closed. Even when the antigen Ag is removed completely, due to the feedback loop the concentration of the responding antibody type i_1 will remain at a higher level than

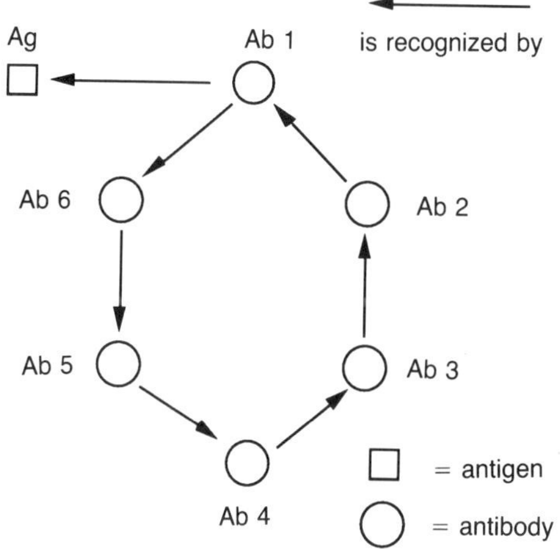

Figure 23.6 A closed loop of interacting antibody types triggered by an antigen Ag. Even if the antigen Ag is removed completely, the loop remains intact and acts as a dynamic memory of the encounter with antigen Ag.

before. The feedback loop acts as a dynamic memory and is shown in Figure 23.6.

This memory may even be associative. Consider an antigen Ag with many different epitopes. Suppose now that different types of antibodies – one for every epitope – have already recognized Ag and responded to it in the past. If, some time later, one of these types *i* again gets activated by Ag then the other types can be activated by *i* due to the network interactions between antibody types even before the other types have encountered Ag.

Finally, the IS has a learning capacity. After a first encounter with a given antigen, Ag, the immune response is much more rapid and larger in size when it is again exposed to the same type of antigen Ag. Due to interactions with other antibody types, the antibodies *i* that respond to the antigen Ag remain at a higher level of concentration than before the first encounter with Ag. As a consequence, the critical concentration needed to mark all newly invading antigens Ag for destruction is reached more rapidly.

Not only is the IS a cognitive system, it is also adaptive, fault-tolerant and its knowledge is distributed. Moreover, these properties of the IS emerge from the network's dynamics and metadynamics. Note that an emergent property is a property of the whole system which none of its components possess. An example is the food seeking behaviour of ant colonies (Deneubourg *et al.* 1983). This global behaviour is adaptive while the behaviour of individual ants is rigid. Such global emergent behaviours can be simulated successfully while only implementing the behaviour of individual ants (see Moyson and Manderick 1988; Manderick and Moyson 1990 for examples). The IS is adaptive since it is capable of responding to an ever-changing antigenic environment. Indeed, new types of antigens invade the organism all the time and most of them are successfully destroyed. The IS is fault-tolerant in the following sense. Suppose that an immune response has been built up against some antigen Ag and that the responding type of antibodies is then removed. Due to the degeneracy of antibody types, other types will now start to respond and get activated. The IS's knowledge about molecular shapes is not localized anywhere but distributed over several components and emerges from the interractions between these components.

In the first half of this century, immunologists discovered that the IS also produces very specific antibodies against all possible man-made substances. The antibodies are specific in the sense that they match only one or sometimes a few particular substances but nothing else. These antibodies could not be present in the organism prior to its exposure to these substances. So one concluded that an instructive process was necessary to transfer information from the outside (antigen) to the inside (antibody) and instructive theories for the working of the IS were proposed. Characteristic of these theories is that (1) external information is used to generate antibodies that recognize the invader, and (2) no pre-existing diversity of antibodies is assumed. For example, Pauling's **directed folding** theory (Pauling 1940) supposes that the IS generates amorphous pre-antibodies. When such a pre-antibody comes in contact with an antigen, it wraps around that antigen and due to its plasticity it retains the imprinting of the antigen's shape. Thus, a pre-antibody becomes an antibody using the antigen as a template.

For these scientists, it was difficult to imagine that a selectionist theory could explain the production of very specific antibodies against any conceivable antigen. It is nowadays generally accepted that the IS is a selectionist system. Only the clonal selection theory was able to correctly explain all known facts (Ada and Nossal 1987).

In this section, we have discussed the cognitive properties of the IS, the difficulties involved in analysing it mathematically and the instructive theories proposed to explain the functioning of the IS.

23.4 THE BRAIN

The brain is probably the most complex adaptive biological system. A lot is known about the working of individual neurons and the way they are interconnected. However, little is known about how large groups of neurons cooperate to perceive stimuli, to categorize these stimuli, to memorize these categorizations and to behave adaptively. As a consequence, only a few biologically realistic theories have been proposed to explain these brain functions. Artificial neural nets studied in artificial intelligence and psychology are no exception to this. Although inspired by biological neural nets, and although an artificial neuron is a reasonable abstraction of a biological neuron, their topologies are biologically unrealistic. For example, in the brain there are no feedforward nets which learn using backpropagation of errors and there are no Hopfield nets as used in the Boltzmann machine.

According to Edelman, the brain's functioning has to be explained by selectionist principles. Below, we sketch the basic principles of the theory of **neuronal group selection**. This theory is also known as **neural Darwinism** due to its similarity to the theory of natural evolution (Edelman 1978, 1981, 1987; Edelman and Reeke 1982; Reeke and Edelman 1984, 1988; Reeke *et al.* 1989).

According to the theory of neuronal group selection (abbreviated as TNGS), the nervous system consists of a population of neuronal groups. A neuronal group consists of hundreds to thousands of strongly interconnected neurons. These groups show a wide variety in anatomical structure. The initial population of groups with which an organism is born is called the **primary repertoire**.

The TNGS makes three fundamental claims:

1. The primary repertoire of structurally different neuronal groups results from a first selectionist process. This selectionist process occurs during the epigenetic development of the brain. The resulting structural diversity is different even for animals belonging to the same species. If we compare corresponding brain regions of two such animals we will find that their connectivity is different.
2. A second selectionist process occurs during the lifetime of an organism and affects the primary repertoire. Neuronal groups whose activities are correlated with input signals arising from the sense organs or from other neuronal groups are selected and modified. In the selected groups, the strengths of synaptic connections within the group and to other groups are modified but the connectivity itself of the group does not change. This modified repertoire of neuronal groups is called the secondary repertoire and is the result of this second selectionist process. This process is the mechanism behind adaptive behaviour.
3. The activities in the different neuronal

groups are coordinated through re-entry. Re-entry means that the response of a neuronal group is fed back to the neuronal groups from which it directly or indirectly receives input. Re-entry is responsible for a coherent global response to the input patterns coming from the external environment.

From now on, we will restrict our discussion to the second selectionist process responsible for adaptive behaviour.

According to the TNGS, neuronal groups as a whole respond to input patterns that arrive at their synapses. Some groups get their input directly from the sense organs and pass their output to other neuronal groups. The latter groups are connected to still other groups and so on. Some of the groups generate the motor output. Even feedback loops are present. This way, we get a whole network of interconnected neuronal groups. Such a network is shown in Figure 23.7.

Neuronal groups getting the same inputs compete to respond to that input. The neural activity in different groups depends on their anatomical structure and the synaptic weights of the interconnections between the neurons of that group. Those groups where the neural activity exceeds some threshold are selected. The selected groups are then amplified so that they will respond faster and make a larger contribution than the unselected groups to the future responses of the organism. This amplification results from modifications of weights of the synaptic connections within the selected group and to other groups. As a result of these modifications, a selected group might capture neurons from other groups. The weights of synaptic connections from a neuron from another group to the selected group may become larger than the weights of connections with neurons of its own group. In the future, that neuron will respond together with the selected group: it is captured by that group. So, according to the TNGS, some groups are selected and amplified according to their usefulness for the organism's behaviour during its lifetime. These amplified groups are called the **secondary repertoire**.

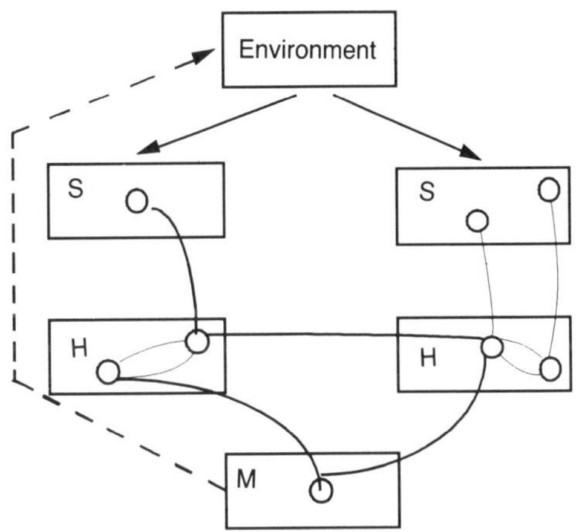

Figure 23.7 The brain according to the neuronal group selection theory (TNGS). The brain consists of different levels of repertoires of neuronal groups. Each repertoire is represented by a rectangular box and groups within the repertoire are represented by circles. The boxes labelled by S are the first-level sensory repertoires receiving their inputs for the sense organs. H represents higher sensory repertoires and M represents a motor repertoire. Only three levels are shown. The connections between neuronal groups, which are unidirectional, are represented by arrows. Some higher level groups are connected with lower level groups or groups at the same level. This property is called re-entry. Re-entry can also operate indirectly through changes in the environment brought about by the organism and is indicated by a dashed arrow. Adapted from Reeke and Edelman (1988).

The diversity among the neuronal groups is large enough for at least one group to respond to every possible input. Also, there is a lot of degeneracy present in the brain, that is, structurally different groups respond in a similar way to the same inputs. Degeneracy is different from redundancy which implies the existence of multiple structurally identical groups. As a consequence of degeneracy, the brain is fault-tolerant: the brain will

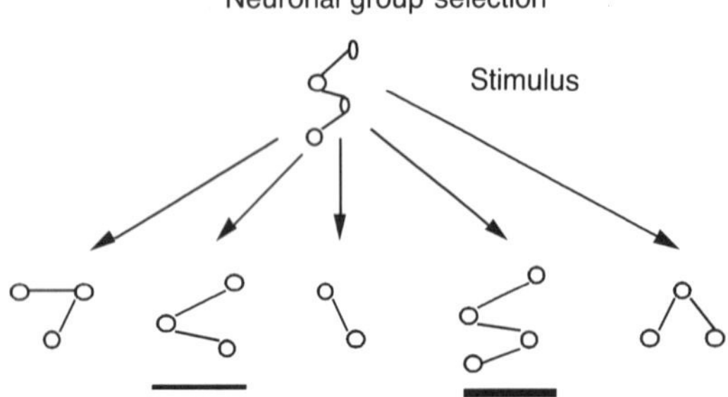

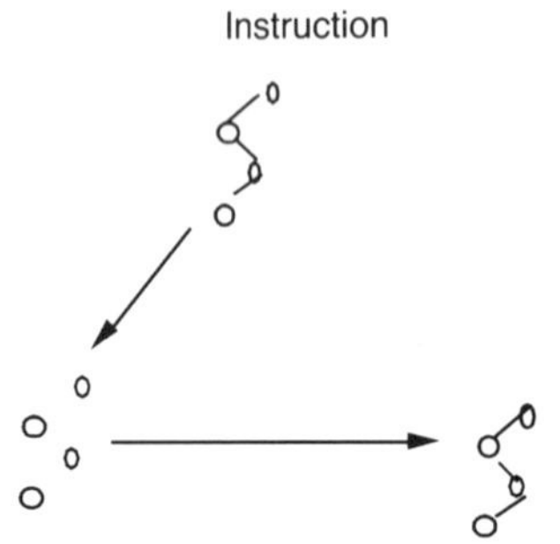

Figure 23.8 At the top, the TNGS is shown schematically. Small circles connected by lines represent predefined patterns of neuronal activity. These patterns are triggered by some of the inputs depending on the structure of that input. In the figure, the triggering depends on the similarity to the input pattern. The neuronal groups where the activity exceeds the threshold are underscored and are said to be selected. The thickness of the underscore is proportional to the response of the neuronal group. The selected groups are subsequentially amplified. At the bottom, instruction as opposed to selection is illustrated. Now, groups do not have predefined patterns of neuronal activity (the small circles within a group are not connected any more) which are triggered by some of the inputs. On the contrary, the pattern of activity is dictated by the interaction with the input pattern. Adapted from Reeke and Edelman (1988).

continue to respond even if some groups are destroyed.

Edelman (1987) sketches how the TNGS accounts for perceptual categorization, memory, learning and adaptive behaviour. Several computer simulations (Reeke and Edelman 1988; Reeke *et al.* 1989) have confirmed that a recognition automaton can be built according to the principles of this theory. The TNGS is illustrated and contrasted with instructive learning in Figure 23.8.

23.4.1 DISCUSSION OF THE TNGS

Today, there is no consensus as to whether the working of the brain should be explained by selectionist theories. A major goal of the TNGS is consistency with the available neurobiological data.

As opposed to neo-Darwinism and the clonal selection theory, the TNGS is not widely accepted yet. It is one of the first attempts to integrate the many neurobiological data into a theory of higher brain functions like perception, categorization, memory and even consciousness. According to Popper (1968), the critical test for a theory is how well it withstands falsification attempts. Edelman (1987) has given a number of predictions which could be used to falsify his theory. However, it is not always clear if experiments can be set up to test these predictions, and how critical these predictions are to a selectionist theory, that is, alternative theories could make the same predictions. Therefore, the theory needs further evidence and confirmation.

A major goal of the TNGS is consistency with the available anatomical data concerning the brains of several species. Some striking facts are (1) organisms belonging to the same species differ in the anatomical details of their brains, and (2) the enormous amount of degeneracy (as opposed to redundancy) in the brain. These facts are consistent with a selectionist theory of the brain where the notions of population and diversity play an important role. In contrast, if, according to the physical symbol system hypothesis (Newell 1980), parts of the brain compute functions, then it seems difficult to see why these parts are different for different organisms of the same species. Moreover, why should there be any degeneracy at all? Redundancy is enough to make the brain fault-tolerant.

Instructive theories have been proposed to explain the functioning of the brain and one of the best known is Hebb's cell assembly theory (Hebb 1949). According to this theory, neurons are assembled in cells and this assembling process is induced by the environment. Connections between neurons are modified in such a way that the response to each given combination of inputs is optimized. An instructive system needs a teacher to impress the desired responses on the system and an instructive system can never bootstrap itself.

However, an organism has to organize stimuli coming from the environment into useful categories. Stimuli are never the same and if the organism is incapable of categorizing these stimuli it would never know what to do since each stimulus would seem novel to it. This categorization has to be done quickly, efficiently and without instruction. Instructive theories cannot explain this. The categorization problem is basic to other cognitive activities like perception, memorization and learning.

Intelligence as computation faces the homunculus problem, that is, to explain the working of an intelligent being one has to presuppose a second intelligent being inside the first one. This leads immediately to infinite regress. According to the physical symbol system hypothesis, intelligence is computation and computation is the manipulation of representations, that is, collections of interrelated symbols, by purely formal rules. These rules do not make any reference to what these symbols represent. The physical symbol system is not aware of the link between a symbol and its referent. So, a physical symbol system at a higher level has to keep track of these links of the first system. But the second system does not know that its representations refer to these links. Therefore, a third system is needed and infinite regress cannot be avoided. In practice, the link between a symbol and its referent exists only in the minds of the designer and the users of the intelligent program. The same problem is known in artificial intelligence (AI) as the **symbol grounding problem** (Harnad 1990), that is, the problem of how a symbol can be linked to its referent in an AI system. It is now receiving a lot of attention.

The focus of the TNGS on the anatomical data concerning the brain makes it difficult for a non-biologist to understand this theory. Although the TNGS is a selectionist theory of the brain we have used the IS as a source of inspiration for the design of the selectionist automaton.

23.5 COMPARISON OF THE THREE SYSTEMS

The natural evolution of species, the pattern recognition capabilities of the immune system and the functioning of the brain are explained by three selectionist theories. These three theories have a lot in common but there are also a number of differences. In all theories, the notion of population is important; the diversity within the population exists prior to any interaction with the environment, the variants in the population are tested by the environment and the fitter variants are selected and amplified. Instructive theories have been proposed for the three biological systems. Differences between the three selectionist theories are in the role of selectionist principles and in the adaptation mechanism. We begin with the differences.

In biology, an important distinction is made between theories explaining the evolution and theories explaining the functioning of structures. Both types of theories are often confounded resulting in confusion between causes of evolution of organisms (ultimate causes) and causes of functioning of organisms (proximate causes) (Mayr 1982). Unawareness of this distinction might also hamper the understanding of adaptation and cognition in terms of selectionist principles.

Selectionist principles play a different role in evolutionary theory on the one hand, and the clonal selection theory and the neuronal group selection theory on the other hand. In evolutionary theory, selectionist principles explain how species become more and more adapted to their environment, that is, it explains the evolution of species. Individuals in a species like monkeys behave adaptively within their lifetime. However, evolutionary theory says nothing about the mechanisms behind this adaptive behaviour. In contrast, it is the functioning of the immune system and the brain that is explained by selectionist principles, that is, these principles clarify the mechanisms responsible for the adaptability of the IS and the brain. This way, the IS is much more important than natural evolution as a metaphor in understanding adaptation in an unknown environment.

In the past, researchers were mostly referring to the theory of natural evolution as a metaphor for adaptation and cognition (cf. Bäck's chapter in this book). This explains much of the resistance against selectionist theories of cognition. Some widely held objections are as follows. Natural evolution operates on a million-year timescale. How could the same principles lie at the basis of behaviour adaptive in real time? Furthermore, evolution theory explains the emergence and evolution of structures like the brain but this does not necessarily mean that the brain functions according to selectionist principles.

Therefore, it is important that we look at selectionist theories like the immune system where selectionism is proposed as the proximate cause. This is why we have paid more attention to the clonal selection theory and neural Darwinism.

Adaptation in dynamical systems comes in two types: adjustment and innovation (Farmer *et al.* 1986). If adaptation is the result of changes of the parameters in the system then it is called adjustment. Learning in neural nets where the weights are the parameters is an example of adjustment. In contrast, if adaptation results from topological changes of the dynamical system then it is called innovation. Adaptation as reflected in the metadynamics of the IS is an example of innovation. The continuous creation of new antibody types changes the IS's network topology (Section 23.3.3). In both the evolution of species and the IS, adaptation is innovation. According to the TNGS, the formation of the primary repertoire is an example of adaptation as innovation (neuronal groups compete for neurons and this affects the topology of the groups) and the formation of the secondary repertoire is adaptation as adjustment (only the synaptic weights of the connections are modified and this does not affect the topology of the groups).

More important than the differences are the similarities between natural evolution, the immune system and the brain. These systems have been explained by selectionist theories: natural evolution by neo-Darwinism, the immune system by the clonal selection theory and the brain by the neuronal group selection theory.

Each system consists of a population of units: a species is a population of genotypes/phenotypes, the immune system is a population of B-lymphocytes and the corresponding antibodies and the brain is a population of neuronal groups.

There is a lot of diversity within the population: no two individuals of a species are identical if twins and the like are excluded. About 10^7 antibody types exist in the immune system. And the neuronal groups are different in their anatomical structure.

The units of the population are tested by either the external or the internal environment: individual organisms are tested by their niche. Antibody types are tested by antigens (the external environment) or antibody types in the immune system (the internal environment). Neuronal groups are tested by signals coming from the sense organs (external environment) or by signals coming from other groups (internal environment).

Due to differences between the units, these units respond differently to their environment and the fitter units of the population are selected by the environment and amplified. In the evolution of species, the units of selection are individuals and better adapted individuals have more offspring than the less well-adapted ones. In the IS, the units of selection are antibodies and the number of activated antibodies increases through the proliferation of the corresponding B-lymphocytes and the release of free antibodies. In the brain, the units of selection are neuronal groups and the groups that respond to an input are amplified through modification of synaptic weights.

The diversity within the population of units is the result of an autonomous process influenced neither directly nor indirectly by the environment. The diversity among the genotypes results from mutation and crossover of the existing genotypes. Mutation and crossover are not affected by the environment. The diversity among the antibody types in the immune system is the result of the recruitment strategy. This strategy is largely unaffected by the external environment: B-lymphocytes activated by an antigen start to mutate at a higher rate than the other B-lymphocytes. This is the only influence of the external environment on the generation of diversity in the immune system. The anatomical differences between neuronal groups are the result of cell movement, cell differentiation and cell death.

A last point of similarity between the three biological systems is that instructive theories have been proposed for (1) explaining natural evolution (Lamarck's theory); (2) the immune system (Pauling's direct folding theory); and (3) the brain (Hebb's cell assembly theory). These instructive theories are either refuted (the first two theories) or questioned (the third theory). The major difference between these instructive theories and their selectionist counterparts is the role of the environment. In selectionist theories, the diversity between the units of the population exists prior to any testing by the environment and this diversity is maintained by a process which is not influenced in any way by that environment. The role of the environment is to select the fitter variants. The priority of the autonomy of the system over its environment is exemplified in Varela's view of the immune system: the role of the immune system is not to eliminate antigens (Varela *et al.* 1988). The immune system is just a self-regulating generator of B-lymphocyte and antibody diversity and self-regulation keeps the concentrations of the several antibody types within acceptable limits. Antigens affect the existing balance and the immune system has to compensate for these imbalances of several sorts. Consequently, the immune response is nothing more than a side effect of the self-regulation of the immune system.

In contrast, instructive theories stress the priority of the environment over the system. In Lamarck's theory, changes in the environment direct changes in the organism so that the organism adapts to these changes. In the direct folding theory, the antigen is used as a template to create matching antibodies. In Hebb's theory, the cell-assembling process is directed by the environment.

The comparison between natural evolution, the immune system and the brain is summarized in Table 23.1.

23.6 SELECTIONIST AUTOMATA

Selectionism has its roots in biology. Selectionist theories have been proposed for

1. the natural evolution of species (neo-Darwinism);
2. the functioning of the immune system (clonal selection theory (Ada and Nossal 1987)); and
3. the functioning of the brain (neuronal group selection theory (Edelman 1987)).

Table 23.1 A comparison between three selectionist theories: (1) the theory of natural evolution of species (left); (2) the clonal selection theory (middle); and (3) the theory of neuronal group selection (right). Here, we have restricted ourselves to the second selectionist process described by the theory of the neuronal group selection

	Natural evolution	*Immune system*	*Brain*
Selectionist theory	Neo-Darwinism	Clonal selection theory	Neuronal group selection theory
Generator of diversity	Mutation and crossover	Recruitment strategy	Cell movement, differentiation and death
Unit of selection	Genotype/phenotype	Antibody type	Neuronal group
Amplification process	Reproduction of the fitter individuals	Secretion of free antibodies and reproduction of the corresponding B-lymphocytes	Modification of the synaptic weights of the connections
Explains	Evolution	Functioning	Functioning
Adaptation	Innovation	Innovation	Adjustment

All these systems solve adaptation problems: natural evolution produces species that are well adapted in form, function and behaviour to their ecological environment. The immune system produces antibodies that recognize and destroy foreign substances called antigens. The brain of a higher organism receives and processes a multitude of environmental stimuli. These stimuli are grouped in categories which have an adaptive value for that organism (Edelman 1987).

These selectionist systems have common properties and for two of these systems – the immune system and the brain – selectionism explains their functioning rather than how these systems are evolved. Moreover, these systems have cognitive properties: the immune system recognizes shapes of antigens, memorizes these shapes and the immune system learns to respond faster to shapes it has recognized in the past. A system that functions according to selectionist principles is called a **selectionist automaton** abbreviated as SA. The immune system and the brain are examples of SAs and will be used to illustrate the key concepts.

In the rest of this section, we define selectionist recognition systems, we discuss the design decisions that have to be taken and we try to determine the scope of selectionist learning.

23.6.1 DEFINITION OF SA

Common properties of selectionist systems are a population of diverse units also called variants. These variants are tested by the environment in parallel. Due to their differences, variants respond differently. And depending on their response, variants are selected and as a result their relative importance in the population is amplified. The diversity in the population exists prior to the testing by the environment and the way this diversity is maintained proceeds independently from the environment in which the system is embedded. The role of the environment is to select the variants that stood the test.

1. A selectionist automaton consists of a population of variants. Examples of such variants are B-lymphocytes in the immune system and neuronal groups in the brain according to the neuronal group selection theory (Edelman 1987). B-lymphocytes

differ in the antibodies that they produce and neuronal groups differ in their anatomical structure. The initial diversity in the immune system results from randomly combining genes from the gene libraries. The recruitment strategy maintains this diversity. The initial diversity in the brain is the result of the first selectionist process described in Section 23.4. The details of how this initial diversity is created and later on maintained are not specified in the theory of neuronal group selection.

2. These variants get the opportunity to respond to inputs from the environment. They are tested by the environment. Different variants respond in a different way to the same input. The response is a function of how well each unit matches that input. A unit recognizes an input if the match with that unit exceeds some threshold.

   B-lymphocytes are in the blood stream with the antigens. If the antibodies on the surface of a B-lymphocyte match a given antigen sufficiently then it is activated. The match depends on the complementarity of the paratope of the antibody and one of the epitopes of the antigens, that is, the affinity of the binding (see Section 23.3 for the details). Different neuronal groups receive the same input signals. Depending on their anatomical structure, different groups respond differently.
3. The selected variants are amplified. This amplification is a function of the response to the current input. Amplification means that the relative importance in the population of the responding variants increases and/or the relative importance of the other variants decreases.

   Some of the matching B-lymphocytes in the IS are activated when they exceed the proliferation threshold and start to proliferate. More B-lymphocytes of the matching type are produced. In contrast, B-lymphocytes that do not get activated die after a few days. If neuronal groups respond sufficiently to an input signal then the weights of the synaptic connections of these groups are modified and these groups will show a higher response in the future to the same or similar input signal.

To design a selectionist automaton the following choices have to be made:

1. Choose the units or variants for your problem.
2. Choose the properties in which these variants differ.
3. Choose the size of the population. The size should be large enough for no inputs to go unmatched. An initial response is needed which can then be improved later on. Otherwise responding variants have to be created in an *ad hoc* way. The size of the population will depend on the environment in which the SA has to operate.
4. Choose a generator of diversity. Although the responding variants are amplified, the diversity within the population has to be maintained for two reasons. First, the SA has to be able to improve on its initial response. That initial response is determined by the responding variants. Amplification of these variants does not allow further improvement. Second, the diversity within the population should remain large enough for the SA to be able to respond to inputs not yet encountered.
5. Choose a function which determines the match between an input and a variant.
6. Choose an amplification mechanism.

The structure of a selectionist automaton is displayed schematically in Figure 23.9.

23.6.2 DISCUSSION OF SAs

A first crucial point in the above definition is that the population consists of diverse units or variants. The diversity should be large enough for at least one variant to respond to every possible input. This way, the selectionist system can adapt itself to an environment of which it has no prior knowledge.

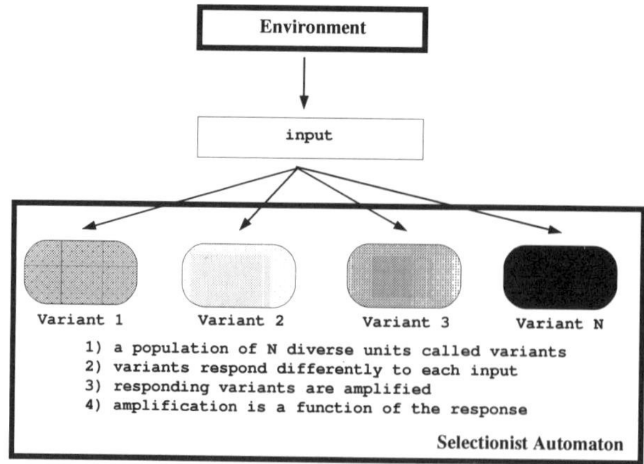

Figure 23.9 A selectionist automaton in its environment.

Also, if a selectionist system is well adapted to an environment there should still be enough diversity among the variants for the system to adapt itself if the environment changes.

A second crucial point in the above definition is that diversity results from some combinatorial process which operates autonomously from the environment. For instance, B-lymphocytes mutate and variants on the original B-lymphocytes are produced by reshuffling genes from the gene libraries. These processes proceed independently of what happens in the environment. The role of environment is to select the fitter variants. The variants that are produced are not prompted by the environment in any way. If the production of variants was correlated with what would be a solution to the environmental problem then the selection step would make no sense at all; the environment would instruct the system rather than select the better adapted variants.

A third point relates to the amplification of selected variants. A common amplification mechanism is differential reproduction. A responding variant produces copies of itself as a function of its response. This way the relative importance of highly responding variants increases in the next population.

23.6.3 SCOPE OF SAs

Symbolic AI has made progress in high-level cognitive areas such as knowledge representation, problem solving and reasoning. However, in low-level cognitive activities such as vision, pattern recognition and adaptive behaviour the situation is completely different. Progress has been slow and difficult to attain. For instance, until recently, work on vision in AI was primarily concerned with the transformation of visual input into an internal symbolic representation of that visual input. This symbolic representation could then be used by a physical symbol system (Marr 1982). The transformation from visual input to internal symbolic representation proved to be a formidable task which has not yet been solved.

The characteristics of high-level and low-level cognitive activities seem so different that Steels (1992) used the terms **behaviour-based intelligence** as opposed to **knowledge-based intelligence** to characterize these two types of intelligence.

In knowledge-based intelligence, everything is centred around the knowledge the system has about the world and about its own internal structure and functioning. Knowledge is represented in such a way that it can be used by reasoning processes like inference and induction. These reasoning processes have been formalized, and this is why this approach has proven to be so successful in the areas of high-level activities like theorem proving, planning, diagnosis and so on.

In contrast, behaviour-based intelligence works on poor models of the world and results from the way very simple modules like behaviours are cooperating and competing to achieve some goal (Brooks 1986; Braitenberg 1986; Tinbergen 1951). Typically, each module has access to a number of sensors and effectors and the sensory information is associated by the module with the behavioural response. No complicated reasoning processes take place at the level of behaviour-based intelligence. Behaviour-based intelligence does not fit the physical symbol system hypothesis very well.

Behaviour-based intelligence uses poor models of the world. Selectionist automata are designed to operate in an environment of which they have no prior environment-specific knowledge. Therefore, it seems reasonable to apply SAs to problems of behaviour-based intelligence. We have deliberately restricted ourselves to two such problems: categorization and adaptive behaviour.

23.7 SUMMARY

In this chapter, we have discussed and compared three biological systems. For two of these systems, the immune systems and the brain, the functioning is explained by selectionist principles. Besides the brain, the IS also has cognitive properties like pattern recognition, memory and learning. Moreover, the IS is fault-tolerant and its knowledge is distributed. The IS is a major source of inspiration for selectionist automata defined in Section 23.6.

Natural evolution, the immune system and the brain solve adaptation problems. Yet, they were not designed in any way to operate in their environment. These biological systems have common properties and these properties are shared by selectionist automata.

ACKNOWLEDGEMENTS

This article was written while the author was on leave from the AI Lab of the Free University, Brussels. This research has been sponsored by the Belgian Government under contract 'Incentive Program for Fundamental Research in Artificial Intelligence, Project: Self-Organization in Subsymbolic Computation'.

REFERENCES

Ada, G. L. and Nossal, G. (1987). The clonal-selection theory. *Scientific American* **257**, 62–9.

Bersini, H. and Varela, F. (1990). Hints for adaptive problem solving gleaned from immune networks. In *Proceedings of the First International Workshop on Parallel Problem Solving from Nature* (ed. H.-P. Schwefel and H. Mühlenbein). Springer Verlag, Berlin.

Bersini, H. and Varela, F. (1991). The immune recruitment mechanism: a selective evolutionary strategy. In *The Proceedings of the Fourth International Conference on Genetic Algorithms*, San Diego, CA. Morgan Kaufmann, Los Altos, CA.

Braitenberg, V. (1986). *Vehicles: Experiments in Synthetic Psychology*. Bradford Books, MIT Press, Cambridge, MA.

Brooks, R. A. (1986). A robust layered control system for a mobile robot. *IEEE Journal of Robotics and Automation*, **1**, 14–23.

Burnet, F. M. (1959). *The Clonal Selection Theory of Acquired Immunity*. Vanderbilt University Press, Nashville, TN.

Darwin, C. (1859). *The Origin of Species by Means of Natural Selection or the Preservation of Favoured Races in the Struggle for Life*. Murray, London.

Dawkins, R. (1986). *The Blind Watchmaker*. Longman, Harlow, UK.

Deneubourg, J. L., Pasteels, J. M. and Verhaeghe, J. C. (1983). Probabilistic behaviour in ants: a strategy of errors?. *Journal of Theoretical Biology*, **105**.

Edelman, G. M. (1978). Cortical organization and the group-selective theory of higher brain functions, in *The Mindful Brain* (eds G. M. Edelman and V. Mountcastle), MIT Press, Cambridge,

MA, pp. 55–96.

Edelman, G. M. (1981). Group selection as the basis for higher brain function, in *The organization of cerebral cortex* (eds F. O. Schmitt, F. G. Worden, G. M. Edelman and S. G. Dennis), MIT Press, Cambridge, MA.

Edelman, G. M. (1987). *Neural Darwinism: The Theory of Neuronal Group Selection*. Basic Books, New York.

Edelman, G. M. and Reeke G. N. Jr. (1982). Selective networks capable of representative transformation, limited generalizations, and associative memory. *Proc. Natl. Acad. Sci. USA*, **79**, pp. 2091–5.

Farmer, J. D., Kauffman, S. A., Packard, N. H. and Perelson, A. S. (1986). Adaptive Dynamic Networks as Models for the Immune System and Autocatalytic Sets. Technical Report LA-UR-86-3287, Los Alamos National Laboratory, Los Alamos, NM.

Harnad, R. (1990). The symbol grounding problem. *Physica D*, **42**(1–3), pp. 335–46.

Hebb, D. O. (1949). *The Organization of Behavior: A Neuropsychological Theory*, Wiley, New York.

Jablonka, Lachmann and Lamb (1992). Evidence, mechanisms and models for the inheritance of acquired characters. *J. Theoret. Biol.* **158**, 245–68.

Jerne, N. K. (1955). The natural selection theory of antibody formation. *Proceedings of the National Academy of Sciences USA* **41**, 849–56.

Jerne, J. H. (1975). *The Immune System: A Web of V-domains*, The Harvey Lectures, Series 70, Academic Press, New York.

Manderick, B. *et al.* (1991). *Selectionism as a Basis for Categorization and Adaptive Behavior*. VUB AI-Lab Technical Report 91–1.

Manderick, B. and Moyson, F. (1990). Self-organization versus programming in massively parallel systems: a case study. In *Proceedings of the International Conference on Parallel Processing in Neural Systems and Computers (ICNC-90)*.

Marr, D. (1982). *Vision: a Computational Investigation into the Human Representation and Processing of Visual Information*. Freeman, San Francisco, CA.

Mayr, E. (1982) *The Growth of Biological Thought: Diversity, Evolution, and Inheritance*, Harvard University Press, Cambridge, MA.

Mitchell, T. M. (1980). *The Need for Biases in Learning Generalizations*. Department of Computer Science, Rutgers University, New Brunswick, NJ.

Moyson, F. and Manderick, B. (1988). *The Collective Behavior of Ants: An Example of Self-Organization in Massive Parallelism*. Proceedings of the AAAI Spring Symposium on Parallel Models of Intelligence, Stanford University, Palo Alto, CA.

Newell, A. (1980). Physical symbol systems. *Cognitive Science* **4**, 135–83.

Paton, R. (1992). *Adaptation and Environment*. Technical Report, Department of Computer Science, University of Liverpool, UK.

Pauling, L. (1940). A theory of the formation of the antibodies. *Journal of the American Chemical Society*, **62**, pp. 2643–57.

Perelson, A. S. (1990). Theoretical immunology. In *1989 Lectures in Complex Systems* (ed. E. Jen). Santa Fe Institute Studies in the Sciences of Complexity, Addison-Wesley, Redwood City, CA.

Popper, K. R. (1968). *The Logic of Scientific Discovery*, Hutchinson, London.

Reeke, G. N. and Edelman, G. M. (1984). Selective networks and recognition automata. *Ann. N.Y. Acad. Sci.*, **426**, pp. 181–201.

Reeke, G. N. and Edelman, G. M. (1988). Real brains and artificial intelligence. *Daedalus*, pp. 143–73.

Reeke, G. N., Finkel, L. H., Sporns, O. and Edelman, G. M. (1989). Synthetic neural modeling: a multilevel approach to the analysis of brain complexity, in *Signal and Sense: Local and Global Order in Perceptual Maps* (eds G. M. Edelman and W. M. Cowan), John Wiley Sons, New York, NY.

Steels, L. (1992). *Kennissystemen*. Addison-Wesley, Amsterdam.

Stewart, J. and Varela, F. J. (1990). Dynamics of a class of immune networks II. Oscillatory activity of cellular and humoral components. *Journal of Theoretical Biology*, **144**, pp. 103–15.

Tinbergen, N. (1951). *The Study of Instinct*. Oxford University Press, New York.

Utgoff, P. E. (1986). Shift of Bias for Inductive Concept Learning. In Michalski, R. S., Carbonell, J. G. and Mitchell, T. M. (1986).

Varela (1991). Discussion at the ERBAS-meeting, Chamonix, January 1991.

Varela, F., Coutinho, A., Dupire, B. and Vaz, N. N. (1988). Cognitive networks: immune, neural, and otherwise. In *Theoretical Immunology*, Vol. II (ed. A. S. Perelson). Santa Fe Institute Studies in the Sciences of Complexity, Addison-Wesley, Redwood City, CA.

Varela, F. J. and Stewart, J. (1990). Dynamics of a class of immune networks I. Global stability of idiotype interactions. *Journal of Theoretical Biology*, **144**, pp. 93–101.

LIFE-LIKE COMPUTING BEYOND THE MACHINE METAPHOR

24

George Kampis

24.1 INTRODUCTION

The question of what models, if any, can serve to represent the complexity of life, is a very important one. The application of biological ideas to novel software or hardware designs, or the seemingly opposite, but in effect closely related, task of using existing computers for the study of life-like phenomena requires an at least partial clarification of what computers can do. The subject of study of this chapter is a fundamental question of this kind.

Following a few earlier writings (Kampis 1991*a*; Kampis 1991*b*) we attempt here to give a short nontechnical summary for the nonspecialist of a set of general ideas about computer modelling, and to present an account of an operational modelling methodology for dealing with models of life and, in particular, with models of evolving systems.

Evolvability is perhaps the most distinctive characteristic of living systems. Many biologists like J. Maynard Smith (1975, 1986) or R. Dawkins (1985) consider this to be the key to life. Evolution produces novelty; so it seems quite natural to say that one of the primary problems of the modelling of life is that of the representation of novelty.

It will be suggested in this paper that in order to represent novelty one has to go beyond what is usually considered to be the computer approach or, rather, the computer metaphor. Afterwards, we shall outline a new model system, which is realizable by computations in an approximative sense, and can possibly serve as a useful tool for the study of novelty.

To describe the idea in a nutshell: we will point out that existing computer-based models of life correspond to encapsulated or enframed systems, the identity of which is subordinated to operations that delimit their complexity and thereby prevent them from incorporating new information. In contrast, the proposed new approach will make it possible to let the model components develop unexpected relations, and to extend their property spaces by means of what will be recognized as a self-modifying behavior.

24.2 RELATIONSHIPS OF COMPLEXITY AND NOVELTY

In an intuitive sense, we may say that a simple system is one for which the characterization does not require much effort. Such a system is readily exhaustible by an analysis which is not computationally intensive. A complex system is, on the other hand, computationally intensive and is not readily exhaustible. Nonlinear phenomena of dynamical systems, as exemplified by chaotic or cellular automata systems, are widely held to be complex, and nicely illustrate the principle which we may call 'complexity-as-nontriviality'.

It is useful to consider more abstract

formalizations of complexity. Kolmogorov complexity, or information complexity (for a good review of this field, see Löfgren (1977) or (1987)) and computational complexity (Garey and Johnson 1979; Wagner and Wechsung 1986) are two precise mathematical concepts that express essentially the same idea. What is deemed complex by these definitions is what requires a large amount of computer resources, that is, where much memory and/or execution time is needed to cope with the given system. Information complexity deals with the length of programs. In terms of information complexity, the most complex objects are those which cannot be described by any computer program shorter than an explicit list of the original object. In terms of computational complexity, the question is how computation speed depends on problem size. Those algorithms which require more than polynomial time for their execution are practically intractable because of their extensive demand of resources.

It is easy to see that these ideas are naturally related to that of novelty. A system that produces and utilizes novelty should become, in general, more complex. This is so because the novel elements that appear in a system also require a description and they also need time and memory if they are to be executed as programs; either way, there will be an increased need for resources to cope with these elements. This reinforces the primary intuition that can be gained from the study of biological evolution. There, the appearance of novelty is usually associated with what is called progress, and this most frequently brings with it the increase of structural and functional complicatedness.

Now, in order to represent systems whose complexity increases we must face the problem of the complexity bottleneck. This notion was introduced and made precise by G. Chaitin (1987). The complexity bottleneck involves the paradoxical issue of producing, in Chaitin's original words, a 'hundred-pound theorem' in a 'ten-pound axiom system': something more complex in a system that is less complex. Chaitin has shown that such a production is not possible and hence the problem is unsolvable in this direct form. For us this suggests that, as is the case with Chaitin's Gödelian systems, in evolution we may reach the limits of formal computability. Just why this is so, and what can we do next is exactly what we shall focus on.

24.3 WHAT IS A COMPUTATION?

Computers are central to our discussion. Therefore, and also in order to fix our ideas, let us characterize computations very briefly. Consider, for instance, the so-called **weak Church thesis**: 'Every effectively definable function is Turing-computable' (e.g., Yasuhara 1971; Rogers 1967). What this means for the modeller is that there is an immediate connection between the more familiar notion of Turing machines (TMs) and the functions of logic and mathematics, and further, that, in a loose sense, all known models of computation are equivalent, and moreover, they are directly equivalent to TMs and to our present-day computers (papers in Herken (1988) discuss various aspects of this curious equivalence).

A TM is a finite-state control system equipped with a reading and writing head as well as a tape, which can be extended without limits in at least one direction and is operated on by the control mechanism (Figure 24.1). All a TM must be able to do, in order to be a universal computer, is to let the head move left and right, to stay where it is and to replace the symbol of the tape under the head by a new one. Perhaps the simplest abstract models of such universal computers were given in Trakhtenbrot (1973).

TMs represent an attractive formalization of computing because the tape-and-controller system offers itself for a relatively easy study; from the Church thesis it follows that we can

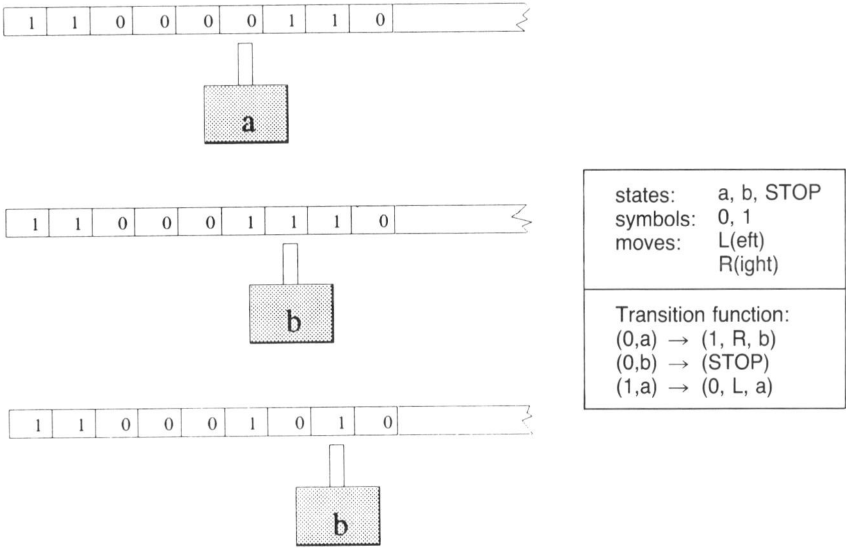

Figure 24.1 Example of a Turing machine.

learn everything about computers by studying nothing but TMs.

Now, if we analyse in detail what the TM model entails, we find that such a machine consists of a set of primitives (such as the permissible symbols of the tape and the states of the controller), a syntax for prescribing how the primitives are related, and a transition scheme to operate on them. Therefore, nothing a TM can ever produce will be different from logical tautology, 'tautology' being a technical word now used for expressions obtained by repeated substitution. Wittgenstein was the first (in his correspondence with Bertrand Russell) to make this point clear; for an account, see Wittgenstein (1939). (In fact Russell believed computation can produce something not contained in the premises and Wittgenstein pointed out that this is not the case.)

The elements that constitute a TM define a one-dimensional or string language. That is, ultimately, every computer is a serial processor, even cellular automata and other complicated systems like computer networks, some of which are intuitively felt to be 'multidimensional' in fact restrict themselves to serial or string-processing modes. This is a somewhat counter-intuitive fact.

The best way to see why every computation is a one-dimensional process is to recall the fact that a computer is, according to one of its many equivalent formulations, a calculating machine or 'number cruncher'. In other words, whatever we do in a computer, it can be expressed for the computer as numbers. Now, numbers inhabit simply the real line; no additional concepts are needed. K. Gödel developed a technique generally called **Gödel numbering** which makes it possible to automatically execute the procedure of transforming everything to numbers and back. In other words, computer programs do nothing but elementary arithmetics – making ones from zeros or back. In this sense, computational processes are extremely simple and rigid. All the complexity a computer can possess is the complexity of the numbers; all the novelty it can represent is that of a 'one' where previously there was a 'nought'.

24.4 THE PROBLEM OF REPRESENTATION

Advanced or adventurous computing style can hide from the eye this underlying structure of a computational process. Unlike the above impoverished picture, on the screen of our desktop computers we find very flexible tools. There are invisible encodings behind them, that do a good part of the job for us, and it is only these encoding procedures, or rather, the programmers who develop them, who must be concerned with the Gödel numbers and the transition functions.

The reason why it is perhaps not pointless to recall these largely trivial facts about computing is that they indicate that it is not a computation in itself but the chosen representation (in other words, the fancy 'blinking' of the screen rather than the numbers in the memory) that carries the relevant information to the user.

Let us consider this question: how is the complexity of a computation, understood in the terms discussed earlier, related to those intuitive complexity sensations which can be obtained by using a computer through the terminal? We can also reformulate the question as this: can the use of suitable interfaces enlarge the capabilities of computers with respect to their ability to deal with complexity, or to produce new complexity? (That the interfaces do increase the strength of computers with respect to their 'handiness' is obvious.)

Unfortunately, the answer must be negative. It is not the case that by changing representations, we can increase complexity; in fact mathematical complexity is defined in a representation-independent way. Or, maybe the question of representation is to be rethought radically.

At the end of this chapter we shall discuss how we can use 'tricks' in order to form new kinds of representations which can in some sense go beyond the complexity which the underlying computers (i.e., the embedding machines and their programs) themselves offer. But the way typical computer models and typical computer embeddings are defined today is rather self-delimiting; a new approach is needed.

Addressed directly, a computer must 'know' in advance, what it should do; that is exactly what present-day programs are for. This gives us all the freedom (because, in a computer, unlike in the physical universe surrounding us, only what we want will happen), but it also imposes a significant constraint. The embedding, or interpretation, of every object that appears in a computer must be pre-specified, either directly, as in a look-up table, or indirectly, by means of some generating function that will be instantiated when necessary. In short, we have to think about every possible embedding before we let it work. We have to invent the future before it will be computed.

Object-oriented programming (e.g., Peterson 1987), a widespread current methodology for dealing with computational objects as objects on their own, provides good examples for what this current modelling strategy implies. Object-oriented languages like C++ or Oberon (Reiser 1991) offer structures similar to those previously developed in AI under the name of frame systems. In both cases, the main idea is to conceive of objects as members of a class represented as a structured template that defines the properties, the inheritance, and the relations of its elements in some invariant scheme (Reichgelt 1991). That is, we end up with an informationally closed, atomistic model.

24.5 NON-COMPUTATIONAL EFFECTS: FORMAL VERSUS PHYSICAL SYSTEMS

Of course, whether there is anything beyond computers is a question many people are interested in, and recent debates about Chinese rooms and emperors' minds put this question in the foreground. But in fact the line of work that deals very seriously with this very question goes back to A. N. Whitehead, the eminent mathematician and philosopher, co-author of the famous *Principia Mathematica* (Russell and Whitehead 1912). What I suggest recalling now owes a lot to him, in particular to his *Process and Reality* (1929).

Expressed in one sentence, what he pointed out was that even computations are not always quite like computations. As a consequence, the usual notion of computation turns out to be too restrictive, a little known idea that requires attention. (To be precise, Whitehead never spoke about computations, because there were no computers in his time. He spoke about using mathematics. However, by the Church thesis, that makes no difference.)

A natural process is, strictly speaking, never reducible to a computation, Whitehead said. Perhaps the best way to show this is by pointing to the physical systems which we use in order to realize the formal computations. A basic fact is this: besides performing a computation, these systems always do something else, as well. One particular form in which they often do something else is what we call 'errors', as if Nature could go wrong. What actually happens when a computer makes an error is that another process interferes with the computation, in a way which was not foreseen (included or accounted for) in the definition of the given computation. That is why we feel the computation breaks down. In other words, every natural process, even if specifically built for doing nothing but computations, does a lot more: there is a potential for further interactions in it.

At the moment when these interactions are 'turned on' we transcend the initial computational framework. But of course nothing prevents us from redefining our initial computation now, so as to include these error-producing processes, as well; in this sense, the computational framework can be saved. (To put it differently, the idea is that the second physical process that embeds the original computation and is responsible for the errors can be incorporated in another, 'bigger', and hence more complicated computation.)

What this all means can be illuminated by using a terminology pertaining not to computer science but to systems research (Klir 1985).

By confining ourselves to an initial computational description of a natural system, we define a model. A model is exactly what we have been taking about so far: it is a kind of abstract construct, which permits a limited characterization which works in a 'typical' case, but for the correctness of which there is no guarantee. That is, strictly speaking, computation is but a fruitful idealization. By enlarging the definition of computation to include more, we define another model, another idealization – presumably, a better one. So the implied question is: how well can we do by using these sequentially refined models for the approximation of physical systems?

What Whitehead and others have noticed is that we encounter an infinite regress here. This refinement game is a race against Nature that cannot be won. No matter how much we have already included in a given mathematical supermodel, what we get is just another computation which would then suffer from new 'side effects', and new 'errors', and so on. Even a quantum description would not be an exception. Unexpected effects can come not only from 'deep down' but from 'the side' or 'far away', as well: no model can be 'foolproof' against everything that affects its original. One can easily imagine such sudden effects as lightning, the entering of hackers into the room, or technicians who alter the hardware of a computer.

'Error' is but one instance of a wide class of unexpected new interactions that can occur within a system, Whitehead recognized. These interactions can drive a system into various new modes of processing we cannot account for in advance. Technically speaking, Whitehead concludes, in every process we deal with a potential infinitude of variables, whereas (expressed in modern terminology) in any model we use but a finite subset.

Whether this infinity (understood in the sense of the number of the potentially relevant variables being 'unbounded') has to be taken seriously is another question. Nature often appears to cooperate. For instance, many systems appear to be 'meaningfully' stratified, and that means that we can split them safely into well-defined and invariant subsystems that correspond to levels of resolution. The level-based approach to mathematical modelling is well-elaborated in Klir (1985) and elsewhere.

It is easy to understand now that the processes we call 'computations' are typically processes confined to one level. That is, usually we do not have to bother with all the variables, other than just a few of them, maybe a handful, which we have selected: these are the ones we include in a computational model or in the definition of a hardware. In particular, if we now consider a Turing machine (such as the one in Figure 24.1) as a physical system, we find that it uses but a fragment of the available information carriers for actually representing information: for instance, in the reading frame defined by the interaction between the head and the memory tape, the majority of the material complexity is omitted (Figure 24.2).

24.6 SHIFTING READING FRAMES

Now, can we find any concrete process of interest, apart from maybe subatomic processes, that does not conform to some version of these reduced computations we have discussed so far? For even the molecular biologist finds it convenient to treat molecules as information processors that can be analysed in terms of computer jargon: the very notions of 'genetic code' and 'genetic program' reflect this.

At the same time, there are various sytems in biology that utilize what can be rightly called shifting reading frames and resemble the noncomputational effects discussed above. That is, there are systems that can change and redefine their own primary use of information carriers. As the choice of information carriers or system variables is what makes computations different from physical systems, the shift of these variables is exactly like crossing the levels or changing the models in the Whiteheadian sense, or using a new reading/writing head in Figure 24.2.

Let us take the example of DNA coding.

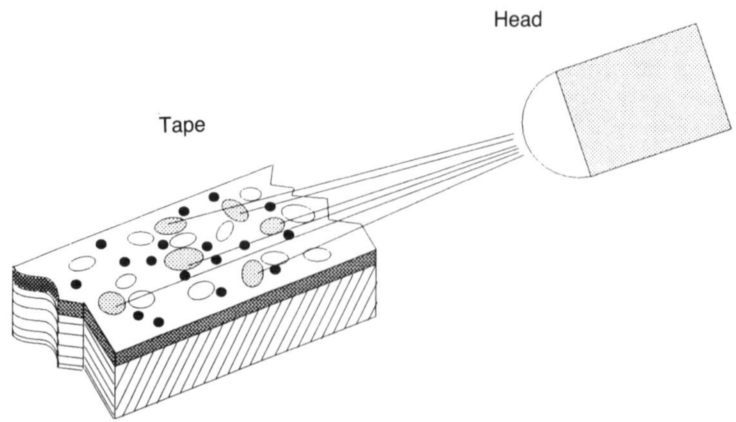

Figure 24.2 Turing machines use a fragment of their physical information potential.

Although the genetic code is universal, DNA or RNA sequences can nevertheless be interpreted in many different ways. For instance, the interpretation of a given DNA chain may depend on where we start to read its sequence. If we have an mRNA subchain ...AGACUG..., it is still an open question what counts as a triplet; is it now GAC, ACU, CUG, or is it something else? The way this selection is made, that is, the way the biological readout frame is matched to the string, determines what amino acid sequence will be produced in turn. But this assignment is not unique, and there is no physical preference for its variants. The selection of this interpretation is biologically controlled and it can be actively changed. As a result, the DNA (or, more precisely, the mRNA formed on the DNA) can use several code systems simultaneously.

Bacteria have a ring-shaped DNA that would offer itself for similar 'autonomous transformations of meaning'. Yet perhaps the most profound example is provided by a virus called ΦX174 in which the same piece of DNA code stands for two distinct structural proteins (Figure 24.3). The choice is mediated by other enzymes that activate this or that pathway. In a permanent readout frame, this information content cannot be expressed.

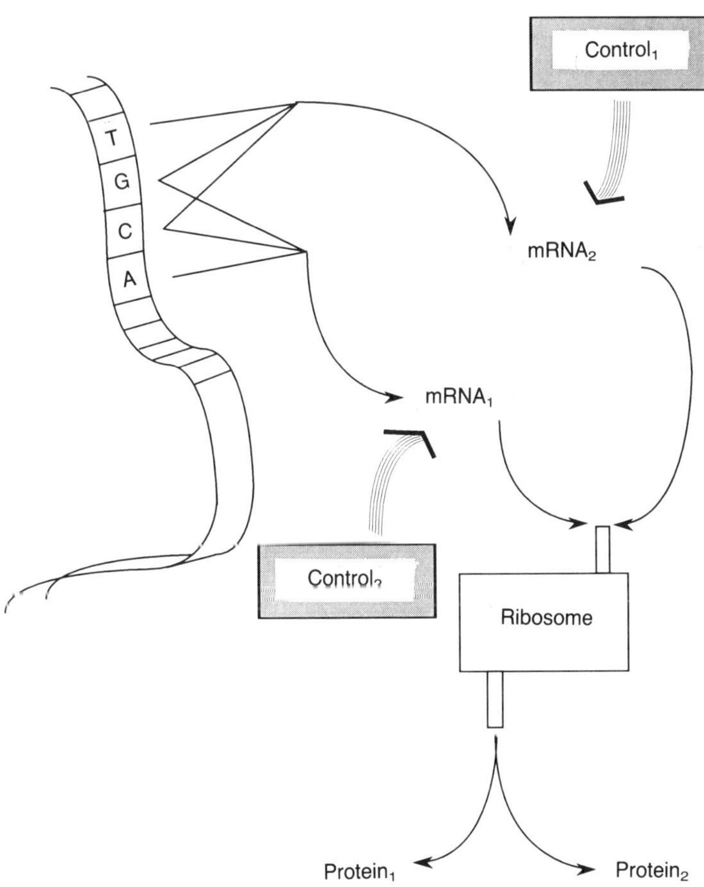

Figure 24.3 The virus ΦX174 with its actively selected coding methods.

The lesson is that codes sometimes overlap and may involve several levels of organization.

Another typical example for the use of shifting frames is provided by catalytic proteins. Allosteric enzymes control biochemical production chains in the following way: they can be 'blindfolded' by specific molecules in the absence of which the enzyme works, and in the presence of which the enzyme is 'silent'. A mechanism called end product inhibition operates by using the same molecule for two different functions (that is, as catalyst and as a blindfold). In other words, this is another example where the structure is unique, but its use is not. How many uses a molecule can have mainly depends on how rich its environment is.

In the organismic realm, we have a perfect analog of these phenomena in the case of what is called the principle of function change. Popularized as 'evolutionary tinkering' by F. Jacob (1981), this mechanism involves the change in the usage of organs. A standard example is the evolutionary formation of the lungs from parts of the digestive tract. In the same way, feet can turn into wings or fins, and so on. It seems everything may become something else in evolution. A remarkable fact is that this transformation can take place without much structural change or sometimes even without any structural change: it is just a matter of starting to use things differently. Later evolution can refine these organs to fit better for the new purpose, but they can be viable without such optimization. Besides being an important evolutionary mechanism, this phenomenon indicates a degree of autonomy of the phenotype in evolution. The phenotype is what is 'visible' from the organism – in other words, what is visible under the given interactions with the internal and external environment.

Classical neo-Darwinism thought that phenotypic function was uniquely determined by the genes and hence that evolution must always operate at the level of the genes. In other words, it was assumed that there is always an invariant and well-defined relation between the phenotype and the genotype, exactly as in the naive sense we assumed that there was a unique relation between a physical system and a computational process. The actual relation turns out to be more blurred, in both cases. In ontogeny, the genotype–phenotype relation can be retransformed or shifted by the environment and by other, partly internal, factors (Goodwin and Saunders 1992).

These mechanisms are important for understanding how life operates and how it produces new complexity, and they are promising candidates to be studied when dealing with biologically motivated computations.

24.7 'NON-TURING MACHINES' AND THE IDEA OF CONTEXT

When a reading frame is shifted a dramatic change occurs. In the cell, a new metabolic pathway is opened. In the computer, a new interaction takes places that invalidates a previous computation.

By this analogy, we find that the biological mechanisms discussed produce novelty that transcends a given set of computational rules. This idea, borrowed from molecules and evolution, is easy to generalize now in terms of Turing machine-like constructs.

We can define a 'non-Turing machine' as one that operates with essentially the same structure as a Turing machine (i.e., it has a finite state transition system, read and write operations, and elementary tape moves), except for one thing: it uses a shifting reading frame instead of a predefined one. It is like traditional computers that permanently produce 'errors', but unlike them does not break down. Instead, it incorporates these errors into its new basic definition. Such a machine could be realized, for instance, by finding a mechanism that develops new reading and

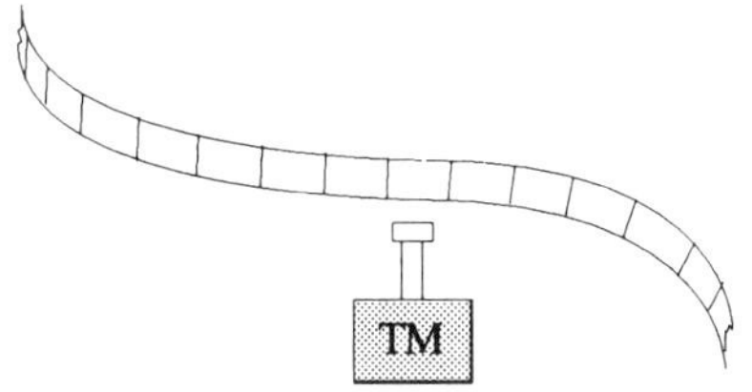

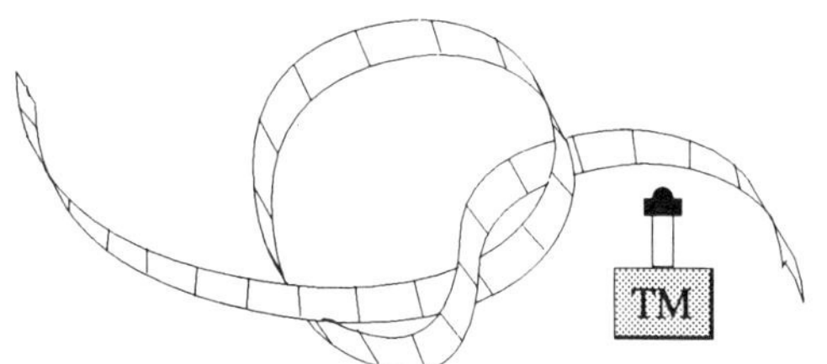

Figure 24.4 Beyond Turing machines: the use of a shifting reading frame.

writing methods (i.e., new tape-handling methods) in runtime.

This possibility has interesting consequences for information theory besides modelling. In effect, this amounts to suggesting that the number of bits stored on a tape can be increased by adding new methods of storage. For instance, we can consider a machine, in the tape of which we tie a knot (Figure 24.4). By developing a new reading method for the knot, the additional information, that is, the existence/nonexistence of the knot, can be accessed. However, and this is very important, every tape has a certain number of knots on it (usually a zero amount as a special case). We are led to the conclusion that the overall information content only depends on the reading frame, and we do not even have to transform the tape. Again, the structure can remain invariant, but the function can change.

24.8 FROM NATURAL TO FORMAL SYSTEMS

For the production of new contexts, and thereby of new information, there is an endless variety of candidates in perhaps every natural system. In actuality these possibilities cannot be explored or even catalogued in advance. Yet we can make an active use of this infinite regress foreseen (but not used) in customary computations. We can use it for the redefinition of the processes that will realize new computations.

We have an equivalent of the modelling process in terms of a hierarchy of Turing machines with various reading heads. There exist Turing machines with multiple heads which may therefore realize multiple reading frames and multiple function modes (that is, one for each head). These machines could try to simulate the shifting systems by simply switching between their predefined modes. But these machines must cling to a limited and predefined variety of frames. Technically speaking, the code sets of these machines must be bounded. There is a better alternative. It consists of organizing the information differently: not to rely on a fixed variety, and not to store the information about the reading methods. Unlike in usual Turing machines we can try to reproduce them dynamically.

24.9 COMPONENT SYSTEMS

We have claimed that certain biological mechanisms realize processes beyond ordinary computation, and that in principle this can be used for the redefinition of computers. The next question is: is there any method for making this idea operational, or should it remain a source of possible criticism against current modelling methods that use a more restricted classical notion of computation?

The biological examples we have considered suggest that systems that define newer and newer properties of their constituting objects can be constructed 'effectively', effectiveness understood in the material sense. Indeed, nothing prevents us from realizing the same biochemical processes as the ones discussed to produce shifts.

In other words, we have examples at hand; the problem is how to generalize them in order to get the definition of a class. Let us turn to this question now.

Work is done on the theory of related systems in various forms (Rosen 1985; Minch 1988; Conrad 1989). One particular formulation is that of component systems (Kampis 1991*a*).

The idea of component systems generalizes the known properties of macromolecular systems. A component system is defined as one that produces and destroys its own components, which form an unlimited pool of component types. By producing a new component, not only a new copy but often also a new type of component will be constructed. The idea is that, by introducing new types of components, new reading frames can be defined. New components can interact with old ones in new ways. That induces new properties in both. In short, the possibility of the emergence of new information processing pathways arises. Therefore, a component system can be recognized as a variant of what we have informally called a 'non-Turing machine'.

The theory of component systems has been developed in detail in Kampis (1991*a*).

24.10 THE REALIZATION OF COMPONENT SYSTEMS ON A COMPUTER

Instead of giving an abstract characterization of component systems, we are going to discuss their possible realizations here. A straightforward method for the approximation of component systems by means of computer models is to use a population of programs that operate on each other, so as to produce new programs, in such a way that these 'interactions' redefine or extend the functionality of the participating programs. In other words, we use programs as components, and we change the properties of components every time they interact with new ones. The interpretation is that we take these variants of a given program to be the context-dependent 'facets' of a bigger, and

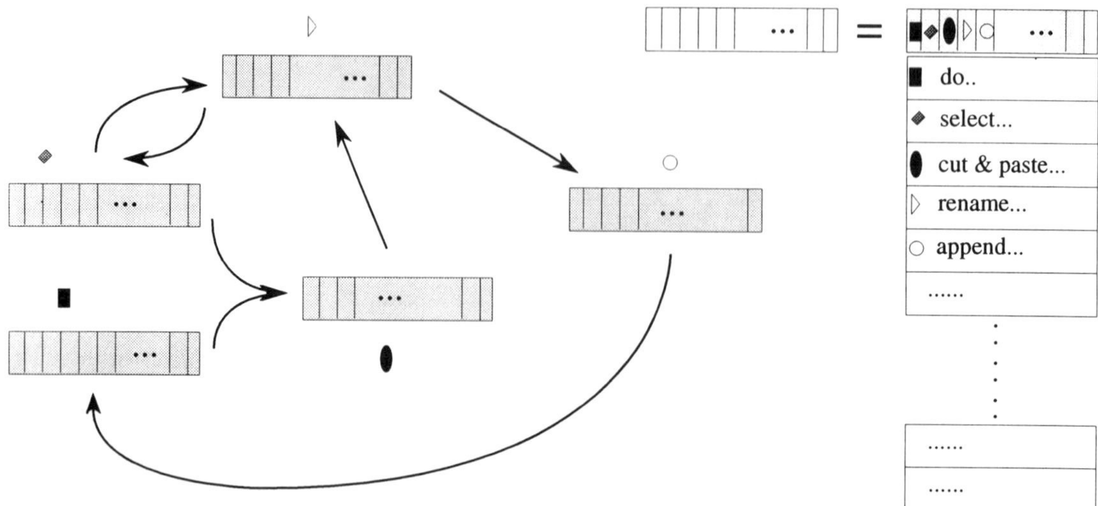

Figure 24.5 The idea of a component system.

invisible, underlying object, and we take the historical envelope of these variants to be the approximation of the object in question. (In this way, objects are fully constructed only at the end of a simulation; this is exactly the opposite of what is usually done today.)

A method by which the redefinition can be achieved is simply by changing program lists during a run. The production of new program code (i.e., the addition or alteration of program statements in a program list) will be made equivalent to the development of new properties for a component. Such a system can be conceived as a modified version of a Markov normal algorithm (Markov 1988) or of a list processor (Figure 24.5). Also Markov algorithms and list processors are programs that can operate on programs.

In a setting like this, it is easy to embed shifting reading frames. Reading frames have to do with object-object interactions. We can now say that an interaction occurs if two programs 'meet' in the sense that one or both emit a specific call for the other. To every call we maintain a separate list of statements defined in the given program (as, for instance, a production system in the sense used in AI), and this list will be activated upon a call. If a given call is new, that is, if it is not yet on the list of the calls of the given program, a new set of program statements can be generated and added (a situation that will certainly occur sooner or later, if we keep on producing new programs that emit their new calls in turn). In this way, by adding self-modification we can simulate Nature's way of utilizing an endless store of 'hidden properties', but without having to construct the systems materially, or having to define these properties in advance.

Although fairly primitive and limited, such a system can already demonstrate the idea of information generation in an otherwise effectively defined system, and, as we shall discuss later, it can produce viable behaviours such as 'natural selection' in a simple form.

24.11 ELEMENTS OF A MODELLING PHILOSOPHY

Let us make a few remarks about the idea that we have just introduced.

Taken literally, one cannot exactly realize a concrete, real-world component system on a computer, because that would require a knowledge of all the system variables and their interactions in advance. That would not be a component system any more, since the component systems are those that produce these modes themselves. What we do is to make computers behave like artificial component systems. By studying these, we can study natural component systems (or at least some of their generic properties) indirectly.

In other words, it will be difficult, if not impossible, to model the exact way a given evolving biological system defines and uses its new variables and properties. In this way, modelling, understood in the strict sense, is abandoned by our methodology. Yet we can build another system, which probably does every detail differently, and where there is no direct correspondence between real-world events and model events but in which we hope some basic characteristics of the original are still retained. In other words, instead of representing one particular system in detail, we can try to build a model for a class.

Accordingly, a method for realizing component systems amounts to the utilization of several independent programs at a time, built with a 'don't care' philosophy (that is, without specifying or pre-designing their interactions by hand). The idea is not to design things beyond what is absolutely necessary, then lean back, and let the system work. This corresponds to the philosophy of having new and previously unknown properties that emerge in a material interaction in a natural system. In a biological system the novelty is bound to the material carrier, but this does not mean that there should be any relation between the old and the novel elements, or that we should consider the latter as preprogrammed: we can just as well think of them as if they were simply random, with respect to each other.

That is, for realizing component systems we can apply some form of artificial chemistry with new properties added randomly whenever new components interact, so that the only thing we need is random property generators which fill the interaction/property matrix and increase its dimensionality – or random program code generators, for that matter. The idea of random addition exemplifies the above 'don't care' philosophy in the simplest way.

Here we understand that the computer can assume a new kind of role, not found in customary simulations. Unlike in traditional computing, where arithmetic and logical abilities of the computer are in the foreground, here the computer, as an embedding framework serves combinatorial purposes. The computer relates different sources of information. What these pieces are, where they come from, and what they do, should not be the given simulation program's concern: that can be pushed 'behind the veil'. What matters is the way these elements are organized and combined.

24.12 AN EARLY COMPUTER MODEL

The first computer model which utilized an early version of a similar philosophy was developed in 1983 and published in 1987 by the author and V. Csányi together (Kampis and Csányi 1987).

The model simulated an evolutionary process producing complex permanent structures in a system with changing composition. The system consisted of a two-dimensional grid, the points of which corresponded to component types in a functional rather than geometrical space. Components could interact with each other along the grid. The interactions were of two kinds. One of them modelled some aspects of chemical catalysis, whereas the other interaction stood for a nonspecific production of new components (in other words, it served for a random redistribution of matter in the phase space). Properties of the components (grid points) corresponded to the existence/nonexistence of a catalytic activity towards their neighbours. These properties could be changed by random errors and by the interactions themselves, according to a simple rule: if a component was produced anew so that its amount changed from zero to a positive number, it was assigned new random properties; in the case of an opposite transition (where the result was the disappearance of a formerly existing component) we removed all its properties, so that the next time it was produced it could assume new ones. In the model, 'catalysis' simply took the form of removing components from randomly selected locations of the functional space, and putting them to the grid point whose production was 'catalized'.

The model was able to produce results that indicated the emergence of self-selecting evolution and showed the spontaneous formation of 'meaningful' stable structures, in spite of the ongoing self-modification (realized as the random change of component properties) and the other noise-like elements. This was the first proof that (1) self-modification of the discussed type is realizable; and (2) that it can lead to 'interesting' behaviour. Permanent catalytic sets (similar to what are called Kauffman sets) were found to form as self-consistent solutions of the self-modifying process on the grid.

Figure 24.6 shows various stages of such a simulation, based on Kampis and Csányi (1987). The model has been significantly developed and extended since then.

24.13 THE SPL SYSTEM

A more recent model, which in fact is still under active development, will now be discussed in detail. It is based on a general

Figure 24.6 A sample run of the self-modifying grid model.

string-processing language called SPL, developed by the author together with M. Vargyas. What will be presented here is a report of the system's basic construction and some current experiments that aim at the study of the principles outlined in the earlier sections.

SPL is part of a complex modelling environment that makes possible the interactive development and simulation of a wide class of biologically relevant computational models, including semi-computational simulations of component systems.

The modelling environment provides tools for the automatic maintenance and the simultaneous execution of a population of programs that operate on each other and on themselves. Besides these new tools, there are several ways of introducing more conventional 'errors' (like noise), and this gives the system the potential for a random search in a functional space. This latter property is shared by many recent evolutionary simulations, and it is one known to be applicable to optimization tasks based on biological mechanisms (as exemplified by genetic algorithms, and other systems). SPL, however, focuses on something else (as in fact there is no inherent need in SPL for self-reproduction, which is the key factor studied in GAs and other evolutionary simulations).

24.13.1 THE STRUCTURE OF SPL

SPL is based on a machine-code-like instruction set designed to allow universal computation besides serving as the substrate for the biological simulation. (Those familiar with the evolutionary model Tierra of Tom Ray or with its several relatives and predecessors, such as the CoreWar systems based on 'Redcode', will recognize SPL as part of a series of efforts towards a general simulation language.)

The SPL instruction set is recapitulated in Figure 24.7 and will be discussed later. A formal definition of the language is given in the Appendix.

An important characteristic of SPL is its coherent string-processing philosophy. It realizes what was anticipated in Figure 24.4. Among other things, this means that, strictly speaking, there is no physical address associated with the program instructions. Rather, there is a pattern-matching process where the arguments of the operations are identified as labels directly on the executing program lists. Informally, this is much like the spatial mechanism by which molecules can find each other directly.

A typical operation of this system is a substitution, in the population of program strings, of the form $A_i \rightarrow B_j$, where A and B are patterns. This is a feature which allows for the use of techniques that are usually excluded from most other programming languages but are of central importance for us. (The actual execution of an SPL substitution takes more than one instruction – it takes at least two, such as a DEL and an INS – but that makes little difference.)

Another feature which needs discussion and which is, indeed, the expression of the central idea behind the whole SPL development, is this. In SPL we introduced two 'metaphysical' operations RND and WISH. The effect of RND is to insert random bytes at a specified place (more precisely, at a given pattern), whereas that of WISH is the insertion of an entire random program, in the same way. The design of the SPL syntax together with the suitable construction of the embedding environment makes every random program functional and executable in this system (that does not mean, of course, that such a random program would necessarily do something 'useful'). This makes random progamming an easy job.

Formally, the introduction of the 'metaphysical' operations renders SPL equivalent to a Turing machine with random inputs. Methodologically, however, we use these random inputs in a novel way. The reason

```
                    String Processing Language

registers (16 bit):                  [stack: STCKSIZE x 16 bit]
S1      address register             S2      address register
I1      index register               I2      index register
D1      data register                D2      data register

instruction set (8 bits, 28 instructions):
------------------------------------------------------------------
control instructions
ADR     find pattern or empty room, write in (S1, I1)
JMP     goto pattern or [S1, I1]
IF      goto pattern or [S1, I1] if D1 != 0
CALL    subroutine call at pattern
RET     return to last CALL
REP     repeat D1 times
EREP    end of repeat block
------------------------------------------------------------------
register instructions
LD      load [S1,I1] or @pattern to D1 (i.e. "read")
DL      download D to [S2, I2] or to pattern (i.e. "write")
PUSH    D1 to stack
POP     stack to D1
MOV     next byte into D1
XCH0    exchange (S1, I1) <--> (S2, I2)
XCH1    exchange D1 <--> I1
XCH2    exchange D1 <--> S1
XCH3    exchange D1 <--> D2
SWP     exchange bytes of D1
------------------------------------------------------------------
arithmetical and logical instructions
ADD     add [S1, I1] or @pattern to D1
SUB     sub [S1, I1] or @pattern from D1
RADD    add stack to D1
RSUB    sub stack from D1
NAND    bw. D1 and stack
SHL     D1 = 2*D1
------------------------------------------------------------------
transfer instructions
COPY    block copy from pattern to [S2, I2]
DEL     delete pattern
INS     insert pattern at [S2, I2]
------------------------------------------------------------------
metaphysical instructions
RND     set D1 random
WISH    insert rnd string of rnd length to [S2, I2]
```

Figure 24.7 The SPL instruction set.

they are called 'metaphysical' is that they can be directly applied to changing the system that uses them, and hence they express the idea of self-modification in the sense discussed earlier. (Note that some primitive form of 'self-modification' is provided by every list-processing language where codes that get altered during execution can be written. But this is different from the mechanism of introducing new, independent code when a program is called by another.)

Besides these, the rest of SPL is concerned with customary computations. A working memory is provided in the form of registers, and we have also introduced the usual arithmetic and logical operations together with instructions for flow of control (like branching or cycles). This part is fairly standard and need not be detailed here.

How these instructions can be used for constructing biological models is briefly discussed next.

24.14 A HIERARCHY OF SIMULATION GOALS IN SPL

Of particular interest is the following hierarchy of goals that can be achieved in this environment. Each step brings us closer to component systems, the last element of the list.

24.14.1 UNIVERSAL TURING MACHINES

A universal Turing machine can be realized in this system in several ways. One is by writing a single special SPL program with no random effects, no self-programming, and no use of the list-processing instructions. The key idea is, in any case, that the pattern-matching operations should be used for establishing and removing a potentially unlimited amount of freely addressable temporal memory in the form of new strings uniquely labelled by patterns. These consitute the extendable tape of the Turing machine.

24.14.2 'DATA EATERS'

Systems nicknamed 'data eaters' are parallel nonrandom programs that use a population of strings for the simultaneous execution of arbitrary tasks and communicate by leaving nonexecutable traces (i.e., labelled data strings) to be picked up by other programs. In this way one can easily realize any parallel

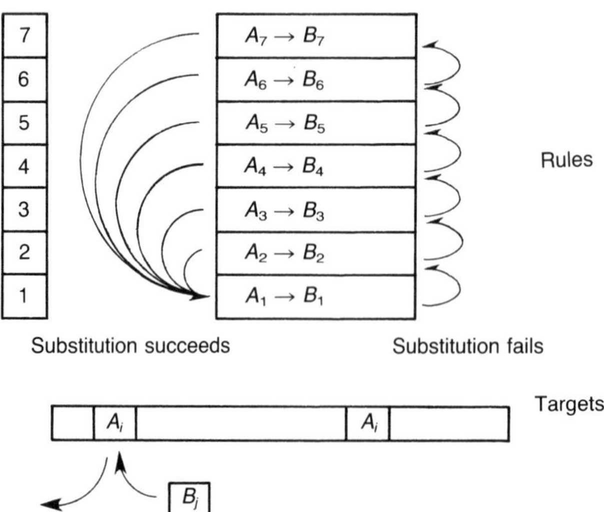

Figure 24.8 Markov normal algorithm.

computation, in a design that shows some similarities to D. Gelernter's Linda system (Carriero and Gelernter 1990).

24.14.3 'PROGRAM EATERS'

This is a more liberal structure where there is no difference between data and program. In SPL such differences can only be maintained by avoiding the use of certain operations; if we program the string system with no restrictions we obtain programs that read and write each other's code deliberately. We call them 'program eaters' because other programs are their 'food', to be transformed ('digested') or to be destroyed ('used up').

As a special example, a wide variety of computer viruses and virus defense mechanisms can be written and tested in SPL, without making harm to the embedding system. Work on the study of this possibility is in progress.

24.14.4 MOLECULAR COMPUTERS

Molecular computers are information processing systems that utilize principles of molecular interactions (Conrad 1985). Of the various formalizations of such systems that can be naturally realized in SPL, we now consider the Liberman molecular computers (Liberman 1979).

A Liberman molecular computer is a pattern-matching and rewriting system with

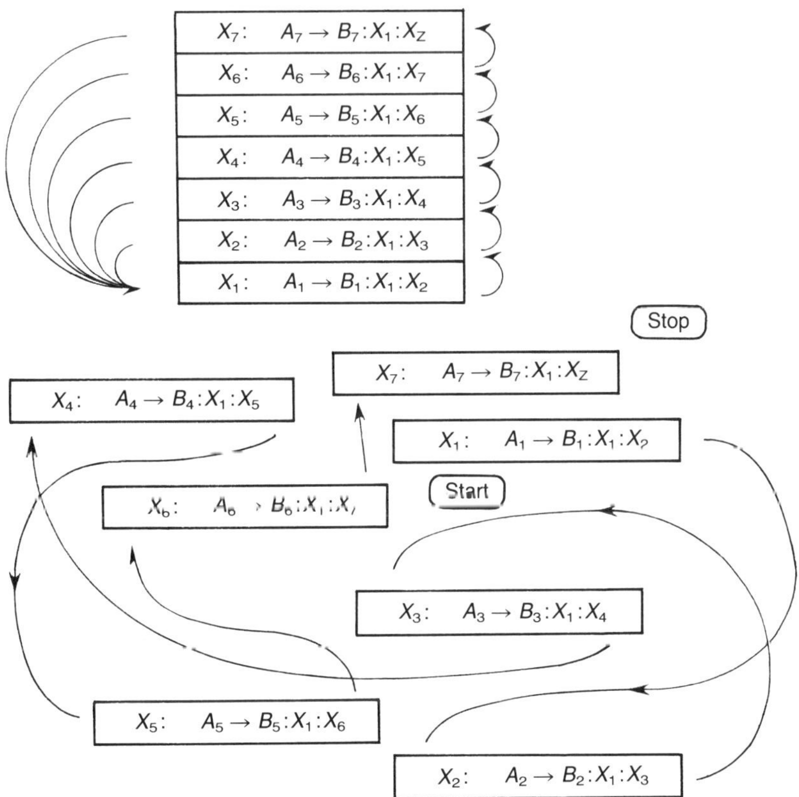

Figure 24.9 A Liberman computer.

a Markov control flow. Recall from Figure 24.8 that the Markov control flow is this: if substitution S_i: $A_i \rightarrow B_i$ fails, go to S_{i+1}; if it succeeds, go back to S_1. In a string processing system this can be realized with a slight modification. A string C_i is supplemented with a header H_i, and the substitutions are executed in strings with the respective headers (S_1 in H_1, S_2 in H_2, etc.). The control flow then takes this form: if S_i is successful, rewrite H_i as H_1; if it fails, rewrite it as H_{i+1} (Figure 24.9). It is easy to see that such a system can be realized with the SPL instruction set: using patterns as headers and using SPLs pattern-directed rewriting rules as substitutions.

24.14.5 TIERRA-LIKE SYSTEMS

Although this was not a goal of the SPL project the system can easily embed evolutionary models of the Tierra type (Ray 1992). In fact in SPL both simpler and more complex self-reproducing programs can exist than in Tierra and, much as in Psoup, another simulation program (de Groot 1992), these can form automatically from an initial population which contains no self-reproducing programs at all. For the sake of example, the simplest SPL program with such Darwinian reproduction and evolution capabilities we have found was

```
"GK ; define header for identification
ADR ; make room for offspring
XCHO ; change registers for COPY
COPY" "GK* ; copy the whole program from
header
```

This is a truly trivial program of no interest, apart from the facts that it exists in SPL and that it can serve as a template for a large family of similar, autonomously developing programs.

24.14.6 EVOLVING 'DATA EATERS'

An interesting class of evolving programs, beyond the capability of the above simple selectionist system where mutation and reproduction are the only controls, is one where the evolving programs can also behave like 'data eaters' at the same time. This makes the autonomous introduction (in runtime) of further selection criteria possible. The executed programs can develop new criteria for their own evolution by letting the reproduction of a given program depend on labelled data strings left behind by some other programs. As these latter programs (just as the evolution criteria themselves) can also be subjects of mutation and selection, we can get an intriguingly complex process where evolution has a choreography mutually defined and changed by the evolving subjects. The same idea can be applied to 'program eaters', too.

24.14.7 EVOLVING PROBLEM SOLVERS

A next logical step is to allow programs to pose problems for each other by means of communication in 'data eater' mode. The target program can find and transform the data, and the source program can evaluate, in turn, the solution. Based on this, it can print out another data string that will be used by the target as an reproductive condition, and so on: there is an endless variety of interaction variants. As this process can proceed on a mutualistic basis, the evolving programs can learn how to solve the problems they have together invented.

Our experiments with evolving 'data eaters' and evolving problem solvers are under way. Early results show that the difficulty is not in the writing of suitable SPL code but in making the program population robust enough to last long enough until something interesting evolves: a problem not uncommon in 'wet' biology either.

24.14.8 WISH PROGRAMMING

There is a possibility for using the WISH and RND instructions for the writing and rewriting of programs. Specifically, the WISH programming philosophy suggests an unusual way of filling the environment with an initial population of random programs. A simple program called 'devil's factory' can do that. As this program works indefinitely, another program is required to kill it after some time has passed.

With the combination of WISH programming with the other methods we discussed it is easy to write very short and surprising programs. As an exercise the reader could write a program which uses WISH or RND instructions for the setting of cycle length in wait cycles (or string length in copy cycles, etc.) to achieve random reproduction. Or, in an evolutionary simulation, these mechanisms can be used for producing 'hot spots' for the evolution. Hot spots are places where random 'errors' occur much above the normal mutation rate. But we can also use similar strategies for more advanced applications.

24.14.9 COMPONENT SYSTEMS

Most importantly, SPL allows for the simulation of component systems by using the strategy of WISH programming.

A method of realizing component systems is combining Liberman computers with WISH elements. When a substitution is not possible, a new header and a new program code will be generated by WISH for the target string. In this way we can obtain a sequence of programs that grow bigger and bigger as they encounter more and more substitutions that correspond to the specific 'calls' we have discussed. Selectively delayed delete instructions can then help these programs to get rid of their never-used program parts, and so on: string programming gives all the flexibility required to such a delicate modelling task. All the programming can be done from within the components, and it can be done autonomously: after an initial startup, no intervention is needed.

24.15 DISCUSSION

The system as described here became functional in February 1993 and has been presented at two conferences since then. The system is equipped with an assembler/disassembler, a visual output generator, and a set of tools such as a dump tracer.

We have early results concerning several forms of the above hierarchy of goals. These results indicate that SPL is a viable tool for experimenting with information-producing systems. As the buildup and analysis of complex simulations is a very time-consuming activity, it will take a long time until the point where 'meaningful' self-organizing structures such as the ones on Figure 24.5 can develop in SPL (and will be understood, once developed).

The SPL interpreter and the programming environment are available upon request from the author in MS-DOS format.

APPENDIX

SYNTACTIC DEFINITION OF THE SPL LANGUAGE

<statement> ::= <single instruction> | <string> | <transfer> | <move>

<string> ::= <string instruction> <global pattern> |
<string instruction> <local pattern> |
<string instruction> <empty pattern>

<transfer> ::= <transfer instruction> <direct pattern> |
<transfer instruction> <indirect pattern>

<global pattern> ::= "<pattern> <eos>

<local pattern> ::= <pattern> <eos>

<empty pattern> ::= "<eos>

<direct pattern> ::= <pattern> <eos>

<indirect pattern> ::= "<pattern> <eos>

<pattern> ::= <character> <pattern> | <character>

<move> ::= MOVE <byte>

<string instruction> ::= ADR | JMP | IF | CALL | LD | DL | ADD | SUB

<transfer instruction> ::= COPY | DEL | INS

<single instruction> ::= RND | WISH | REP | EREP | RET | PUSH | POP | XCH0 |
XCH1 | XCH2 | XCH3 | SWP | RADD | RSUB | NAND | SHL |

<character> ::= ' ' | '!' | .. | '}' | '~'

<byte> ::= 0 | 1 | .. | 255

<eos> ::= 0

ACKNOWLEDGEMENTS

This chapter was written with the support of research grant OTKA 2314 obtained from the Hungarian Academy of Sciences. The support is gratefully acknowledged. Part of the work was carried out during the author's stay at the Department of Theoretical Chemistry, University of Tübingen, Germany. The author wishes to thank Professor O. E. Rössler for his hospitality and cooperation. He thanks Mr M. Vargyas for his work on SPL and related matters. This chapter uses figures and excerpts from other, noncopyrighted manuscripts of the author.

REFERENCES

Carriero, N. and Gelernter, D. (1990). *How to Write Parallel Programs: a First Course*. MIT Press, Cambridge, MS.

Chaitin, G. J. (1987). *Information, Randomness and Incompleteness*. World Scientific, Singapore.

Conrad, M. (1985). On design principles of a molecular computer. *Comm. ACM* **28**, 464–80.

Conrad, M. (1989). The brain-machine disanalogy. *BioSystems* **22**, 197–213.

Dawkins, R. (1986). *The Blind Watchmaker*. Norton, New York.

Garey, M. R. and Johnson, D. S. (1979). *Computers and Intractability: A Guide to the Theory of NP-Completeness*. Freeman, San Francisco, CA.

Goodwin, B. and Saunders, P. T. (eds.) (1992). *Theoretical Biology: Epigenetic and Evolutionary Order from Complex Systems*. Johns Hopkins University Press, Baltimore, MD.

de Groot, M. (1992). *Psoup Manual*. Obtainable from marc@os.com

Herken, R. (ed.) (1988). *The Universal Turing Machine: A Half-Century Survey*. Oxford University Press.

Jacob, F. (1981). *The Possible and the Actual*. University of Washington Press, Seattle, WA.

Kampis, G. (1991*a*). *Self-Modifying Systems in Biology and Cognitive Science: A New Framework for Dynamics, Information and Complexity*. Pergamon, Oxford, p. 543.

Kampis, G. (ed.) (1991*b*). *Creativity in Nature, Mind and Society*. Gordon and Breach, New York. (Also available as Special Issue of World Futures: *The Journal of General Evolution*, **32**, 2–3.)

Kampis, G. and Csányi, V. (1987). A computer model of autogenesis. *Kybernetes* **16**, 169–81.

Klir, G. J. (1985). *Architecture of Systems Problem Solving*. Plenum, New York.

Liberman, E. A. (1979). Analog-digital cell computer. *BioSystems* **11**, 111–24.

Löfgren, L. (1977). Complexity of descriptions of systems: a foundational study. *Int. J. General Systems* **3**, 197–214.

Löfgren, L. (1987). Complexity of systems. In *Systems and Control Encyclopedia* (ed. M. Singh). Pergamon, Oxford, pp. 704–9.

Markov, A. A. (1988). *The Theory of Algorithms*. Kluwer, Dordrecht.

Maynard Smith, J. (1975). *The Theory of Evolution*. Penguin, London.

Maynard Smith, J. (1986). *The Problems of Biology*. Oxford University Press.

Minch, E. (1988). Representation of Hierarchical Structure in Evolving Networks. Ph.D. Dissertation, Department of Systems Science, SUNY at Binghamton, NY.

Peterson, G. E. (ed.) (1987). *Tutorial, Object-oriented Computing*. Computer Society Press of the IEEE, Washington, DC.

Ray, T. (1992). Tierra and Tierra documentation. (Although there have been numerous writings on Tierra in the press, it is hard to find a quality publication. Hence, currently the texts that accompany the PD versions of the software are of primary importance. The whole package can be pulled down from the Net through: life.slhs udel.edu)

Reichgelt, H. (1991). *Knowledge Representation: an AI Perspective*. Ablex, Norwood, NJ.

Reiser, Martin (1991). *The Oberon System: User Guide and Programmer's Manual*. ACM Press, New York.

Rogers, H. (1967). *Theory of Recursive Functions and Effective Computability*. McGraw-Hill, New York.

Rosen, R. (1985). *Anticipatory Systems: Philosophical, Mathematical, and Methodological Foundations*. Pergamon, Oxford.

Russell, B. and Whitehead, B. (1912). *Principia Mathematica*. Cambridge University Press.

Trakhtrenbrot, B. A. (1973). *Finite Automata; Behavior and Synthesis*. North-Holland, Amsterdam.

Wagner, K. and Wechsung. G. (1986). *Computational Complexity*. Reidel, Dordrecht.

Whitehead, A. N. (1929). *Process and Reality*. Free Press, New York.

Wittgenstein, L. (1976). *Lectures on the Foundations of Mathematics*. Wittgenstein's lectures on the foundations of mathematics, Cambridge, 1939 (from the notes of R. G. Bosanquet, Norman Malcolm, Rush Rhees, and Yorick Smythies, edited by Cora Diamond), Cornell University Press, Ithaca, NY.

Yasuhara, A. (1971). *Recursive Function Theory and Logic*. Academic Press, New York.

NATURE'S MACHINE: MIMESIS, THE ANALOG COMPUTER AND THE RHETORIC OF TECHNOLOGY

James M. Nyce

Of importance for cybernetics was its setting in the flesh . . . For me it proved that brains do not secrete thought as the liver secretes bile, but that they compute thoughts the way computing machines calculate numbers.

Warren McCulloch (1974) p. 10.

25.1 INTRODUCTION

At its most uncritical, history often appears to be a series of foregone conclusions. Nowhere perhaps is this more true than in the history of computing. As K. W. Smillie put it: 'the history of computing remains largely . . . outside history [and historiography]' (1992, p. 69). What scholars of science and engineering have not done is pay much attention to the cultural idioms at work in these rhetorics or how these idioms sustain particular arguments and make them credible and persuasive to others. Unfortunately, in the social studies of science culture, rhetoric and argument tend to be much too quickly reduced to individual or historical events. In short, it is necessary to look at the cultural resources that have informed the arguments and practices we call science and engineering.

What we want to do here is look at two influential, but quite different, sets of analogies that have helped constitute for scientists and engineers both the brain and the computer. To do this, we will begin with the 1942–52 Macy Conferences on cybernetics. The Josiah Macy Jr Foundation funded these ten conferences to bring together biologists, engineers, mathematicians and social scientists to apply cybernetic models to biology and neurobiology particularly the individual and human society. At the conferences, discussion and debate were shaped by two different ways of 'making sense' of the brain and the computer. The first, now less understood and known, emerged from pre-war analog technology and theory. The other, supported and informed by digital theory and technology, is the one on which our understanding of computational machines and the brain primarily rests today. Furthermore, the latter is both so pervasive and taken for granted that its history and the circumstances which favoured its acceptance have received little attention. In fact, Conrad has to remind us of 'the untenability of a machine-like picture of the brain [and that this] is a consequence of . . . evolution, structure and intelligence' (1989, p. 197).

Looking at these conferences reminds us

(and the conference transcripts make it quite clear) that digital analogy 'won' over competing, but now largely forgotten, models of the brain. Rather than discuss, as others have done, the digital analogy and the role it has had in supporting the Macy Conferences and informing late-twentieth-century discourse (Heims 1975, 1977, 1991; Heims 1980), we want to look at what progress and success, as we now define it, has left behind. This chapter will concern itself with a set of models, analogies, metaphors and theory that for the most part has been lost or forgotten.

Despite demonstrated differences between biology and machines, the digital metaphor continues, in powerful ways, to inform research in areas where mind, brain and computers intersect (West and Travis 1991*a*, p. 69; Paton 1992). However, to decide whether this metaphor obscures, diverts or supports particular research agendas requires that at the very least it be the subject of discussion. For the most part, recent attempts to do this either promote (or inventory) alternative models or they treat this metaphor as though it is just a figure of speech (see, for example, West and Travis 1991*a*, 1991*b*).

However if this metaphor is to be dealt with in any serious, reflective fashion, strategies of this kind will not be sufficient. In short, lexical analysis cannot answer questions about what supports and sanctions metaphors like this. What is required in fact is a stronger analytic program – one that can take things like practice, ideology and history (particularly institution building) into account. In particular what scholars of scientific practice have not done is pay much attention to how culture and cultural idiom sustain particular arguments and projects and makes them creditable to others. This chapter is an attempt to move the study of computional metaphors in this direction.

What brings biology, machinery and metaphor together here are a set of parallels (strong family resemblances) thought to exist between the brain and the nervous system, and certain classes of machines that can carry out mathematical and logical operations (Heims 1975, p. 371). In other words, there could or can be enough parallels drawn between brain and machine that the same logic could be used to describe them both. As a result, logical theory can be used to describe the brain and any act or mechanism of machine intelligence (Heims 1977, p. 143). Thus from this analogy it has become possible to claim that there are fundamental identities between brain and machine. What made this analogy so compelling is that it holds out the promise of helping us to better understand certain classes of machine action. Further, this analogy was attractive because it was believed that models derived from it could yield more adequate, complete models of brain structure and organization.

What supported this analogy was the belief that it was possible to both perceive and draw salient parallels between biological structures and processes and electrical circuitry and systems (Heims 1980, p. 173). At the Macy Conferences, von Neumann, for example, applied a formalism originally intended to describe the organization of the human brain to the design of computing machines (Heims 1980, p. 183). Von Neumann believed he had identified common features between the nervous system and basic units of logic. What held the two together was their '*prima facie* digital character' (von Neumann 1958, p. 44). This led (and permitted) von Neumann to move back and forth between brain and automata. Moving between essentially dissimilar systems to gain insight and to derive models is hardly a unique strategy. In fact, Stephen Pepper (1942) and others have argued that intellectual, if not scientific, discourse and practice rests upon seeking and working through just these correspondences.

25.2 THE MACY CONFERENCES

To tell part of this story takes us to the Macy Conferences. Essentially (and this is common knowledge) a digital set of analogies provided a kind of common ground for these conferences. What has not been acknowledged is the role that these meetings had in validating this analogy set. At the time turning to digital theory to formally model brain structure and function (today this is so accepted, it is seldom even questioned) was relatively novel. It can be argued that the success this analogy has had in framing subsequent conversations about brain and machine is at least in part because it emerged from and was supported by the Macy Conferences. 'The advent of cybernetics', Coulter reminds us, 'brought with it . . . the idea that the brain is a computer' (1974, p. 10). In short, the Macy Conferences helped to validate digital theory and logic as legitimate ways to model and to think about the brain and the machine. The participants' direct involvement in (and commitment to) logical machines, for example, servomechanisms, of the time helped confirm a digital model of the brain and machine (Edwards 1985, pp. 36, 40).

The McCulloch and Pitts model of the brain was essentially a logical construct in which idealized neurons, often in series or sets, formed its basic units. The model's descriptive strength led von Neumann, among others, to seek not just parallels but identities between the brain and specific kinds of electrical circuitry (Heims 1980, p. 211). As a result, in time, this logical construct, this model of the brain, would essentially become the thing itself. The fact that not only the model but these parallels and identities could be expressed in formal (mathematical and logical) terms was what made this model particularly attractive. From it too, the digital computer emerged (Heims 1975, p. 371). At the time of the Macy Conferences however, the McCulloch and Pitts model had not established the hegemony it was later to have over research, technology and discourse.

From the first, questions were raised about the role continuous variables (both chemical and electrical) had in brain function and whether variables like these could either be accounted for or adequately handled in terms of binary (all-on, on-off) digital theory (Heims 1991, p. 21). As the Macy Conferences took up questions about brain and machine, the nature of each and what one could tell us about the other, the digital model was challenged. For example, at the seventh meeting Gerard noted that 'much of the electrical action of the brain is analogical . . . Brain waves themselves, the spontaneous electrical rhythmic beats of individual neurons are analogical [and] . . . represent a continuously variable potential, not . . . discontinuous spikes' (1951, p. 13). Digital devices cannot handle value or ranges of value like this directly: they have to be encoded into finite steps or sets (Bromley 1990, p. 195). Gerard argued that this is not as unproblematic as it seems (1951, p. 17). For Gerard, the idea that one can move from artifact or event, experimental or not, to variable to range to binary code relatively easily preempts and obscures a number of fundamental issues in neurophysiology.

Throughout the Macy Conferences, there were discussions about whether and which brain mechanisms and processes could be better described or modelled in analog or digital terms (Heims 1991, p. 92). Gerard put it this way: 'Although it remains true that nerve impulses may be atomic in character and that they move or don't move, I think it dangerous to . . . conclude that functioning of the nervous system can be expressed essentially in terms of digital mechanisms' (1951, p. 17). Gerard was not alone: a number of papers given at the conferences raised the issue of whether the brain itself might function more 'analogically' than 'digitally' (Heims 1991, p. 74). Gregory Bateson even raised the stakes: at one point he spoke of 'the body as a whole as a possible analogic calculating machine' (Heims 1991, p. 153).

All this threw the McCulloch–Pitts model into question because it basically represents a set of idealized neurons whose formal properties had much in common with a digital computer (Heims 1975, p. 371). Handling information that had been transformed and coded in binary, a model of this kind would not have been able to represent certain kinds of continuous variation that help determine brain function. For the same reason, a binary model cannot support or handle direct parallels or isomorphisms between the brain and itself. The issue here was whether a model has to resemble the thing itself more than, say, words resemble what they describe.

What was at the heart of, or at least supported, these critiques of McCulloch and Pitts was another set of assumptions and theories about computers and computation. While it is clear that computer–brain comparisons informed the Macy Conferences (Heims 1991, p. 92), what have not been looked at closely are the intellectual resources sceptics had available to them to call into question the digital analogy. Next we will trace out what some of these resources were.

25.3 COMPUTATION, TECHNOLOGY AND THEORY AT MID-CENTURY

By the 1950s, to a large extent scientific work had shifted to digital machines (Aspray 1990, p. 255).* Before this the practice of science was supported and informed by a different set of computing and calculating technologies. Analog devices belong to a long tradition of scientific instruments, starting in the seventeenth century, that 'made visible what could not be seen' (Hackmann 1989, p. 31). Unlike most scientific instruments, however, analog devices supported both understanding (literally by measurement and number, like an astrolabe) and investigation for they, like an orrery, were 'models' of phenomena. Often analog machines were also heroic devices (Hackmann 1989, p. 32): their sheer size, and because they represented the limits of what at the time was technologically possible, gave these devices (and the results derived from them) a certain rhetorical power. What also made them persuasive is that they were both statements about and direct imitations of the things they represented. Mimesis is 'hidden' or absent in digital machines: analog machines represent phenomena vividly and directly. In other words, what these machines do is bring together, bind together and extend two strong, distinct descriptions of the world – the declarative and the figurative.

Throughout World War II, analog theory and technologies, both mechanical and electrical, represented the state of the art in computers (Burks and Burks 1988, pp. 263, 272). By the end of the war Smith tells us: 'the analog computer was accepted both as a practical mathematical approximator and as a practical mechanism for the control and simulation of physical forces, whether the computer was mechanical . . . with cams and wheels, or electrical . . . or a mixture of both' (1989, p. 11).

The analog tradition of engineering, technology and theory, while once widely known, accepted and used (cf. Hartley 1962), has largely been forgotten. However, because this tradition and its devices provided at least in part the vocabulary and evidence used to critique Pitts and McCulloch and what has become today largely unchallenged assumptions about the brain and the machine, they deserve another look. This takes us to Vannevar Bush (1890–1974) who was deeply committed to analog machines (Shurkin 1984, pp. 96, 145).

Before World War II Bush developed the

*The transition from analog to digital machines was neither as absolute nor abrupt as someone like Foucault would have portrayed it. For example, even before World War II, Smith tells us that a MIT student would be 'in a position to learn both about differential analysers and . . . to explore the logical properties of binary arithmetic . . . [in particular] how . . . binary arithmetic, alias yes-no logic, could be embodied in electronic circuits' (1989, p. 11). At this time, in other words, when one technology was chosen over the other, this reflected more pragmatic choice and calculation than a victory for a particular epistemology.

differential analyser and a series of other analog machines and after the war he continued to work with and improve this technology. It was this technology and the arguments that supported it that was a counterfoil to McCulloch and Pitts. Bush found analog machines attractive because they literally modelled phenomena (Nyce and Kahn 1991, p. 61). He believed that analog machines could, with only a little simplification, directly mirror the thing itself. (They also provide a vocabulary and a correspondence set that helped make this possible.) Unlike digital or logical machines, analog devices did not reduce phenomena to any set of general principles and then treat those principles as the thing itself. In effect then, they both follow and obey the same laws. Consequently, an analog machine, without any intervening steps, can represent either abstract or actual phenomena.

As Owens (1991) argues, Bush's machines were not simply elegant reproductions of phenomena. They also, he notes (1991, p. 75), kinetically act out and so represent those processes that constitute and define the thing itself. In Bush's machines, for example, physical events (action and relation) are depicted as similar continuous variables. Here, process and event are neither abstracted, reduced nor defined as a set of principles. In other words, these machines represent a triumph over a tendency in science to formalize and typify phenomena. Offering an alternative to describing phenomena in general terms, analog machines and theory raised the question whether the brain could be modelled or even understood in idealized terms. This, of course, challenged the McCulloch and Pitts model whose sets and neurons were logical constructs and whose operations had been reduced to code or information exchange along binary principles.

Bush wanted to design machines that were not bound to mathematical or logical operations. While he wrote to Edward Weeks at the *Atlantic Monthly* that 'a great deal of our brain cell activity is closely parallel to the operation of relay circuits' (Bush 1944*a*, p. 1), Bush saw no necessary reason to reduce the operations and mechanisms of either to any formal or principled terms. Nor did he use languages of this kind to make statements about identity between them. What he concluded instead was that one can explore 'this parallelism' as well as 'the possibilities of . . . making something out of it almost indefinitely' (Bush 1944*a*, p. 1). In effect, for Bush, analog machines were not valued because law, principle or general statements could be derived from them.

25.4 BUSH'S MEMEX: THE MACHINE AS RHETORIC

To argue that science rests on figurative language has become commonplace (Latour and Woolgar 1979; Knorr-Cetina 1981; Lynch 1985). However, little attention has been given to the role culture and cultural idiom has in framing and validating scientific argument. As C. Wright Mills reminds us, working vocabularies are not just individual or institutional dialects, they are in important ways culturally grounded (1940, p. 909). It is not enough to focus on any one laboratory, the trajectory of a particular experiment or the 'career' of a set of experimental results. It is also necessary to take into account the cultural resources and idioms that support and sustain the arguments and practices we call science and engineering. Using Bush's Memex machines we will attempt to do this here.

Bush never built his Memex machines (for some reasons for this, see Burke (1991)). Perhaps because the machines were never an endpoint on a development cycle and thus never tested against experience, for Bush they remained, as Levi-Strauss said about totemism, something good to think with. What we have here is a series of rhetorical machines that both inform and comprise argument. Not just icons, emblems or pictures, rhetorical machines like Bush's interrogate, constrain and make statements about mechanical possibility. However, not only do they frame argument, action and object, they define what these possibilities might be. This is both more and less inclusive than what is generally meant by 'metaphor'. Because the Memex machines were only embodied in words, arguments and rhetoric, taking a close look at these machines opens a window on how arguments about brain and machine can be put together and what makes some of these arguments seem more 'correct' than others.

A single premise underlies Bush's writings on Memex: to solve human problems, technology must imitate nature and natural systems.* In this passage, Bush makes his position quite clear (1959, pp. 2–3).

> In bringing machines to man's aid, we have thus far built them on patterns which fitted the technical elements at hand and our habitual ways of doing things, rather than to cause them to imitate and extend the actual processes by which the brain functions. Many years ago I described a machine, called the Memex, which I conceived of as a device that would supplement thought directly . . . Thanks to the psychologists and the neural physiologists we now have a clearer conception of how the brain actually functions, and hence a better chance of joining it with a machine.

In his work, Bush extends this argument. For example, he goes on to argue that if information technology is to be effective, it has to simulate elements of human nature and functioning (Bush 1945, p. 106).

> Our ineptitude in getting at the record is largely caused by the artificiality of systems of indexing. When data of any sort is placed in storage, then filed alphabetically or numerically, and information is found (when it is) by tracing it down from subclass to subclass . . . The human mind does not work that way . . . The speed of its action, the intricacy of trails, the details of mental pictures, is awe inspiring beyond all else in nature . . . Man cannot hope fully to duplicate this mental process artificially, but he certainly ought to be able to learn from it.

Conventional indexing systems, then, do not work because they are artificial and by definition inadequate. These systems represent human conventions of work, labour and technology. They are cultural artifacts. In the end, they fail for just these reasons. These systems break down because they neither duplicate the mind nor its processes. In short, to solve such problems, Bush argues, one has to turn to natural or biological systems. Here the brain, an artifact of biology and nature, provides the most adequate model for man to emulate. As we shall see, Bush would later extend this argument.

Moving between natural and cultural systems like this is hardly unique to Vannevar Bush. Others have used this rhetorical strategy to gain insight and derive models (Lienhardt 1961; Tambiah 1969). What makes Bush's rhetoric so attractive is the next step he takes. Baldly put, it is man's technology and knowledge that enables him to bridge gaps between essentially dissimilar systems.

To bridge the gap between nature and culture, Bush uses the principle of analogy. Indeed, Bush's Memex machines emerge from analogies between biological and mechanical systems. Analogy allows for similarity without necessarily having to argue for identity. An analogy between natural and mechanical systems, while incommensurable, makes an argument about how much they have in common. However, analogy, as a trope, does not require that each term resemble the other in every respect. All an analogy establishes is a link or links between different terms. Bush's Memex was never to be some kind of literal mechanical analog to the brain. In fact, for

*Nyce and Kahn (1991) have published or reprinted much of this material.

Bush, there were never to be exact, point by point, identity between Memex and the brain. In short, what is sought here is not identity but a series of parallels.

Analogy makes Bush's technological projects (and solutions) possible. Through analogy, we can build mechanical systems or structures that simulate the actions of important natural systems. Further, through analogy, the promise Bush holds out to us will eventually be realized. We can learn to move between nature and culture.

Reading Bush's work on Memex reveals the kinds of analogies and the analogy set he uses. Ordered by scale and complexity, Bush builds this analogy set up level by level. The first is nature:culture. From this analogy, another central pair is derived (biology: technology). What holds these pairs together is electricity. For example, electricity is both a natural force and created and controlled by humans.

How important is electricity? Bush is quite clear on this. 'In the outside world all forms of intelligence, whether of sound or light, have been reduced to . . . currents in an electrical circuit . . . Inside the human frame exactly the same sort of process occurs' (Bush 1939, p. 42). For Bush, an electrical engineer, electricity is a universal force. It is common to both biology and technology and it drives them both.

For Bush, it is electricity that links all the other analogies. For example, the next is nerve:wire. The nerve:wire, cable analogy, like the others, is quite explicit. Bush writes (1945, p. 108):

> We know that when the eye sees, all the consequent information is transmitted to the brain by means of electrical vibrations in the channel of the optic nerve. This is an exact analogy with the electrical vibrations which occur in the cable of the television set: they convey the picture from the photocells which see it to the radio transmitter from which it is broadcast.

nature (biology)	culture (technology)
nerve	wire, cable
brain	relays, circuits
memory trails (brain)	associative trails (Memex)
human judgement, experience	machine judgement (Memex II)

Figure 25.1 Bush's analogy set.

To aid the reader, this system of analogies will be presented as a table. Level by level, it would look much like Figure 25.1.

For all these pairs, electricity remains the common denominator. It not only joins the terms but makes it possible to move from one set to another.

The next pair Bush uses is brain:relays, circuits. As Bush wrote to Edward Weeks 'the chances are that a great deal of our brain cell activity is closely parallel to the operation of relay circuits . . . One can explore this parallelism and the possibilities of making something out of it almost indefinitely' (1944*a*, p. 1). Through relays and circuits, then, it might be possible to imitate, not replicate, basic biological processes in the brain.

With Memex, Bush takes the next step. If basic processes in the brain can be reproduced mechanically, why not its more complex ones? Here, Bush turns to memory. Memory, he argues, 'operates by association. With one item in mind it snaps instantly to the next that is suggested . . . in accordance with some intricate web of trails carried by the cells of the brain' (Bush 1939, pp. 33–4). Memex essentially mechanized this. With Memex, Bush writes, any item 'can be . . . followed by the next, instantly and automatically, wherever it might be. There is formed an associative trail through the material. It is closely analogous to the trails formed in the cells of the brain' (1944*b*, p. 5). Like all Bush's analog machines, Memex was to mirror, not imitate, memory. This is an important point. Memex was not, and perhaps not intended to be, a direct translation of a complex biological system(s) into mechanical terms. Instead it represented a transposition and borrowing from analogy to analogy. The set now reads; nerve:wire; brain:relays, circuits; memory trails(brain):associative trails (Memex).

25.4.1 THE RHETORIC SHIFTS: MEMEX II

Late in his life, Bush's rhetoric changes and he begins to describe Memex in very different terms from those which we have presented here. For example, Memex, Bush writes in his autobiography *Pieces of the Action*, 'is an extended, physical supplement for man's mind, and seeks to emulate his mind in its associative linking of items' (1970, p. 190). While Bush often describes Memex as supplementing the mind, he talks of it here as extending and emulating the mind. Prior to this, Bush had on occasion described Memex as representing the mind itself (see 1967, pp. 76–7). No longer is nature the model to imitate. Nor is the task of technology just to bridge the gaps between nature and culture. Here for the first time, Bush suggests technology might be used differently. The ends to which it will be put now, as we shall see, will distance Bush from the rhetoric he used earlier.

In an unpublished manuscript 'Memex II' (1959), Bush writes about machines, like Shannon's mouse, that learn from experience. Here, Bush discusses linking Memex (and Memexes) to other machines to improve their performances (1959, p. 24). The result would be, he feels, a Memex machine that could not only imitate, but improve upon the brain itself (Bush 1959, p. 21). This shift is reflected in the analogies he now draws; human judgement, experience:machine judgement (Memex II). With Memex II, the analog machine could imitate human memory and experience and it would be capable of judgement. For the first time, Bush equates machine and human judgement (1959, p. 24).

Bush's rhetoric has changed. Technology, it seems, should not just try to imitate (human) nature, it should also help to perfect it. What will this technology accomplish? What will be the result? With Memex II, Bush tell us (1959, pp. 20–1).

> Going beyong the extension and ordering of man's memory, it can also touch those subtle processes of the mind, its logical and rational processes, its ability to form judgements in the presence of incomplete and contradictory data, as these become facilitated by better memory . . . It . . . provides a memory which does not fade . . . This use of Memex II, in turn, remolds the trails of the user's brain, as one lives and works in close interconnection with machine . . . For the trails of the machine become duplicated in the brain of the user, vaguely as all human memory is vague, but with a concomitant emphasis by repetition, creation and discard, refinement as the cells of the brain become realigned and reconnected, better to utilize the massive explicit memory which is its servant.

Essentially Memex II would join the brain and the machine (Bush 1959, p. 3).

This in some ways is a logical outcome of Bush's faith in technology. Like many Americans, Bush strongly believed that technology could improve the human condition (Bush 1945; Nyce and Kahn 1991, p. 42). For example, by linking human and machine, Memex II could solve problems that humans alone could not. With Memex II, Bush simply extended his own rhetoric. No longer would this rhetoric and these analogies be used just to bridge the gap between nature and culture. With Memex II, Bush's project now was to merge the two and so improve them both.

25.5 CONCLUSION

While today analog machines and theory are often just subjects of historical interest, at the time of the Macy Conferences this was not the case. Neither digital computation nor the procedures it seemed to validate about how to study human and machine yet had the upper hand. In short, the paradigm within which we work and argue today about these subjects had not yet been fixed. What the analog machines held out was a possibility that for the brain or the machine to be understood, neither necessarily has to be reduced to any one set of laws or logical principles.

For Bush, analog machines gave his writing and thinking direction. Bush's analogy set is still good to think with. It not only handles key parallels, for example, nature:culture and biology:machine, it can also link them. Bush's Memex machines, like all analog machines, reproduce and 'work' through similarity and identity. With these machines, Bush attempted to bridge, if not bring together, nature and culture – for Americans opposed irreconcilable categories. Whether Bush's machines are seen as figures of speech or models to be implemented, we need better to understand Bush's rhetoric and the promises it holds out to us. This is because it offers us an opportunity better to understand how we have come to see objects and subjects, like brain and machine, as we do. Further it reminds us, as history often does, that there may be equally valid alternatives to the way we frame and define these subjects today.

ACKNOWLEDGEMENTS

This paper is based on a SIGBIO paper (Nyce 1992), a paper presented at the 1991 meeting of the American Society for Cybernetics in Amherst, Massachusetts and a lecture given at the Department of Computer and Information Science, Linköping University in 1989. I would like to thank Gail Bader, Geoffrey Bowker, Steve Heims, Roland Hjerppe, Magnus Johansson, Paul Kahn, Jonas Löwgren, Ray Paton and Toomas Timpka, MD for their comments on drafts of the paper.

REFERENCES

Aspray, W. (ed.) (1990). Epilog. In *Computing Before Computers*. Iowa State University Press, Ames, IA, pp. 251–6.

Bromley, A. G. (1990). Analog computing devices. In *Computing Before Computers* (ed. W. Aspray). Iowa State University Press, Ames, IA, pp. 156–99.

Burke, C. (1991). A practical view of Memex: the career of the Rapid Selector. In *From Memex to Hypertext: Vannevar Bush and the Mind's Machine* (ed. J. Nyce and P. Kahn). Academic Press, Boston, MA, pp. 145–64.

Burks, A. R. and Burks, A. W. (1988). *The First Electronic Computer: The Atanasoff Story*. University of Michigan Press, Ann Arbor, MI.

Bush, V. (1939). Mechanization and the Record. Bush Papers, Library of Congress, typescript.

Bush, V. (1944*a*). Letter to Edward Weeks (6 November). Bush Papers, Library of Congress.

Bush, V. (1944*b*). Letter to Edward Weeks (20 November). Bush Papers, Library of Congress.

Bush, V. (1945). As we may think. *Atlantic Monthly* **176**, 101–8.

Bush, V. (1959). Memex II. Vannevar Bush Papers, MIT Institute Archives and Special Collections, typescript.

Bush, V. (ed.) (1967). Memex revisited. In *Science Is Not Enough*. William Morrow, New York, pp. 75–101.

Bush, V. (1970). *Pieces of the Action*. William Morrow, New York.

Conrad, M. (1989). The brain-machine disanalogy. *Biosystems* **22**, 197–213.

Coulter Jr, N. A. (1974). Mind: the software of the brain. *ASC Forum* **6**, 10–12.

Edwards, P. N. (1985). *Technologies of the Mind: Computers, Power, Psychology and World War II*. Silicon Valley Research Group Working Paper, No. 2, University of California, Santa Cruz.

Gerard, R. W. (1951). Some of the problems concerning digital notions in the central nervous system. In *Transactions – Conference on Cybernetics*, Vol. 7 (ed. H. von Foerster). Josiah Macy, Jr Foundation, New York, pp. 11–52.

Hackmann, W. D. (1989). Scientific instruments: models of brass and aids to discovery. In *The Uses of Experiment: Studies in the Natural Sciences* (ed. D. Gooding, T. Pinch and S. Schaffer). Cambridge University Press, pp. 31–65.

Hartley, M. G. (1962). *An Introduction to Electronic Analogue Computers*. Methuen, London.

Heims, S. J. (1975). Encounter of behavioral sciences with new machine-organism analogies in the 1940s. *J. Hist. Behav. Sci.* **11**, 368–73.

Heims, S. J. (1977). Gregory Bateson and the mathematicians: from interdisciplinary interaction to societal functions. *J. Hist. Behav. Sci.* **13**, 141–59.

Heims, S. J. (1980). *von Neumann, J. and Wiener, N.: From Mathematics to the Technologies of Life and Death*. MIT Press, Cambridge, MA.

Heims, S. J. (1991). *The Cybernetics Group*. MIT Press, Cambridge, MA.

Knorr-Cetina, K. D. (1981). *The Manufacture of Knowledge*. Pergamon, New York.

Latour, B. and Woolgar, S. (1979). *Laboratory Life: The Social Construction of Facts*. Sage, Beverly Hills, CA.

Lienhardt, G. (1961). *Divinity and Experience: The Religion of the Dinka*. Oxford University Press, Oxford.

Lynch, M. (1985). *Art and Artifact in Laboratory Science*. Routledge, London.

McCulloch, W. S. (1974). Recollections of the many sources of cybernetics. *ASC Forum* **6**, 5–16.

Mills, C. W. (1940). Situated actions and vocabularies of motive. *American Sociological Review* **5**, 904–13.

Nyce, J. (1992). Analogy or identity: brain and machine at the Macy Conferences on cybernetics. *ACM SIGBIO* **12**, 32–7.

Nyce, J. and Kahn, P. (eds.) (1991). A machine for the mind: Vannevar Bush's Memex. In *From Memex to Hypertext: Vannevar Bush and the Mind's Machine*. Academic Press, Boston, MA, pp. 38–66.

Owens, L. (1991). Vannevar Bush and the differential analyzer: the text and context of an early computer. In *From Memex to Hypertext: Vannevar Bush and the Mind's Machine* (ed. J. Nyce and P. Kahn). Academic Press, Boston, MA, pp. 3–38.

Paton, R. C. (1992). Towards a metaphorical biology. *Biology and Philosophy*, **7**, 279–94.

Pepper, S. (1942). *World Hypotheses*. University of California Press, Berkeley, CA.

Shurkin, J. (1984). *Engines of the Mind: A History of the Computer*. Norton, New York.

Smillie, K. W. (1992). Review of inventing accuracy: a historical sociology of nuclear missile guidance, by Donald MacKenzie (MIT, 1990). *Annals of the History of Computing* **14**, 69–70.

Smith, T. M. (1989). Some perspectives on the early history of computers. In *Perspectives on the Computer Revolution*, 2nd edn (ed. Z. Pylyshyn and L. Bannon). Ablex, Norwood, NJ, pp. 5–12.

Tambiah, S. J. (1969). Animals are good to think and good to prohibit. *Ethnology* **8**, 424–59.

von Neumann, J. (1958). *The Computer and the Brain*. Yale University Press, New Haven, CT.

West, D. R. and Travis, L. E. (1991*a*). From society to landscape: alternative metaphors for artificial intelligence. *AI Magazine*, Summer, 69–83.

West, D. R. and Travis, L. E. (1991*b*). The computational metaphor and artificial intelligence: a reflective examination of a theoretical falsehood. *AI Magazine*, Spring, 64–79.

COMPUTING WITH BIOLOGICAL METAPHORS – SOME CONCEPTUAL ISSUES

26

Ray Paton

26.1 INTRODUCTION

This book is about the two-way transfer of ideas between biology and computing. Hopefully it has been demonstrated that biological models can be the inspiration for engineering designs (e.g., Holcombe, Winter, Arbib, Bäck, Koza, Horswill). Computational thinking is also invaluable for modelling biological systems (e.g., Thomas and Thieffry, Marijuan, Vertosick). This is not to say that biological and computational systems should in any way be treated as the same ontological category but rather that in a number of ways they share common properties.

Many biologically inspired ideas have been discussed in this book such as biologically motivated neural computing (Arbib, Carpenter and Grossberg), computational ecologies (Huberman), enzyme repair analogues in evolutionary algorithms (Bäck), interplay of genetic and socio-ecological metaphors (e.g., Koza, Bergman, Davidor), selectionist systems in genetics, immune systems and brains (Manderick, Bersini and Varela) and the application of genetic operators for simulating dendritic growth (Hamilton). Each of these (and many other) topics are fascinating on their own. However, the picture becomes even more exciting when a synthesis is attempted across topics. Some readers may have done this already, but consider a simple example to demonstrate the point. Space, time and computation are very important for Zajicek's cellular automata and also feature in Davidor's ECO framework and in Welch's field metaphor. Here we see the emergence of common ideas from cells to ecosystems and the potential for transferring ideas across even these three chapters is considerable.

In order to probe aspects of this synthesis further a number of topics will now be explored, namely, the nature of the models we construct, the need to accept the pluralistic nature of some aspects of the subject and a detailed look at an alternative biological source.

26.2 WORKING WITH CONCEPTS – MODELS AND METAPHORS

At a general level we may describe a model as the representation of one thing (either verbal, symbolic, mathematical or physical) in terms of something else. A plastic skeleton which a student of anatomy could use for study purposes may be a very good replica but it is still a model. In this case the source of the model (a bony skeleton) and the subject (a plastic replica) are very similar. Often, however, sources and subjects can be quite different. For example, some very good models of the mammalian blood circulatory system are based on the notion of a closed hydrodynamic circuit. The modelling language is used to describe the heart as a pump, the aortic arch as a compression chamber, the arteries and arteriole as pre-capillary resistance vessels and so forth. It is also interesting to note that aspects of blood system functionality provided the metaphorical source for computational systems called systolic arrays.

As we see from this book, a large biocomputational vocabulary has emerged which overlaps both the biological and the computational sciences. It includes such terms as code, cognition, automata, self-organization, competition and cooperation, ecology, emergence, environment, evolution, fault tolerance, parallel distributed processing, pattern recognition and selection. There are a number of ways of managing this cross-disciplinary vocabulary. For example, Farmer (1990) introduces 'the Rosetta Stone of connectionism' in order to demonstrate common ideas between neural, immune and autocatalytic networks. In contrast, with this mathematical approach, we shall explore common ground looking at a set of pervasive metaphors connected with systemic properties which can be displaced between domains. The nature and importance of these are investigated more fully in Paton (1992, 1993).

Current thinking about systems has developed because it rightly applies to the modelling requirements of complex real world situations. For the purpose of analysing systemic metaphors, each system has a number of general properties: interacting parts, organization and whole-system (collective) behaviour. These three properties can be used to describe a set of systemic metaphors which include machine, organism, society, text and circuit (Paton 1992). Each systemic metaphor has a number of associated properties which, depending on context, may or may not be unique to the particular system being described. The philosophical basis underlying this approach can be found in Paton *et al.* (in press). Figure 26.1 shows some properties associated with particular systemic metaphors.

The three biological organizational levels on the left of Figure 26.1 share common features associated with each of the five systemic metaphors. For example, in attempting to understand the openness of a cell we may wish to apply ecosystem thinking. In so doing, a number of society-based properties may be displaced to the subject (cell) domain such as openness, zonation and spatial arrangement, history (succession in ecological

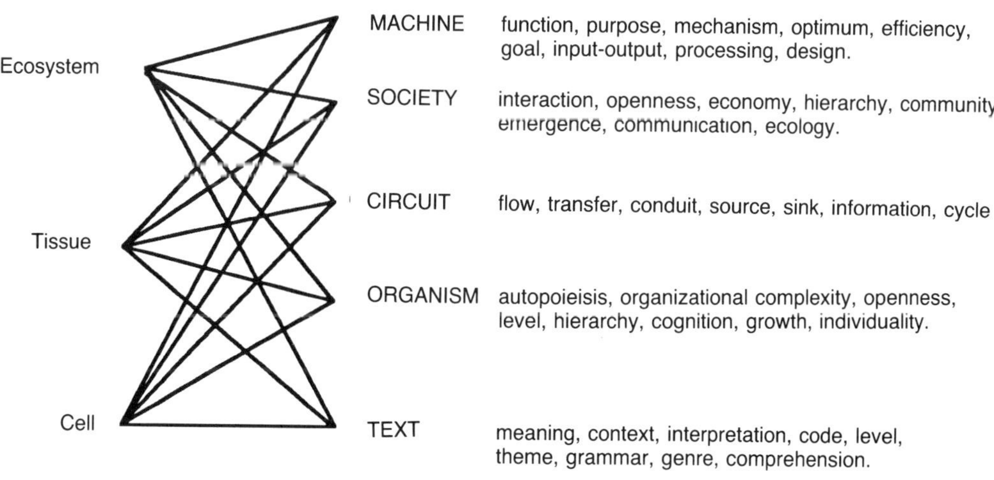

Figure 26.1 Relations between organizational levels and systemic metaphors.

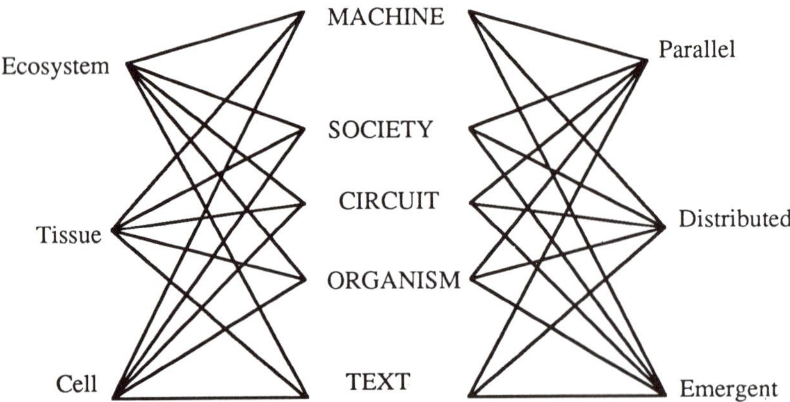

Figure 26.2 Interaction of biological and computational ideas through systemic metaphors.

terms), regime and economy. Alternately, in seeking to appreciate the contextual relations between subcellular organizational levels we may wish to take account of text and the part-whole relations in a tissue. This will focus attention on issues such as the emergence of meaning and the nature of information processing in a hierarchy. The common ideas shared across different domains reflects the common ontology which these general metaphors can provide (Harré 1986; Paton 1992).

This analytical scheme can now be extended to include computational ideas. Figure 26.2 includes some very general approaches to computation (at the right of the figure) and follows the categorizations of Forrest (1990). Let us consider how this scheme can help us appreciate the displacement of a number of concepts in distributed or loosely coupled systems from chapters in this book. For example, Staniford makes use of society and organismic thinking to develop a model of how a set of interacting agents would be organized in a distributed AI system and we find a hierarchical organization (prescribed rather than emerging) and a clear division of labour. On the other hand, Stark makes use of machine and circuit thinking to provide a view of an artificial tissue as a society with important implications for communication.

We now apply the scheme summarized in Figure 26.2 to anticipate some possible future developments in artificial neural networks. Over the past 40 years the biological neuron has variously been modelled as a switch, transistor, multiplexer and microprocessor. As computer technology has progressed so the biological analogue has become more sophisticated. This kind of thinking illustrates how computing has inspired neurobiology. We may also consider the way neurobiology can inspire computing. The chapter by Carpenter and Grossberg provides a fascinating example with the synapse-inspired artificial system called ART 3, and Arbib looks at how the modulatory effects of hormones could be applied to artificial neural networks in his chapter. It can be argued that each organizational level of a biological neuron should be modelled as a parallel distributed processing unit. This should not come as a surprise. For example, Shepherd (1990) has argued that the basic unit of neuronal computation should be the synapse. Indeed, he looks at processing at a number of neuronal organizational levels. For the purposes of applying biology to computing, the result of such an approach would be a set of nested systems each subsuming all the complexity of underlying levels. This could be achieved,

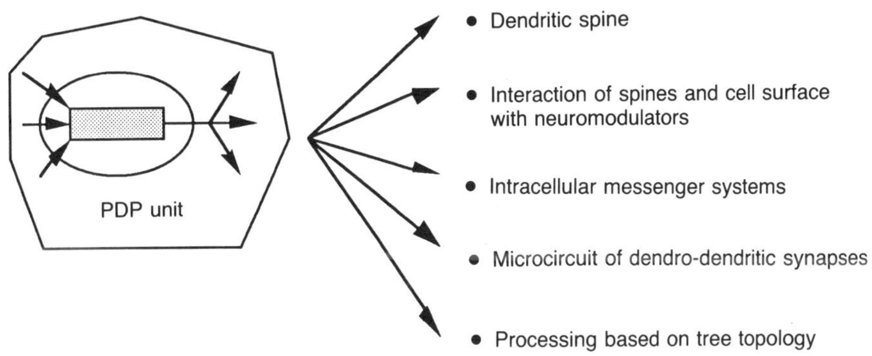

Figure 26.3 Dendritic tree as a nested hierarchy of PDP units.

for example, by producing more complicated transfer functions.

An appeal to greater analogous processing mechanisms would also be required. Figure 26.3 summarizes some of the ways in which artificial neural network processing unit functionality could be extended by nesting functionality at increasing levels of dendritic organization. We return to issues connected with hierarchies in the next two sections.

26.3 SOME CONCEPTS ARE TOO BIG – THE CASE OF BIOLOGICAL INFORMATION

It should be clear from reading this book that biological information has been a common and important term (see, for example, the chapters by Kampis, Holcombe, Marijuan, Stark and Manderick). The purpose of this section is to bring together a number of key thoughts and demonstrate that, at present, a pluralistic account of the concept is required if its richness is to be appreciated. In saying this, it is important to acknowledge that we are dealing with a general concept of information rather than the ideas which are specific to information theory (Shannon 1948). Although information has a technical meaning in communication theory we are faced with a problem that its meaning and use in biology and computing is much bigger. The application of the technical (Shannonian) idea of information brings with it the need to measure uncertainty and associate with it ideas from thermodynamics. These ideas have been applied to mechanisms of biomolecular computation (Schneider 1991) and enable researchers to anticipate the way a subcellular machine is controlled and how input–output relations can be formulated into predictions.

Information theory provides a rigorous mathematical system for reducing certain biological phenomena to the processing of binary information. This is acceptable if we acknowledge limits to the computational metaphor and that Shannon's information theory has a purely syntactic nature (see also, Rosen (1988)). A number of questions then appear. For example, can all biological phenomena be reduced to some kind of syntactic theory? At a deeper level, it is pertinent to ask whether a reductionist account can always provide the necessary and sufficient conditions by which predictive and explanatory hypotheses about the nature of systems can be made. Consider the following example. Kincaid (1990) seeks to demonstrate that the signal hypothesis for intracellular protein transport could not be accounted for (i.e., reduced to a syntactic form alone) without an understanding of the context in which a code (protein sequence) is executed. The meaning of the code (a kind of sentence) required the cellular context (a kind of text). We examine semantic issues associated with information below. There are inadequacies with the concept of cellular information based solely on thermodynamic models, which talks in terms of negative entropy and physical constraints (see Kampis's chapter and Kampis (1992)). A greater descriptive account is needed drawing on complementary metaphors.

The idea of a text metaphor for articulating biosystem organization has been seen in recent times in the writings of Polanyi (1968) and Pattee (1977). Both writers seek to demonstrate that biosystems require two descriptive modalities one related to biosystem-as-machine and the other to biosystem-as-text. For Polanyi this enabled him to demonstrate the need for appreciating boundary conditions which demarcate one level of organization from another. The text metaphor provided Pattee with a non-reductionist means of accounting for certain autonomous properties of the system which he describes as its linguistic mode (see comment on Varela later in this section).

A common notion which appears in the description of biological information based on the text metaphor is the idea of 'glue'. For example, Albrecht-Buehler (1990) assesses the importance of the glue concept in his analysis of the nature of cellular information. In this case he defines cellular information as the glue which holds the cell together. Using the notion of text he considers whether the information is in the text of the cell or in the letters (note also the earlier comment on Kincaid's argument). Clearly the metaphor is rather loose in this form. When a cell is decomposed into molecular letters its meaning is destroyed. In this case cellular information is not only context-dependent (i.e., the cell provides context for the meaning of its parts) but it is also non-reductive (i.e., meaning is lost when the whole is decomposed into parts). Another example from a very different area of biology is that of Eldredge (1985) who uses the glue metaphor to describe cohesion between the replicator and interactor levels in their respective hierarchies. Replicators are entities which pass on their structure directly in replication and their hierarchy is: germ line → organisms → demes → species → monophyletic taxa. Interactors are entities that directly interact as cohesive wholes with their environment in such a way that replication is differential. Their hierarchy is: soma → organisms → avatars → ecosystems → biosphere. The replicator hierarchy is a hierarchy of information transfer and the glue which unites the subparts to give a higher level is the 'more making' of the lower level. Further articulation of information transfer as glue in this hierarchy could be of value in the design and identification of emergent processes and hierarchical relations in evolutionary computation, for example, ecosystem effects (interactors' hierarchy) in relation to deme → species emergence (see chapters by Collins, Sumida and Hamilton, Koza, Davidor).

The notion of information as text has a close association with the idea of information as form. Historically we can see a conceptual bridge between the two in the classical work of D'Arcy Thompson (1942) who notes how 'the form of an object is a "diagram of forces"' (p. 16) and that the living organism 'represents or occupies a field of force' (p. 30). Welch applies similar thinking in his chapter on the application of the field construct to cell biophysics. It is also important to appreciate that the spatio-temporal partitioning of biosystems, from cells to ecosystems, is a major aspect of their computational capacity and it is instructive to note how spatial factors are becoming increasingly important with respect to understanding parallel and distributed computation (Feldman and Shapiro 1988). From the point of view of topology, Thom (1975) sought to equate information as form. For him information is a topological complexity for which the scalar measures of entropy and energy in thermodynamics should be interpreted. We see this kind of idea applied by both biologists and computer scientists. Young (1978), following Thom, proposes that the whole organism is a coded representation of its environment and Jefferson (1991), from studies of simulated evolution, suggests that certain aspects of an animal's environment become encoded into an animal's genome. In these two cases the scientist acts as an interpreter of biological reality, a reality conceived in holistic terms with respect to form. It is also crucial to note that biosystems themselves have an interpretative capacity. Thus, Brandon (1990) describes a phytometer as a plant which is used as a measuring instrument or interpreter of an environment. These views on the relation between information and form raise the question of representation. In the example of Young and Jefferson the system and its environment require the scientist to act as interpreter whereas Brandon's phytometer is its own interpreter. So how does this help our appreciation of bio-information? Not least, it can help to clarify the bootstrap problem described in Manderick's chapter and help to provide a modelling framework for some of the system-environment issues raised by Horswill. We shall see that when a system is described in cognitive terms there is a necessary condition in which it generates meaning itself.

The value of the glue metaphor (see above) in understanding information can be related to a class of objects in mathematical logic called functors (Curry 1977). In the language of mathematical linguistics, these functors bind objects through relations as in the case when verbs bind noun phrases. Thus, it should come as no great surprise that when Barwise (1984) develops a theory of situation semantics, verbs are the glue which hold together nouns and other parts of speech. This aspect of meaning as applied to verbs, as applied to bio-informational glue, is invaluable, not so much in formulating testable hypotheses about biosystem information processing but rather in pointing to one way forward in promoting our understanding of the relational nature of biological information. In describing this dimension we need go no further than reflect on the important contributions made by Rashevsky (1973) and Rosen (1973). Rashevsky introduced the principle of relational invariance to describe sets of commonly occurring invariances among living things. Following on from the discussion about glue, we may consider the role of verbs as functors to be similar to that of invariant relations in abstract models of biosystems. For example, the relation locomotion → catching prey → ingestion is invariant across many predatory species. For Rosen, these and other invariant modelling relations are subsumed within a category theoretic framework and Rosen (1988) describes the idea of an informational framework to take account of the emergent, primarily semantic nature of bio-information and this leads us to consider its cognitive nature.

The cognitive capacities of biosystems have been developed by a number of theoreticians such as Varela (1979) and Kampis (1992). Of great importance to their work is the relation between the biological system and its environment. In order to demonstrate some basic ideas, consider the following example from physics. It concerns the interaction between the gain medium and the radiation field in a laser system. According to Atmanspacher and Weinberger (1990) this system exhibits cognitive properties, as the two components influence each other. The self-organized appearance of complexity in this system is associated with a self-generation (construction) of meaning (in-formation). We may also note another striking feature about viewpoints on such a system. A photon ages internally (intrinsically) and externally. As such there is a perspective on time in terms of the internal structure of the photon (the endosystem view) and there is also a external perspective on the photon related to dynamical behaviour (the exosystem view). It will not be possible to discuss endo/exo system viewpoints in detail although it should be clear even from this inanimate analogy that the self-construction of meaning is a cognitive capacity of many systems (see also, Kampis (1992) and his chapter in this book).

According to Varela (1979) biological systems are autonomous devices because their emergent behaviours and internal self-organizing processes define what counts as relevant interactions. This autopoietic description provides a contrast with machine thinking associated with control and the transformation of inputs into outputs. Indeed, Varela draws the important distinction between a heteronomous device which is defined according to a set of instructions and the related control mechanisms acting on it and an autonomous device which is defined according to its internal self-organizing processes, emergent behaviours and operational closure. Varela points out that information in biological systems is not simply related to the way behaviour can be adequately represented independent of the systems structure (what he calls the representational view of information), it is also about the way in which the system constructs information, namely, the instructional view. (Further aspects of Varela's approach can be found in the chapters by Bersini and Varela and by Manderick.)

A further dimension related to the cognitive and textual dimensions could be called the sociology of information. Emmeche and Hoffmeyer (1991) examined several ways in which biological information can be described using a basic metaphorical assumption of nature-as-language. For them, information as

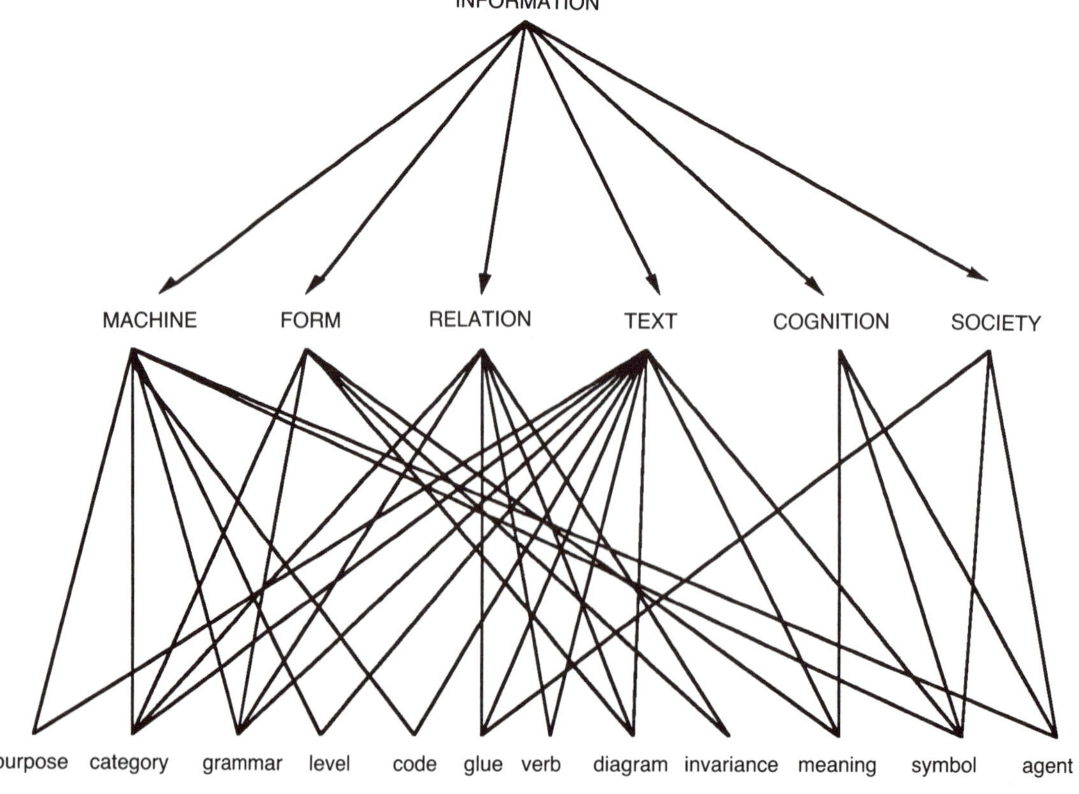

Figure 26.4 Alternative attributes of information.

articulated within the mathematical theory of information is a category which is much less comprehensive than information exchanged between people. They argue that biological information must be understood as embracing the semantic openness that is characteristic of information exchange in human communication. Consequently, information is inseparable from a subject to whom the information makes sense, so that, for example, the meaning of the genetic message is only discernible in the concrete process of development. This context dependency is particularly noticeable when it comes to identifying and modelling the different kinds of currencies operating in the economics of ecosystems (Mansson and McGlade, in press).

Figure 26.4 summarizes some features related to the multidimesional nature of the concept of information as it can be applied to biological systems. In this case a non-exhaustive list of six information metaphors (machine, form, etc.) is given for which the relation can be described as 'information_as'. Beneath this list is another collection of attributes which may be associated with the metaphors in various ways. It should be clear from this that many ways are available for describing information. This can promote dialogue between scientists but unfortunately it may also raise intra- and interdisciplinary barriers, hence the need to accept a pluralistic approach.

26.4 EXPLOITING SYSTEMIC PROPERTIES – TISSUE-LEVEL COMPUTATION

The purpose of this section is to draw together a number of valuable ideas about tissue-level computation based on new source models. Most computational ideas associated with the tissue/organ level of biological organization deal with networks of processors or agents such as artificial neural networks, immune networks and parallel classifier systems. The choice of these systems reflects their ability to adapt, their high degree of parallelism and their common mathematical properties (Farmer 1990). A number of chapters in this book cover various aspects of these approaches including Arbib, Carpenter and Grossberg, Vertosick, Bersini and Varela, and Manderick. Some other chapters address aspects of tissue computation not usually discussed including Stark and Zajicek.

We shall consider an idealized tissue to be a set of physically connected cells of the same structural type together with a blood supply which allows materials (data) to be transported to and from the cells. In computational terms both cells and blood are computing devices. (The reader should be reminded that some of the information-processing capacities of blood are discussed in the chapters by Vertosick, Bersini and Varela, and Manderick.)

Figure 26.5 presents an idealized representation of a simple computational tissue. Such components as basement membrane and capillary endothelium are omitted. Let us consider this tissue to consist of a string of cells with blood flowing in one direction at right angles to one side of this string. The two-way exchange of materials (data) between cell and adjacent blood depends on a number of processes including: diffusion of materials, active transport, receptor binding and signal transduction. Some of the data which are exchanged between cell and blood will be of an instructional nature, for example, primary messengers, the biological chemicals which are transferred between cells via body fluids such as blood. Other data will include the substrates of intermediary metabolism, ionic species and trace elements.

The variety of primary messengers includes hormones, circulating synaptic transmitters, cytokines, chemotactic factors and metabolic products. Two ways primary messenger analogues could act as sources for computational systems are in modulation and priming. The idea of a connection weight not only applies to artificial neural networks but

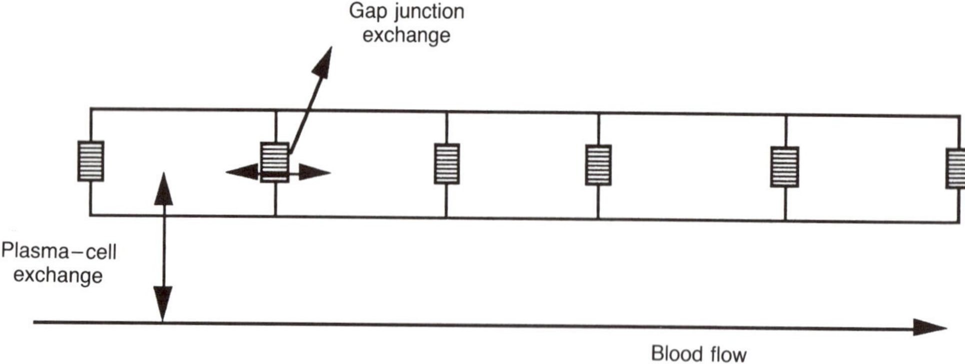

Figure 26.5 An idealized representation of a computational tissue.

also to immune networks. However, certain limitations are imposed by weighted connections in artificial neural networks. For example, they do not take account of symmetric connections in some circuits (such as dendro-dendritic and axo-axonic two-way synapses) and they do not involve neuromodulators and other chemical messengers. Artificial neural networks have hardly begun to exploit many of these features as Arbib makes clear in this chapter. An example of non-hormonal modulation can be seen with the activity of lymphokines which are a group of intercellular signalling molecules of particular importance to the immune system. Vertosick discusses some implications for the operation of these substances in PDP network models in his chapter.

The priming action of hormones is discussed by Brooks (1991) who describes a hormone-motivated approach to the behaviour-based integration in robot design and the coordination of activated and deactivated behavioural schemes. For Brooks, the hormone system provides a 'global repository of state' in the robot such that the system can be primed to enable appropriate behaviours to become active. Behavioural building blocks were specified which allow the functionality of the robot to then emerge. This kind of hormonal system, based on 'releasers', seems to have been particularly useful in facilitating the coordination of the behavioural building blocks.

In addition to cell–blood–cell communication, there is also cell–cell communication via gap junctions. Gap junctions (plasmodesmata in plants) are regions of the plasma membrane which provide cytoplasmic continuity between adjacent (touching) cells. A group of cells coupled together in this way forms a compartment within which small ions and molecules are freely exchanged (MacDonald 1985). This direct intercellular coupling allows for the exchange of metabolites and the transfer of secondary messenger control signals such as cyclic AMP and Ca^{2+} resulting in such activities as metabolic cooperation, coordination of tissue activities and synchronization of cellular behaviour. Stark's chapter elaborates a distributed information processing model of an idealized tissue based on gap junctional integration. We shall now look at examples of computational tissue units which make use of primary messenger and gap junctional communication based on the liver (see also Paton *et al.* 1992).

The reasons for choosing this multifunctional structure are many and a number of computational features at various levels of organization can be described including: parallel organization of processing units,

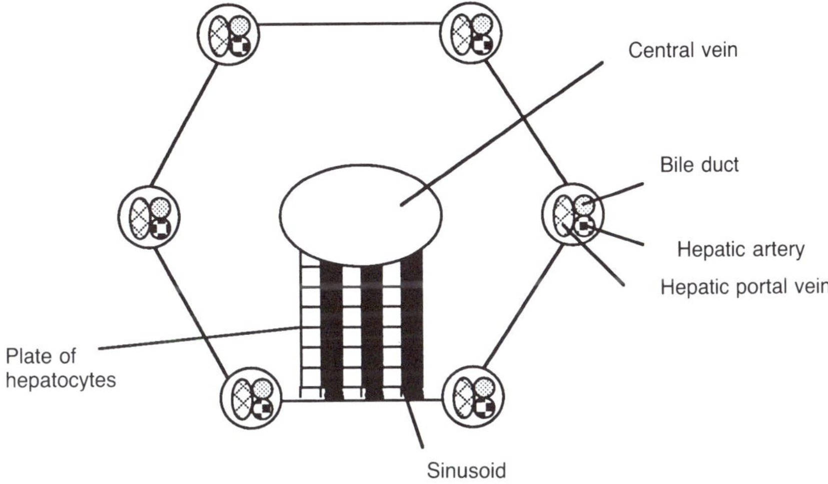

Figure 26.6 Highly simplified cross-section through a lobule.

asynchronous processing, anonymous communication, partial global control and partial local control, distributed processing, local interactions, fault tolerance, demand-driven processing, pattern recognition, self-organization, data flow and pipelining. The liver has a striking internal construction, consisting of about one million polyhedral prismatic lobules which are approximately 2 mm long by 0.7 mm (see Figure 26.6).

It has two blood supplies one rich in oxygen and the other rich in materials absorbed from the gut. The architecture of the hepatic vascular tree provides a very effective mixing so that blood entering any one lobule will be constitutionally very similar to blood entering any other. Blood flows along the sinusoid from the portal end (the periphery) of the lobule to the venous (central vein) end. As blood flows along the sinusoid, its chemical composition changes. This is due to the action of the hepatocytes (and other cells not considered here) lining the sinusoid. Indeed, the cells at the portal end of the sinusoid are in an environment which can be richer in oxygen, glucose, lipids, amino acids, toxins and hormones whereas the cells at the venous end will be poorer in oxygen and richer in carbon dioxide and certain metabolic waste products. As a result of this, cells are adapted to the particular microenvironment in which they are found and there is a striking division of labour, often referred to as metabolic zonation. Periportal cells are involved in bile formation, glucose mobilization and catabolism of fatty acids and amino acids. Perivenous cells are involved in glucose utilization, ketogenesis and detoxification of xenobiotics.

The idealized liver functional unit we shall consider in some detail is a two-dimensional sheet of cells (called a plate or lamina) which is one cell thick. Either side of this sheet is a space called a sinusoid which is filled with blood as shown in Figure 26.7. The laminae are arranged radially around a central vein as shown in Figure 26.6.

A number of other organizational levels can be described beginning with the basic liver cell, the hepatocyte. The next level in the functional hierarchy after the hepatocyte is a compartment consisting of four cells: hepatocyte, Ito cell, Kupffer cell and sinusoid lining cell. This quartet makes up a functional unit (Dioguardi 1989) and is the basic unit of the recognized physiological functional unit, the acinus (Paton *et al.* 1992). Figure 26.8

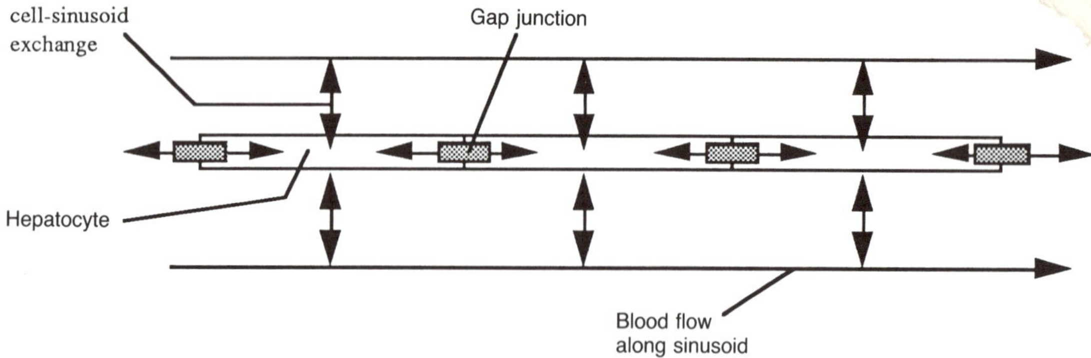

Figure 26.7 Schematic cross-sectional model of part of a hepatocyte plate. (Note: blood flow is left to right, that is, from periportal to perivenous zone.)

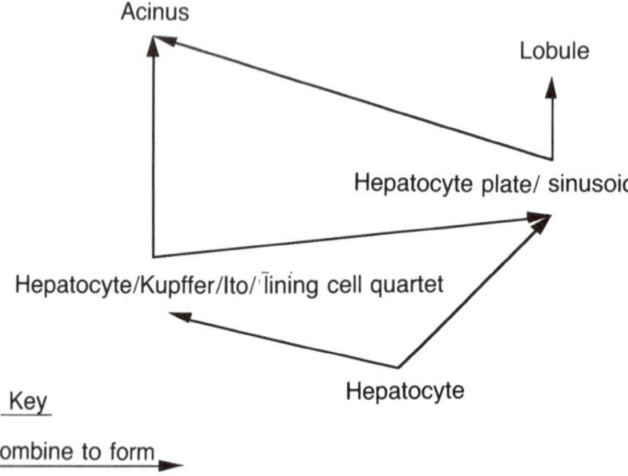

Figure 26.8 Some component interrelations in the lobule.

summarizes some of the hierarchical interrelationships between hepatic processing units.

Following Ward and Halstead's (1990) extension to Flynn's well-known taxonomy of microprocessors we may identify a number of features of the computational liver which exhibit MIMD (Multiple Instruction Stream, Multiple Data Stream) or NIMD (No Instruction Stream, Multiple Data Stream) machines. Consider the hepatocyte. In computational terms this cell type approximates to a MIMD machine (or complex of nested MIMD machines) with respect to its input–output relations in the operating sinusoid and also to a NIMD machine (or complex of nested NIMD machines) with respect to its internal organization. Its NIMD capabilities can be seen, for example, in the way it acts as a complex pattern recognizer of input patterns including hormones, metabolites, nerve impulses and oxygen as well as the integrating effects of gap junction cooperation. The hepatocyte is not only like a machine (Holcombe; Thomas and Thieffry) it is also like a society (Welch and, by displacement, Stark), an ecology in Huberman's terms and a cognitive system from Kampis's point of view. The many loosely coupled processes in the cell contribute to its multifunctionality, pattern-recognition capability and distributed memory. These processes, which are interrelated in numerous ways include amplification of genes, gene regulation of enzyme levels, distribution of membrane receptors, pathway regulation, action of secondary messengers, intracellular signalling system, organization of the trabeculum, compartmentation and pooling.

The sinusoid/hepatocyte plate unit acts in many ways like a MIMD machine but also as a somewhat loosely coupled system. Control is partly global, in the case of primary messengers and nerve impulses, and also local in the case of gap junctional communication. The advantages of this kind of multimodal approach to control, memory and communication require further investigation. However, the role of the external factors should not be seen in a simplistic sense as global control because hepatocytes in one zone can transform some hormones so that their effect downstream is considerably different. Partial computation takes place when some periportal transformations are completed in the perivenous zone. For example, thyroxine is transformed by periportal cells to the more potent tri-iodothyronine (T3). Not only this, some processing upstream can be reversed downstream. For example, glucose can be mobilized in periportal cells only to be converted back to glycogen in perivenous cells. Marking cells for particular roles is partly related to the set of membrane receptors which they produce – again this seems to be a zone-dependent feature. The configuration of the receptor set is an ecological issue as it changes depending on the microenvironment of the cell. The implications for this are yet to be explored although there seems to be a valid case for marking separate processors with a differential set of 'receptors' within a multiprocessor system. Figure 26.9 summarizes some of these processes.

The high degree of communication is achieved through local intracellular broadcasts via gap junctions or even by gaseous diffusion (should the role of nitric oxide in the liver be clarified). Multimodal communication provides a processing unit with the ability to deal with asynchronous changes and anonymous messages. One could argue that each cell has the same basic algorithm (genome) which is differentially expressed within the plate according to spatio-temporal location and the zonal microenvironment. The MIMD qualities that can be described using this comparison deal with the differential expression of the same code according to the demands placed on individual processing units.

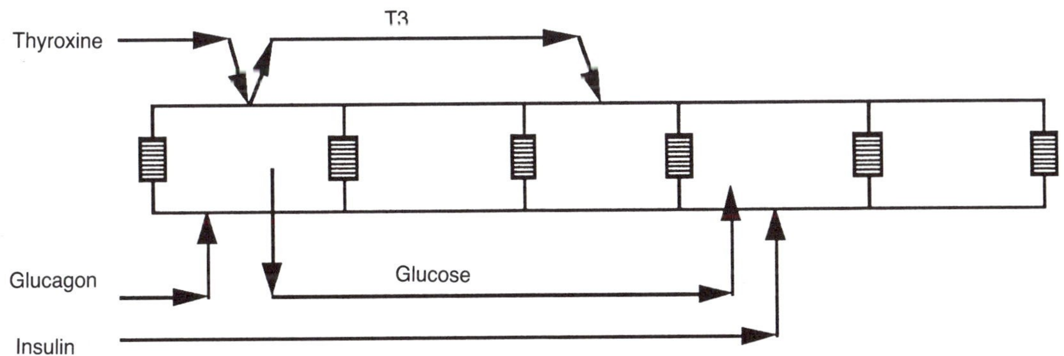

Figure 26.9 Some downstream effects and partial computations.

26.5 FINAL COMMENT – A PLEA FOR DIALOGUE

In concluding this chapter and this book I wish to elaborate on one issue namely, the need for dialogue. This is pertinent not only for cross-disciplinary encounters but specializations within disciplines can also make it difficult to exchange ideas. It is the hope of the writer that the reader will have become convinced from reading this book that dialogue can be a very fruitful activity, even if the reader needs to take a 'foreign discipline phrase book' and risk the chance of sometimes misunderstanding colleagues and being misunderstood. If scientific communities are even a little like ecosystems, then fitness, competition and cooperation should be expected; so should new niches and invader and colonizer as well as stabilizer species!

It seems to me that we may look at two approaches to a multidisciplinary basis to biocomputing. Firstly we may seek a common language or secondly we may entertain a plurality of views. The first approach is likely to have problems associated with reduction and over-simplification and the second with fuzzy concepts, unclear hypotheses and over-complication. So maybe, at present, we need both. This book gives both!

REFERENCES

Albrecht-Buehler, G. (1990). In defense of 'non-molecular biology'. *International Review of Cytology* **120**, 191–241.

Atmanspacher, H. and Weinberger, E. D. (1990). Dualities, context and meaning. *Proceedings of NATO ASI on Information Dynamics*. Irsee, FRG, June.

Barwise, J. (1984). *The Situation in Logic*. CSLI Report, Stanford University, CA.

Brandon, R. (1990). *Adaptation and Environment*. Princeton University Press, NJ.

Brooks, R. A. (1991). Integrated systems based on behaviours. *SIGART Bulletin* **2**(4), 46–9.

Curry, H. B. (1977). *Foundations of Mathematical Logic*. Dover, New York.

Dioguardi, N. (1989). The liver as a self-organising system I. *Res. Clin. Lab.* **19**, 281–99.

Eldredge, N. (1985). *Unfinished Synthesis. Biological Hierarchies and Modern Evolutionary Thought*. Oxford University Press.

Emmeche, C. and Hoffmeyer, J. (1991). From language to nature: the semiotic metaphor in biology. *Semiotica* **84**(1/2), 1–42.

Farmer, J. D. (1990). A Rosetta stone for connectionism. *Physica D* **42**, 153–87.

Feldman, Y. and Shapiro, E. (1988). Spatial Machines: Towards a More Realistic Approach to Parallel Computation. Unpublished report TR CS88-05, Department of Applied Mathematics and Computer Science, The Weizmann Institute of Science, Israel.

Forrest, S. (1990). Emergent computation: self-organizing, collective and cooperative phenomena in natural and artificial computing systems. *Physica D* **42**, 1–11.

Harré, R. (1986). *Varieties of Realism*. Blackwell, Oxford.

Jefferson, D. R. (1991). Evolution as a theme in artificial life. From abstract of a paper presented to the American Mathematical Society workshop on Biologically Motivated Computing, Florida, March.

Kampis, G. (1992). Process, information and the creation of systems. In *Evolution of Information Processing* (ed. K. Haefner). Springer, Berlin, pp. 83–102.

Kincaid, H. (1990). Molecular biology and the unity of science. *Philosophy of Science*, pp. 575–93.

Levins, R. (1984). The strategy of model building in population biology. In *Conceptual Issues in Evolutionary Biology* (ed. E. Sober). Bradford/MIT Press, Cambridge, MA, pp. 18–27.

MacDonald, C. (1985). Gap junctions and cell–cell communication. *Essays in Biochemistry* **21**, 86–118.

Mansson, B. A. and McGlade, J. M. (in press). Ecology, thermodynamics and H. T. Odum's conjectures. *Oecologia*.

Paton, R. C. (1992). Towards a metaphorical biology. *Biology and Philosophy* **7**, 279–94.

Paton, R. C. (1993). Some computational models at the cellular level. *BioSystems* **29**, 63–75.

Paton, R. C., Nwana, H. S., Shave, M. J. R. and Bench-Capon, T. J. M. (1992). Computing at the tissue/organ level (with particular reference to the liver). In *Towards a Practice of Autonomous Systems* (ed. F. J. Varela and P. Bourgine). Bradford/MIT Press, Cambridge, MA, pp. 411–20.

Paton, R. C., Nwana, H. S., Shave, M. J. R. and Bench-Capon, T. J. M. (in press). An examination of some metaphorical contexts for biologically motivated computing. *British Journal for the Philosophy of Science*.

Pattee, H. H. (1977). Dynamic and linguistic modes of complex systems. *International Journal of General Systems* **3**, 259–66.

Polanyi, M. (1968). Life's irreducible structure. *Science* **160**, 1308–12.

Rashevsky, N. (1973). The principle of adequate design. In *Foundations of Mathematical Biology*, Vol. III, *Supercellular Systems* (ed. R. Rosen). Academic Press, New York, pp. 145–75.

Rosen, R. (ed.) (1973). Is there a unified mathematical biology?. In *Foundations of Mathematical Biology*, Vol. III, *Supercellular Systems*. Academic Press, New York, pp. 361–93.

Rosen, R. (1986). Causal structures in brains and machines. *Int. J. General Systems* **12**, 107–26.

Rosen, R. (1988). Information and complexity. *Journal of Computational and Applied Mathematics* **22**, 211–18.

Schneider, T. D. (1991). Theory of molecular machines: sequence logos, machine/channel capacity, Maxwell's demon, and molecular computers. Paper presented to Second Foresight Conference on Nanotechnology, November 7–9, Palo Alto. To appear in *Nanotechnology*.

Shannon, C. E. (1948). A mathematical theory of communication. *Bell System Tech. Journal* **27**, 379–423, 623–56.

Shepherd, G. M. (1990). The significance of real neural architectures for neural network simulations. In *Computational Neuroscience* (ed. E. L. Schwartz). Bradford/MIT Press, Cambridge, MA, pp. 82–96.

Thom, R. (1975). *Structural Stability and Morphogenesis*. Benjamin, Reading, MA.

Thom, R. (1990). *Semio Physics: A Sketch*. Benjamin, Reading, MA.

Thompson, D'Arcy W. (1942). *On Growth and Form – A New Edition*. Cambridge University Press.

Varela, F. J. (1979). *Principles of Biological Autonomy*. North-Holland, New York.

Ward, S. A. and Halstead, R. H. (1990). *Computation Structures*. MIT Press, Cambridge, MA.

Young, J. Z. (1978). *Programs of the Brain*. Oxford University Press.

AUTHOR INDEX

SUBJECT INDEX